D1050068

Nonaqueous Solution Chemistry

Nonaqueous Solution Chemistry

OREST POPOVYCH

Department of Chemistry
City University of New York, Brooklyn College
Brooklyn, New York

REGINALD P. T. TOMKINS

Department of Chemical Engineering and Chemistry
New Jersey Institute of Technology
Newark, New Jersey

A WILEY-INTERSCIENCE PUBLICATION

JOHN WILEY & SONS

New York • Chichester • Brisbane • Toronto

Library of Congress Cataloging in Publication Data:

Popovych, Orest.
Nonaqueous solution chemistry.

"A Wiley-Interscience publication."
Includes bibliographies and index.
1. Solution (Chemistry) 2. Nonaqueous solvents. I. Tomkins, Reginald P. T., joint author. II. Title. [DNLM: 1. Chemistry. 2. Solutions. 3. Solvents. QD544.5 P829n]

QD541.P58 541.3′423 80–21693
ISBN 0–471–02673–5

Printed in the United States of America

10 9 8 7 6 5 4 3 2 1

Preface

This book is designed to provide the student with a broad background of the scope of the chemistry and physical properties of nonaqueous solvent systems and, we hope, will serve as a springboard to both graduate level courses or specialized courses in this subject. We have tried to relate the various approaches to this topic in order that the reader will appreciate the need for the different methods used to shed light on problems such as ion–solvent and ion–ion interactions, acid–base equilibria, and mechanisms of chemical and electrochemical reactions in nonaqueous solvents.

The first three chapters are key chapters, as they provide a firm foundation for a thorough understanding of the more specialized topics in subsequent chapters. These early chapters focus on the numerous solute–solvent interactions, the nature of solvation, the classification of solvents and the physicochemical properties of some typical solvents. Later chapters concentrate on thermodynamic and transport properties, correlation of properties in different solvents, acid–base chemistry, spectroscopy, electrode processes, and kinetics. The final chapter discusses some specialized applications of nonaqueous solvents.

In each chapter, we have attempted to present a concise account of the key principles and equations governing each topic. Derivations of equations are not given, but adequate references are provided for those interested in pursuing the theory in greater detail. A general discussion for each topic is followed by specific examples for a range of selected nonaqueous solvents. An attempt has been made to cover a varied approach to the topics with emphasis on the more recent advances. A fairly extensive coverage of actual data and/or graphs is included to illustrate the main points of the discussion.

For the purposes of further study, a set of general references is given at the end of each section. In addition, a glossary of solvents and various solvent symbols is included.

One of the authors (RPTT) would like to acknowledge the helpful discussions provided by G. J. Janz (Rensselaer Polytechnic Institute).

OREST POPOVYCH
REGINALD P. T. TOMKINS

Brooklyn, New York
Newark, New Jersey
January 1981

Contents

Glossary of Nonaqueous Solvents Mentioned in This Book

Name of solvent	Chemical formula	Common abbreviation
Acetic acid	CH_3COOH	HAc, HOAc
Acetic anhydride	$(CH_3CO)_2O$	
Acetone	$(CH_3)_2CO$	Me_2CO
Acetonitrile	CH_3CN	MeCN, AN
Acetyl chloride	CH_3COCl	
Acrylonitrile	$H_2C{=}CHCN$	
2-Aminoethanol	$H_2NCH_2CH_2OH$	
Ammonia	NH_3	
Amyl alcohol (see 1-Pentanol)		
Aniline	$C_6H_5NH_2$	
Antimony trichloride	$SbCl_3$	
Arsenic tribromide	$AsBr_3$	
Arsenic trichloride	$AsCl_3$	
Arsenic trifluoride	AsF_3	
Benzene	C_6H_6	ϕH
Benzonitrile	C_6H_5CN	
Benzoyl chloride	C_6H_5COCl	
Benzyl alcohol	$C_6H_5CH_2OH$	
Benzyl cyanide	$C_6H_5CH_2CN$	
Bromobenzene	C_6H_5Br	
1-Bromonaphthalene	$C_{10}H_7Br$	
Bromine trifluoride	BrF_3	
1-Butanol n-Butanol	$CH_3(CH_2)_3OH$	*n*-BuOH
2-Butanol	$CH_3CH_2CH(OH)CH_3$	
i-Butanol	$(CH_3)_2CHCH_2OH$	*i*-BuOH
Butanone	$C_2H_5COCH_3$	

Name of solvent	Chemical formula	Common abbreviation
t-Butyl alcohol	$(CH_3)_3COH$	
n-Butylamine	$CH_3(CH_2)_3NH_2$	
tert-Butylamine	$(CH_3)_3CNH_2$	
Butyrolactone	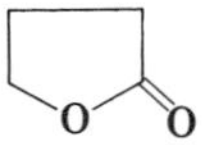	
iso-Butyronitrile	$(CH_3)_2CHCN$	
n-Butyronitrile	$CH_3CH_2CH_2CN$	
Carbon disulfide	CS_2	
Carbon tetrachloride	CCl_4	
Chlorobenzene	C_6H_5Cl	
Chloroform	$CHCl_3$	
Chlorosulfuric acid	HSO_3Cl	
Cyclohexane	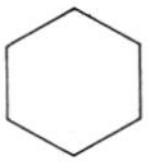	
Dibutyl carbitol	$C_4H_9OCH_2CH_2OCH_2OC_4H_9$	
1,1-Dichloroethane	CH_3CHCl_2	
1,2-Dichloroethane	$ClCH_2CH_2Cl$	
Dichloromethane	CH_2Cl_2	
(N,N)-Diethylacetamide	$CH_3CON(C_2H_5)_2$	
Diethyl ether	$(C_2H_5)_2O$	
(N,N)-Diethylformamide	$HCON(C_2H_5)_2$	
1,2-Dimethoxyethane	$CH_3OCH_2CH_2OCH_3$	
(N,N)-Dimethylacetamide	$CH_3CON(CH_3)_2$	DMAC, DMA
Dimethyl ether	$(CH_3)_2O$	
(N,N)-Dimethylformamide	$HCON(CH_3)_2$	DMF
Dimethyl methylphosphonate	$(CH_{30})_2P(O)CH_3$	DMMP
(N,N)-Dimethylthioformamide	$HCSN(CH_3)_2$	
Dimethylsulfoxide	$(CH_3)_2SO_2$	DMSO
(1,4)-Dioxane	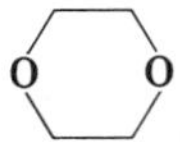	
Diphenylphosphinic chloride	$(C_6H_5)_2P(O)OCl$	
Ethanol	C_2H_5OH	EtOH, E (as superscript)
N-Ethylacetamide	$CH_3CONH(C_2H_5)$	
Ethyl acetate	$CH_3CO_2C_2H_5$	EtOAc
Ethylamine	$C_2H_5NH_2$	
Ethyl ether (see Diethyl ether)		
Ethylene carbonate	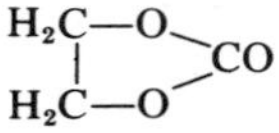 H_2C—O CO H_2C—O	$EnCO_3$

Name of solvent	Chemical formula	Common abbreviation
Ethylenediamine	$H_2NCH_2CH_2NH_2$	
Ethylene dichloride (see 1,2-Dichloroethane)		
Ethylene glycol	$(CH_2OH)_2$	
Ethylene sulfite		
Fluorobenzene	C_6H_5F	
Formamide	$HCONH_2$	
Formic acid	$HCOOH$	
Glycerol	$CH_2OHCHOHCH_2OH$	
Hexamethylphosphoramide or Hexamethylphosphorotriamide	$[(CH_3)_2N]_3PO$	HMPA HMPT
Hydrazine	H_2NNH_2	
Hydrogen bromide	HBr	
Hydrogen chloride	HCl	
Hydrogen fluoride	HF	
Hydrogen iodide	HI	
2-Hydroxycyanoethane	$NCCH_2CH_2OH$	
Iodobenzene	C_6H_5I	
Isobutyl alcohol (see *i*-Butanol)		
Isooctane	$(CH_3)_2CHCH_2C(CH_3)_3$	
Isopentanol (or *i*-Pentanol)	$(CH_3)_2CHCH_2CH_2OH$	*i*-PeOH
Isopropanol (or Isopropyl alcohol)	$(CH_3)_2CHOH$	*i*-PrOH
Isopropyl amine	$(CH_3)_2CHNH_2$	
Methanol	CH_3OH	MeOH, M (as superscript)
N-Methylacetamide	$CH_3CONHCH_3$	NMA
Methyl acetate	$CH_3CO_2CH_3$	
Methylene chloride (see Dichloromethane)		
N-Methylformamide	$HCONHCH_3$	NMF
Methyl formate	HCO_2CH_3	MF
Methyl isobutyl ketone	$(CH_3)_2CHCH_2COCH_3$	MIBK
N-Methylpropionamide	$CH_3CH_2CONHCH_3$	NMP
N-Methyl-2-pyrrolidone (or 1-Methyl-2-pyrrolidone)		

Name of solvent	Chemical formula	Common abbreviation
3-Methylsulfolane	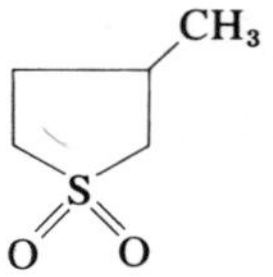 CH_3 (ring) S, O O	
2-Methyltetrahydrofuran	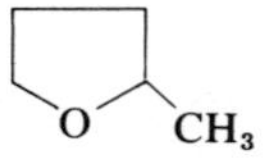 (ring) O CH_3	
Nitric acid	HNO_3	
Nitrobenzene	$C_6H_5NO_2$	
Nitroethane	$C_2H_5NO_2$	
Nitrogen tetroxide	N_2O_4	
Nitromethane	CH_3NO_2	$MeNO_2$
Nitrosyl chloride	NOCl	
Octanol	$CH_3(CH_2)_7OH$	
n-Pentanol	$CH_3(CH_2)_4OH$	*n*-PeOH
t-Pentyl alcohol	$C_2H_5C(CH_3)_2OH$	
Phenylphosphinic dichloride	$C_6H_5POCl_2$	
Phenylphosphinic difluoride	$C_6H_5POF_2$	
Phosgene	$COCl_2$	
Phosphorus oxychloride (or Phosphoryl chloride)	$POCl_3$	
Propanediol-1,2-carbonate (see Propylene carbonate)		
1-Propanol	$CH_3CH_2CH_2OH$	1-PrOH
2-Propanol	$CH_3CHOHCH_3$	2-PrOH
iso-Propyl alcohol (see Isopropanol)		
n-Propyl alcohol (see 1-Propanol)		
Propylene carbonate	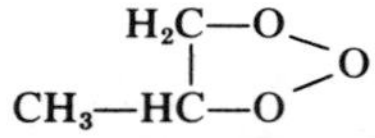 $H_2C—O$, $CH_3—HC—O$ (C=O)	PC
Propylene glycol	$CH_3CH(OH)CH_2OH$	
Selenium oxychloride	$SeOCl_2$	
Sulfolane	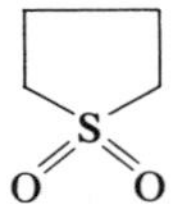 (ring) S, O O	SL, TMS
Sulfur dioxide	SO_2	
Sulfuric acid	H_2SO_4	
Sulfuryl chloride	SO_2Cl_2	
Tetrahydrofuran	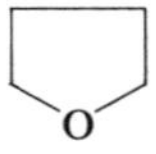 (ring) O	THF

Name of solvent	Chemical formula	Common abbreviation
Tetramethylenesulfone (see sulfolane)		
1,1,3,3-Tetramethylguanidine	$(CH_3)_2NC(=NH)N(CH_3)_2$	
Tetramethylurea	$(CH_3)_2NCON(CH_3)_2$	
Thionyl chloride	$SOCl_2$	
Tributyl phosphate	$[CH_3(CH_2)_3O]_3PO$	TBP
Trichloroacetic acid	CCl_3COOH	
Triethyl phosphate	$(C_2H_5O)_3PO$	
Trifluoroethanol	CF_3CH_2OH	TFE
Trimethyl phosphate	$(CH_3O)_3PO$	TMPA

Nonaqueous
Solution Chemistry

CHAPTER ONE

Introduction

There is no mystery why in college curricula the study of solution chemistry has been largely confined to aqueous solutions. Water is plentiful, cheap, and happens to be an excellent solvent for the majority of inorganic salts, acids and bases, and for a large number of polar organic substances as well. Traditionally, water has been *the* solvent in which physicochemical properties and chemical reactions were investigated. Indeed, the first modern concept of acidity and basicity—that of Arrhenius—was designed solely for aqueous solutions. Also, until recently the physicochemical data available in the literature for aqueous electrolyte solutions have been more precise.

This is not to say that solvents other than water are new to chemistry. Organic chemists always resorted to nonaqueous media either to achieve adequate solubility, or to increase the reaction rates, or because of the incompatibility of some organic reagents with water. Biochemists found that many water-insoluble proteins, enzymes, and polysaccharides could be dissolved in hydrofluoric acid and then recovered from it unchanged. Physical and analytical chemistry research in nonaqueous media dates back at least to 1863, referring to Weyl's pioneering work in liquid ammonia, followed by Raoult's studies of cryoscopy in glacial acetic acid in 1884. In the early 1900s, we note the fundamental work in liquid ammonia by E. C. Franklin and the determination of the standard potentials of the alkali–metal electrodes in water via emf measurements in ethylamine. In the 1920s, Walden and Hantzsch made their pioneering contributions to solution chemistry in such media as sulfuric acid, acetonitrile, chloroform, benzene, and ethyl ether. Around 1930, Conant, Hall, and Werner introduced glacial acetic acid as a medium for acid–base titrations. At about the same time, systematic studies of electrolytic conductance in nonaqueous solvents were initiated by Kraus and continue to this day by numerous investigators.

However, it was not until the last three decades that both fundamental research and practical applications involving nonaqueous media have experienced an exponential growth. Investigators are too numerous to mention here, but their names will be found in chapters on specialized topics. One milestone comes to mind, however; it is the detailed interpretation of acid–base equilibria in acetic acid (Kolthoff and Bruckenstein).

Probably no branch of chemistry has been affected by the nonaqueous revolution as much as analytical. Traditionally, analytical chemists took advantage of organic solvents to shift solubility equilibria in the desired direction, such as in solvent extraction processes. Later, application of nonaqueous (including mixed) solvents has expanded enormously the range of acidic and basic strengths that can be titrated and differentiated. Acid–base titrations in nonaqueous solvents are the subject of a recent monograph by Fritz and of annual review articles in *Analytical Chemistry*. pH scales, originally established for aqueous solutions, are now extended to several mixed solvents. Extension of potentiometry and polarography to nonaqueous media has opened up whole new areas of inquiry, such as to the meaning of ion activities and electrode potentials in different solvents and the magnitudes of liquid-junction potentials at the interfaces of different media. Practical analytical questions such as these brought about a boom in fundamental research on the energetics of individual ions in solution and have advanced significantly our knowledge of solute–solvent interactions. One index of these decades of intensified activity is the large body of physicochemical data determined in nonaqueous solvents, such as compiled in the two-volume handbook by Janz and Tomkins (1) and discussed in major treatises and monographs, for example, Coetzee and Ritchie (2), Covington and Dickinson (3), Kolthoff and Elving (4), Lagowski (5), and Gordon (6).

Despite the wealth of theoretical and applied knowledge available today on nonaqueous systems, little of it manages to diffuse into our undergraduate or graduate textbooks. The typical chemistry graduate is indoctrinated exclusively in aqueous chemistry, so that his view of solution chemistry is both limited and distorted. As a result the average chemist is generally unaware of the potentialities offered by the use of nonaqueous solvents and is unprepared to tackle practical problems that might involve them. One reason for this state is the fact that information on nonaqueous chemistry is scattered throughout the literature in specialized articles and chapters. The objective of this book is to provide a single concise and yet comprehensive source of information on nonaqueous chemistry for the advanced student and for the practicing chemist as well. This book is not meant to be a comprehensive review for the specialist. It is an introductory text to nonaqueous solvent chemistry, and, because of space limitations, it must be selective in the choice of material; therefore the selections are bound to be somewhat arbitrary, reflecting the interests of the authors. However, a sufficiently extensive bibliography is provided to offset these limitations.

It should be stressed that by "nonaqueous" we mean not only the pure liquids but any one of their mixtures with each other or with water. Thus we are talking about a virtually infinite number of solvent media and one in which a continuous variation of physicochemical properties can be realized. Studies in molten salts are not included in this text as there exist several authoritative treatments of the chemistry in fused salt media.

We mentioned earlier that the almost exclusive emphasis on aqueous chemistry now prevailing in the training of chemists contributes to shortcomings in their understanding of solution chemistry. Why does knowledge limited to aqueous solutions leave one with a distorted view of chemistry? To answer this question let us review briefly the unusual, indeed unique, properties of water as a solvent—properties that tend to modify profoundly the nature and the behavior of substances dissolved in it. Water has a higher dielectric constant than most solvents (78.5) and an appreciable dipole moment (1.85 D). Furthermore, water forms strong hydrogen bonds, especially to smaller-sized anions, and actually donates electron pairs to form coordinate bonds with transition-metal cations. As a result, most ionophores (substances that in their pure state exist in the form of ions) readily dissolve in water and are completely dissociated in aqueous solution. Dipolar nonionic substances are also appreciably soluble in water, especially if they can also act as hydrogen-bond acceptors.

All acids and many bases are so-called ionogens. In the pure state they are not composed of ions, but will ionize by undergoing a proton-transfer reaction with solvents that possess basic or acidic properties. Water is both an acid and a base. Consequently, in aqueous solution many acids and bases ionize, and the resulting ions remain completely dissociated due to the high dielectric constant of water. Surprisingly, water can also accommodate limited concentrations of nonpolar substances in the interstitial voids of its loose three-dimensional structure.

We have chosen to enumerate the most important properties of water as a solvent so as to establish a relative framework in which nonaqueous solvents can be discussed. None of the nonaqueous solvents possesses all of the outstanding solvent and ionizing properties of water. Therefore, if a chemist with an "aqueous education" wishes to predict or to rationalize the chemistry in a given nonaqueous medium, it might be easiest for him to begin by asking how the given solvent differs in its relevant properties from water. What is the dielectric constant of the solvent? Does it have a dipole moment and if so, how large? Does it possess pronounced acidic or basic properties? Is it likely to act as a donor or acceptor of hydrogen bonds, or of electron pairs? What are its structural characterstics?

To provide a framework for discussion of these and related questions, Chapter 2 is devoted to a consideration of the various intermolecular forces that exist between ions and solvent molecules and between ions themselves. A fairly detailed treatment of the nature of the solvation process, including the solvation of uncharged molecules and ions, is also presented in this chapter. One of the initial problems encountered in studying nonaqueous solvents is that of classifying solvents according to various properties, and Chapter 3 attempts to focus on this problem. The classification used in this text divides solvents into (1) organic and amphiprotic inorganic solvents, which include protophilic, protogenic, neutral, dipolar aprotic, and inert solvents, and (2) inorganic aprotic solvents. To set the stage for later chapters, attention is given to the physicochemical properties of specific solvents that are representative of each of the enumerated solvent classes. These are acetic acid, liquid ammonia, acetonitrile, benzene, liquid sulfur dioxide, arsenic trichloride, and, for the purposes of comparison, water. Properties considered include self-dissociation, acid–base chemistry, solubility relationships, electrochemistry, and solvent structure.

The subject of thermodynamic properties of nonaqueous electrolyte solutions is the topic of Chapter 4. The material comprises equilibrium emf measurements, vapor-pressure studies, cryoscopy, solubility, heats of solution calorimetry, and density studies. Enough background on theoretical principles is provided to enable the reader to understand how the thermodynamic data are evaluated and interpreted.

A feature of special interest over the past two decades has been that of the correlation of properties in different solvents. Thermodynamic transfer functions have formed the central theme of these discussions. In Chapter 5 both the determination of thermodynamic transfer functions and the estimation of transfer functions for single ions are discussed, together with applications such as the correlation of ion-activity scales and emf series in different solvents.

Acid–base chemistry is the subject of Chapter 6. Various concepts of acidity are reviewed and the interpretation of pH scales in aqueous and nonaqueous solvents is highlighted. A special section is devoted to acid–base titrations and the quantitative treatment of acid–base equilibria in selected solvents.

Transport properties such as electrical conductance, diffusion, viscosity, and transference numbers provide key information on the structure of electrolyte solutions, and parameters such as diffusion coefficients and transference numbers are often needed for evaluation of transport data. In addition, electrical conductivity and viscosity are normally important criteria when assessing technical applications of nonaqueous solutions. Chapter 7 discusses the properties of electrical conductance, transference numbers, diffusion, and viscosity with selected examples in a range of solvents. An attempt has been made to trace the historical development in treating the data, especially with regard to electrical conductivity, so that the reader is also made aware of the more recent analytical methods that have become available using fairly sophisticated computer programs.

Chapter 8 of this text contains sections on the applications of spectroscopy with particular reference to the providing of evidence for interactions between solutes and solvents and the identification of specific species in solution. The application of ultraviolet and visible spectroscopy in determining the strength of coordination between transition-metal ions and nonaqueous solvent molecules and examination of the nature of ion pairing is featured in this chapter. The study of solute–solvent complexes using infrared and Raman spectroscopy is discussed, and the use of infrared spectroscopy for identification of ion pairs, the study of hydrogen bonding, and the measurements of acid–base strength is included. Studies of ionic solvation, using nuclear magnetic resonance and the solvent effects on free radicals and radical ions in solution, are another topic discussed in this chapter.

Electrode processes in nonaqueous solvents is the subject of Chapter 9. For example, the use of voltammetry in studying the rates and mechanisms of some organic reactions and voltammetric studies of ionic solvation are the main points of interest. Electrical double-layer studies in nonaqueous solvents such as alcohols and amides are also discussed.

It is fairly well known that solvents play a significant role in chemical reactions. The rates and mechanisms of both organic and inorganic reactions are featured in Chapter 10. Studies of fast reactions using NMR and relaxation methods are also presented in this chapter.

Finally, Chapter 11 concentrates on some specialized applications of nonaqueous solvents. In particular attention is given to hydrometallurgical processes, nonaqueous electrolyte batteries, electrodeposition from nonaqueous solvents, and solvent extraction.

LITERATURE CITED

1 Janz, G. J., and R. P. T. Tomkins, *Nonaqueous Electrolytes Handbook*, Academic Press, New York, Vol. I, 1972; Vol. II, 1973.

2 Coetzee, J. F., and C. D. Ritchie, Eds., *Solute–Solvent Interactions*, Marcel Dekker, New York and London, Vol. I, 1969; Vol. II, 1976.

3 Covington, A. K., and T. Dickinson, Eds., *Physical Chemistry of Organic Solvent Systems*, Plenum Press, London and New York, 1973.

4 Kolthoff, I. M., and P. J. Elving, Eds., *Treatise on Analytical Chemistry*, Part I, Vol. 1, Chap. 12, and Part I, Vol. 2, Chap. 19, 2nd ed., Wiley-Interscience, New York, 1978–1979.

5 Lagowski, J. J., Ed., *The Chemistry of Non-Aqueous Solvents*, Vols. I–V, Academic Press, New York and London, 1966–1978.

6 Gordon, J. E., *The Organic Chemistry of Electrolyte Solutions*, Wiley, New York, 1975.

CHAPTER TWO

Solvent–Solute Interactions

2.1 INTERMOLECULAR FORCES

A crystal of sodium chloride dissolves readily in water but not in benzene, while just the opposite is true for a crystal of naphthalene. Evidently the forces that bind sodium and chloride ions together in the solid state are easily overcome by the forces of interaction between the ions and the water molecules, while the water–naphthalene interactions are too weak to compete with the cohesive forces between naphthalene molecules. Benzene, on the other hand, seems to interact much more effectively with the naphthalene molecules than with the sodium and chloride ions.

What are these competing intermolecular forces that determine the solubility relationships as well as the nature of solvent–solute interactions in solution? Qualitatively, most of the forces that act between stable atoms, molecules, and ions can be viewed as essentially electrostatic in nature, so that the interaction energies arising from them can be calculated approximately from electrostatic models. Below we consider briefly the most important types of the nonspecific electrostatic interactions between molecules and ions in the order of their decreasing energies. In addition, many molecules and ions interact by association via hydrogen bonds, which is a widely occurring phenomenon in amphiprotic liquids. Finally, there are specific chemical interactions, such as Brønsted and Lewis acid–base reactions, coordination, and charge transfer.

2.1.1 Ion–Ion Interactions The strongest intermolecular forces exist between ions. The attractive energy between two oppositely charged ions of charges Z^+e and Z^-e, separated by distance r *in vacuo*, is given by Coulomb's law:

$$E_{\text{ion-ion}} = \frac{Z^+Z^-e^2}{r} \tag{2.1a}$$

In a medium of dielectric constant ϵ, the energy for the same pair of ions becomes lower:

$$E_{\text{ion-ion}} = \frac{Z^+Z^-e^2}{\epsilon r} \tag{2.1b}$$

Throughout this section, the interaction distance r is in centimeters, e is the electronic charge equal to 4.80×10^{-10} electrostatic units of charge (esu), and Z^+ and Z^- are ionic valencies with appropriate sign. Thus the energies are calculated directly in the units of erg molecule^{-1}. More common in chemical practice are the energy units of kcal mol^{-1} and kJ mol^{-1}. The conversion factors are 1 erg molecule^{-1} = 1.44×10^{13} kcal mol^{-1} and 1 kcal = 4.18 kJ. Obviously the energy in Eqn. 2.1 is negative only for pairs of opposite charges; for like charges, it is positive (repulsion). Coulombic forces are omnidirectional and, being inversely proportional to the square of the distance r (since E = force $\times$ r), operate over a longer range than other intermolecular forces, which are functions of the inverse interaction distance raised to higher powers. Coulombic forces are primarily responsible for the large lattice energies and the high melting points of ionic crystals.

Coulomb's law in its simplest form represents the energy of interaction between two point charges in a vacuum. Two modifications are required before it can be applied to a pair of real ions situated inside a crystal lattice, that is, to the calculation of the lattice energy of an ionic crystal. In a crystal, the interaction between a pair of oppositely charged ions is influenced by all the attractions and repulsions due to neighboring ions. The net effect of these interactions depends on the geometry of the crystal lattice and is quantitatively expressed in the form of the so-called *Madelung constant, A*, which must be introduced in the numerator of Eqn. 2.1a. For example, in a crystal of sodium chloride the Madelung constant is 1.75, but it can attain much higher values for other crystal lattices. Another necessary addendum when considering the energy of a pair of ions *in vacuo* or in a crystal lattice derives from the fact that real ions are not point charges but possess electron clouds that repel each other even when the overall interaction is one of attraction between oppositely charged ions. The energy of repulsion has the general form $E_R = Br^{-n}$, where B and n are constants. Depending on the electronic configuration of the ion and the model employed for the energy, the exponent n (known as the Born exponent) may assume values of 5, 7, 9, 10, and 12. What is important for our purposes is not their precise numerical values but that energies of repulsion are negligible at large distances, increasing very rapidly when electron clouds of the ions begin to penetrate each other. It is clear that in the absence of repulsive energy there could be no stable ionic lattice: if attraction were the only governing force, ions of opposite charge would simply fuse together and annihilate each other.

While the coulombic and the repulsion forces account for all but a few percent of the lattice energy of ionic crystals, three additional minor energy components should be mentioned for the sake of completeness. They are the dispersion interaction (discussed in Section 2.1.7), the zero-point energy and the correction for heat capacity. The zero-point energy is the vibrational energy of the ions in the crystal at absolute zero, while the correction for heat capacity represents the energy change resulting from raising the temperature of the ions from absolute zero to the given value.

The electrostatic model has been remarkably successful in that the calculated ionic-lattice energies agreed with the corresponding measured values to within a few percent. For uni-univalent salts, the lattice energies typically range from 6.94×10^{-12} to 13.9×10^{-12} ergs molecule^{-1}, which corresponds to 100–200 kcal mol^{-1}. This success in predicting experimental values arises from the fortunate circumstance that the model representation of an ionic lattice as one consisting of rigid (nonpolarizable) ionic charges in a fixed and known geometrical configuration corresponds very closely to reality in the case of the smaller ions. As we will see, other modelistic representations of intermolecular interactions are not that lucky.

2.1.2 Ion–Dipole Interactions Many molecules that carry no net electrical charge are nevertheless characterized by electrical dissymmetry—their centers of positive and negative charge do not coincide. The simplest example of this type is provided by the molecule of HCl, where the partial negative charge resides on the chlorine atom and the partial positive charge, on the hydrogen atom. As a result, a molecule of HCl can be pictured as equivalent to an electric dipole, an arrangement of two equal charges q^+ and q^- separated by the distance l (Fig. 2.1). A polar molecule will have a dipole moment μ, given by

$$\mu = ql \tag{2.2}$$

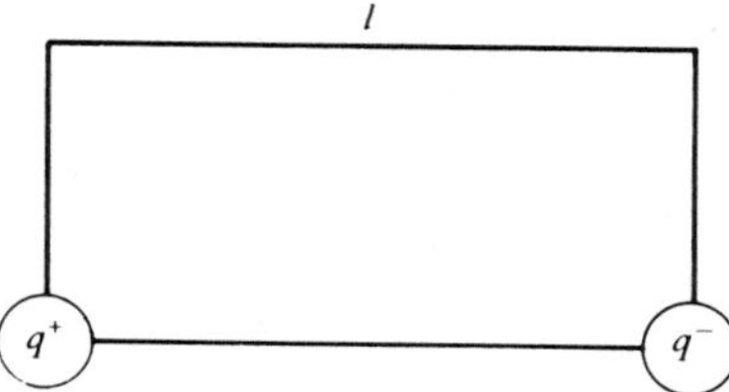

Figure 2.1 An electric dipole.

Because the charge q is of the same order of magnitude as that of an electron (4.80×10^{-10} esu), while the interaction distances are of the order of 10^{-8} cm, the dimensions of a molecular dipole moment are of the order of 10^{-18} esu cm. This unit of dipole moment has been named after the Nobel Prize winner Peter Debye, who contributed greatly to our knowledge of solution chemistry, including studies of dipole moments. Thus, 10^{-18} esu cm ≡ 1 Debye.

When a dipole is placed in the field of the electrical charge of an ion Ze, it will tend to orient itself in such a way as to have its end with charge opposite to that of the ion directed toward the ion. For the general case, without preferred orientation, depicted in Fig. 2.2, the energy of an ion–dipole interaction is given by

$$E_{\text{ion-dipole}} = \frac{-Ze\mu \cos \theta}{r^2} \tag{2.3}$$

where r is the distance between the center of the ion and the midpoint of the polar axis of the dipole and θ is the angle between the polar axis and the ion–dipole axis, as shown in Fig. 2.2. The most favorable interaction occurs when the attractive end of the dipole is pointing to the ion, as shown in Fig. 2.3. The value of $\cos \theta$ in Eqn. 2.3 for this case would be unity.

Ion–dipole interactions are somewhat weaker than ion–ion interactions, since they are functions of r^{-2} rather than r^{-1} and because the charges in

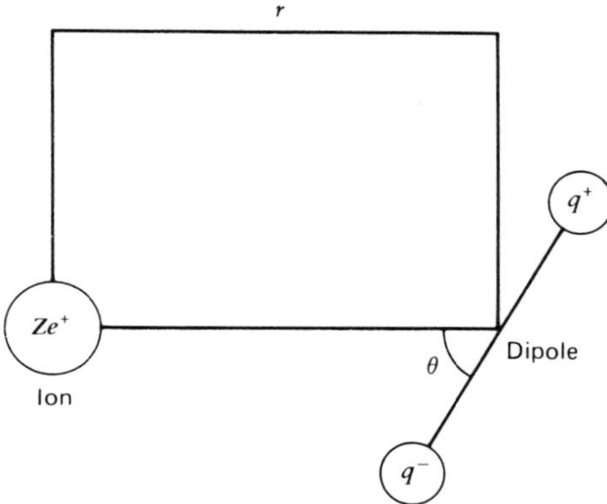

Figure 2.2 A general case of ion–dipole interaction (without preferred orientation).

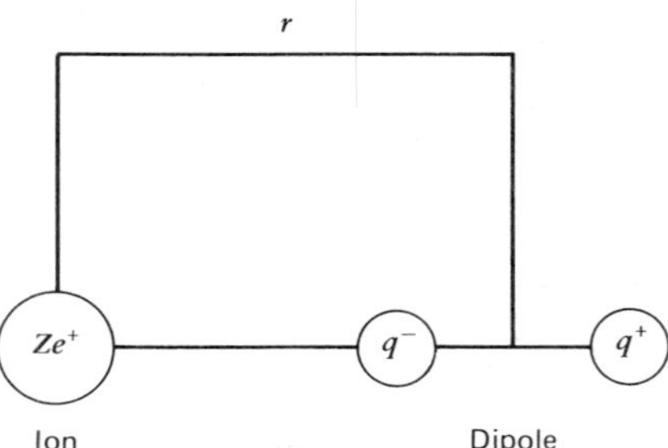

Figure 2.3 The most favorable ion–dipole interactions (minimum energy).

dipoles are usually considerably less than the full electronic charge. As we shall see later, ion–dipole forces contribute significantly to the solvation of ionic substances in polar solvents. Actual calculations of the interaction energy between an ion and the solvent dipoles require the knowledge of the interaction distance r, which is taken as the sum of the radii of the ion and the solvent molecule when the two are in contact. Furthermore, the number of solvent dipoles interacting with one ion must be known and that factor introduced in the numerator of Eqn. 2.3.

2.1.3 Ion–Quadrupole Interactions A molecule that has more than two charge centers can be represented by an electrical quadrupole, an arrangement of four charges of equal magnitude, two positive and two negative. Quadrupolar representations can be made both for molecules that possess permanent dipoles (e.g., water, methanol) as well as for molecules that have zero dipole moments as a result of compensating effects of oppositely directed dipoles in their structures (e.g., carbon tetrachloride, benzene). When the electric field of an ion of charge Ze and a quadrupole assume the most favorable relative orientation, as shown in Fig. 2.4, the interaction energy is given by

$$E_{\text{ion-quadrupole}} = \frac{-Ze\mu}{r^2} \pm \frac{Ze\phi}{2r^3} \tag{2.4}$$

The quadrupole moment ϕ is defined as $\Sigma_i q_i l_i$, where the q's are the charges of the quadrupole and the l's are their distances from the center of the quad-

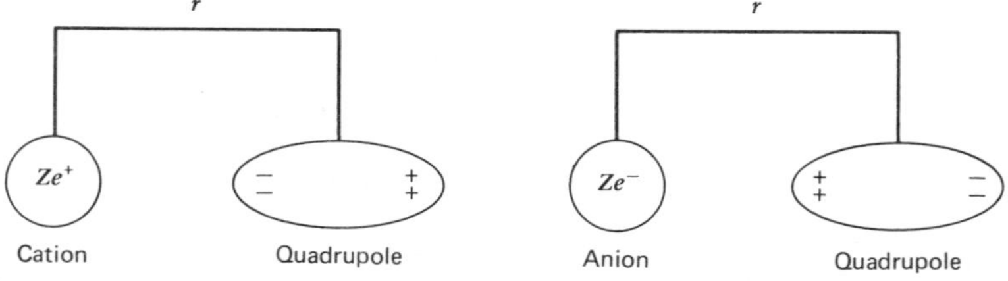

Figure 2.4 The most favorable ion–quadrupole interactions.

rupole. In Eqn. 2.4 the first term is the familiar expression for the maximum attractive energy of an ion–dipole interaction (Eqn. 2.3), while the new term specific to ion–quadrupole interactions is $\pm Ze\phi(2r^3)^{-1}$. The ion–quadrupole term is positive for cations and negative for anions, accounting for the previously unexplained fact that hydration energies of anions and cations of equal radii are not equal. Because of their dependence on r^{-3}, ion–quadrupole energies are often neglected, but this is hardly justified (see Table 2.1).

2.1.4 Ion-Induced Dipole Interactions When an uncharged, nonpolar molecule, such as an atom of xenon or a molecule of benzene, is placed in an electric field, its electron cloud is displaced from the normal position and the molecule assumes polar character. We say that a dipole has been induced in the nonpolar molecule. The extent to which the electron cloud will be displaced by a given electric field is determined by the *polarizability* of the molecule, α. If the electric field is due to an ion of charge *Ze*, the net interaction energy between the ion and the dipole induced by it will be

$$E_{\text{ion-induced dipole}} = \frac{-\tfrac{1}{2}\alpha Z^2 e^2}{r^4} \tag{2.5}$$

This energy is always one of attraction, since the induced dipole assumes the most favorable orientation along the molecule–ion axis, as shown in Fig. 2.5.

2.1.5 Dipole–Dipole Interactions The most favorable attractive interaction (minimum energy) between two dipoles is achieved when they are placed "head-to-tail," as shown in Fig. 2.6. For this most favorable case, the energy is given by

$$E_{\text{dipole–dipole}} = \frac{-2\mu_A\mu_B}{r^3} \tag{2.6}$$

Other orientations will lead to higher (smaller negative or positive) energy values. Because this energy depends on the relative orientation of the dipoles, it is also known as the "orientation effect." Obviously thermal agitation is going to interfere with any fixed orientation of the dipoles, and at sufficiently high temperatures the dipole–dipole interactions will vanish al-

Table 2.1 Interaction Energies of Na^+ and Cl^- Ions with Water Molecules[a]

Ion	Ion–Dipole term	Ion–Quadrupole term	Ion–Induced dipole term
Na^+	−94.2	+42.6	−32.7
Cl^-	−50.3	−16.6	−9.4

[a] Data of Bockris and Reddy (1). All energies are in kcal mol^{-1} calculated for interaction with four water molecules.

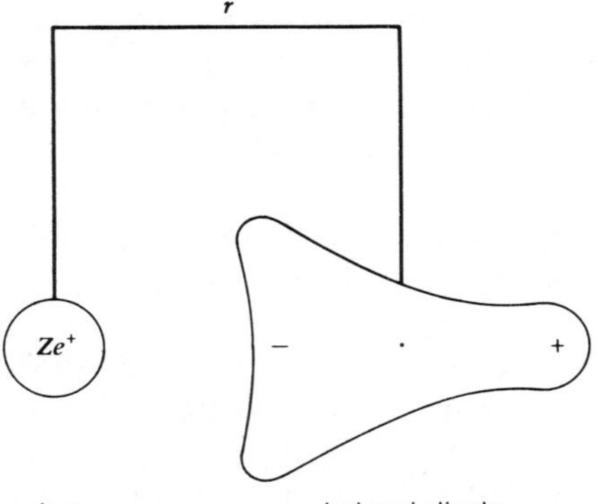

Figure 2.5 Ion-induced dipole interaction.

together. Therefore, Eqn. 2.6 holds only under conditions where the attractive dipole–dipole energy is larger than thermal energy. At room temperature the thermal energy $RT = 0.6$ kcal mol^{-1}, while estimates for the dipole–dipole energy of common polar molecules, such as NH_3, H_2O, and CH_3OH, are in the range of 1–5 kcal mol^{-1}. As expected, dipole–dipole interactions are weaker than ion–dipole interactions and, being a function of r^{-3}, diminish more rapidly with the distance. They are responsible in part for the association of polar liquids and for the fact that polar substances in the gaseous state condense into liquids and solids at lower temperatures.

2.1.6 Dipole-Induced Dipole Interactions As was the case with an ion, a permanent dipole can also induce another dipole in a normally nonpolar molecule. The maximum energy of attraction between such a dipole and the dipole induced by it is

$$E_{\text{dipole-induced dipole}} = \frac{-\alpha\mu^2}{r^6} \tag{2.7}$$

where μ is the permanent dipole moment and r is the distance between the midpoints of the inducing and the induced dipole.

2.1.7 Dispersion, or London, Interactions In molecules that have no permanent dipole moment, the electron cloud is on the time average symmetrically distributed about the nucleus. At any given instant, however, the

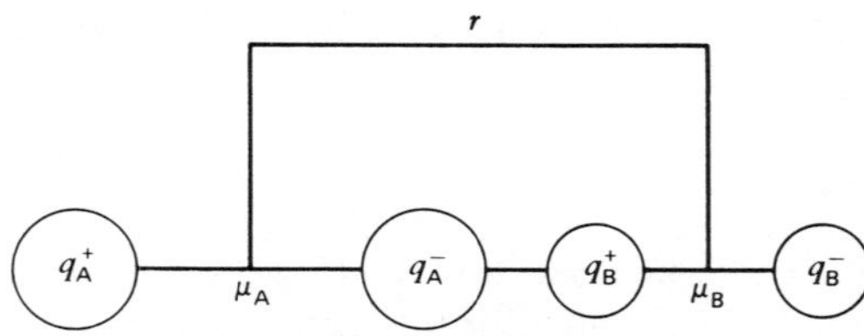

Figure 2.6 The most favorable dipole–dipole interaction.

electron distribution about the nucleus need not be symmetrical, so that momentary dipoles arise. Such instantaneous, fluctuating dipoles induce similar dipoles in neighboring molecules, resulting in attractive dipole–dipole interactions between them. The energy of these always attractive interactions is known as the London, or dispersion, energy, given by

$$E_{\text{dispersion}} = -\frac{3}{2}\frac{\alpha_1\alpha_2}{r^6}\left(\frac{I_1 I_2}{I_1 + I_2}\right) \qquad (2.8a)$$

when the interaction occurs between two different species having polarizabilities α_1 and α_2 and ionization potentials I_1 and I_2, respectively. When applied to two molecules of the same substance, the above equation reduces to

$$E_{\text{dispersion}} = -\frac{3}{4}\frac{\alpha^2 I}{r^6} \qquad (2.8b)$$

Dispersion forces are the weakest of all chemical forces of attraction. They are temperature independent and increase with the molecular volume as well as with the number and the looseness of electrons. Dispersion forces are responsible for the condensation of nonpolar substances into liquids and solids as well as for the solubility of nonpolar substances in nonpolar liquids. Being a function of r^{-6}, dispersion energies operate over extremely short ranges, which is the reason why nonpolar solids generally melt at lower temperatures than do ionic crystals, where the binding energy is a function of r^{-1}.

It is interesting at this point to compare the numerical values of some of the electrostatic interaction energies. In Table 2.1 we compare the relative magnitudes of the ion–dipole, ion–quadrupole, and ion-induced dipole energy terms in the interaction of Na^+ and Cl^- ions with water molecules. As we can see, the energy values in Table 2.1 vary qualitatively as expected from their functional dependence on the interaction distance r. The sign of the ion–quadrupole term reflects the sign of the ionic charge. All the tabulated energy terms are appreciable, but when comparing them with the energies of ion–ion interactions (100–200 kcal mol^{-1} for univalent ions) one should remember that the tabulated values were computed for an interaction with *four* water molecules. The energies of interaction with *one* water molecule (corresponding to Eqns. 2.3–2.5) are, of course, smaller by a factor of 4.

The interaction energies between pairs of water molecules in ice were calculated to be -4.75 for the dipole–dipole terms, -1.09 for the dispersion term, and -0.317 for the dipole–induced dipole term, in kcal mol^{-1} (2). Evidently interactions between uncharged molecules are weaker than ion–molecule interactions by at least one order of magnitude.

2.1.8 Hydrogen Bonding So far we have limited our discussion of intermolecular interactions to those arising from chemically nonspecific electrical forces. A more specific type of interaction and yet one which is characteristic of a large class of molecules known as amphiprotic is hydrogen bonding. A hydrogen bond is formed when a proton donor HA associates with a species containing unshared electron pairs X as follows: X ⋯ HA. Hydrogen bonding is responsible for the extensive self-association of water and alcohols as well as the favorable solvation of anions by these and other amphiprotic liquids. The energy associated with a hydrogen bond is only of the order of a few kcal mol^{-1}. However, since many hydrogen bonds are made or broken in the course of a given solution process, the overall energies associated with H-bonding can be very appreciable. The role of H-bonding in solvation and acid–base equilibria is discussed in greater detail in Section 3.1.2.

2.1.9 Specific Chemical Interactions An extreme version of a hydrogen bond is the complete transfer of a proton from an acid to a base. This represents a Brønsted acid–base reaction, which is one of the most powerful types of intermolecular interaction. Examples are the reactions between protonic acids and liquid ammonia or amines. Similarly strong and widespread are a variety of coordination reactions between electron donors and acceptors, including Lewis acid–base reactions and charge-transfer processes. For example, the extensive solvation of the Ag^+ ion by acetonitrile and hexamethylphosphoramide can be viewed as a Lewis acid–base interaction. Similarly, the enhanced solvation of the I^- ion by liquid sulfur dioxide is thought to be a function of strong charge-transfer interactions. These and many other examples of specific chemical effects on solvation are discussed in Section 2.2.3.

Summary If one neglects the energies of repulsion, which operate only at extremely short distances, the sum total of the energies of electrostatic intermolecular interactions can be represented as a power series in the reciprocal interaction distance r. In this series, the terms range from one in r^{-1} for the ion–ion energies (strongest) to one in r^{-6} for the dispersion energies (weakest).* It is this functional dependence on the distance as well as the order of magnitude of the energy terms, rather than their exact computation, that is emphasized here. One should never lose track of the fact that the equations for the interaction energies presented here are based on electrostatic models, which attempt to reproduce the most essential features of intermolecular interactions in a simplified way. In real chemical systems, the possibly complex structure of the interacting species and a variety of competing effects must be taken into account for any reliable calculation of the energies.

Furthermore, nonspecific electrostatic energy terms do not account for all intermolecular interactions. In many systems, there are significant energy

* This includes only the most important terms. Rarely considered are the following interaction energies, which are also inverse functions of r, as indicated: dipole–quadrupole, r^{-4}; ion-induced quadrupole and quadrupole–quadrupole, r^{-5}; induced dipole-induced quadrupole, r^{-8}; induced quadrupole-induced quadrupole, r^{-10}.

contributions from H-bonding and from the more specific chemical interactions, such as Brønsted and Lewis acid–base reactions and charge transfer.

2.2 THE NATURE OF SOLVATION

2.2.1 The Process of Solution When a substance in its solid or liquid state dissolves in a solvent, two processes take place. Bonds are broken between the ions or molecules of the pure substance, and the detached solute species are then dispersed homogeneously throughout the solvent medium. The solute species becomes "solvated." For example, when solid sodium chloride dissolves in methanol, bonds between sodium and chloride ions are broken, and the ions are enveloped by methanol molecules. Similarly, when a liquid like benzene dissolves in acetone, the benzene–benzene bonds must be broken by acetone molecules. For dissolution to occur, the forces that bind the pure substance together must be overcome by the forces of solvent–solute interactions. Dissolution of gaseous substances also depends on solute–solvent interactions, except that the intermolecular forces in the gas phase that must be overcome in the process are very much weaker than those in the condensed phases.

What are the forces that compete in the process of solution? Consider the dissolution of a crystal of sodium chloride in methanol. The ions in a crystal of NaCl are held together by a lattice energy of 186 kcal mol^{-1}, dominated by powerful coulombic forces. Molecules of methanol, with their dipoles pointing to ions of opposite charge, must pry the sodium and chloride ions loose from their positions in the solid, eventually dismantling the crystal lattice (Fig. 2.7). Once in solution, the ions are surrounded by molecules of methanol—they are solvated. The interactions between the polar methanol molecules and the ions must be predominantly of the ion–dipole, ion–quadrupole, and ion-induced dipole type (Eqns. 2.3, 2.4, and 2.5) with hydrogen bonding between the solvent and the chloride ions playing an important auxiliary role. We can see now why benzene, which has no permanent dipole and cannot form hydrogen bonds, would be unable to dissolve a significant quantity of sodium chloride. Evidently the dispersion and the ion-induced dipole forces are too weak to break up the lattice of NaCl. In liquid benzene, however, the binding forces are almost exclusively of the weak dispersion type (Eqn. 2.8b) and the dissolution of benzene in acetone must be a result of the dispersion and the dipole-induced dipole interactions between the acetone and the benzene molecules, which are strong enough to dismember the structure of liquid benzene.

When we speak of solvent–solute interactions, we include not only the ion–dipole, ion–quadrupole, ion-induced dipole, and dispersion energies already discussed, but also the Born charging energy for ions and a variety of nonelectrostatic and specific chemical interactions, which are discussed

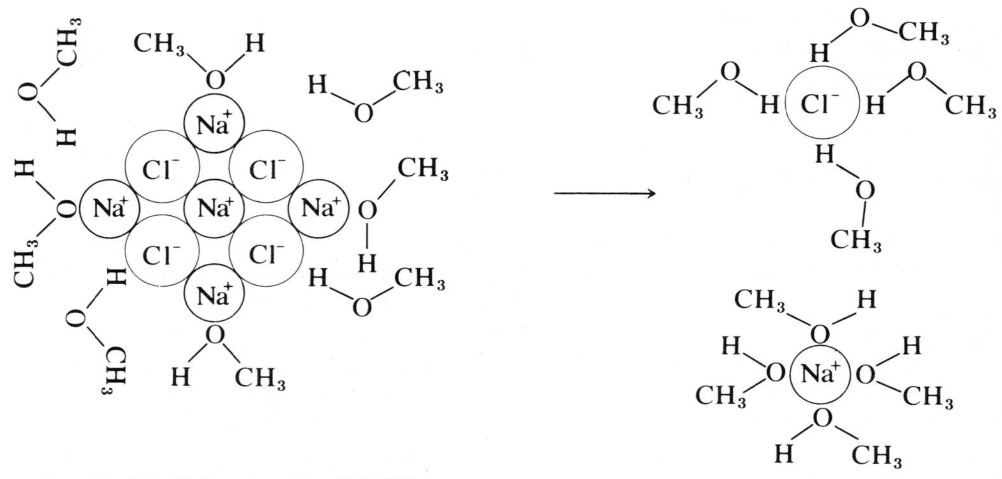

Crystal of NaCl Enveloped by CH_3OH Molecules

Ions Solvated by CH_3OH

Figure 2.7 Dissolution of a crystal of sodium chloride in methanol.

later in this chapter. The sum total of all the solvent–solute interactions is reflected quantitatively in the *solvation energy* of the dissolved substance. To represent solvent–solute interactions only, the solvation free energy, ΔG° solv., must be so defined as to exclude the energy that binds the pure substance together, such as the crystal lattice energy. It is therefore defined as the standard free energy change for the transfer of the gaseous substance S from vacuum (not from its solid or liquid state) into solution:

$$\Delta G^\circ\text{solv.:} \qquad S(\text{g})(\text{vacuum}) \longrightarrow S(\text{solv.})(\text{solution}) \tag{2.9}$$

Solvation energy should not be confused with the free energy of *solution*, which refers to the process:

$$\Delta G^\circ\text{soln.:} \qquad S(\text{s, or l}) \longrightarrow S(\text{solv.})(\text{solution}) \tag{2.10}$$

The two differ, of course, by the lattice energy, ΔG° latt., corresponding to the process*

$$S(\text{s, or l}) \longrightarrow S(\text{g})(\text{vacuum}) \tag{2.11}$$

Consequently,

$$\Delta G^\circ\text{solv.} = \Delta G^\circ\text{soln.} - \Delta G^\circ\text{latt.} \tag{2.12}$$

* In the literature the lattice energy is frequently defined in terms of the process which is the reverse of that shown in Eqn. 2.11 and designated by the symbol U. Also the free energies of solution and solvation are usually approximated by the corresponding enthalpies, ΔH°.

In Fig. 2.8 the relationships between the solvation, solution, and lattice energy are shown by means of the customary Born–Haber cycle diagram. Finally, the solubility product, K_{sp} (or the solubility in case of uncharged molecules), is related to the standard free energy of solution by the expression

$$\Delta G°\text{soln.} = -RT \ln K_{sp} \tag{2.13}$$

The magnitude of the solvation energy and the solubility of a substance as well as the nature of the resulting solution are determined by the properties of the solute and the solvent. In general, the old rule of thumb that "like dissolves like" is helpful, but one must first learn to recognize the relevant properties that determine the "likeness" between solvents and solutes.

What are the properties of a solute that are relevant to solvation? We have learned so far that it matters whether a solute is ionic, polar, or nonpolar and, in the last type, the extent to which it is polarizable. An ionic substance could be an ionophore or an ionogen. Ionophores (sometimes called "true electrolytes") are composed of ions in their pure state, examples being the majority of salts and metal hydroxides. Ionogens do not consist of ions in the pure state, but do ionize partially or completely as a result of proton-transfer reactions with solvents or solutes that possess acidic or basic properties. They are sometimes referred to as "potential electrolytes." All Brønsted acids as well as bases other than metal hydroxides are ionogens. For example, the Brønsted acid HCl is a covalent molecule, but in aqueous solution it ionizes 100% via proton-transfer reaction with water:

$$HCl + H_2O \longrightarrow H_3O^+ + Cl^- \tag{2.14}$$

Similarly, a covalent Brønsted base like methylamine will ionize partially by accepting a proton from water in aqueous solution:

$$CH_3NH_2 + H_2O \rightleftharpoons CH_3NH_3{}^+ + OH^- \tag{2.15}$$

We learned in the preceding section that when a substance is not ionic, but its centers of positive and negative charge do not coincide, it has a dipole moment and would be referred to as a polar or a dipolar solute. Common examples of polar substances are water, alcohols, ketones, nitriles, ammonia, and hydrogen halides. Finally, a solute with no permanent dipole

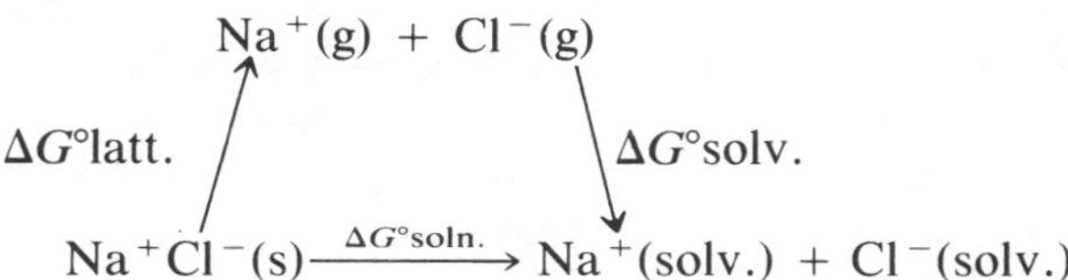

Figure 2.8 The relationship between solvation, solution, and lattice energies.

moment would be termed nonpolar. Benzene, cyclohexane, carbon tetrachloride, and the noble gases are examples of nonpolar substances.

What are the properties of a solvent that are relevant to solvation? In Chapter 3 solvents are classified in reference to their proton-transfer properties and the presence or absence of a dipole in their molecules. A fundamental distinction between amphiprotic and aprotic solvents is emphasized. One quantitative index of the effectiveness of an amphiprotic solvent is the magnitude of the equilibrium constant for its proton-transfer reaction with an acidic or a basic solute. For example, when comparing the proton-accepting ability (basicity) of water vs that of ammonia, we can compare the value of the ionization constant of acetic acid in the two solvents (1.8×10^{-5} in water vs complete ionization in ammonia). More generally applicable parameters that determine the magnitude of solvent–solute interactions are the dielectric constant, the dipole moment (Eqn. 2.2), the quadrupole moment (Eqn. 2.4), and the polarizability (Eqn. 2.8) of the solvent. Another property that may have a profound effect on the solvation energy, but one that is not easy to formulate quantitatively, is the structure of the solvent.

Water is the most structured liquid and one whose structure has been thoroughly investigated. Although the precise nature of the structure of water still remains the subject of some controversy, it is generally accepted that liquid water has the short-range order of hydrogen-bonded, tetrahedrally coordinated structure of slightly expanded ice, with interstitial cavities filled by monomeric water (3). That order is believed to persist on a time scale of about 10^{-11} sec. While water as well as di- and polyhydric alcohols are structurally exceptional in that their hydrogen-bonded networks are three-dimensional, other amphiprotic solvents are also associated via hydrogen bonds, but they form either linear polymeric chains (simple alcohols, *N*-alkylamides) or rings (HF). Chain structures are also believed to exist in *N,N*-dimethylformamide and dimethylsulfoxide.

Even more complex is the structure of mixed solvents. When small amounts (below 10 wt %) of alcohols, acetone, dioxane, or tetrahydrofuran are added to water, the water structure becomes actually reinforced, as in the case when nonpolar solutes, solid, liquid, or gaseous, are dissolved. For this solvent composition, maxima are observed for such properties as viscosities, Walden product, and heats of solution. Additional quantities of the organic liquid, however, cause progressive disruption of the water structure until a minimum is reached somewhere in the vicinity of the equimolar region for the two components. A variety of physicochemical properties, such as the Gibbs free energy of mixing and the Hammett acidity functions, exhibit maxima or minima in the vicinity of that composition. Presumably, the extrema correspond to a maximum breakdown (depolymerization) of the water structure by the nonaqueous component. A similar structure-breaking effect is observed when *N,N*-dimethylformamide, which is not associated via H-bonding, is added to the H-bonded *N*-methylacetamide. Such effects are not observed when none of the components is highly associated in the pure state. In the case of alcohol–water mixtures, however, the depolymerization of water is followed by the formation of strong water–alcohol bonds at higher alcohol contents (4).

To summarize, the effectiveness of a liquid solvent must be evaluated in terms of a superimposition of its macroscopic parameters (dielectric constant, equilibrium constants) and the microscopic properties of the individual solvent molecules (dipole moment, quadrupole moment, polarizability, H-bonding, coordination ability) as well as its structure. Generally, ionic substances exhibit higher solubilities in solvents of high dielectric constants, large dipole moments, and significant hydrogen-bonding or coordination ability. Polar uncharged substances dissolve preferentially in polar solvents. Hydrogen bonding and specific chemical interactions, such as acid–base reactions between the solute and the solvent, may play a decisive role here. Nonpolar substances prefer solvents of high polarizability. We shall return to some of these topics when we discuss the solvation of uncharged molecules and of ions in the immediately following sections.

2.2.2 Solvation of Uncharged Molecules What kind of energy changes are involved when an uncharged nonpolar molecule is inserted into a solvent medium? Compared to the energetics of the solvation of ions and dipolar molecules, this topic has received little attention. However, it is generally accepted that energy terms corresponding to the following processes must be considered:

1 A cavity must be formed within the liquid, large enough to accommodate the solute molecule plus any solvent molecules that might be associated with it. Depending on the size of the solute species, this process is likely to involve the breaking of solvent–solvent bonds and would tend to be energetically disfavored, particularly in highly structured liquids, such as water. In the specific case of water, the dissolution of nonpolar substances is disfavored for another reason. Here the introduction of nonpolar solutes causes actually an enforcement of the water structure via formation of so-called "icebergs" or "cages" around the solute species. The result is a decrease in the entropy of the solution and an increase in the free energy.

2 Once in the cavity inside the solvent, a nonpolar substance, in the absence of chemical reaction, interacts with the solvent by means of the London, or dispersion, forces. The energy of dispersion interactions was given by Eqn. 2.8. It follows from the equation that dispersion interactions in solution are a function of the polarizabilities of the solvent and the solute—the extent to which their electron clouds can be mutually distorted.

3 Finally, there is an energy term associated with the volume change resulting from the transfer of the solute species from the gas phase to the

solution phase. Of course, when solvation in two solvents is compared this volume term drops out.

The above solvation scheme applies only to those molecular substances that (a) have no permanent dipole, and (b) experience no specific chemical interaction with the solvent. Naturally, when the uncharged molecule has a permanent dipole and the solvent is also dipolar, additional interactions will occur, namely, between the oppositely charged ends of their dipoles. The energy of interaction between two molecules possessing permanent dipoles was given by Eqn. 2.6. Furthermore, hydrogen bonding and other specific chemical interactions, such as proton transfer or coordination, are often the determining factors in the solvation of uncharged molecules. In Chapter 3 we cite many examples where proton-transfer (acid–base) interactions between solutes and solvents lead to enhanced solubilities. The importance of hydrogen bonding in solvation and chemical reactivity is also stressed there. We will see that the coordination ability of the solvent molecules is sometimes so pronounced as to be chosen as a criterion for classifying organic solvents.

Interaction between electron donors and electron acceptors to form complexes is a widespread phenomenon. When the resulting complex shows a characteristic absorption spectrum due to intermolecular charge-transfer transition, where an electron is transferred from the donor to the acceptor, it is properly called a *charge-transfer complex* (5, 6). Any electron-rich species can serve as a donor substance. Distinction is made between donors of n-electrons (lone pairs), such as the amines, amine oxides, ethers, alcohols, sulfoxides, and phosphines, and π-electron donors, such as the aromatic hydrocarbons. Typical electron acceptors include the Lewis acids, such as BX_3, AlX_3, SnX_4, I_2, and Br_2 as well as aromatic hydrocarbons containing electron-withdrawing groups, such as NO_2 or X (where X is a halogen). Condensed polycyclic hydrocarbons may act both as donors and acceptors. Obviously substances that form complexes with the solvent molecules will readily dissolve in those solvents.

A criterion used occasionally to correlate solubility of nonpolar substances is the square root of the "internal pressure," $[\Delta E_{vap} (\Delta V)^{-1}]^{1/2}$, where ΔE_{vap} is the energy of evaporation and V is the molal volume (7). Internal pressure is related to the binding forces of the liquid and, in the absence of complications, substances with similar internal pressures will tend to dissolve in each other.

To what extent are the above modelistic views of solvation borne out by experiment? Within the same solvent a correlation between the solvation energies and the solubilities for a series of solid or liquid substances can be made only if the lattice energies of the substances are known, or can be assumed to be either small or comparable within the series (see Eqn. 2.12). Fortunately, in the case of gases we do not have to worry about the complications from lattice energies. In Table 2.2 we compare the solubilities of

Table 2.2 The Effect of Polarizability on the Solubility of Noble Gases

Gas	Atomic polarizability $\times 10^{24}$ cm^3 →	Water (0°C) 1.48	Ethanol (0°C) 5.29	Acetone (0°C) 6.59	Benzene (7°C) 10.87
	↓	Gas Solubility in $10^4 \times k$ [a]			
He	0.20	0.177	0.599	0.684	0.55
Ne	0.39	0.174	0.857	1.15	0.87
Ar	1.63	0.414	6.54	8.09	8.66
Kr	2.46	0.888	—	—	28.9
Xe	4.00	1.94	—	—	140.7
Rn	5.40	4.14	211.2	254.9	638.1

[a] k is Henry's law constant, where solubility $= kP$ (P = pressure of gas). Data taken from the compilation by M. Davies, *Some Electrical and Optical Aspects of Molecular Behavior,* Pergamon Press, Oxford and New York, 1965, p. 169.

the noble gases in water, ethanol, acetone, and benzene, where the polarizability of the solvent molecules increases in that order. In general, the solubility of the noble gases increases with the polarizability of the solvent. Also, within a given solvent, the solubility of the noble gases increases as their own size and polarizability increases. Both trends attest to the predominance of the dispersion interactions as the determining factor in the solubility of nonpolar solutes. For a given substance, we can follow the *changes* in the solvation energy in different solvents from the corresponding changes in the solubility: the lower the relative solvation energy, the greater the solubility. The solvation-energy changes calculated from the solubilities of noble gases and other nonpolar solutes, such as alkyl and aryl compounds, were found to be roughly a linear function of the cube of the solute radius (8, 9). Indeed, this seems to confirm not only the predominance of the dispersion interactions, which are a function of the solute volume, but also possibly our first assumption that a cavity must be formed in the liquid to accommodate the solute. The energy required to form such a cavity is likely to be proportional to the volume of the solute molecule. The previously mentioned formation of ordered water structures around nonpolar solutes, which leads to a diminution of their solubility, has been deduced from a variety of spectroscopic and transport properties. We will have occasion to refer to them again.

In Chapter 3 we point out that nonpolar substances are generally more soluble in dipolar aprotic solvents and in liquid ammonia than in water. This is attributed primarily to the smaller polarizability of the water molecule as compared to the molecules of the other solvents (see Table 2.2), confirming the importance of dispersion interactions in determining the solubility of nonpolar substances. Dipolar substances incapable of H-bonding (nitrobenzene, nitromethane) are slightly soluble in water, but the solubility increases significantly when the solute is able to form hydrogen bonds (aniline, phenol). It seems that the importance of hydrogen bonding outweighs that of dipole–dipole interactions.

2.2.3 Solvation of Ions Despite the enormous literature on the subject of ionic solvation, the magnitude of solvation energy of ions can be predicted with semiquantitative success only for aqueous solutions. For nonaqueous solutions, we are just beginning to advance methods of estimation, using extensions and analogies based on aqueous models.

Equations for the solvation energy of ions are derived from physical models, which are intended to represent certain selected features of an ion–solvent interaction in a simplified, idealized manner. In modelistic terms, it is convenient to divide ion–solvent interactions into electrostatic, nonelectrostatic, and specific chemical. The last may include hydrogen-bonding, coordination, and acid–base reactions.

The simplest model for the electrostatic component of the ion–solvent interaction energy was introduced by Born in 1920 (10). In the Born model, an ion of crystallographic radius r and charge Ze is identified with a rigid sphere of the same radius and charge. The solvent, which in reality is usually structured and, in the vicinity of the ion, polarized, is represented by a structureless continuum of uniform dielectric constant ϵ, corresponding to its bulk value. The solvation energy, $\Delta G°$(Born), is computed as the net electrostatic work of discharging the sphere in vacuum $(-Z^2e^2(2r)^{-1})$ and then recharging it to the same charge as the ion in a medium of dielectric constant ϵ, $[Z^2e^2(2r\epsilon)^{-1}]$:

$$\Delta G°(\text{Born}) = \frac{-NZ^2e^2}{2r}\left[1 - \frac{1}{\epsilon}\right] \qquad (2.16a)$$

(per mole)

$$= \frac{-166}{r}\left[1 - \frac{1}{\epsilon}\right] \quad \text{in kcal mol}^{-1} \qquad 2.16b)$$

where N is Avogadro's number and r is in Ångstroms.

Although constantly (and justly) criticized for its oversimplification of reality and other shortcomings, the Born equation provides an invaluable order-of-magnitude handle on the relative values of ionic solvation energies, a starting point in the evolution of more sophisticated approaches, and one that chemists have not been able to do without. There are two obvious flaws in the Born model. In the vicinity of an ion, the solvent may be oriented by the electric field of the ion, and its dielectric constant there will be much lower than the bulk value used ordinarily in Eqn. 2.16. This effect is likely to be minimal only for very large ions ($r \geq 5$ Å in water). Furthermore, it is not clear that it is legitimate to use crystallographic radii for ions in solution. Corrections for both the solvent dielectric constant and the ionic radii in the Born equation have been tried with some success.

Nevertheless, even the uncorrected Born equation provides useful information. Table 2.3 lists values of the Born solvation energies calculated for hypothetical ions of different radii as a function of the dielectric constant of some common solvents. What qualitative conclusions can be drawn from this table? First of all, the fact that the Born solvation energy is always negative means that all ions prefer to exist in the solvated state rather than in vacuum. Of course, it does not follow that all salts dissolve in all liquids, because in many cases the solvation energy is not great enough to overcome the lattice energy (see Eqn. 2.12). Second, the smaller the ion, the more negative its solvation energy. Indeed, experiment tends to confirm that salts of the smallest metal cation, lithium, generally exhibit the highest solubilities. Of course, this type of correlation cannot be exact, because the solubility is determined not only by the solvation energy but also by the lattice energy. Third, for a given ion, the solvation energy decreases (becomes more negative) with an increase in the dielectric constant of the solvent. This is in line with the common observation that the solubilities of salts generally tend to be higher in solvents of higher dielectric constant. It should be noted, however, that the absolute value of the Born energy is a direct function of the inverse radius of the ion but is comparatively insensitive to changes in the dielectric constant of the medium. Eventually the effect of the dielectric constant on the solvation energy begins to level off somewhere between $\epsilon = 30$ and 60 (depending on the size of the ion), which means that beyond

Table 2.3 Born Solvation Energies Calculated for Ions with Radii of 1–5 Å in Some Common Molecular Solvents, $-\Delta G°(\text{Born})$ in kcal mol^{-1}

Solvent	ϵ at 25°C	$r = 1$	1.5	2	3	4	5(Å)
N-Methylformamide	186.9	165	110	82.6	55.0	41.3	33.0
Formamide	109.5	165	110	82.5	54.8	41.2	33.0
Water	78.5	164	109	82.0	54.7	41.0	32.8
Propylene carbonate	64.4	163	109	81.7	54.3	40.9	32.6
Dimethylsulfoxide	48.9	163	108	81.3	54.2	40.6	32.5
Nitromethane	38.6	162	108	80.8	53.9	40.4	32.3
70 wt % Ethanol–H_2O	38.0	162	108	80.8	53.9	40.4	32.3
N,N-Dimethylacetamide	37.8	162	108	80.8	53.9	40.4	32.3
N,N-Dimethylformamide	36.7	161	108	80.7	53.8	40.4	32.3
Acetonitrile	36.0	161	108	80.7	53.8	40.3	32.3
Methanol	32.6	161	107	80.4	53.6	40.2	32.2
Hexamethylphosphoramide	29.6	160	107	80.2	53.5	40.1	32.1
Trifluoroethanol	26.7	160	106	79.9	53.3	39.9	32.0
Ethanol	24.3	159	106	79.6	53.1	39.8	31.8
Acetone	20.7	158	105	79.0	52.7	39.5	31.6
Acetic acid (glacial)	6.2	139	93	70	46	35	28

a certain threshold value all solvents should do an equally good job of dissolving a given salt if Born energy were the only governing factor.

Admittedly, the Born model offers an incomplete picture of electrostatic ion–solvent interactions. As was mentioned earlier, ions tend to orient the molecules of a dipolar solvent at their surfaces, resulting in the formation of what is known as a *primary solvation shell*. In water, the primary solvation shell of completely oriented water molecules is followed by a *secondary solvation shell*, where the water molecules are only partly oriented, also known as the *structure-broken region*. This region is disordered due to the competing influences of the spherically symmetrical field of the ion and the normal forces of tetrahedral symmetry characteristic of liquid water. Beyond this disordered region, lies the water continuum unaffected by the electric field of the ion. Dipolar nonaqueous solvents are also believed to form primary solvation shells around ions (more on the subject of solvation in Chapter 5).

The predominant short-range forces operating in the primary solvation shell are thought to be of the ion–dipole type, where the energy for a single ion–dipole interaction is given by Eqn. 2.3. For a mole of solvated ions that energy has been formulated as $(-NnZe\mu \cos\theta r^{-2})$, where n is the number of solvent molecules in the primary solvation shell, N is Avogadro's number, and the other symbols have been defined in conjunction with Eqn. 2.3 and Fig. 2.2. (Actually the above formulation is strictly valid for vacuum as the medium, and some authors prefer to include in the formula the reciprocal dielectric constant of the solvent medium.) Application of the ion–dipole models of Bernal and Fowler (11) and Eley and Evans (12) in combination with the Born term has led to a fair agreement between calculated and experimental values for the hydration energies of electrolytes. Further improvement was achieved when the ion–dipole model was replaced by the ion–quadrupole model of Buckingham (13). Here, the water molecule in the primary solvation shell is represented by a quadrupole—a system of four charges of equal magnitude, two positive charges at the hydrogen atoms and two negative charges near the oxygen atom. Based on the relationship already stated in Eqn. 2.4, the maximum interaction energy for a mole of ions with solvent molecules having a quadrupole moment ϕ consists of the previously mentioned ion–dipole term plus an additional energy term $(\pm NnZe\phi r^{-3})$.

Both the ion–dipole and the ion–quadrupole energy terms are quite significant in comparison with the Born energy. For example, in the case of the K^+ ion in water, the interaction energies calculated using a hydrated radius are (in kcal mol^{-1}) -40.8 for the Born term, -69.7 for the ion–dipole term, and $+27.1$ for the ion–quadrupole term (1). The ion–quadrupole model has been applied also in the calculation of ionic solvation energies in methanol (9) and in *N*-methylformamide (14). However, calculations in these and other nonaqueous media are presently hampered by the uncertainties as to the values for the quadrupole moments and the coordination numbers n in the primary solvation shells. These difficulties are likely to be gradually alleviated in the future as more information on the properties of nonaqueous solvents and on the structure of their solutions is accumulated.

Additional electrostatic components in the solvation energy of ions are those representing ion-(induced dipole), ion-(induced quadrupole), and dispersion interactions. The last was already discussed in connection with the solvation of uncharged substances. These terms, however, are functions of higher powers of reciprocal ionic radius (see Eqns. 2.5 and 2.8) and are therefore comparatively small.

In addition to those interactions that result from ionic charge, the solvation energies of ions include energy terms similar to those that hold for uncharged substances (see preceding section). Certainly a cavity must be formed in the liquid to accommodate an ion and its primary solvation shell, if any, and a certain number of solvent–solvent bonds are likely to be broken in the process. These and other *nonelectrostatic* interactions between an ion and a solvent are thought to be similar to those of an uncharged molecule having the same size and structure as the ion. However, the two cannot be identical (except perhaps for the largest ions) because the electric field of an ion modifies the solvent structure at its surface in a way that an analogous uncharged molecule does not. In water, which is a three-dimensional network connected by hydrogen bonds, the "structure-breaking" energy required to form a cavity was computed in terms of the energy required to break a certain net number of hydrogen bonds. It was estimated to be 20 kcal mol^{-1} for cations and 30 kcal mol^{-1} for anions (1). Obviously we need to know much more about the structure of liquids other than water before such computations become possible for nonaqueous solutions.

Specific chemical interactions between solvents and solutes are responsible for some of the largest values of solvation energies known. However, because in the case of ions the specific interactions always occur between a solvent and an *individual* ionic species (not with a pair of oppositely charged ions), the assignment of energy values to such interactions must be based on extrathermodynamic assumptions (see Chapter 5). The most common and powerful specific interactions are acid–base reactions both of the Brønsted and the Lewis type. We mentioned earlier that when hydrogen chloride dissolves in water, the proton-transfer reaction from the acid to water proceeds 100% (Eqn. 2.14). Not surprisingly, in liquid ammonia the solvation energies of hydrogen halides are even lower than in water by about 25 kcal mol^{-1}, indicating more favorable solvation. Because the corresponding solvation-energy differences for the alkali salts are small, the more favorable solvation of the acids in ammonia relative to water must be a function of the powerful specific interaction between ammonia and the proton. Unfortunately, this apparently obvious conclusion cannot be confirmed by thermodynamics, which provides information only about electrically neutral combinations of ions, such as $(H^+ + Cl^-)$ or $(H^+ - K^+)$, but not about individual ions. Energy changes for individual ions can be estimated by so-called extrathermodynamic assumptions, discussed in Chapter 5. According to one such estimate about 23 kcal mol^{-1} (of the total of 25 kcal mol^{-1}) in the solvation-energy change accompanying the transfer of hydrogen halides from water to ammonia is attributable to the more favorable solvation of the proton by ammonia (15).

Solvents that can act as electron-pair donors (Lewis bases) undergo specific interactions with metal cations that act as electron-pair acceptors (Lewis acids). In extreme examples of this type, rather strong complexes form, for example, between acetonitrile and Ag^+, Cu^+, and Au^+ ions, between ammonia or hydrazine and Ag^+, Cu^{2+}, Zn^{2+}, Cd^{2+}, Pb^{2+}, and Hg^{2+} ions, or between water and Cu^{2+} or Al^{3+} ions. The silver ion forms complexes also with *N,N*-dimethylformamide, dimethylsulfoxide, and hexamethylphosphoramide. Dipolar aprotic solvents containing basic oxygen atoms, such as the amides, dimethylsulfoxide, hexamethylphosphoramide, and *N*-methyl-2-pyrrolidone, solvate cations more strongly than does water. One practical consequence of these specific interactions is a large increase in the relative solubilities of the salts containing the interacting cations. Another consequence is that these cations become more difficult to reduce electrolytically in the complex-forming media, that is, their reduction potentials are shifted in the negative direction, relative to their position in water.

Solvation of certain anions is enhanced in solvents that can act as electron acceptors. If the polarizability of the anion is sufficiently large and its ionization potential sufficiently small, its interaction with the solvent molecules may become strong enough so that charge-transfer complexes develop (see preceding section). For example, solutions of iodides and thiocyanates in liquid sulfur dioxide have an intense yellow color that is independent of the cation. The absorption responsible for the yellow color is a charge-transfer transition involving transfer of an electron from the anion to an SO_2 molecule. Obviously, complexation of this type leads to abnormally high solubilities for electrolytes containing these anions.

Specific chemical interactions manifest themselves in the form of *preferential*, or *selective, solvation* of certain ions by one of the components of a solvent mixture. For example, it was shown from measurements of Hittorf transference numbers that in acetonitrile–water mixtures, the silver ion is preferentially solvated by acetonitrile and the nitrate, by water (16). Similarly, Grunwald and his associates (17), by applying an extrathermodynamic assumption (see Chapter 5), concluded that in dioxane–water mixtures most anions were preferentially solvated by water, while the alkali cations were, rather surprisingly, solvated appreciably (though not exclusively) by dioxane. We discuss preferential solvation further in Chapter 8.

It is clear that the relative solubilities of ionic substances in different solvents depend on the extent of solvation of both the anion and the cation, though frequently one of the ions may experience solvation effects of such magnitude as to overshadow the other.

Can we apply some of the principles discussed so far to predict or explain the relative solubilities of electrolytes? As an example, consider the solubility products K_{sp} of potassium chloride, a salt relatively free of specific interactions, tabulated as a function of solvent dielectric constant in Table 2.4 and plotted in the form of pK_{sp} vs ϵ^{-1} in Fig. 2.9. An instructive picture emerges. For protic solvents, the pK's are roughly a linear function of the reciprocal dielectric constant of the solvent, but for the dipolar aprotic solvents, they form a scatter diagram. Apparently when the comparison is restricted to solvents of like type, the relationship in Fig. 2.9 may reflect the Born energy term because the solvation-energy components that are not functions of ϵ^{-1} happen to be small in comparison with it.

This phenomenon where solubilities of electrolytes do vary linearly with ϵ^{-1} of the medium, but only within a series of related solvents, has been known for a long time. For example, Izmaylov and Chernyi (18) reported that the straight-line plots of the logarithm of solubility of several electrolytes vs ϵ^{-1} had different slopes for a series of alcohols and ketones, respectively. They attributed the change in the slope to a difference between the dipole moments of the two types of solvents, formulating the energy of ion–dipole interactions as including the dielectric constant of the medium in the denominator of Eqn. 2.3.

Although our understanding of the above discrepancies is far from complete, the fact that KCl has the same pK_{sp} in propylene carbonate (ϵ = 64.4) as in l-butanol (ϵ = 17.1) and in acetonitrile (ϵ = 36.0) as in isopropanol (ϵ = 18.3) should constitute a dramatic warning of the limitations of the dielectric constant as a parameter for predicting the relative magnitude of any ion–solvent interactions, including the solubility relationships in media of diverse nature.

The drastic differences between the solubilities of KCl in protic and di-

Table 2.4 Solubility Products of KCl as a Function of Solvent Dielectric Constant (Molal Scale)

Solvent	ϵ (25°C)	ϵ^{-1}	pK_{sp}
Formamide	109.5	0.00913	0.6
Water	78.5	0.0127	−0.90
Propylene carbonate	64.4	0.0155	6.80
N,N-Dimethylacetamide	37.8	0.0265	1.84
N,N-Dimethylformamide	36.7	0.0272	5.7
Acetonitrile	36.0	0.0278	7.04
Methanol	32.6	0.0307	2.98
Trifluoroethanol	26.7	0.0374	4.0
Ethanol	24.3	0.0412	5.22
1-Propanol	20.1	0.0498	6.4
Isopropanol	18.3	0.0546	7.0
1-Butanol	17.1	0.0585	6.8
2-Butanol	15.8	0.0633	7.9

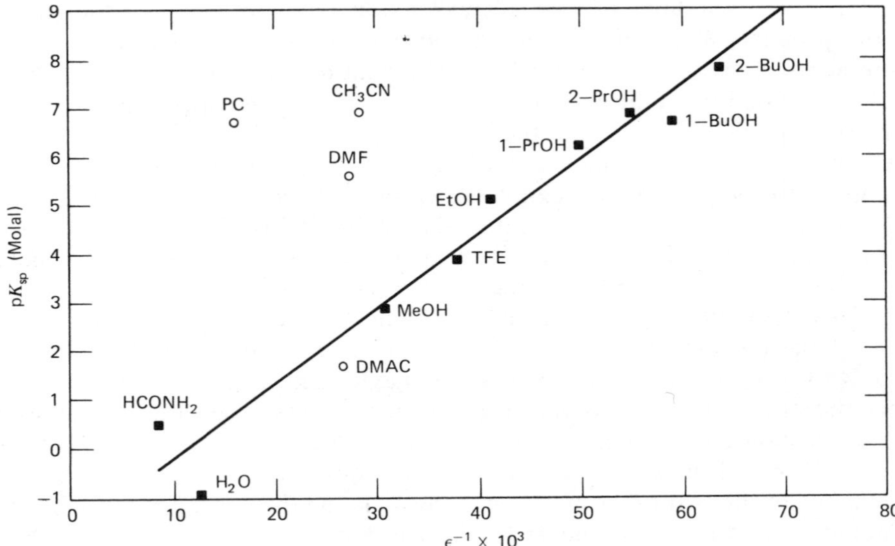

Figure 2.9 Solubility products, pK_{sp} (Molal) of KCl as a function of solvent dielectric constant. (■) Protic solvents: (○) dipolar aprotic solvents; PC = propylene carbonate; DMF = dimethylformamide; DMAC = dimethylacetamide; TFE = trifluoroethanol.

polar aprotic media are primarily a function of the relative H-bonding abilities of these classes of solvents. In protic solvents, the solvation of anions is favored due to formation of hydrogen bonds from the solvent to the anion. In a given solvent, hydrogen bonding is strongest for the smallest anions and those with localized charge, such as F^-, OH^-, Cl^-, and the carboxylates. It becomes progressively weaker as the size of the anion increases. The solvation of anions by protic solvents decreases in the order: OH^-, $F^- \gg Cl^- > Br^- > N_3^- > I^- > SCN^- >$ Picrate $> BPh_4^-$. The reverse order of preference holds for the solvation of anions by dipolar aprotic solvents (19). It is the inability to form hydrogen bonds that accounts for the poor solubility of the salts with smaller anions in dipolar aprotic solvents, as compared to protic solvents. This apparently offsets the contribution of the ion–dipole terms, which must be quite appreciable if one considers that the dipole moments of the dipolar aprotic solvents are much greater than those of water and the alcohols (Table 3.2). Larger ions, on the other hand, are more polarizable and they are preferentially solvated by the molecules of dipolar aprotic solvents by a combination of dispersion and ion–dipole interactions.

Some striking examples of the consequences of anion solvation on the solubility are provided in Table 2.5. In considering the solubility relationships for the thallium(I) salts in acetonitrile we observe that they indeed conform to the progression in anionic solvation outlined above. In this connection it is noteworthy that the solubility of TlCl in 71 wt % aqueous ethanol, which has a dielectric constant close to that of acetonitrile, is more than 2000 times greater than in acetonitrile (7.31×10^{-4} *M*). However, as was stated earlier in this section, we should not always expect a direct correlation between solvation energy and solubility, as the latter depends also on the lattice energy. This ambiguity can be removed by considering the differences between the pK_{sp} values in water and a given dipolar aprotic solvent for a series of salts containing a common cation. In such comparisons any complications due to inequality of crystal lattice energies are eliminated and what remains are the differences between the relative solvation of the anions by water and the nonaqueous solvent. In Table 2.5 we compare the ΔpK_{sp} values (acetonitrile–water) for a series of silver salts. The more negative the ΔpK_{sp}, the greater the solvation of the anion by acetonitrile, relative to its hydration. Positive ΔpK_{sp} values mean that hydration is preferred to nonaqueous solvation. Clearly, the ΔpK_{sp} trends in Table 2.5 essentially confirm the order of preference in anionic solvation stated earlier.

When an electrolyte is composed of large organic ions, the above relationship is qualitatively reversed. For example, the solubility of tetraphenylarsonium picrate (Ph_4AsPi) is 7.99×10^{-2} *M* in acetonitrile, but only 4.23×10^{-3} *M* in 71 wt % ethanol. Evidently, dispersion and nonelectrostatic interactions begin to predominate here. Salts composed of organic ions in which the charge is virtually "buried" inside the organic residue behave almost as if they were composed of a pair of analogous uncharged molecules. The best example of this type of electrolyte is tetraphenylarsonium tetraphenylborate (Ph_4AsBPh_4), and one of its uncharged analogs is tetraphenylmethane (Ph_4C). It is indeed remarkable that in dipolar aprotic solvents the values of pK_{sp} for Ph_4AsBPh_4 fall in general within one log unit of the

Table 2.5 Effect of Anion Solvation

The solubilities of thallium(I) salts in acetonitrile (20)		Differences between the solubility products of silver salts in water and acetonitrile (21)	
Salt	Solubility, molar	Salt	$\Delta pK^a \equiv {}_s pK_{sp} - {}_w pK_{sp}$
TlCl	3.19×10^{-7}	AgCl	3.3
TlBr	4.25×10^{-7}	AgN_3	1.4
TlI	8.07×10^{-7}	AgBr	1.2
TlSCN	9.04×10^{-5}	AgI	−1.7
$TlNO_3$	5.03×10^{-4}	AgSCN	−1.9
$TlClO_4$	3.31×10^{-2}	AgPi	−5.4
		$AgBPh_4$	−9.8

[a] The subscripts s and w denote acetonitrile and water, respectively. The ΔpK_{sp} is a measure of the free energy for the transfer of the electrolyte from water to acetonitrile (see Chapter 5).

Table 2.6 The Predominance of Nonelectrostatic Solvation: A Comparison of the Solubilities of Tetraphenylmethane and Tetraphenylarsonium Tetraphenylborate[a]

Solvent	$pK_{sp}(Ph_4AsBPh_4)$	$2pS(Ph_4C)$[b]
Acetonitrile	5.8	6.4
N-Methyl-2-pyrrolidone	3.3	4.0
N,N-Dimethylformamide	3.9	5.0
N,N-Dimethylacetamide	4.0	4.8
Hexamethylphosphoramide	3.7	4.6
Formamide	8.8	9.8
Dimethylsulfoxide	3.6	4.8

[a] Calculated from the data of Alexander et al. (22).
[b] $pS \equiv -\log S$, where S = solubility in moles liter^{-1}.

corresponding values for $2pS(Ph_4C)$, where S is the molar solubility and $pS = -\log S$ (Table 2.6). The electrolyte behaves essentially as two molecules of tetraphenylmethane. As was mentioned earlier, salts containing bulky organic ions, such as tetraalkylammonium and tetraaryl ions, have low solubilities in water because of entropy losses resulting from the formation of rigid solvent structures around these solutes. Their solubilities increase significantly in aprotic solvents, where their entropy increases and where the contribution of the dispersion forces becomes more significant due to the generally greater polarizabilities of the aprotic solvents.

Summary In general, ionic substances are most soluble in protic solvents of high dielectric constant and moderately so in dipolar aprotic solvents. In all polar solvents, cations are solvated by interactions with the negative end of the solvent dipoles and by the formation of coordinate bonds with the lone pairs of electrons from oxygen, nitrogen, and sulfur atoms. In all polar solvents, anions are solvated via interaction with the positive end of the solvent dipoles. In addition, in protic solvents, anions are solvated via H-bonding (strongest for the smallest ions) and in dipolar aprotic solvents, by dispersion interactions (strongest for the largest ions). Overall, in dipolar aprotic solvents, anions are less solvated than cations. Salts composed of large organic ions with buried charges approach the behavior of the corresponding uncharged molecules in solution.

LITERATURE CITED

1 Bockris, J. O'M., and A. K. N. Reddy, *Modern Electrochemistry*, Vol. I, Plenum Press, New York, 1971, Chap. 2.

2 Davies, M., *Some Electrical and Optical Aspects of Molecular Behavior*, Pergamon Press, Oxford and New York, 1965, pp. 173–4.

3 Lee, T. S., and O. Popovych, "Chemical Equilibrium and the Thermodynamics of Reactions," in I. M. Kolthoff and P. J. Elving, Eds., *Treatise on Analytical Chemistry*, Part I, Vol. I, 2nd ed., Wiley-Interscience, New York, 1978, Chap. 9, p. 552.

4 Franks, F., and D. J. G. Ives, *Quart. Rev.*, **20,** 1 (1966).

5 Foster, R., *Organic Charge-Transfer Complexes*, Academic Press, London and New York, 1969.

6 Person, W. B., and R. S. Mulliken, *Molecular Complexes*, Wiley, New York, 1969.

7 Hildebrand, J. H., and R. L. Scott, *The Solubility of Nonelectrolytes*, 3rd ed., Reinhold, New York, 1950.

8 Alfenaar, M., and C. L. deLigny, *Rec. Trav. Chim.*, **86,** 929 (1967).

9 Bax, D., C. L. deLigny, and M. Alfenaar, *Rec. Trav. Chim.*, **91,** 452 (1972).

10 Born, M., *Z. Physik.*, **1,** 45 (1920).

11 Bernal, J. D., and R. H. Fowler, *J. Chem. Phys.*, **1,** 515 (1933).

12 Eley, D. D., and M. G. Evans, *Trans. Faraday Soc.*, **34,** 1093 (1938).

13 Buckingham, A. D., *Disc. Faraday Soc.*, **24,** 151 (1957).

14 DeLigny, C. L., H. J. M. Denessen, and M. Alfenaar, *Rec. Trav. Chim.*, **90,** 1265 (1971).

15 Izmaylov, N. A., *Dokl. Akad. Nauk SSSR*, **126,** 1033 (1959).

16 Strehlow, H., and H. M. Koepp, *Z. Elektrochem.*, **62,** 373 (1958).

17 Grunwald, E., G. Baughman, and G. Kohnstam, *J. Amer. Chem. Soc.*, **82,** 5801 (1960).

18 Izmaylov, N. A., and V. S. Chernyi, *Zhur. Fiz. Khim.*, **33,** 127 (1959); **34,** 319 (1960).

19 Parker, A. J., *Quart. Rev.*, **16,** 163 (1962).

20 Coetzee, J. F., and J. J. Campion, *J. Amer. Chem. Soc.*, **89,** 2517 (1967).

21 Popovych, O., "Transfer Activity Coefficients (Medium Effects)," in I. M. Kolthoff and P. J. Elving, Eds., *Treatise on Analytical Chemistry*, Part I, Vol. 1, 2nd ed., Wiley-Interscience, New York, 1978, Chap. 12, p. 722.

22 Alexander, R., A. J. Parker, J. H. Sharp, and W. E. Waghorne, *J. Amer. Chem. Soc.*, **94,** 1148 (1972).

GENERAL REFERENCES

Coetzee, J. F., and C. D. Ritchie, Eds., *Solute–Solvent Interactions*, Marcel Dekker, New York and London, Vol. I, 1969; Vol. II, 1976.

Gordon, J. E., *The Organic Chemistry of Electrolyte Solutions*, Wiley, New York, 1975.

CHAPTER THREE

General Features and Characteristics of Nonaqueous Solvents

3.1 CLASSIFICATION OF SOLVENTS

Solvents have been classified on the basis of various criteria, some more general, others reflecting the particular interests and objectives of a specific branch of chemistry. Any one classification is bound to be arbitrary to some extent and as such will not satisfy all chemists. Moreover, the boundaries between solvent classes are not always sharp, and in the literature the same liquid is sometimes listed in different solvent classes, depending on the author or a particular application. Therefore, the placing of solvents in one category or another should be viewed more as an organizational aid to discussion, rather than some absolute rigid truth.

In this book, solvents are divided into two broad categories: (1) *organic and amphiprotic inorganic solvents*, and (2) *inorganic aprotic solvents*. The first category includes all organic solvents as well as those inorganic solvents that are capable of donating or accepting protons (e.g., H_2SO_4, NH_3). The virtue of such grouping is in part historical, since all the liquids in the first category can be discussed within the framework of Brønsted's concept of acidity and were in fact included in his original scheme of solvent classification. The second category covers molecular SO_2 and amphoteric inorganic solvents. The latter are liquids that exhibit a measurable degree of self-ionization (e.g., $AsCl_3$, BrF_3). Truly ionic liquids—the molten salts—are not discussed in this book.

After introducing the general features of each class of solvents, we select a specific liquid as a typical representative of the class and discuss in some detail its physicochemical properties, offering a condensed overview of the chemistry of its solutions. Because solvents must be classified and discussed in terms of phenomena each of which is also a subject of a separate chapter (acid–base, solvation, electrochemical, kinetic, spectroscopic, etc.), the treatment of these subjects in the present chapter will necessarily have an introductory quality.

3.1.1 Organic and Amphiprotic Inorganic Solvents In this category, the solvents are divided into groups depending on their behavior with respect to proton transfer. Thus, their role is interpreted within the framework of the Brønsted concept of acidity, which defines an acid as a proton donor and a base as a proton acceptor. Brønsted himself proposed a classification scheme for solvents based on the relative magnitude of three solvent properties: the dielectric constant, the acidic strength, and the basic strength. By assigning a plus sign where a property was relatively predominant and a minus sign where it was relatively weak or absent, Brønsted came up with the eight combinations of solvent properties that defined the eight solvent classes as shown in Table 3.1. Water was given as the example of class 1 solvent and benzene, as the example of class 8. Examples of other classes were not given by Brønsted, and there may be slight variations in the way that later authors have completed his table. In particular, the lower limit for the dielectric constants that merit a plus sign has been variously set anywhere between 15 and 30.

Current classification schemes for organic and amphiprotic inorganic solvents are generally derived from Brønsted's prototype but incorporate some modifications. A typical modern breakdown differentiates between the following types of solvents: (1) *Amphiprotic* (sometimes called simply "protic"), which means capable of accepting and donating protons. These are subdivided into (a) protophilic (predominantly basic), (b) protogenic (predominantly acidic), and (c) neutral. (2) *Aprotic*, which means incapable

Table 3.1 Brønsted's Classification of Solvents

Class no.	Diel. const.	Relative acidity	Relative basicity	Modern term	Examples
1	+	+	+	Neutral amphiprotic	Water, methanol
2	+	+	−	Protogenic	H_2SO_4, HF, formic acid
3	+	−	+	Protophilic	*N*-methylpropionamide, tetramethylurea
4	+	−	−	Dipolar aprotic	Acetonitrile, acetone
5	−	+	+	Neutral amphiprotic	Alcohols
6	−	+	−	Protogenic	Acetic acid
7	−	−	+	Protophilic	Most amines
8	−	−	−	Inert	Benzene, alkanes, CCl_4, CS_2

of transferring protons to any appreciable extent. Aprotic solvents are subdivided into (a) dipolar aprotic, and (b) inert (nonpolar). Within the dipolar aprotic class Kolthoff makes a further fine distinction between liquids that are protophilic (slightly basic) and protophobic (without basic properties). The physical properties of many solvents covered by the above categories are collected in Table 3.2. Let us discuss each type in some detail.

A quantitative index of the amphiprotic nature of a solvent is a well-defined and reproducible autoprotolysis constant. For a generalized amphiprotic liquid SH, the autoprotolysis (or self-dissociation) process can be represented as

$$2SH \rightleftharpoons SH_2^+ + S^- \tag{3.1}$$

It is really a combination of two reactions, the first expressing the intrinsic acidity and the second, the intrinsic basicity of the solvent:

$$SH \rightleftharpoons H^+ + S^- \quad \text{(solvent acting as an acid)} \tag{3.2a}$$

$$SH + H^+ \rightleftharpoons SH_2^+ \quad \text{(solvent acting as a base)} \tag{3.2b}$$

(The above equations are not meant to imply necessarily complete dissociation into ions.) The equilibrium constant for the reaction 3.1 is the *autoprotolysis constant* of the solvent, K_s, expressed as the ion-activity product, $K_s = (SH_2^+)(S^-)$. The magnitude of the autoprotolysis constant is a function of both the intrinsic acidity (Eqn. 3.2a) and the intrinsic basicity (Eqn. 3.2b) of the solvent as well as its dielectric constant. Numerical values of the autoprotolysis constants of some solvents are given in Table 3.2. Although apparent values of K_s have been determined experimentally for such nominally aprotic liquids as acetonitrile ($pK_s = 32.2?$) and dimethylformamide ($pK_s = 21?$), there is the danger that their magnitude might be

Table 3.2 Physical Properties of Selected Organic and Amphiprotic Inorganic Solvents

Solvent	Diel. const.	μ, D[a]	BP, °C	FP, °C	d, g ml^{-1}	pK_s
			Amphiprotic Solvents			
H_2SO_4	101	—	~270	10.4	1.8267	3.6
HF	84 (0°)	1.83	19.5	−83.4	1.002 (0°)	9.7, 11.7
H_2O	78.5	1.84	100	0	0.9970	14.0
HNO_3	50 ± 10 (14°)	2.16	82.6	−41.6	1.504	1.7 (−41°)
CH_3OH	32.6	1.70	65.0	—	0.7914	16.7
C_2H_5OH	24.3	1.69	78.3	—	0.7895	19
NH_3	23 (−33.4°)	1.49	−33.4	−77.7	0.6900 (−40°)	29.8
HCl	11.3 (188.1 K)	—	−85.1	−114.2	—	—
HBr	7.33 (187.1 K)	—	−66.8	−86.9	—	—
HAc	6.194 (18°)	0.83	117.7	16.635	1.04365	14.45
HI	3.57 (−45°)	—	−35.4	−50.8	—	—

Dipolar Aprotic Solvents

Solvent	Diel. const.	μ, D	BP, °C	d, g ml^{-1}
N-Methylformamide (NMF)	182.4	3.82	183	0.9988, 0.9976
Formamide	109.5	3.25	193	1.1296, 1.12918
Propylene carbonate (PC)	64.9	4.94	242	1.0257
Dimethylsulfoxide (DMSO)	46.7	4.3	189	1.1014
Nitromethane	38.6	3.44	101	1.1354
N,N-Dimethylformamide (DMF)	36.7	3.82	152	0.9445
Acetonitrile (AN)	36.0	3.84	80.1	0.7856
Hexamethylphosphoramide (HMPA)	29.6	5.37	235	1.0253
Acetone	20.7	2.88	56.2	0.792

Inert Solvents

Solvent	Diel. const.	BP, °C	d, g ml^{-1}
1,1-Dichloroethane	9.90	57.28	1.1757
Chloroform	4.81 (20°)	61.7	1.4832
Benzene	2.27	80.1	0.87865
Carbon tetrachloride	2.23	76.54	1.5940
1,4-Dioxane	2.21	101	1.0337
Cyclohexane	2.202	80.74	0.77855

Unless otherwise stated, all temperatures are in °C and all properties were determined at 25°C. BP = boiling point. FP = freezing point.

[a] Dipole moment in Debye units.

seriously distorted due to the presence of traces of acidic and basic impurities, particularly water and carbon dioxide. Their true pK_s values are likely to be much higher, if at all measurable.

Protophilic Protophilic (basic) solvents are those with a pronounced tendency to accept protons from solutes. They are generally much stronger bases and much weaker acids than water. Examples of common protophilic solvents are the amines and liquid ammonia. Ethers, including dioxane and tetrahydrofuran, are sometimes listed in this category, although they are comparatively weaker bases and can be also found classified as aprotic. This is one of many instances where the boundaries between classes of solvents become vague. Depending on the dielectric constant, protophilic solvents could fall into Brønsted's classes 3 or 7. When a generalized protophilic solvent S reacts with a Brønsted acid HA, the product is the solvated proton, or the lyonium ion, SH^+:

$$\underset{\text{(basic solvent)}}{S} + \underset{\text{(acidic solute)}}{HA} \rightleftharpoons \underset{\text{(lyonium ion)}}{SH^+} + A^- \tag{3.3}$$

The ions SH^+ and A^- may be completely dissociated or partially associated in solution, depending mainly on the dielectric constant of the solvent. Presumably they would tend to be dissociated in solvents of class 3 and associated in solvents of class 7.

Any substance that can protonate the solvent to a measurable extent acts as an acid in it. Moreover, since the strongest acid that can exist in a basic solvent is the protonated solvent itself, SH^+, all those acids that are intrinsically stronger than SH^+ will be converted to it completely by reaction with the solvent. They are known as the *strong acids* in the given medium. Consequently, all strong acids in a protophilic solvent appear to be of the same strength; they are indistinguishable in terms of their acid–base behavior. We say that strong acids are *leveled* down to the same strength, which is that of the lyonium ion SH^+. Those acids that react with the solvent incompletely are called *weak acids* in the given medium.

Protogenic Protogenic (acidic) solvents are those with a pronounced tendency to donate protons to solutes. They are much stronger acids and much weaker bases than water. The most common protogenic solvents that can be used at room temperature and without special apparatus are glacial acetic acid and concentrated sulfuric acid. Most of the others, such as HCl, HBr, and HI, are liquids only at very low temperatures (see Table 3.2 for their liquid ranges), while liquid HF obviously requires special apparatus and handling. In a protogenic solvent, all solutes that accept a proton from the solvent to a measurable degree act as bases:

$$\underset{\text{(basic solute)}}{B} + \underset{\text{(solvent)}}{SH} \rightleftharpoons BH^+ + \underset{\text{(solvent lyate ion)}}{S^-} \tag{3.4}$$

In Eqn. 3.4, a generalized protogenic solvent SH protonates the solute base B to form the protonated base and the solvent lyate ion. The ions may be partially associated in some solvents.

As protophilic solvents bring out the acidity of a solute, protogenic solvents allow manifestaton of a solute's basic strength, ultimately exerting on it a leveling effect. Since the strongest base that can exist in a protogenic solvent is the solvent lyate ion (solvent molecule minus one proton), all bases stronger than that will be converted completely to the solvent lyate ion. Thus in a protogenic solvent all *strong bases* will appear to be of equal strength, or leveled. Again, as in the case of acids, the direction of leveling is downward, to a lower basic strength. Bases that react incompletely with the solvent are called *weak bases*.

Neutral Neutral (amphiprotic) solvents possess both acidic and basic properties; they can act either as proton donors or as proton acceptors, depending on the properties of the solute. The term "neutral amphiprotic" is generally reserved for solvents whose acidic and basic strengths are roughly balanced and similar to those of water. The alcohols and of course water belong to this class.

Depending on the intrinsic acidity or basicity of a solute, a neutral amphiprotic solvent will act as an acid or as a base toward it. If the solute is acidic enough to protonate the solvent, the latter will behave in effect as a protophilic solvent (Eqn. 3.3). If the solute is basic enough to accept a proton from the solvent, the latter will act as a protogenic solvent (Eqn. 3.4). Ultimately a neutral solvent will level all the strong acids by converting them 100% to its lyonium ion SH_2^+ and will level all strong bases by converting them 100% to its lyate ion S^-. Thus, neutral amphiprotic media impose limits on the measurable strengths of both acids and bases. On the other hand, they are generally very good solvents for a large number of electrolytes and polar substances and are the media of choice in which most of the available physicochemical data have been determined.

Dipolar Aprotic Dipolar aprotic solvents exhibit no appreciable tendency to participate in the transfer of protons but are moderately good solvating and ionizing media due to their dipolar nature (dipole moments range from 2.7 to 4.7) and their intermediate (20–40) dielectric constants. In Brønsted's scheme they fall into class 4. Ketones, nitriles, amides, sulfoxides, and nitro compounds are examples of dipolar aprotic liquids. Kolthoff makes the finer distinction between dipolar *protophilic* solvents (dimethylformamide, dimethylsulfoxide, dioxane, pyridine, tetrahydrofuran), and dipolar *protophobic* solvents (ketones, nitriles, nitromethane).

In contrast to amphiprotic solvents, which mask and distort the intrinsic acidic and basic properties of solutes, aprotic solvents allow a considerable range of these properties to be exhibited. As opposed to the leveling effect on acids and bases characteristic of the amphiprotic solvents, the behavior of aprotic solvents toward acids and bases is termed *differentiating*. Dipolar aprotic solvents are moderately good solvents for electrolytes and have come into focus in recent decades because of their ability to increase the rates of certain organic reactions by many orders of magnitude (see Chapters 5 and 10).

Inert Inert solvents are nonpolar aprotic liquids, usually characterized by very low (10–2) dielectric constants. Hydrocarbons and halogenated hydrocarbons are typical examples of inert solvents. Again, the term "inert" as applied here is only relative. Benzene and other aromatic hydrocarbons are known to be π-electron donors and can accept protons under drastic conditions (e.g., in liquid HF). Similarly, some hydrocarbons lose their hydrogens in strongly basic media, forming carbanions. Still, whatever acidities or basicities hydrocarbons may possess are many orders of magnitude below those of typical protogenic or protophilic solvents.

While offering a differentiating, noninterfering medium for chemical reactions, even better in this respect than their dipolar cousins, inert solvents are handicapped by their low solvation ability for most substances of interest, particularly those that are ionic or polar. Thus, their use for physicochemical and analytical purposes is rather limited.

3.1.2 The Role of Hydrogen Bonding If one were asked to point to the single most significant feature that distinguishes amphiprotic from aprotic solvents, the choice would have to be hydrogen bonding. Amphiprotic liquids are characterized by their ability to self-associate via hydrogen bonding and to solvate anions and protophilic uncharged species by donating hydrogen bonds to them. Aprotic liquids are incapable of either of these interactions. This difference in H-bonding ability has many far-reaching consequences and lies at the root of the difference between solution chemistry in amphiprotic and aprotic media.

Despite its widespread recognition in chemistry, the hydrogen bond has eluded a precise definition, which has been the subject of considerable controversy. There is agreement, however, that a hydrogen bond exists when there is evidence of molecular association that involves bonding to a hydrogen atom already bonded to another atom. When a compound contains an atom X that is more electronegative than hydrogen, it will be capable of forming a hydrogen bond, usually depicted as $AH\cdots X$, where AH is a Brønsted acid, or proton donor, and X has an unshared pair of electrons and acts as the base. Thus, in a hydrogen bond, the hydrogen atom is formally divalent. On Pauling's electronegativity scale, atoms with electronegativities greater than hydrogen (2.1) are Se (2.4), I (2.5), Br (2.8), N (3.0), Cl (3.0), and F (4.0).

Hydrogen bonding determines much of the nature of the pure liquids as well as of their solutions. A water molecule can act simultaneously as a hydrogen-bond donor and acceptor and liquid water is associated into a three-dimensional network by H-bonding. Liquid alcohols are associated by hydrogen bonds into linear polymers. In mixtures of water and alcohols, hydrogen-bonded complexes of both species are formed. In amphiprotic solvents, anions with localized charge and uncharged basic solutes are stabilized by hydrogen bonds donated by the solvent, while acid solutes are stabilized via hydrogen bonds donated from the acid to the solvent. In the latter case, the stability of the solvates decreases as the basicity of the liquid decreases. In water, both H_3O^+ and OH^- ions are believed to be associated with three water molecules by hydrogen bonds. The abnormally high electrolytic conductance of both these ions in water is in part a function of the extensive hydrogen bonding. The ability or inability to form hydrogen bonds has a profound effect on solvation and solubility (see Chapter 2).

In amphiprotic solvents, the process of ionic dissociation of acids and bases can be visualized as occurring with the intermediate formation of hydrogen-bonded complexes. When an uncharged acid HA is dissolved in a protophilic solvent S, the reaction can be assumed to proceed in three stages (Eqn. 3.5): (1) an uncharged hydrogen-bonded complex is formed, followed by (2) transition to a hydrogen-bonded ion pair characterized by partial proton transfer, and finally (3) complete proton transfer with varying degrees of dissociation into the free (solvated) ions, depending on the dielectric constant of the medium:

$$\mathrm{AH} + \mathrm{S} \rightleftharpoons \mathrm{AH}\cdots\mathrm{S} \rightleftharpoons \mathrm{A}^-\cdots\mathrm{HS}^+ \rightleftharpoons \mathrm{A}^- + \mathrm{SH}^+ \tag{3.5}$$

An analogous sequence can be formulated for the ionic dissociation of a base in a protogenic solvent. The two intermediate hydrogen-bonded complexes are not detectable in aqueous solutions but may become the predominant species in aprotic and low-dielectric media.

When the solvent is aprotic and thus incapable of forming hydrogen bonds, hydrogen bonding among the solute species becomes prevalent in solution. In other words, in aprotic media those solutes that are capable of hydrogen bonding begin to "solvate" each other. Carboxylic acids, phenols, and primary and secondary amines, as well as alcohols and water are associated in aprotic media. Self-association in acids occurs through formation of hydrogen bonds of the type O—H···O. Carboxylic acids are known to form dimers both in the vapor state and in aprotic solvents:

$$2\mathrm{RCOOCH} \rightleftharpoons (\mathrm{RCOOH})_2 \tag{3.6}$$

$$K_{1,2} = \frac{(\mathrm{RCOOH})_2}{(\mathrm{RCOOH})^2} \tag{3.7}$$

The dimerization constants, $K_{1,2}$, can be quite large, those in benzene ranging from 10^1 to 10^3. Similarly, phenols in aprotic solvents self-associate into dimers and possibly trimers, while in the case of phenols substituted in the *ortho* position to the OH group, intramolecular hydrogen bonding (chelation) occurs. Primary and secondary amines undergo extensive self-association through the formation of NH···N bonds. This occurs in the solid and liquid states as well as in concentrated solutions in aprotic solvents.

In aprotic media, anions with localized charge, such as Cl^- and the carboxylates, are stabilized through hydrogen bonds donated by solute acids. When anion A^- interacts with its conjugate acid HA, the product HA_2^- is appropriately termed a *homoconjugate anion* and the complexation process is known as *homoconjugation* (1):

$$A^- + HA \rightleftharpoons HA_2^- \tag{3.8}$$

$$K_f^{HA_2} = \frac{(HA_2^-)}{(A^-)(HA)} \tag{3.9}$$

where $K_f^{HA_2}$ is the formation constant of the homoconjugate anion. It is generally assumed that the activity coefficients of the ions in Eqn. 3.9 are equal and cancel, so that concentrations can be legitimately substituted for activities. Sometimes the homoconjugate anion can be further associated, forming higher complexes having the general formula $A(HA)_n^-$.

Complexation between an anion A^- and a nonconjugate acid HR is known as *heteroconjugation* (2):

$$A^- + HR \rightleftharpoons AHR^- \tag{3.10}$$

and the complex AHR^- is called a *heteroconjugate anion*. Predictably, the formation constants of homo- and heteroconjugate anions are much greater in protophobic (acetonitrile, nitrobenzene) than in protophilic (dimethylsulfoxide, dimethylformamide) dipolar aprotic solvents. The latter, being more basic, compete more successfully with the anion for the acid. For example, values of $K_f^{HA_2}$ for carboxylate anions are of the order of 10^4–10^5 in nitrobenzene (least basic), 10^3 in acetonitrile, 10^2 in DMF and 10^1 in DMSO (most basic). Competition between a basic solvent and an anion has an effect on the degree of dissociation of hydrogen-bonded ammonium salts in aprotic media. The ion pair $R_3NH^+ \cdots A^-$, which is stabilized by hydrogen bonding, dissociates less than a corresponding quaternary ammonium salt $R_4N^+A^-$, which cannot hydrogen-bond. However, the dissociation of $R_3NH^+ \cdots A^-$ will increase as the basicity of the solvent increases, as the basic solvent can successfully compete with the anion for the solvation of the R_3NH^+ cation. For example, triethylammonium 3,5-dinitrobenzoate has a dissociation constant of only 1.2×10^{-5} in acetonitrile but is practically a strong electrolyte in DMSO. The small increase in dielectric constant (36 to 49) cannot begin to account for a difference of this magnitude.

Consider the effect of conjugation on the solubility of a slightly soluble salt M^+A^-, where A^- is an anion with localized charge. The solubility in an aprotic medium can be increased by addition of acids HA or HR, because the equilibrium

$$M^+A^-(s) \rightleftharpoons M^+ + A^- \tag{3.11}$$

is driven to the right by the formation of conjugate anions HA_2^- or AHR^-. Also the degree of dissociation of salts of the type BH^+A^- and B^+A^- can be increased when the anion A^- is stabilized by conjugation. Increases in solubility due to homoconjugation are known to occur even in neutral amphiprotic solvents, but there the formation constants for the homoconjugate ions are very small. For example, it has been known for a long time that the solubility of benzoic acid is higher in concentrated aqueous alkali benzoate solutions than in pure water. What was traditionally called the "salting in" of benzoic acid is now known to be promoted by the formation of HA_2^- ions between benzoic acid and the benzoate anion. Conjugation effects are important only for anions with localized charge. If the anionic charge is delocalized, as in the case of the picrate ion, conjugation effects become negligible.

Cationic acids, BH^+, can form *homoconjugate cations* $BH^+\cdots B$ and *heteroconjugate cations* $BH^+\cdots B'$ by hydrogen bonding to bases. In aprotic solvents the degree of dissociation of salts of the BH^+A^- type increases slightly because of the homoconjugation of BH^+ with excess base B.

In general, conjugation increases the solubilities of salts, acids, and bases, affects dissociation equilibria and manifests itself in peculiar shapes of potentiometric and conductometric acid–base titration curves (Chapter 6). We will have repeated occasion to return to the subject of conjugation and hydrogen bonding in general both later in this chapter and throughout this book. Specific examples are discussed in the sections devoted to the chemistry of acetonitrile and benzene solutions in this chapter.

3.1.3 Other Classification Schemes for Organic Solvents The proton-oriented scheme for the classification of organic and amphiprotic inorganic solvents just presented is particularly popular and useful in acid–base and electrochemistry. It focuses on the acid–base properties, the dielectric constant, and the dipole moment of the solvent. On the other hand, we know that other solvent properties, such as its structure, polarizability, and coordination ability may in many instances play an important role and might form the basis for other classification schemes. For example, in their efforts to interpret solvent effects on the rates of organic reactions, kineticists have proposed a number of scales in which organic solvents are ranked according to properties rather vaguely described as "polarity," solvating power," or "ionizing power." The quantitative measures for such solvation parameters are variously derived from spectral changes observed as a function of the solvent, from the heats of certain reactions, or from the analysis of the solvent dependence of the rate constants. Reference will be made to several of these scales in the chapters on kinetics (Chapter 10), acid–base chemistry (Chapter 6), and spectroscopy (Chapter 8). Here we will mention only one such scale for the "ionizing power" of solvents, established by Kosower (3). Each solvent is characterized by a "Z-value," which corresponds to the spectral maximum (in kcal mol^{-1}) for the charge-transfer transition in the ion pair 1-ethyl-4-carbomethoxypyridinium iodide. Because the transition occurs between an ionic ground state and a nonionic excited state, the *Z*-values are expected to be higher for those solvents that better stabilize an ionic state. Indeed, the *Z*-value is 94.6 for water, 71.3 for acetonitrile, and 60.1 for isooctane. A variety of other scales based on the solvent dependence of visible and NMR spectra have been proposed.

Coordination chemists are likely to classify solvents on the basis of their ability to donate electron pairs to solutes. Several scales for the coordination power of the solvents are known. In one of them the solvents are ranked according to the extent of ligand-field splitting in the *d*-levels of the Ni^{2+} ion caused by the coordination with the solvent molecules. On this scale ammonia ranks at the top (1080 cm^{-1}), compared to acetonitrile (1026 cm^{-1}) and water (860 cm^{-1}).

The best-known scheme for classifying solvents as electron-pair donors is Gutmann's scale of "donor numbers" (DN) (see General References and also Table 6.2). The donor number is defined as the negative enthalpy of the reaction between the solvent and antimony pentachloride, $SbCl_5$. It really amounts to a scale of Lewis basicity in which $SbCl_5$ serves as the reference acid (see Chapter 6). On Gutmann's scale, the dipolar aprotic solvent hexamethylphosphoramide, $[(CH_3)_2N]_3PO$, ranks at the top, with DN = 38.8, water has a DN = 18.0, acetonitrile, DN = 14.1, and the DN is negligible for 1,2-dichloroethane.

Another similar scheme classifies solvents according to the enthalpies of formation of adducts between the solvents and either phenol or iodine (see Table 6.3). Of course, the donor numbers do not correlate with either the dielectric constant or the dipole moment of the solvent. Classifications of this type seem to be useful for certain limited specialized purposes.

3.1.4 Inorganic Aprotic Solvents Inorganic solvents that contain no hydrogen atoms and do not accept protons can be divided conveniently into two subclasses: (a) molecular solvents, which exhibit no measurable self-ionization, and (b) amphoteric solvents, which are appreciably self-ionized.

Molecular Solvents The only inorganic liquid for which the existence of even slight self-ionization was actually disproved is liquid sulfur dioxide. Its properties will be discussed in greater detail in the following section. Its closest colleague seems to be liquid dinitrogen tetroxide, N_2O_4, for which the proposed self-ionization scheme $N_2O_4 \rightleftharpoons NO^+ + NO_3^-$ must occur to a very slight extent (if at all), considering the negligible specific conductance of the liquid (2.36×10^{-13} mho cm^{-1}).

Amphoteric Solvents Most inorganic liquids exhibit a measurable degree of self-ionization (autoionization) into well-defined cations and anions, according to the general scheme:

$$nS \rightleftharpoons s^+ + s^- \tag{3.12}$$

Although these solvents are aprotic, so that no proton transfer is involved, the above equilibria are analogous to the autoprotolysis of amphiprotic solvents. By analogy, an amphoteric aprotic solvent, S, is interpreted to behave both as an acid and as a base within the framework of so-called *solvent acidity*. It produces a characteristic lyonium ion s^+ (the solvent cation), which is the carrier of acidic properties, and a characteristic lyate ion s^- (the solvent anion), which is the carrier of basic properties in the given solvent. The key difference between solvent acidity and Brønsted acidity lies in the fact that the former has no single ion common to all acid–base systems that could play the unifying role analogous to that of the proton in the Brønsted concept.

Within each system, however, a solvent acid would be a substance that produces in solution the characteristic solvent lyonium ion s^+ and a solvent base would be a solute that yields the characteristic solvent lyate ion s^-. Some examples of the self-ionization of amphoteric aprotic solvents follow:

$$COCl_2 \rightleftharpoons COCl^+ + Cl^- \tag{3.13a}$$

$$2POCl_3 \rightleftharpoons POCl_2{}^+ + Cl^- \tag{3.13b}$$

$$2AsCl_3 \rightleftharpoons AsCl_2{}^+ + AsCl_4{}^- \tag{3.13c}$$

$$2BrF_3 \rightleftharpoons BrF_2{}^+ + BrF^- \tag{3.13d}$$

In the following section we shall discuss the properties of $AsCl_3$ as a representative of amphoteric aprotic solvents. The physical properties of some inorganic aprotic solvents are collected in Table 3.3.

Table 3.3 Physical Properties of Selected Aprotic Inorganic Solvents[a]

Solvent	Diel. const.	BP, °C	FP, °C	d, g ml^{-1}
Sulfur dioxide, SO_2	15.4 (0°)	−10.02	−75.46	—
Nitrogen tetroxide, N_2O_4	2.42 (18°)	21.3	−12.3	1.49
Nitrosyl chloride, NOCl	19.7 (−10°)	−5.5	−64.5	1.59 (−6°)
Phosphoryl chloride, $POCl_3$	13.9 (22°)	1	108	1.71 (0°)
AsF_3	5.7 (−6°)	63	−6	2.45 (20°)
$AsCl_3$	12.6 (17°)	130	−13	2.16 (20°)
$AsBr_3$	8.8 (35°)	220	35	3.33 (50°)
$SbCl_3$	33.0 (75°)	219–223	73	2.44 (178°)
BrF_3	—	126	9	2.8

[a] Unless otherwise stated, all temperatures are in °C and all properties were measured at 25°C. BP = boiling point. FP = freezing point.

3.2 PHYSICOCHEMICAL PROPERTIES OF TYPICAL SOLVENTS

3.2.1 Water—A Neutral Amphiprotic Solvent Except in the field of organic reactions, the chemistry of solutions that we learn in traditional academic courses is the chemistry of *aqueous* solutions. Therefore, for most of us it might be easiest to discuss any given nonaqueous solvent in terms of its similarities and differences relative to water. To provide such a comparative framework, we single out for discussion water, rather than an alcohol, as the representative of neutral amphiprotic liquids, so that we can review some of the key properties of that familiar solvent.

Self-Dissociation The autoprotolysis constant of water ($\sim 10^{-14}$ at 25°C), though not one of the largest, indicates that the self-dissociation reaction

$$2H_2O \rightleftharpoons H_3O^+ + OH^- \tag{3.14}$$

proceeds to a measurable extent. In aqueous solution the carrier of acidic properties is the hydronium ion, usually represented by the simplified formula H_3O^+. Actually, in dilute aqueous solutions the proton is believed to be associated with four water molecules in the primary solvation shell, corresponding to a formula $H_9O_4{}^+$.

The carrier of basic properties in aqueous solutions is the hydroxide ion OH^-, believed to be hydrated by three water molecules in its primary solvation shell. Both the hydronium and the hydroxyl ions are further hydrated through hydrogen bonding. In pure water the law of electroneutrality requires that the concentrations of H_3O^+ and OH^- ions be equal, so that $(H_3O^+) = (OH^-) \cong 10^{-7}\ M$. Variations in ion activities are most conveniently expressed on a logarithmic scale. We can thus define

$$\mathrm{p}a_\mathrm{H} \equiv -\log a_\mathrm{H} \tag{3.15a}$$

and

$$\mathrm{p}a_\mathrm{OH} \equiv -\log a_\mathrm{OH} \tag{3.15b}$$

Obviously, in pure water, $\mathrm{p}a_\mathrm{H} = \mathrm{p}a_\mathrm{OH} \cong 7$. The aqueous scale is arbitrarily considered to extend from $\mathrm{p}a_\mathrm{H} = 0$ to $\mathrm{p}a_\mathrm{H} = 14$, the extreme values corresponding to solutions of a strong acid and a strong base of unimolar activity, respectively.

Acids Any solute that increases the concentration of H_3O^+ ions in aqueous solution beyond its value in pure water ($\sim 10^{-7}\ M$) acts as an acid. The ionic dissociation (often incorrectly called "ionization") of a generalized Brønsted acid HA in water can be regarded as proceeding in two steps. In the first step water acting as a base accepts a proton from the acid, resulting in the formation of the hydronium ion and the acid anion:

$$HA + H_2O \underset{\text{(ionization)}}{\rightleftharpoons} H_3O^+ \cdots A^- \tag{3.16}$$

(The ion pair in Eqn. 3.16 is sometimes written as $H_2OH^+ \cdots A^-$, in order to emphasize the fact that the anion is hydrogen bonded to the hydronium ion. Here we chose to retain the familiar identity of H_3O^+.)

This first step is properly called "ionization" because it involves the formation of ions from covalent reactants. In the second step the ion pair $H_3O^+ \cdots A^-$ dissociates into the free ions:

$$H_3O^+ \cdots A^- \rightleftharpoons H_3O^+ + A^- \tag{3.17}$$

Because of the high dielectric constant of water (78.5), the ion pairing represented by Eqn. 3.16 is generally not detectable in aqueous solutions, and the process of ionic dissociation of an acid is commonly simplified to

$$HA + H_2O \rightleftharpoons H_3O^+ + A^- \tag{3.18}$$

In other words, the extents of ionization and dissociation are equal in aqueous solutions. However, as the dielectric constant of the medium decreases, the ionization process begins to play an increasingly important role in comparison with the dissociation, ultimately acquiring a predominant importance.

Thus, in aqueous solution, the strength of an acid is measured quantitatively by the equilibrium constant for the reaction 3.18:

$$K_a = \frac{(H_3O^+)(A^-)}{(HA)} \tag{3.19}$$

where the quantities in parentheses are the activities. The extensive hydration of all the species in these equations has been omitted for the sake of clarity.

Acids characterized by finite values of K_a are called weak acids. A list of the more common weak acids and their pK_a's in water is shown in Table 3.4. Solutes that are converted 100% to the hydronium ion by the protolysis reaction (Eqn. 3.18) are the strong acids. In water, examples of the strong acids are $HClO_4$, HCl, HBr, HI, HNO_3, and H_2SO_4 (first hydrogen). In aqueous solution, strong acids are leveled, that is, they appear to be of the same strength, and when any of their mixtures are titrated a single titration curve is obtained.

The neutralization process for all strong acids with a strong base is the same:

$$H_3O^+ + OH^- \rightleftharpoons 2H_2O \tag{3.20}$$

For weak acids the neutralization reaction

$$HA + OH^- \rightleftharpoons H_2O + A^- \tag{3.21}$$

liberates the basic anion of the acid A^-, whose basicity increases as the acid strength decreases. Water, acting as an acid, tends to protonate the anion of a weak acid A^-, thus reversing the titration reaction. Because of this acidic property of water the lower limit of acid strengths that can be titrated in water corresponds to about a $pK_a = 9$.

Bases Any solute that increases the concentration of OH^- ions in aqueous solution beyond the value of 10^{-7} *M* acts as a base. As in the case of acids, the ionic dissociation of bases can be formulated as a two-step process of ionization followed by dissociation:

$$B + H_2O \rightleftharpoons BH^+OH^- \qquad \text{(ionization)} \tag{3.22}$$

$$BH^+OH^- \rightleftharpoons BH^+ + OH^- \qquad \text{(dissociation)} \tag{3.23}$$

Again, since ion pairs of the type BH^+OH^- are not detectable in aqueous solution, the ionic dissociation of a base is usually simplified to

$$B + H_2O \rightleftharpoons BH^+ + OH^- \tag{3.24}$$

In the above reactions, water acting as a Brønsted acid protonates the base B and releases an equivalent amount of OH^- ions in the process. The quantitative measure of the strength of a base in water is the equilibrium constant for the reaction 3.24, expressed in terms of activities:

$$K_b = \frac{(BH^+)(OH^-)}{(B)} \tag{3.25}$$

Weak bases, those with finite values of K_b as well as all of the acids, are so-called *ionogens*. In the pure state they exist as covalent molecules and ionize only in the course of a proton-transfer reaction, such as reaction 3.24. Strong bases, on the other hand, are not ionogens but *ionophores*. They contain OH^- ions in their crystal structures in the pure state and so require no solvent for ionization. Thus, the strong bases such as NaOH and R_4NOH (R_4N^+ = tetraalkylammonium) simply dissociate into OH^- ions in water without having to undergo a preliminary ionization via proton transfer.

When a weak base is titrated in water with a strong acid, the neutralization is essentially

$$B + H_3O^+ \rightleftharpoons BH^+ + H_2O \tag{3.26}$$

Just as the range of the pK_a's of acids that can be titrated in water is limited by the acidity of water, the range of pK_b's that produce detectable endpoints when titrated with strong acids in water is limited by the basicity of water. Acting as a base, water tends to reverse the reaction in Eqn. 3.26 by competing with base B for the proton. Of course, the weaker the base, the less complete the neutralization reaction. The limit of titrability of bases in water is about $pK_b = 9$. Ionic dissociation constants of some bases in water are shown in Table 3.4.

Solubility The high dielectric constant and the appreciable dipole moment of water are primarily responsible for the fact that it is an excellent solvent

Table 3.4 Ionic Dissociation Constants of Some Organic Acids and Bases in Water

Acid	pK_a	Base	pK_b
Salicylic	2.98	Butylamine	3.39
Formic	3.77	Ammonia	4.76
Benzoic	4.20	Hydrazine	6.01
Acetic	4.76	Pyridine	8.81
Propionic	4.88	Quinoline	8.94
Phenol	9.95	α-Naphthylamine	10.08

for many ionic and polar substances. Furthermore, the high dielectric constant is the main reason why electrolytes in aqueous solution remain dissociated into ions. Cations are hydrated by interactions with the negative ends of the water dipoles and sometimes as a result of the electron-donor abilities of the water molecule. Anions and other electron-rich species are hydrated by interactions with the positive ends of the water dipoles and, very importantly, by accepting hydrogen bonds from the water molecules.

Acids and bases dissolve in water because of the extensive proton-transfer reactions discussed in the immediately preceding sections. The resulting ions are further hydrated as discussed above. The low polarizability of water makes it a poor solvent for organic nonpolar substances, although small amounts of such solutes can be accommodated in the interstitial voids of the loose three-dimensional structure of liquid water. For a fuller discussion of solvation phenomena the reader is referred to Chapter 2.

Electrochemistry In addition to its excellent solvating and ionizing properties, what makes water such a desirable medium for electrochemistry is the large potential range for oxidation and reduction reactions that can be accommodated by aqueous solutions. Even without the added benefits of overvoltage, the strength of oxidizing agents or the value of potential that can be applied to inert electrodes in water is quite appreciable, being limited only by the redox reaction

$$O_2 + 4H^+ + 4e \longrightarrow 2H_2O \qquad E^\circ = 1.23\ \text{V} \tag{3.27}$$

Oxidizing agents stronger than oxygen in acid solution cannot (thermodynamically) exist in aqueous solution, as they would simply oxidize water according to the reverse of reaction 3.27. Similarly, the limit for reducing agents that can exist in water is set by the potential of the half-reaction

$$2H_2O + 2e \longrightarrow H_2 + 2OH^- \qquad E^\circ = -0.81\ \text{V} \tag{3.28}$$

We will see later that not all solvents are that accommodating with respect to oxidizing and reducing agents.

We have referred to the fact that, as a consequence of the high dielectric constant of water, ions are very rarely associated in aqueous solutions. This simplifies the calculation of ionic concentrations from total (analytical) solubilities of electrolytes. A less widely known consequence of the high dielectric constant of water is its effect on the values of activity coefficients of electrolytes and ions. The simplest approximation for the value of an ionic acitivity coefficient γ is given by the Debye–Hückel limiting law: $\log \gamma = -AI^{1/2}$, where I is the ionic strength (equal to molar concentration for 1:1 electrolytes), and the constant $A = 354.4\epsilon^{-3/2}$ at 25°C. For aqueous solutions at 25°C, $A = 0.509$, and the value of γ calculated for a 10^{-3} M 1:1 electrolyte from the Debye–Hückel limiting law is 0.96. As we can see, the activity correction is relatively small and would become even smaller for more dilute solutions.

Thus, the high dielectric constant of a solvent simplifies matters in two ways: not only can we equate the molar solubility of an electrolyte to ionic concentration, but in dilute solution we can further use it as an approximation of ionic activity. As a rule, neither of these simplifications can be assumed when dealing with solvents of lower dielectric constant.

The Structure of Water Water is the most structured liquid, comprising a three-dimensional network of molecules highly associated through hydrogen bonds. Although the literature on the structure of water is enormous, experts still disagree on some of the fine points in the models for liquid water. What is not in dispute is that the short-range order in liquid water is tetrahedral, where each oxygen atom is coordinated to four other oxygen atoms by means of hydrogen bonds. The network of water molecules associated through hydrogen bonds, described by some as clusters, exists in a dynamic equilibrium with unassociated, monomeric water that fits into the interstitial regions of the open water structure. The effects of water structure on solvation are discussed in Chapter 2.

Liquid alcohols do not possess the three-dimensional structure of water. Each oxygen atom of an alcohol molecule forms only two hydrogen bonds (as opposed to four in water) and the resulting structures are linear polymers.

3.2.2 Acetic Acid—A Protogenic Solvent Glacial acetic acid is the most common acidic solvent used in the laboratory, mainly for the titration of bases, and one that has been extensively investigated both from the applied and theoretical standpoint. Outstanding in this regard is the quantitative treatment of acid–base equilibria in acetic acid by Kolthoff and Bruckenstein, from which the key equations are presented in Chapter 6.

Self-Dissociation Although primarily a protogenic solvent, acetic acid does possess some protophilic tendencies manifested by accepting protons from strong acids and by limited self-protonation:

$$2\mathrm{HAc} \rightleftharpoons \mathrm{H_2Ac^+} + \mathrm{Ac^-} \tag{3.29}$$

where H_2Ac^+ is, of course, $CH_3COOH_2^+$.

As liquid acetic acid is highly associated into dimers and ions exist in it mainly as ion pairs, the self-ionization process 3.29 has been depicted by structural formulas such as

$$\mathrm{H_3C{-}C}\begin{matrix}\diagup\!\!\!\diagup \mathrm{O\cdots H{-}O} \diagdown \\ \diagdown \mathrm{O{-}H\cdots O} \diagup\!\!\!\diagup\end{matrix}\mathrm{C{-}CH_3} \rightleftharpoons \mathrm{CH_3{-}C^{\oplus}}\begin{matrix}\diagup \mathrm{O{-}H\cdots O} \diagdown \\ \diagdown \mathrm{O{-}H\cdots O} \diagup\!\!\!\diagup\end{matrix}\mathrm{{}^{\ominus}C{-}CH_3} \tag{3.30}$$

Historically, investigators found it difficult to agree on the numerical value for the autoprotolysis constant of acetic acid. Today, the most reliable value is considered to be $K_s = 3.5 \times 10^{-15}$ ($pK_s = 14.45$), reported by Bruckenstein and Kolthoff (4).

A value of K_s approximately equal to K_w does not mean that glacial acetic acid is water-like. Water possesses a roughly matched acidity and basicity and a high dielectric constant, which favors the dissociation of the ions formed in the self-ionization process. Acetic acid has a low dielectric constant (6.2) and a very low basicity, so that its K_s value must be dominated by the high proton-donating tendency.

Acids If one were to apply the aqueous criterion of ionic dissociation, no acid could qualify as being "strong" in acetic-acid medium. Furthermore, those acids which appear to be equally strong (leveled) in aqueous solutions have finite and different acidity constants in acetic acid—they are differentiated. These are some of the obvious consequences of the low basicity and the low dielectric constant of acetic acid. Due to its low basicity, acetic acid can be protonated measurably only by the strongest acids, in a process known as *ionization*:

$$\underset{\text{(acid)}}{\mathrm{HA}} + \underset{\text{(solvent)}}{\mathrm{HAc}} \rightleftharpoons \underset{\text{(ion pair)}}{\mathrm{H_2Ac^+\cdots A^-}} \tag{3.31}$$

$$K_i^{HA} = \frac{(\mathrm{H_2Ac^+\cdots A^-})}{(\mathrm{HA})} \tag{3.32}$$

For strong acids, values of K_i^{HA} may be of the order of unity. Because of the low dielectric constant of acetic acid, the ion pair $H_2Ac^+\cdots A^-$ will experience only slight dissociation into the free ions:

$$\mathrm{H_2Ac^+\cdots A^-} \rightleftharpoons \mathrm{H_2Ac^+} + \mathrm{A^-} \tag{3.33}$$

$$K_d^{HA} = \frac{(\mathrm{H_2Ac^+})(\mathrm{A^-})}{(\mathrm{H_2Ac^+A^-})} \tag{3.34}$$

Thus, in acetic acid, a "strong" acid will exist primarily in the form of the ion pair $H_2Ac^+A^-$ in dilute solution and in the form of triple ions and even higher ionic aggregates at higher concentrations. Obviously, the extent of ionic dissociation is not a good measure of the strength of an acid in acetic-acid medium, as it is in water. Instead, the relative strengths of acids are best expressed by the equilibrium constants K_i^{HA}, which govern the extent of ionization (or ion-pair formation) between the acid and the solvent (Eqn. 3.32).

Values of K_i^{HA} and K_d^{HA} are not easy to determine experimentally. The common potentiometric or conductometric techniques, which determine ionic activities or concentrations, respectively, yield a combination of K_i^{HA} and K_d^{HA}, known as the *overall dissociation constant* of the acid, K_{HA}:

$$K_{HA} = \frac{(H_2Ac^+)(A^-)}{[(HA) + (H_2Ac^+A^-)]} = \frac{K_i^{HA}K_d^{HA}}{1 + K_i^{HA}} \tag{3.35}$$

Since the degree of dissociation in acetic acid is always very small, the quantity $[(HA) + (H_2Ac^+A^-)]$ equals, to a good approximation, the total analytical concentration of the acid, C_{HA}.

Bruckenstein and Kolthoff determined the overall dissociation constants of a number of acids and bases (Table 3.5). It is significant that perchloric, sulfuric, and *p*-toluenesulfonic acids, all of which are strong and leveled in water, have clearly differentiated strengths in acetic acid, their pK_{HA} values being 4.87, 7.27, and 8.44, respectively. Evidently, in terms of ionic dissociation, no acids in acetic-acid medium are stronger than acetic acid itself is in water.

Bases In acetic acid, the carrier of basic properties is the acetate ion—the strongest base that can exist in that medium. Any solute capable of increasing the acetate concentration beyond its value in pure acetic acid ($pa_{Ac} \cong 7.2$), acts as a base. Obviously, solutes that happen to be acetates to begin with

Table 3.5 Some Overall Dissociation Constants of Acids and Bases in Acetic Acid

Acid	pK_a	Base	pK_b
Perchloric	4.87	Tribenzylamine	5.4
Sulfuric	7.24	Pyridine	6.10
p-Toluenesulfonic	8.44	Potassium acetate	6.15
Hydrochloric	8.55	Dimethylaminoazobenzene	6.32
		Ammonia	6.40
		Sodium acetate	6.68
		Lithium acetate	6.79
		2,5-Dichloroaniline	9.48
		Urea	10.24
		Water	12.53

Reprinted with permission from *Treatise on Analytical Chemistry*, Part I, Vol. 5, I. M. Kolthoff and P. J. Elving, Eds., Wiley-Interscience, New York, 1966, p. 502 of the Interscience Reprint. Table courtesy of John Wiley & Sons, Inc.

act as strong bases in acetic acid, analogous to hydroxides in aqueous solution. Covalent bases, such as amines, which are ionogens, form the acetate ion by accepting a proton from acetic acid, which is the process of *ionization*:

$$\underset{\text{(base)}}{\text{B}} + \underset{\text{(solvent)}}{\text{HAc}} \rightleftharpoons \underset{\text{(ion pair)}}{\text{BH}^+ \cdots \text{Ac}^-} \tag{3.36}$$

$$K_i^B = \frac{(\text{BH}^+\text{Ac}^-)}{(\text{B})} \tag{3.37}$$

Because of the high acidity of acetic acid, bases that are moderately strong or weak in water (e.g., amines with up to $pK_b \cong 9.3$) are protonated completely by acetic acid and leveled. Weaker bases, including many substances that exhibit no basic properties in water, experience partial protonation and act as bases of various strengths. As in the case of acids, it is the ionization step that determines the relative strengths of bases, and the values of K_i^B for some bases in acetic acid are quite high ($K_i^B \leq 10^5$). However, the dissociation of the ion pairs BH^+Ac^- is as slight as that of any electrolyte in acetic acid:

$$\text{BH}^+\text{Ac}^- \rightleftharpoons \text{BH}^+ + \text{Ac}^- \tag{3.38}$$

$$K_d^B = \frac{(\text{BH}^+)(\text{Ac}^-)}{(\text{BH}^+\text{Ac}^-)} \tag{3.39}$$

For bases, as for acids, the quantity best accessible experimentally is the overall dissociation constant K_B:

$$K_B = \frac{(\text{BH}^+)(\text{Ac}^-)}{[(\text{B}) + (\text{BH}^+\text{Ac}^-)]} = \frac{K_i^B K_d^B}{1 + K_i^B} \tag{3.40}$$

The quantity $[(B) + (BH^+Ac^-)]$ is very nearly equal to the analytical concentration of the base, C_B.

The negligible dissociation of all electrolytes in acetic acid and its powerful protogenic quality combine to produce many, sometimes subtle, consequences in the behavior of more complicated acid–base equilibria, such as apply to titrations, indicator, and buffer systems. For example, when a base is titrated in acetic acid with the strong acid $HClO_4$, the relevant product of the reaction is not the solvent, but the ion pair $BH^+ClO_4^-$. Further discussion of acid–base equilibria in acetic acid will be deferred to Chapter 6, where their detailed formulation will be presented. Here it will suffice to stress that all acids and bases in acetic acid are "weak" in the sense of being slightly dissociated into ions. Thus, the hydrogen-ion concentration in acetic acid is a function of the square root of acid or base concentration. Nominally, the pH scale in acetic acid should extend from 0 to 14.45 pH units ($pK_s = 14.45$). However, as even the strongest acid, $HClO_4$, is incompletely dissociated in acetic acid, a pH of zero is never reached in practice. For example, a solution of 1 *M* $HClO_4$ has a pH of 2.4. Consequently, the pH scale in acetic acid is probably no longer than 10 pH units.

Solubility On the basis of its low dielectric constant and dipole moment, one would expect acetic acid to be a poor solvent for ionic and polar substances. On the other hand, its high acidity and the ability to form hydrogen bonds should make it a promising solvator for anions and other electronegative species. In fact, some salts are soluble in acetic acid to a surprisingly large extent. This is especially true of the acetates, where such solubilities as 23.46 mol % (NH_4Ac) and 19.72 mol % ($Pb(Ac)_2$) were reported. Most soluble, however, seem to be $ZnCl_2$ (40 mol % at 30°C) and $LiClO_4$ (38.0 mol % at 25°C). Generally, most nitrates and chlorides also exhibit good to moderate solubilities. A number of solvates of acetic acid have been isolated.

The slight ionic dissociation of all electrolytes in acetic acid exerts some peculiar effects on the solubility behavior as well. For example, the solubility of KBr can be increased by addition of electrolytes without a common ion. Since the changes in ionic strength involved are negligible, the increase in solubility is attributed to the exchange of ion pairs of the type $K^+Br^- + M^+X^- \rightleftharpoons K^+X^- + M^+Br^-$, which drives the reaction from left to right. The solubilities of some sparingly soluble acetates, such as $Zn(Ac)_2$ and AgAc, can be increased by addition of soluble acetates, which form soluble complexes.

Of course, any molecular substance that possesses even the slightest basic property will dissolve in acetic acid with the aid of the proton-transfer reaction discussed earlier. Often, molecular adducts form between the basic solutes and acetic acid; a number of higher complexes have been reported between acetic acid and pyridine as well as other amines. Weak 1:1 complexes are believed to form between acetic acid and such weakly basic substances as acetonitrile, dioxane, dimethylformamide, tetrahydrofuran, and water.

Interestingly, acetic acid also forms complexes with the Lewis acids, such as $SnCl_4$, $SnBr_4$, $SbCl_3$, and BF_3. The products are strong acids postulated to have formulas such as $H_2SnCl_4(Ac)_2$.

Electrochemistry When contrasting the electrochemistry in aqueous solutions and in solvents of low dielectric constant, the key factor to remember is that electrochemical measurements by and large involve the properties of *ions*. Since ions are in short supply in media of low dielectric constant, electrochemical measurements in such solvents as acetic acid are likely to be difficult and their interpretation, less reliable.

One case in point is electrolytic conductance. All electrolytes are "weak" in acetic acid and the observed values of equivalent conductance Λ are generally below unity. In dilute solution, plots of log Λ vs log C are linear, typical of extensive ion pairing, but the straight lines begin to curve at concentrations higher than about 10^{-2} *M*, indicating the formation of triple ions. Extrapolations of such data to infinite dilution to obtain the limiting equivalent conductances, Λ_0, are not very reliable.

It is not surprising that even small amounts of water can have a drastic effect on the ionizing behavior in acetic-acid solutions. For example, addition of 1% water increases the specific conductance of acetic acid about tenfold. Much of the discrepancy among the conductance and other results in acetic acid reported by different investigators may be due to different degrees of contamination by trace amounts of water.

Nothing like a series of standard electrode potentials is available in acetic acid, and even the few values for the $E°$'s of cells without transference that are known are sometimes disputed. In low-dielectric anhydrous media such factors as electrode preparation, poisoning of the electrode by impurities, contamination by water and even the method of extrapolation to arrive at the standard potentials provide ample cause for irreproducible results as well as for questioning any one of them. Nevertheless, we have for the cell Pt, $H_2(g)|H_2SO_4|Hg_2SO_4(s)$, Hg an $E° = 0.338$ V (at 25°C) and for the cell Pt, $H_2(g)|HCl|AgCl(s)$, Ag, an $E° = -0.62084$ V (at 35°C).

More common have been potential measurements on cells with liquid junction, mainly employed for the determination of pH and for acid–base titrations. Historically, the first of this type was the chloranil electrode of Conant and his associates (6): Pt$|C_6Cl_4O_2$ (satd.), $C_6Cl_4(OH)_2$ (satd.), HX$|$, which was found to respond in Nernstian fashion to changes in hydrogen-ion activities in acetic acid, forming the basis for the establishment of a scale of relative pH values in that medium. For a chloranil electrode the half-cell reaction is

$$C_6Cl_4O_2 + 2H^+ + 2e \longrightarrow C_6Cl_4(OH)_2 \qquad (3.41a)$$

and the corresponding potential E is given by

$$E = E° + 0.05916 \log a_H \quad \text{at } 25°C \qquad (3.41b)$$

An aqueous calomel electrode was generally used as a reference, so that the measurements contained an unknown, hopefully reproducible, liquid-junction potential at the interface of acetic acid and water. Of course, since all acids in acetic acid are weak, a tenfold change in the concentration of an acid changes the potential of the chloranil electrode by 29.6 mV.

A subject almost never mentioned is that of ionic activity coefficients in acetic acid or in other solvents with dielectric constants that low. Because the concentration of ions is low to begin with, corrections for activity coefficients are universally neglected. But are they really negligible? We mentioned earlier that for a 10^{-3} M solution of a 1:1 electrolyte (e.g., NaCl) in water, the activity coefficient calculated from the Debye–Hückel limiting law would be almost unity (0.96), or that the activity of each of the ions would equal 9.6×10^{-4} M, which is within a few percent of their molar concentration. In acetic acid, however, the Debye–Hückel expression for the activity coefficient is $\log \gamma = -23\, I^{1/2}$. Accordingly, the activity coefficient in a solution of 10^{-3} *M ions* would be about 0.19. Thus, the activity correction in a situation typical of the strongest electrolytes in acetic acid (e.g., a 10^{-1} M solution of a salt having a $K_d = 10^{-5}$) would be quite significant.

It goes without saying that all of our preceding discussion pertained to *anhydrous* acetic acid because even trace amounts of water would introduce changes in the observed physicochemical properties.

3.2.3 Liquid Ammonia—A Protophilic Solvent

Acid–Base Chemistry Although ammonia exists as a liquid only at low temperatures (−33 to −78°C for the pure liquid), it has been studied more than any other nonaqueous solvent, including those that are liquid at room temperature. In many ways the chemistry of ammonia and ammoniacal solutions is remarkably like that of water and aqueous solutions. To begin with, liquid ammonia undergoes an autoprotolysis reaction analogous to that of water:

$$2NH_3 \rightleftharpoons NH_4^+ + NH_2^- \qquad (3.42)$$

However, the low specific conductance of pure ammonia (10^{-11} mho cm^{-1}) and the low autoprotolysis constant (1.9×10^{-33} at −50°C) indicate that the above dissociation must proceed only to a very minute degree. Undoubtedly the low dielectric constant of ammonia and its low acidity are responsible for the negligible self-dissociation. Nevertheless, the NH_4^+ and the NH_2^- ions can be identified in ammonia as the carriers of acidic and basic properties, respectively. The strongest base in ammonia is the NH_2^- ion; hydroxide and the alcoholates are weaker. The pH of a 10^{-2} *M* solution of amide is about 30, but that of a 10^{-2} *M* hydroxide, only about 24.5. Nominally, the pH scale in ammonia stretches from 0 to 33 units, the neutrality point being at pH = 16.5. At that point, $pa_{NH_4} = pa_{NH_2}$.

Many of the chemical reactions of the NH_4^+ and NH_2^- ions in ammonia are analogous to the reactions of H_3O^+ and OH^- ions in water. Examples are acid–base reactions:

$$NH_4Cl + KNH_2 \rightleftharpoons KCl + 2NH_3 \quad \text{(in ammonia)} \qquad (3.43a)$$

and

$$HCl + KOH \rightleftharpoons KCl + H_2O \quad \text{(in water)} \qquad (3.43b)$$

amphoteric reactions:

$$\left.\begin{array}{l} ZnCl_2 + 2KNH_2 \rightleftharpoons Zn(NH_2)_2\downarrow + 2KCl \\ Zn(NH_2)_2 + 2KNH_2 \rightleftharpoons K_2\,Zn(NH_2)_4 \end{array}\right\} \text{in ammonia} \qquad (3.44a)$$

and

$$\left.\begin{array}{l} ZnCl_2 + 2KOH \rightleftharpoons Zn(OH)_2\downarrow + 2KCl \\ Zn(OH)_2 + 2KOH \rightleftharpoons 2K^+ + Zn(OH)_4^{2-} \end{array}\right\} \text{in water} \qquad (3.44b)$$

displacement of acidic hydrogen by active metals:

$$Na + NH_4Cl \rightleftharpoons \tfrac{1}{2}H_2 + NaCl + NH_3 \quad \text{in ammonia} \qquad (3.45a)$$

and

$$Na + H_3O^+ + Cl^- \rightleftharpoons \tfrac{1}{2}H_2 + NaCl + H_2O \qquad \text{in water} \tag{3.45b}$$

precipitation reactions, where NH_2^- in ammonia acts as OH^- in water:

$$AgNO_3 + KNH_2 \rightleftharpoons AgNH_2\downarrow + KNO_3 \tag{3.46a}$$

$$PbI_2 + 2KNH_2 \rightleftharpoons PbNH\downarrow + 2KI + NH_3 \tag{3.46b}$$

and ammonolysis of esters (catalyzed by NH_4^+) as compared to hydrolysis of esters (catalyzed by H_3O^+):

$$CH_3COOC_2H_5 + NH_3 \rightleftharpoons CH_3CONH_2 + C_2H_5OH \qquad \text{(in ammonia)}$$

and (3.47a)

$$CH_3COOC_2H_5 + H_2O \rightleftharpoons CH_3COOH + C_2H_5OH \qquad \text{(in water)}$$

(3.47b)

These analogies would be complete if it were not for the fact that, in ammoniacal solutions, electrolytes are largely associated into ion pairs, triple ions, and higher ionic aggregates.

Liquid ammonia is much more basic than water, so that many substances that exhibit little or no acidity in aqueous solution act as acids in ammonia. These include amides, sulfonamides, amines, and even some hydrocarbons. Sulfamic acid, SO_2NH_2OH, which is a strong monobasic acid in water, acts as a dibasic acid in ammonia. Most of the so-called weak acids in water (e.g., acetic acid) appear to be strong in ammonia and leveled due to completeness of the proton-transfer reaction:

$$\underset{\text{(solute)}}{HA} + NH_3 \rightleftharpoons NH_4^+A^- \tag{3.48a}$$

$$K_i^{HA} = \frac{(NH_4^+A^-)}{(HA)} \tag{3.48b}$$

The ionization constant K_i^{HA} is probably the best measure of the strength of an acid in liquid ammonia. However, for a more complete measure, the dissociation of the ion pair must also be taken into account:

$$NH_4^+A^- \rightleftharpoons NH_4^+ + A^- \tag{3.49}$$

$$K_d^{HA} = \frac{(NH_4^+)(A^-)}{(NH_4^+A^-)} \tag{3.50}$$

An example from the literature (7) where both K_i^{HA} and K_d^{HA} have been determined for a pair of acids is shown in Table 3.6. It is clear that, while both the ionization and the dissociation step are significant in determining the strength of these acids, the ionization is the predominant process.

Very few acidity constants have been determined in liquid ammonia and

Table 3.6 Ionization and Dissociation Constants of *o*- and *p*-Nitroacetanilide in Liquid Ammonia at $-55.6°C$[a]

Acid	$K_i \times 10^2$	$K_d \times 10^4$
o-NO_2-C_6H_4-$NHCOCH_3$	2.2 ± 1.4	2.2 ± 1.4
p-NO_2-C_6H_4-$NHCOCH_3$	9.3 ± 0.6	0.89 ± 0.06

[a] Data of Cuthrell, Fohn, and Lagowski (7).

the equations governing acid–base equilibria have not been developed to the same extent as for some of the other solvents. This neglect probably stems from the fact that ammonia is not a liquid at room temperature and therefore found no application as an analytical titration medium for acids.

Solutions of Metals One property of liquid ammonia that finds no parallel in water is its ability to form solutions of alkali- and alkaline-earth metals in which solvated electrons exist as reasonably stable species. Dilute solutions of these metals in ammonia have identical absorption spectra, independent of the nature of the metal, with an absorption maximum in the vicinity of 15,000 Å. The metal solutions are blue because the tail of the absorption band extends into the visible region. They show an abnormally high electrolytic conductance (Λ_0 = 1022!) and a molar magnetic susceptibility approaching that of a mole of free electrons. These facts are interpreted by assuming that in dilute ammoniacal solutions the metals dissolve with the formation of cations and solvated electrons situated in solvent cavities. The absorption, paramagnetism, and high electrolytic conductance, all of which are independent of the metal, are accounted for by the existence of solvated electrons. At higher concentrations, the metal cations and the electrons recombine, forming a bronze-colored solution of the metal, and the electrolytic conductance of these solutions does approach that of the pure metal.

Metal–ammonia solutions are very strong reducing agents in which the oxidation–reduction reactions can be easily followed both conductometrically and spectrophotometrically. Ammonium ions are reduced by these solutions to hydrogen gas and ammonia.

Solubility When compared to water, liquid ammonia has a lower dielectric constant, dipole moment, and acidity but is much more basic and polarizable. Thus, ion–solvent interactions in ammonia are expected to be weaker than in water, except with ions that are acidic or highly polarizable. Indeed, the solubility of ionic substances in ammonia is generally much lower than in water. Salts of doubly and triply charged anions are practically insoluble because their lattice energies are too high. In general, the highest solubilities are exhibited by ammonium salts, probably as a result of specific interactions with the structurally similar solvent. Since ammonia is more polarizable than water, the dispersion (mutual polarizability) interactions account for the phenomenon that polarizable nonpolar substances (such as benzene) are more soluble in ammonia than in water. In general, covalent organic substances have greater solubilities in ammonia than in water. The high polarizabilities of the iodide and the thiocyanate ions is also the reason why their salts are very soluble in ammonia.

Polar organic compounds, such as alcohols, amides, amines, aldehydes, ketones, and esters dissolve readily in ammonia probably as a result of dipole–dipole interactions aided wherever possible by the formation of hydrogen bonds from the solute to ammonia (e.g., for alcohols, primary and secondary amines). Interactions via hydrogen bonding from ammonia to the solute are also known (e.g., ethers, tertiary amines). Only saturated hydrocarbons and dicarboxylic acids are virtually insoluble in ammonia.

Ammonia is, of course, a good solvent for Brønsted acids because of the ability to accept protons and it also reacts with Lewis acids, which act as electron-pair acceptors. Thus, complexes known as ammonates (analogous to hydrates in water) are formed with such metal ions as Ag^+, Cu^+, Cu^{2+}, Co^{3+}, Cr^{3+}, Fe^{2+}, Hg^{2+}, and Pt^{2+}. It is the specific interaction between ammonia and Ag^+ ion that causes the reverse of an elementary reaction in water solution to occur in ammonia:

$$Ba(NO_3)_2 + 2AgCl \longrightarrow BaCl_2\downarrow + 2AgNO_3 \qquad (3.51)$$

In Table 3.7, we compare the solubilities of some salts in ammonia and in water.

Electrochemistry In terms of ionic dissociation, all electrolytes in ammonia are weak. Up to about 10^{-3} M, they exist predominantly as ion pairs, such as M^+A^-; at higher concentrations, triple ions, such as $A^-M^+A^-$ and $M^+A^-M^+$ tend to form. Nevertheless, the dissociation constants of electrolytes in liquid ammonia are large enough to produce an appreciable concentration of free ions in dilute solutions. Values of K_d range from 10^{-6} to 10^{-2}. A typical salt that undergoes no specific interactions with the solvent (e.g., NaCl) has a $K_d \approx 10^{-3}$, which means that for a 10^{-1} M analytical concentration of NaCl the concentration of each of the free ions would be of the order of 10^{-2} M. This is well within the range of reliable electrochemical measurements, and it is no wonder that electrolytic conductances as well as electrode potentials have been determined for a large number of systems in liquid ammonia. In Table 3.8 we compare the limiting equivalent conductances for a number of ions in ammonia and in water. In general, ionic conductances are higher in ammonia than in water, suggesting a lower degree of solvation by ammonia. It is significant that no abnormally high conductance is observed for the NH_4^+ and NH_2^- ions in ammonia, in contrast to their analogs H_3O^+ and OH^- in water. In aqueous solutions H_3O^+ and OH^- ions are not actually transported through the solution; instead, their conductance is believed to be the result of a series of proton jumps from H_3O^+

Table 3.7 A Comparison of the Solubilities of Salts in Liquid Ammonia and in Water[a]

Salt	Solubility in NH_3 ($g/100$ g of NH_3) at 25°C[b]	Solubility in H_2O ($g/100$ ml of H_2O)[c]
NH_4Cl	102.5	29.7
NH_4Br	237.9	59.8
NH_4I	368.4	154.2
NH_4SCN	312.0	128
NH_4ClO_4	137.9	10.74
NH_4NO_3	390.0	118.3
NH_4Ac	253.2	148 (4°C)
$(NH_4)_2CO_3$	0.0	100 (15°C)
AgCl	0.83	8.9×10^{-5}
AgBr	5.92	(10°C)
AgI	206.84	8.4×10^{-6}
NaF	0.35	3×10^{-7}
NaCl	3.02	4.22 (18°C)
NaBr	137.95	35.7
NaI	161.9	79.5
KCl	0.04	158.7
KBr	13.5	34.7 (20°C)
KI	182	53.48
		127.5

[a] Data in liquid ammonia were taken from the compilation by Waddington (8). Data for water were taken from the *Handbook of Chemistry and Physics,* 40th ed., Chemical Rubber Publishing Co., Cleveland, Ohio, 1958–1959.
[b] At elevated pressure.
[c] Unless otherwise specified, the temperature was 0°C.

Table 3.8 A Comparison of the Limiting Ionic Conductances in Ammonia[a] and in Water

Ion	Ammonia (−33.5°C)	Water (25°C)
Na^+	158	50.1
NH_4^+	142	73.5
H_3O^+	Does not exist	349.8
NO_3^-	177	71.4
NH_2^-	166	Does not exist
OH^-	—	198.3
Br^-	170	78.2

[a] From a compilation by Waddington (8).

ions to favorably oriented adjacent water molecules and from water molecules to adjacent OH^- ions. Apparently, this proton-jump mechanism believed to be responsible for the abnormally high conductance of the H_3O^+ and OH^- ions in water is lacking in liquid ammonia.

A number of standard electrode potentials have been determined in ammonia (Table 3.9). No direct comparison can be made between these potentials and their counterparts in water, because all of them are referred to the arbitrary zero of the standard hydrogen electrode *in ammonia*. For extrathermodynamic methods of correlating electrode potentials in different solvents, see Chapter 5.

Oxidation–reduction chemistry in liquid ammonia is thermodynamically extremely limited by the tiny range of oxidation–reduction potentials that can be tolerated by the solvent. The strongest reducing agent that can exist in ammonia is hydrogen, and the strongest oxidizing agent, nitrogen. Stronger redox agents would decompose the solvent according to the reverse of the following reactions:

$$NH_4^+ + e \longrightarrow \tfrac{1}{2}H_2 + NH_3 \qquad E^\circ = 0 \tag{3.52}$$

$$\tfrac{1}{2}N_2 + 3NH_4^+ + 3e \longrightarrow 4NH_3 \qquad E^\circ = 0.04 \tag{3.53}$$

These limiting reactions provide a thermodynamic potential range of only 0.04 V. Fortunately, both the hydrogen and the nitrogen couples in ammonia are subject to high overvoltage, which extends the available potential range beyond this short thermodynamically stable value.

Table 3.9 Standard Electrode Potentials in Liquid Ammonia[a]

Electrode	E°, V (molal scale)
Li^+/Li	−2.24
K^+/K	−1.98
Cs^+/Cs	−1.95
Rb^+/Rb	−1.93
Na^+/Na	−1.85
Zn^{2+}/Zn	−0.53
Cd^{2+}/Cd	−0.20
H^+/H_2	0
Pb^{2+}/Pb	+0.32
Cu^+/Cu	+0.41
Cu^{2+}/Cu	+0.43
Ag^+/Ag	+0.83

[a] Data of Pleskov and Monossohn (9) at −35°C.

3.2.4 Acetonitrile—A Dipolar Aprotic Solvent Acetonitrile, CH_3CN, has an intermediate dielectric constant (36.0), a large dipole moment (3.84), very low basicity, and virtually negligible acidic properties. Although its basicity is much lower than that of water, there can be no doubt that its lyonium ion, CH_3CNH^+, is a stable species in solution. Thus $HClO_4$ behaves as a strong acid in acetonitrile,* while in a medium of 99% H_2SO_4, acetonitrile is reportedly 50% protonated. On the other hand, the conjugate base of acetonitrile, CH_2CN^-, appears to be unstable and attempts to produce appreciable concentrations of it by reacting acetonitrile with alkali metals lead to polymerization of the solvent. As a result, the quantitative aspects of the proposed self-dissociation

$$2CH_3CN \rightleftharpoons CH_3CNH^+ + CH_2CN^- \tag{3.54}$$

are somewhat controversial. Indeed, the original estimate of the autoprotolysis constant of acetonitrile [$pK_s = 28.5$ (1)] was later revised upward to $pK_s = 32.2$ (10), with the qualification that the true value might be even higher.

It must be stressed that the determination of an autoprotolysis constant in an essentially aprotic medium can be frustrated by the presence of trace amounts of acidic and basic impurities. In acetonitrile, such impurities would be likely to include primarily carbon dioxide and water as well as the hydrolysis products of the solvent, namely, acetamide, ammonium acetate, ammonia, and acetic acid. While every precaution is taken in the preparation of the solvent to reduce the levels of the above contaminants to a negligible value, complete elimination of contaminants is impossible, particularly in the course of subsequent reactions and measurements. Nevertheless, the pK_s of 32.2 can be taken as an effective or operational value of the autoprotolysis constant of acetonitrile under the best conditions attainable experimentally. Thus the practical pH scale in acetonitrile would extend from 0 to 32.2 units, with neutrality at a pH of 16.1. The long pH scale means that, compared to water, acetonitrile is a much more differentiating solvent toward acids and bases.

Acids As in any solvent with even the slightest protophilic properties, acids in acetonitrile undergo measurable ionization followed by dissociation:

$$HA + CH_3CN \rightleftharpoons CH_3CNH^+\cdots A^- \quad \text{(ionization)} \tag{3.55}$$

$$CH_3CNH^+\cdots A^- \rightleftharpoons CH_3CNH^+ + A^- \quad \text{(dissociation)} \tag{3.56}$$

* This has been the prevailing conclusion in earlier acid–base literature, derived from conductometric and potentiometric evidence, but it was disputed since by Kinugasa et al. [*J. Phys. Chem.*, **77**, 1914 (1973)] on the basis of an infrared study. The latter showed that perchloric acid is not a strong electrolyte in *pure* acetonitrile, but appears to be one only in the presence of acetic acid in acetonitrile, due to formation of protonated acetic acid. The point is that in all those acid–base studies where $HClO_4$ was observed to be a strong acid in acetonitrile, its solutions were prepared by dehydration with acetic anhydride, thus introducing acetic acid to all solutions. Of course, as a practical matter, the $HClO_4$–HAc mixtures, which are commonly employed as "$HClO_4$" in acetonitrile, nevertheless represent solutions of strong acid.

As was mentioned earlier, the only acid that seems to be completely dissociated in acetonitrile is $HClO_4$ in solutions containing also acetic acid. When processes 3.55 and 3.56 predominate, we have the familiar simple dissociation, governed by the overall dissociation constant:*

$$K_d^{HA} = \frac{(CH_3CNH^+)(A^-)}{(CH_3CNH^+A^-) + (HA)} \tag{3.57}$$

However, in acetonitrile the above equation applies only to acids containing anions with delocalized charge, such as picric and 2,6-dihydroxybenzoic acid. More common is the case where the anion A^- has a localized charge, for example, the chloride or carboxylates. Such anions are not stabilized by acetonitrile, which is incapable of hydrogen bonding, but by their conjugate acids HA, which act as hydrogen-bond donors. Thus, in the majority of cases the dissociation of an acid in acetonitrile is governed by the equilibria of dissociation as well as homoconjugation (see Eqn. 3.8):

$$HA + CH_3CN \rightleftharpoons CH_3CNH^+ + A^- \tag{3.58}$$

and

$$A^- + HA \rightleftharpoons HA_2^- \tag{3.59}$$

$$2HA + CH_3CN \rightleftharpoons CH_3CNH^+ + HA_2^- \tag{3.60}$$

Then, the overall dissociation constant becomes†

$$K_2^{HA} = \frac{(CH_3CNH^+)(HA_2^-)f^2}{(HA)^2} = K_f^{HA_2} K_d^{HA} \tag{3.61}$$

$K_f^{HA_2}$ is the formation constant of the homoconjugate anion HA_2^- and was defined earlier in Eqn. 3.9.

The formation constants $K_f^{HA_2}$ have been determined potentiometrically on mixtures of HA with the corresponding tetraethylammonium salts using the glass electrode, which behaves reversibly in acetonitrile with respect to changes in hydrogen-ion activity. Independently, in some systems the formation constants were also derived from the variation of the solubility of the carboxylate salt as a function of carboxylic-acid concentration (see Eqn. 3.11 and its discussion). The dissociation constants K_d^{HA} are derived from potentiometric, spectrophotometric, and conductometric data. In Table 3.10 the values of $K_f^{HA_2}$, K_d^{HA}, and K_2^{HA} (in logarithmic form) for some of the acids are compiled. It can be seen that the formation constants for the homoconjugate anions in acetonitrile are generally very large (10^3–10^4), so that the overall dissociation equilibria for most acids correspond to reaction 3.60, rather than to simple dissociation. As we mentioned earlier, the simple

* In the original literature the superscripts and subscripts on constants for acetonitrile solutions are reversed, e.g., K_{HA}^{d}. Here, we are being consistent with the symbols used for equilibria in acetic acid and ammonia. Ionic charges have been dropped from subscripts and superscripts to avoid confusion with + and − signs in equations.

† The ambiguous literature symbol K_2(HA) was replaced by K_2^{HA}.

Table 3.10 Dissociation Constants of Acids in Acetonitrile

Acid	pK_d^{HA}	$\log K_f^{HA_2}$	pK_2^{HA}
HCl	8.94	2.23	6.71
HNO_3	8.89	2.30	6.59
HBr	5.51	2.45	3.06
H_2SO_4	7.29	3.03	4.26
3,5-Dinitrobenzoic	16.9	4.0	12.9
Salicylic	16.7	3.3	13.4
p-Nitrobenzoic	18.7	3.8	14.9
m-Bromobenzoic	19.5	3.7_5	15.7
p-Hydroxybenzoic	20.8	3.0_5	17.7
Benzoic	20.7	3.6	17.1
Phenol	26.6	5.76[a]	
o-Nitrophenol	22.0		
p-Nitrophenol	20.7	4.93	
2,4-Dinitrophenol	16.0		
Picric	11.0	0.3	

[a] Corresponding to the equilibrium $A^- + 2HA \rightleftharpoons (HA)_2A^-$.
Data of Kolthoff, Chantooni and Bhowmik (12) and Coetzee and Padmanabhan (13).

dissociation scheme (Eqn. 3.57) prevails only for picric and 2,6-dihydroxybenzoic acids, where homoconjugation is negligible.

Bases Acetonitrile is such a weak acid that the characteristic dissociation of a base

$$B + CH_3CN \rightleftharpoons BH^+ + CH_2CN^- \tag{3.62}$$

is barely detectable in acetonitrile solutions. A conductometric estimate of the equilibrium constant

$$K_d^B = \frac{(BH^+)(CH_2CN^-)}{(B)} \tag{3.63}$$

for 1,3-diphenylguanidine ($pK_b = 4.00$ in water) in dilute solutions was reported to be $\sim 2 \times 10^{-11}$ (11). At base concentrations greater than $5 \times 10^{-2}\ M$, the predominant equilibrium seems to involve the formation of a homoconjugate cation between the ammonium ion and the base $BH^+\cdots B$

$$2B + CH_3CN \rightleftharpoons (BHB)^+ + CH_2CN^- \tag{3.64}$$

$$K_2^B = K_d^B\, K_f^{BHB} \tag{3.65}$$

The overall dissociation constant corresponding to Eqn. 3.65 (equilibrium 3.64) was estimated to be 4×10^{-10}, from which the value for the formation constant of $(BHB)^+$ is found to equal 20. Of course,

$$K_f^{BHB} = (BHB^+)[(B)(BH^+)]^{-1}$$

The formation constants of homoconjugate cations in acetonitrile range from zero to 35. Thus they are much smaller than the formation constants for the homoconjugate anions, reflecting the relative preference of acetonitrile to solvate cations as opposed to anions. The combination of low dissociation and low homoconjugation constants means that the overall dissociation constants of bases in acetonitrile must be much lower than those of acids. The data in Table 3.11 confirm this.

An acid–base titration in acetonitrile involves fundamentally the formation of the salt, for example, $BH^+ClO_4^-$, which may be partially associated, so that its dissociation constant must be known for equilibrium calculations. However, the titrations are generally more complicated due to

Table 3.11 Dissociation Constants of Protonated Monoamines in Acetonitrile

Amine	K_f^{BHB} [a]	$pK_d^{BH^+}$ [b]
Ammonia	11	16.46
Methylamine	35	18.37
Dimethylamine	31	18.73
Trimethylamine	6	17.61
Ethylamine	25	18.40
Diethylamine	2	18.75
Triethylamine	0	18.46
n-Propylamine	19	18.22
Tri-*n*-propylamine	0	18.10
n-Butylamine	26	18.26
Di-*n*-butylamine	0	18.31
Tri-*n*-butylamine	0	18.09
t-Butylamine	20	18.14
Isobutylamine	—	17.92
Diisobutylamine	0	17.88
Triisoamylamine	0	18.04
Piperidine	26	18.92
Pyrrolidine	32	19.58
Pyridine	4	12.33
Aniline	0	10.56
p-Toluidine	0	11.25
Benzylamine	15	16.76
Monoethanolamine	24	17.53
Morpholine	10	16.61
1,3-Diphenylguanidine	0	17.90

Reprinted with permission from *J. Amer. Chem. Soc.*, **87**, 5007 (1965). Copyright 1965. American Chemical Society.

[a] $K_f^{BHB} = (BHB^+)/(B)(BH^+)$.

[b] $K_d^{BH^+} = (H^+)(B)/(BH^+)f_1$, where f_1 is the activity coefficient of BH^+ calculated from the Debye–Hückel limiting law.

homoconjugation, which leads to the appearance of maxima in conductometric titration curves and to additional inflections in potentiometric titration curves, both at 50% neutralization points. The discussion of titration equilibria is in Chapter 6.

Solubility Acetonitrile is a good solvent for covalent substances and for ionic compounds containing highly polarizable anions, such as picrate, thiocyanate, triiodide, and perchlorate, a property evidently determined by its own high polarizability. On the basis of its large dipole moment, one would expect acetonitrile to have strong interactions with all ions, but generally this does not seem to be the case. The poor solvation ability of acetonitrile toward anions with localized charge seems to be primarily a function of its inability to act as a hydrogen-bond donor. The solvation ability of acetonitrile for anions is known to increase as the hydrogen-bonding ability of the anion decreases. Although, as is pointed out repeatedly in Chapter 2, the solubility of electrolytes is determined not only by the solvation energy, but also by the crystal lattice energy, the solubilities of salts in acetonitrile do tend to increase in the series OH^-, $F^- \ll Cl^- < Br^- < N_3^- < I^- < SCN^- <$ picrate $< BPh_4^-$. For example, the pK_{sp} values in acetonitrile are 7.20, 5.60, 4.0, and 2.75 for KCl, KBr, KPi, and $KBPh_4$, in that order. This is the order in which the polarizability of the anions and hence their dispersion interactions with acetonitrile molecules increase. In this series, the most polarizable anions are picrate and tetraphenylborate and they are responsible for the highest solubilities. When a large organic cation with dispersed charge is combined with the picrate ion, the effect on the solubility is predictably even greater: for tetraphenylarsonium picrate the pK_{sp} in acetonitrile is 2.5. Additional examples of the effect of anion solvation on the solubility of salts are cited in Table 2.5.

Silver and mercury halides dissolve in the presence of excess halide due to formation of complex ions of the type AgX_2^- and HgX_3^- or higher complexes. Evidently, the more polarizable complex ions are stabilized by acetonitrile relative to X^- ions because of enhanced dispersion interactions.

Earlier in this chapter we mentioned that the solubility of salts in aprotic solvents may be strongly influenced by conjugation equilibria. Thus, in pure acetonitrile, the molar solubility of potassium 3,5-dinitrobenzoate is 3.14×10^{-4}, but in the presence of 3.47×10^{-2} *M* 3,5-dinitrobenzoic acid it is 7.38×10^{-3} *M*, an increase of more than twentyfold, brought about by homoconjugation. Solubility in acetonitrile can be increased also by heteroconjugation with such hydrogen-bond donors as water, alcohols, or acetic acid. An example of analytical importance is the solubilization of alkali carboxylates by small amounts of acetic acid.

Electrochemistry Ionophores are extensively dissociated in acetonitrile (association constants range typically from 0 to 20) so that potentiometric, conductometric, and other electrometric studies are generally not hampered by low concentrations of ions. Instead, obstacles to the determination of standard electrode potentials referred to the usual zero of the hydrogen electrode come from a different source. In amphiprotic solvents the $E°$'s are classically derived from the electromotive force of cells without transference, using primarily the hydrogen (gas) electrode, silver–silver halide, and mercury–mercurous chloride (calomel) electrodes. None of these electrodes are compatible with acetonitrile. The hydrogen electrode behaves irreversibly and gives irreproducible potentials, probably because platinum black catalyzes the hydrogenation of acetonitrile. (Fortunately, the glass electrode does behave reversibly toward hydrogen-ion activity and has been employed successfully.) The halide electrodes dissolve in the presence of excess halide due to formation of complexes, while the calomel electrode specifically disproportionates and dissolves as well.

Because of these difficulties, $E°$'s in acetonitrile have been determined from the emf of cells with salt bridges and liquid junctions. By far the most popular reference electrode has been the silver–metal electrode dipping in a solution of 0.01 *M* $AgNO_3$. The salt bridge is usually a 0.1 *M* solution of tetraethylammonium perchlorate (Et_4NClO_4), although the picrate has also been used. Some of the more reliable standard potentials determined in this manner are collected in Table 3.12.

3.2.5 Benzene—An Inert Solvent The most inert liquids are saturated aliphatic hydrocarbons. Unfortunately, their extreme inertness toward solutes also means extremely low solubility for most substances of interest. Thus, as a practical matter, very few chemical reactions can be studied in those media that would least interfere with them. Benzene, on the other hand, is only relatively inert. It is known to be a π-electron donor and therefore a

Table 3.12 Standard Electrode Potentials in Acetonitrile[a]

Electrode[b]	Standard potential $E°$, V
Ag^+–Ag	0.13_3
Hg_2^{2+}–Hg	0.49_0
Hg^{2+}–Hg_2^{2+}	0.64_4
Rb^+–Rb	−3.282
Tl^+–Tl	−0.648
Ferricinium–ferrocene	0.074
Ferroin(III–II)	0.846

[a] The data in this table are from Coetzee and Campion (15) and Coetzee, Campion, and Liberman (16).

[b] The reference electrode in each case was $Ag|Ag^+$ 0.01 $M|$0.1 M $Et_4NClO_4|$.

Lewis base, but, in comparison with bona fide bases such as the amines or even ethers, benzene is so weakly protophilic that it can be safely considered aprotic. With its low dielectric constant (2.27), zero dipole moment, and virtually no tendency to participate in proton transfer or in hydrogen bonding, benzene is a very poor solvent for ionic and dipolar substances. Any capacity for dissolving that it does possess derives from its polarizability and the slight π-electron basicity. In fact, the complexes that form between benzene and such Lewis acids as iodine and iodine monochloride are rather strong and belong to the charge-transfer type. Aside from iodine, elemental sulfur, phosphorus, and hydrocarbons are examples of substances with appreciable solubilities in benzene. Of the ionophores, only $AgClO_4$ and tetraalkylammonium salts dissolve to any extent in benzene.

Whenever solvent–solute interactions are weak, the solute species will tend to interact more with each other. Indeed, the chemistry of benzene solutions is dominated by a variety of association equilibria.

Electrolytes in Benzene The complexity of electrolyte equilibria in benzene has been demonstrated in the classical studies by Kraus and his associates using the techniques of cryoscopy, molecular polarization, and electrolytic conductance. According to Kraus (17), salts in benzene solution can exhibit three types of behavior, depending on the relative sizes of the ions and the dipole moment of the salt. In type 1, the salt is composed of two large ions, which are hydrogen bonded. For this type, both the ion-pair dissociation constant K_d and the dipole moment μ are relatively small. In type 2, a large cation is combined with a smaller anion. Here the dissociation constant and the dipole moment are relatively large. In type 3, both ions are large but not hydrogen bonded; both the dissociation constant and the dipole moment are somewhat larger than for type 2.

In Fig. 3.1 the three types of electrolyte behavior are exemplified by the shapes of the conductance–concentration curves for triisoamylammonium picrate (type 1), tetraisoamylammonium thiocyanate (type 2), and tetraisoamylammonium picrate (type 3). The straight-line portions of curves 2 and 3, corresponding to the lowest concentration range (10^{-5}–10^{-6} M) represent equilibria between the ion pairs and the free ions: $M^+X^- \rightleftharpoons M^+ + X^-$. In this region, ion pairs are the predominant species, since their dissociation constants in benzene are extremely small (e.g., for i-$Am_4N^+Pi^-$, $K_d = 2.18 \times 10^{-17}$). As the concentration increases, the conductance decreases at first, due to increasing association of ions to form ion pairs, until a conductance minimum is reached. When the concentration is increased beyond this minimum, the conductance rises again due to formation of triple ions, $M^+X^-M^+$ and $X^-M^+X^-$, which carry a net charge. Further increase in the concentration causes another drop in conductance (clearly visible in curve 2), apparently as a result of association into nonconducting ion quadrupoles. Even higher ionic aggregates can be deduced from curve 2.

The extent of aggregation in the above systems has been determined

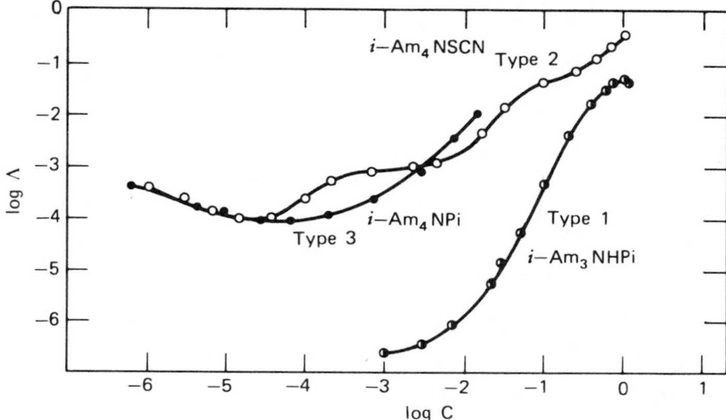

Figure 3.1 Conductance–concentration plots for three types of salts in benzene (17). Reprinted with permission from *J. Chem. Educ.*, **35**, 330 (1958). Copyright 1958. Division of Chemical Education, American Chemical Society.

quantitatively from cryoscopic studies. The type 1 salt, *i*-Am_3NHPi (K_d $= 4 \times 10^{-21}$, $\mu = 13.3$ D), exhibits the least association: at the concentration of 1×10^{-2} *M*, it is only 6% associated to quadrupoles. Somewhat greater degree of association occurs for the type 3 salt, *i*-Am_4NPi ($K_d = 2.18 \times 10^{-17}$, $\mu = 18.3$ D), which exists 40% in the quadrupole state at the concentration of 3.8×10^{-3} *M*. Also the conductance minimum occurs at a lower concentration for *i*-Am_4NPi as compared to *i*-Am_3NHPi. By far the largest aggregates are formed by type 2 salt, exemplified in Fig. 3.1 by *i*-Am_4NSCN. Here, association to quadrupoles is appreciable in solution as dilute as 4×10^{-5} *M* and the maximum association number* reached by that salt is 26 (!) at 0.14 *M*. Type 2 behavior is typical of salts composed of quaternary ammonium (large) cations and smaller anions, such as the halides or perchlorate, but the association maxima differ in magnitude and are reached at different concentrations, depending on the nature of the salt. Evidently the ionic association equilibria in benzene are very sensitive to the size and the structure of the ions.

Acids and Bases In contrast to amphiprotic solvents, benzene can act neither as a reference acid nor as a reference base toward solutes. To define and measure the relative strength of an acid or a base in any truly aprotic medium, we must dissolve in it another solute that can act as a reference substance. Such reference acids or bases are usually indicators, which enable us to follow the extent of acid–base reactions spectrophotometrically. As

* The association number is the apparent molecular weight of a salt, as derived from measurements of freezing-point depression, divided by the formula weight.

the concentration of free ions in benzene is vanishingly small, electrometric methods are of little use in the study of acid–base equilibria.

In general, an acid–base reaction in benzene is one of association between an acid HA and a base B:

$$B + HA \rightleftharpoons BHA \tag{3.66}$$

The nature of BHA can range from a loose hydrogen-bonded complex B···HA, in the case of weak-acid–weak-base interactions, to a more common hydrogen-bonded ion pair $BH^+ \cdots A^-$, forming between moderately strong reactants, to a purely electrostatic ion pair, such as $BH^+ClO_4^-$, which results from reaction between a strong base and a strong acid. There is actually believed to be a whole spectrum of hydrogen bonds intermediate between B···HA and $BH^+ \cdots A^-$, in which the equilibrium position of the proton is varying according to the basicity of B and the acidity of HA.

Many investigators have shown that the simple association represented by Eqn. 3.66 is indeed the predominant acid–base equilibrium in dilute benzene solutions and that the correct measure of the relative strength of an acid HA toward a base B (or vice versa) is the association constant governing that process:

$$K_{\text{assn.}} = \frac{(BHA)}{(B)(HA)} \tag{3.67}$$

A simple test of the applicability of Eqn. 3.67 to an acid–base equilibrium is to plot the quantities $\log[(BHA)(HA)^{-1}]$ vs $-\log(B)$. If Eqn. 3.67 is obeyed, the plot will be a straight line with the slope of -1. Davis and her associates made particularly extensive contributions in the spectrophotometric studies of acid–base equilibria in benzene (18). In Fig. 3.2 we show some of their data, for the reaction of triethylamine with isomeric dinitrophenols, which obey the linear relationship described above. In this series 2,6-dinitrophenol is the most acidic and 2,5-dinitrophenol the least acidic of the isomers.

While a 1:1 complex between an acid and a base is the likely product of reaction in dilute benzene solution, higher complexes form at higher reactant concentrations. Using the differential vapor pressure method, Bruckenstein and Saito (19) have demonstrated that when trichloroacetic acid is added to benzyl-*N,N*-dimethylamine the first product to form is BHA, but that further addition of the acid yields the acid salt BHA·HA. The initial complex, however, is not always monomeric. In the same study it was shown that 0.01 *M* dibenzylamine reacts with trichloroacetic acid to form initially the dimeric complex $(BHA)_2$; then excess acid produces the acid salt BHA·HA. Homoconjugation is likely to be a factor in the formation of these acid salts. In fact, homoconjugation is quite evident from the titration curves of trichloroacetic acid with amines, in which intermediates having the formula $BH^+ \cdots A^- \cdots HA$ have been postulated. Many other acids, including

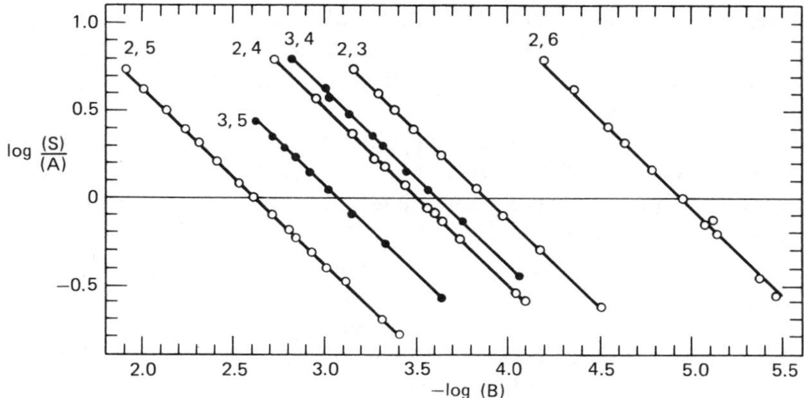

Figure 3.2 Graph constructed from spectral absorbance data and stoichiometric concentrations for mixtures of triethylamine with the isomeric dinitrophenols in benzene at 25°C. The reaction assumed is addition of base and acid, forming a salt consisting of hydrogen-bonded ion pairs. Symbols enclosed in parentheses signify equilibrium concentrations of acid, base, and salt. Reprinted with permission from *J. Amer. Chem. Soc.*, **84,** 3624 (1962). Copyright 1962. American Chemical Society.

HCl and HNO_3 (but not $HClO_4$), are known to form homoconjugate anions in benzene.

3.2.6 Liquid Sulfur Dioxide—A Molecular Inorganic Solvent Sulfur dioxide exists as a liquid in the temperature range of −10.08°C to −75.52°C. It is probably the only inorganic liquid that is truly molecular in nature, as the existence of the once-proposed self-dissociation $2SO_2 \rightleftharpoons SO^{2+} + SO_3^{2-}$, however attractive from the viewpoint of solvent acidity, was actually disproved by experiments of isotopic exchange. Thus it turned out that $SOCl_2$ and $SOBr_2$, which would act as acids in the hypothetical self-ionization scheme, failed to exchange radioactively labeled sulfur, ^{35}S, or oxygen, ^{18}O, with the solvent. Nor is there any evidence for the existence of SO^{2+} ions based on electrolytic conductance.

Solubility In general, covalent substances are more soluble than salts in liquid SO_2. Most organic compounds, with the exception of saturated aliphatic hydrocarbons, are very soluble. Dipole–dipole and dispersion interactions as well as the ability of SO_2 to act as an electron acceptor are responsible for the solvation by SO_2. Of these, the dispersion interactions seem to be particularly dominant. Thus the dispersion potential, which is a measure of the attractive field due to the polarizability of the molecule, was calculated to be 294 erg cm^6 for liquid SO_2, as compared to only 47 erg cm^6 for water (20).

Many organic substances that are capable of donating electrons (e.g., amines, ethers, alcohols, olefins) form molecular addition compounds with SO_2, where the latter acts as an electron acceptor. The amine complexes are intensely colored and are believed to be of the charge-transfer type. Occasionally SO_2 acts as an electron donor through an oxygen atom, but only to the most powerful Lewis acids, such as BF_3 or SbF_5.

Liquid SO_2 has a relatively low dielectric constant (15.4 at 0°C), but an appreciable dipole moment (1.62D). Predictably it is in general a poor solvent for ionic substances. Nevertheless, the solubilities of many salts in liquid SO_2 are surprisingly high, particularly those of alkali- and tetraalkylammonium halides. Stable solvates are known to form between SO_2 and these salts (e.g., $NaI{\cdot}2SO_2$, $RbI{\cdot}4SO_2$, $(CH_3)_4NBr{\cdot}2SO_2$), attesting to strong interactions. The exceptionally high solubilities of the iodides and the thiocyanates betray interactions between these anions and SO_2 more powerful than mutual polarization. Their solutions in liquid SO_2 have an intense yellow color, which is believed to arise from the charge-transfer transition in the anion–SO_2 complex.

Even more surprising is the ability of liquid SO_2 to cause appreciable ionization of some covalent halides, such as triphenylchloromethane, which ionizes and partially dissociates into Ph_3C^+ and Cl^- ions. Again, it has been suggested that a specific interaction involving charge transfer between the solvent and the halide ions may be responsible both for the ionization of triphenylchloromethane and for the appreciable solubilities of ionic halides.

Electrochemistry The electrolytic conductance of solutions in liquid SO_2 follows a pattern typical of low-dielectric media. Below analytical concentrations of about 10^{-2} M, salts exist in the form of ion pairs and some free ions. At higher concentrations, the conductance due to triple ions takes over, leading to the well-known minimum in log Λ vs log C curves, which in the case of SO_2 appears at the concentration of about 10^{-1} M. Above this concentration the conductance is due primarily to the triple ions. Limiting equivalent conductances and overall dissociation constants for some electrolytes in liquid SO_2 are compiled in Table 3.13.

In principle liquid SO_2 is a good medium for potentiometry because it generally acts as an inert carrier for oxidation–reduction reactions. The iodide ion is oxidized quantitatively by such oxidizing agents as $FeCl_3$ and $SbCl_5$, while iodine is known to oxidize sulfite to sulfate. Some systematic determinations of standard electrode potentials in liquid SO_2 have been carried out and the known E°'s are listed in Table 3.14.

Reactions For some time, SO^{2+} and $SO_3{}^{2-}$ were believed to be the carriers of acidic and basic properties, respectively, in liquid SO_2. This interpretation was supported by apparent acid–base reactions such as

$$Cs_2SO_3 + SOCl_2 \rightleftharpoons 2CsCl + 2SO_2 \quad (3.68)$$

Table 3.13 Electrolytic Conductance and Dissociation Constants in Liquid SO_2 [a]

Ionophore	Λ_0	$K_d \times 10^4$
LiBr	189	0.27
NaBr	265	0.48
KCl	243	0.74
KI	244	3.0
Me_4NCl	243	10.3
Me_4NBr	236	11.8
Me_4NI	234	13.9
Me_4NClO_4	218	8.4
$(C_6H_5)_3CCl$	207	0.41

[a] At 0°C. Values selected from the compilation by Burow (20).

and

$$K_2SO_3 + SO(SCN)_2 \rightleftharpoons 2KSCN + 2SO_2 \tag{3.69}$$

as well as apparent amphoteric reactions, such as the dissolution of aluminum sulfite by excess sulfite:

$$Al(SO_3)_3(s) + 3[(CH_3)_4N]_2SO_3 \rightleftharpoons 2[(CH_3)_4N]_3(SO_3)_3 \tag{3.70}$$

Now these are interpreted as double-displacement (metathesis) reactions. Unambiguous metathesis reactions are also known, for example, the following reaction employed for the preparation of anhydrous HI:

$$KI + HCl \rightleftharpoons HI + KCl\downarrow \tag{3.71}$$

A solvolysis reaction in which SO_3^{2-} or SO^{2+} ions were formed appeared to be reminiscent of acid–base "hydrolysis" in water:

$$2NH_4Ac + 2SO_2 \rightleftharpoons (NH_4)_2SO_3 + SO(Ac)_2 \tag{3.72}$$

However, other types of solvolytic reactions are known to occur in liquid SO_2, for example,

$$PCl_5 + SO_2 \rightleftharpoons POCl_3 + SOCl_2 \tag{3.73}$$

$$NbCl_5 + SO_2 \rightleftharpoons NbOCl_3 + SOCl_2 \tag{3.74}$$

$$Zn(C_2H_5)_2 + SO_2 \rightleftharpoons ZnO + (C_2H_5)_2SO \tag{3.75}$$

Liquid SO_2 possesses a number of advantages as an inert medium for organic reactions. Thus, it has been used as a solvent for carrying out Friedel–Crafts reactions because the solubilities of the organic reactants and of the catalysts ($AlCl_3$, $FeCl_3$) are appreciable, for example,

$$C_6H_6 + (CH_3)_3CCl \xrightarrow{AlCl_3} C_6H_5C(CH_3)_3 + HCl \tag{3.76}$$

Table 3.14 Electrode Potentials in Liquid SO_2 [a]

Cell	Potential, V
$Pb\|PbCl_2$(satd.), $Hg_2Cl_2\|Hg$	0.43
$Zn\|ZnBr_2$(satd.), $Hg_2Cl_2\|Hg$	0.37 to 0.40
$Cd\|CdI_2$(satd.), $Hg_2Cl_2\|Hg$	0.42 to 0.45
Pb, $PbCl_2$(s)$\|Cl^-$ $Hg_2Cl_2\|Hg$	$E° = 0.36$
Ag, AgCl(s)$\|Cl^-$ $Hg_2Cl_2\|Hg$	$E° = 0.04$ (initial)
Ag, AgBr(s)$\|Br^-$ $Hg_2Cl_2\|Hg$	$E° = 0.048$ (initial)

[a] From a compilation by Burow (20).

Halogenation and sulfonation reactions have also been carried out in liquid SO_2. The low boiling point of SO_2 makes it easy to remove the solvent from a reaction mixture by evaporation.

3.2.7 Arsenic Trichloride—An Amphoteric Inorganic Solvent Arsenic trichloride exists as a liquid between −13 and +130°C. It is believed to undergo the self-ionization:

$$2AsCl_3 \rightleftharpoons AsCl_2^+ + AsCl_4^- \tag{3.77}$$

which is corroborated by a measurable specific conductance (1.4×10^{-7} mho cm^{-1}) and by the nature of some of the chemical reactions that occur in it. We stated earlier that within the framework of solvent acidity, the cation $AsCl_2^+$ would act as the carrier of acidic properties, while the anion $AsCl_4^-$ would act as the carrier of basic properties in $AsCl_3$. Any substance that increases the $AsCl_2^+$ concentration acts as a solvent acid in that medium. Indeed, such Lewis acids as antimony pentachloride and tellurium tetrachloride seem to fit the bill by forming solutions that are believed to contain $AsCl_2^+$ and $SbCl_6^-$ ions in the first case and $AsCl_2^+$ and $TeCl_6^{2-}$ ions in the second case. Similarly, a basic substance should increase the concentration of the $AsCl_4^-$ ions in the $AsCl_3$ medium. Pyridine acts as a base via the reaction:

$$C_5H_5N + 2AsCl_3 \rightleftharpoons C_5H_5N{\cdot}AsCl_2^+ + AsCl_4^- \tag{3.78}$$

Chloride ion added in the form of $(CH_3)_4N^+Cl^-$ also appears to act as a base, forming the salt $(CH_3)_4N^+AsCl_4^-$.

Solutions of such solvent acids and bases have high conductivities in $AsCl_3$ and can be titrated conductometrically, for example,

$$(CH_3)_4N^+AsCl_4^- + AsCl_2^+SbCl_6^- \rightleftharpoons (CH_3)_4N^+SbCl_6^- + 2AsCl_3 \tag{3.79}$$

The resulting salt has been isolated from solution and identified.

Analogous results are obtained when solutions of pyridine in $AsCl_3$ are titrated conductometrically with acids such as $SnCl_4$ or VCl_4. The titration curves show endpoints at the base-to-acid ratio of 2 : 1, and the salts produced have the compositions that can be formulated as $(C_5H_5N{\cdot}AsCl_2^+)_2SnCl_6^{2-}$ and $(C_5H_5N{\cdot}AsCl_2^+)_2VCl_6^{2-}$, respectively.

Given the low dielectric constant (12.6), it is not unexpected that $AsCl_3$ is a very poor solvent for ionic substances. The only easily soluble salts are the tetraalkylammonium halides, which form well-defined solvates that have been isolated. Solvates of KCl and RbCl are also known. Very soluble are those metal halides that behave as acids in this medium, such as $AlCl_3$, $FeCl_3$, $SnCl_4$, and VCl_4. Nonmetals such as sulfur, iodine, and phosphorus dissolve readily.

SUMMARY

In this chapter, solvents are divided essentially into two major categories depending on whether they are capable of H-bonding and proton transfer (amphiprotic solvents) or incapable (aprotic solvents). These categories are subdivided further and the solution chemistry of a specific solvent representative of each of the subcategories is discussed in some detail, with particular emphasis on such areas as solubility, acid–base, and electrochemistry. Thus within the class of amphiprotic solvents, which are subdivided into neutral, protogenic, and protophilic, we highlight the chemistry in water (neutral), glacial acetic acid (protogenic), and liquid ammonia (protophilic). Within organic aprotic solvents, acetonitrile and benzene serve as illustrative of dipolar aprotic and inert solvents, respectively. Within inorganic aprotic solvents, the chemistry in liquid sulfur dioxide and in arsenic trichloride is singled out as representative of molecular and amphoteric solvents, respectively. In aqueous solutions, virtually all electrolytes are completely dissociated into ions and there exist many strong acids and bases. On the other hand, in organic solvents, which generally have lower dielectric constants and poorer solvation ability for ions, ion pairing and association into higher ionic aggregates is the rule. In terms of ionic dissociation, all acids and bases are weak and the proper measure of acid–base strength is ionization (ion-pair formation) rather than dissociation. Particularly high ionic aggregates dominate the chemistry of benzene solutions. The existence of H-bonding in amphiprotic media and its absence in aprotic media accounts for much of the difference between their respective solvation and dissociation equilibria. Amphiprotic solvents solvate preferentially small and carboxylate anions. Dipolar aprotic solvents stabilize the large polarizable anions and those cations which participate in specific donor–acceptor interactions with the solvent molecules.

LITERATURE CITED

1 Kolthoff, I. M., S. Bruckenstein, and M. K. Chantooni, Jr., *J. Amer. Chem. Soc.*, **83**, 3927 (1961).

2 Kolthoff, I. M., and M. K. Chantooni, Jr., *J. Amer. Chem. Soc.*, **85,** 2195 (1963).

3 Kosower, E. M., *J. Amer. Chem. Soc.*, **80,** 3253 (1958); **82,** 2188 (1960).

4 Bruckenstein, S., and I. M. Kolthoff, *J. Amer. Chem. Soc.*, **78,** 2974 (1956).

5 Kolthoff, I. M., and S. Bruckenstein, "Acid–Base Equilibria in Nonaqueous Solutions," in I. M. Kolthoff and P. J. Elving, Eds., *Treatise on Analytical Chemistry*, Part I, Vol. 5, Wiley, New York, 1966, Chap. 13.

6 Conant, J. B., and N. F. Hall, *J. Amer. Chem. Soc.*, **49,** 3062 (1927); Conant, J. B., and T. H. Werner, *J. Amer. Chem. Soc.*, **52,** 4436 (1930); Hall, N. F., *J. Amer. Chem. Soc.*, **52,** 5115 (1930); Hall, N. F., and J. B. Conant, *J. Amer. Chem. Soc.*, **49,** 3047 (1927); Hall, N. F., and T. H. Werner, *J. Amer. Chem. Soc.*, **50,** 2367 (1928).

7 Cuthrell, R. E., E. C. Fohn, and J. J. Lagowski, *Inorg. Chem.*, **5,** 111 (1966).

8 Waddington, T. C., *Non-Aqueous Solvents*, Appleton, New York, 1969, pp. 17–18.

9 Pleskov, V. A., and A. Monossohn, as cited by H. Strehlow in "Electrode Potentials in Non-Aqueous Solvents," in J. J. Lagowski, Ed., *The Chemistry of Non-Aqueous Solvents*, Vol. I, Academic Press, New York, 1966, Chap. 3.

10 Kolthoff, I. M., and M. K. Chantooni, Jr., *J. Phys. Chem.*, **72,** 2270 (1968).

11 Muney, W. S., and J. F. Coetzee, *J. Phys. Chem.*, **66,** 89, (1962).

12 Kolthoff, I. M., M. K. Chantooni, Jr., and S. Bhowmik, *Anal. Chem.*, **39,** 1627 (1967).

13 Coetzee, J. F., and G. R. Padmanabhan, *J. Phys. Chem.*, **69,** 3193 (1965).

14 Coetzee, J. F., and G. R. Padmanabhan, *J. Amer. Chem. Soc.*, **87,** 5005 (1965).

15 Coetzee, J. F., and J. J. Campion, *J. Amer. Chem. Soc.*, **89,** 2513 (1967).

16 Coetzee, J. F., J. J. Campion, and D. R. Liberman, *Anal. Chem.*, **45,** 343 (1973).

17 Kraus, C. A., *J. Chem. Educ.*, **35,** 324 (1958).

18 Davis, M. M., "Brønsted Acid–Base Behavior in "Inert" Organic Solvents," in J. J. Lagowski, Ed., *The Chemistry of Non-Aqueous Solvents*, Vol. III, Academic Press, New York, 1970, Chap. 1.

19 Bruckenstein, S., and A. Saito, *J. Amer. Chem. Soc.*, **87,** 698 (1965).

20 Burow, D. F., "Liquid Sulfur Dioxide," in J. J. Lagowski, Ed., *The Chemistry of Non-Aqueous Solvents*, Vol. III, Academic Press, New York, 1970, Chap. 2.

GENERAL REFERENCES

Audrieth, L. F., and J. Kleinberg, *Non-Aqueous Solvents*, Wiley, New York, 1953.

Covington, A. K., and T. Dickinson, Eds., *Physical Chemistry of Organic Solvent Systems*, Plenum Press, London and New York, 1973.

Gutmann, V., *Coordination Chemistry in Non-Aqueous Solutions*, Springer, New York, 1968.

Kolditz, L., "Halides of Arsenic and Antimony," in V. Gutmann, Ed., *Halogen Chemistry*, Vol. II, Academic Press, London and New York, 1967.

Kolthoff, I. M., *Anal. Chem.*, **46,** 1992 (1974).

Kolthoff, I. M., and P. J. Elving, Eds., *Treatise on Analytical Chemistry*, Part I, Vol. 5, Wiley, New York, 1966.

Lagowski, J. J. Ed., *The Chemistry of Non-Aqueous Solvents*, Vols. I–V, Academic Press, New York and London, 1966–1978.

Payne, D. S., "Halides and Oxyhalides as Solvents," in T. C. Waddington, Ed., *Nonaqueous Solvent Systems*, Academic Press, New York and London, 1965, Chap. 8.

Tremillon, B., *Chemistry in Non-Aqueous Solvents*, D. Reidel, Dordrecht, Holland/Boston, Mass., 1974.

Waddington, T. C., Ed., *Non-Aqueous Solvent Systems*, Academic Press, New York, 1965.

Waddington, T. C., "Liquid Sulfur Dioxide," in T. C. Waddington, Ed., *Non-Aqueous Solvent Systems*, Academic Press, London and New York, 1965, Chap. 6.

Zingaro, R. A., Nonaqueous Solvents, Heath, Lexington, Mass., 1968.

CHAPTER FOUR

Thermodynamic Properties

Thermodynamic properties of nonaqueous electrolyte solutions are often useful in discussing the structure of the solutions with particular reference to ion–solvent and ion–ion interactions. In addition, the free energies of chemical reactions are frequently obtained from equilibrium emf measurements in which the overall electrochemical cell reaction is the one under investigation. The enthalpies and free energies of transfer of electrolytes from one medium to another have been widely used to test various theoretical relationships and models, such as the Born equation and modifications of the Born equation.

The various thermodynamic investigations include equilibrium emf measurements, which are used to determine standard electrode potentials and mean ionic activity coefficients, heats of solution and dilution calorimetry and heat capacities, solubility data, cryoscopic studies, and density measurements leading to partial molal volumes.

To provide the necessary background to discuss the thermodynamic properties of solutions, it is important to review some basic definitions and general relationships.

4.1 DEFINITIONS AND GENERAL RELATIONSHIPS

The property of reactants and products that determines the position of chemical equilibria is the *partial molal free energy*, or the *chemical potential*, $\bar{G}$, defined for species i by Eqn. 4.1:†

$$\bar{G}_i = \left(\frac{\partial G}{\partial n_i}\right)_{n_j, n_k \ldots T, P} \tag{4.1}$$

where n is the number of moles of the species denoted by the subscript, and T and P are temperature and pressure, respectively. Activity and free energy are related by Eqn. 4.2:

$$\bar{G}_i = \bar{G}_i^\circ + RT \ln a_i \tag{4.2}$$

where a_i is the activity and $\bar{G}_i^\circ$ is the *standard free energy* of species i. For solutes, $\bar{G}_i^\circ$ is identical with the solvation energy discussed in Chapter 2. Of course, at unit activity, $\bar{G}_i = \bar{G}_i^\circ$. In an ideal solution, that is, one where solute–solute interactions are absent, $\bar{G}_i^\circ$ at a given temperature and pressure depends only on the nature of the solute and the solvent, but not on the concentration. This condition represents a most desirable reference state to which measurements at other concentrations could be compared. Unfortunately, in order to be "ideal," a solution must be infinitely dilute and therefore cannot have unit activity. As a result, the conventional reference state for solutes is not a real solution but a *hypothetical* ideal solution of unit concentration. It follows from this choice of reference state that as the molality $m \to 0$, $(a/m) \to 1$. The activity and molality are related by means of the salt-effect activity coefficient γ, which for electrolytes is identical with the mean ionic activity coefficient $\gamma_\pm$:

$$a_i = m_i \gamma_i \tag{4.3}$$

Thus, the molal activity coefficient γ becomes unity at infinite dilution in the given solvent. Of course, the free energies and activities can be formulated also on the molar scale:

$$a_i = c_i y_i \tag{4.3a}$$

† The symbol μ is often used instead of $\bar{G}$. Also, in many cases the bar notation is omitted.

and the mole-fraction scale:

$$a_i = X_i f_i \tag{4.3b}$$

where c_i and X_i are the molar concentration and the mole fraction of species i, respectively, and y_i and f_i are its activity coefficients.† In each case, the numerical value of $\overline{G}_i^\circ$ will be different, depending on the units of a_i. For dilute solutions, the relationship between the activity coefficients on the three scales is given by

$$\gamma = \frac{cy}{md_0} \tag{4.3c}$$

and

$$f = \gamma(1 + 0.001\nu Mm) \tag{4.3d}$$

In the above equations, d_0 is the density of the solvent, M is the molecular weight of the solvent, and ν is the number of ions into which an electrolyte dissociates.

4.2 EQUILIBRIUM EMF MEASUREMENTS

The advantage of using a galvanic cell to obtain thermodynamic information is that the measurements are carried out under conditions of zero current flow and hence approximate to a truly thermodynamic reversible situation. The decrease in the free energy of the electrochemical cell reaction, ΔG, at constant temperature and pressure is given by the well-known relation

$$\Delta G = -nFE \tag{4.4}$$

where n is the number of equivalents of reactants converted into products, that is, the number of electrons involved in the overall cell reaction, E is the emf of the cell, and F is the Faraday.

In addition, by undertaking emf measurements as a function of temperature, both the entropies and enthalpies of reaction can be obtained directly using the equations

$$\Delta S = nF\left(\frac{\partial E}{\partial T}\right)_P \tag{4.5}$$

and

$$\Delta H = -nFE + nFT\left(\frac{\partial E}{\partial T}\right)_P \tag{4.6}$$

† Ionic charges are generally omitted from the subscripts to the symbols for activities, activity coefficients, free energies, etc. No confusion should arise from this simplification, as one hardly ever deals with solutions of such *uncharged* species as H, Cl, or Ag.

In theory, it might appear reasonably easy to obtain a large amount of thermodynamic data for electrolytes in nonaqueous media. However, in practice, one encounters several problems in selecting electrodes that will behave both reversibly and reproducibly in the medium. Some aspects of this problem will be discussed later.

Some of the earlier investigations in pure organic solvents involved a study of the behavior of the hydrogen halides and, where possible, cell 4-I was used.

$$\text{Pt; H}_2 \mid \text{HX} \mid \text{AgX} \mid \text{Ag} \tag{4-I}$$

where X = Cl or Br (solutions of HI are often easily oxidized and buffered cells are preferable).

As a review of the approach used to determine standard electrode potentials and mean ionic activity coefficients from equilibrium emf measurements, consider the reversible cell 4-II without liquid junction:

$$\text{Pt; H}_2 \mid \underset{\text{(solvent)}}{\text{HBr}(m)} \mid \text{AgBr} \mid \text{Ag} \tag{4-II}$$

The overall cell reaction is:

$$\tfrac{1}{2}\text{H}_2 + \text{AgBr} = \text{H}^+ + \text{Br}^- + \text{Ag} \tag{4.7}$$

Using the Nernst equation, the standard emf of this cell, E_m°, on the molal scale (molar and mole-fraction scales are also used) is given by

$$E = E_m^\circ - \frac{2RT}{F}\ln m - \frac{2RT}{F}\ln \gamma_\pm \tag{4.8}$$

where $\gamma_\pm$ is the mean ion activity on the molal scale, E is the potential of the silver–silver bromide electrode with respect to the hydrogen electrode, and F is the Faraday. In order to correct E to the standard state from any other pressure of hydrogen, the expression

$$\Delta E = \frac{RT}{2F}\,\frac{760}{P_{\text{bar}} - P_{\text{soln.}}} \tag{4.9}$$

is used, where P_{bar} is the barometric pressure and $P_{\text{soln.}}$, the vapor pressure of the solution. ΔE is added to the observed emf if the hydrogen electrode is the negative terminal and subtracted if it is the positive terminal.

In order to obtain the standard electrode potential, several extrapolation procedures have been proposed and are clearly discussed by Ives and Janz (1). Almost all the methods involve some form of the Debye–Hückel interionic attraction equation for the mean ionic activity coefficient:

$$\log \gamma_\pm = -\frac{A\sqrt{md}}{1 + \mathring{a}B\sqrt{md}} + B'm - \log(1 + 0.002\,Mm) \tag{4.10}$$

where A and B are the Debye–Hückel constants, calculable from fundamental constants, dielectric constant of the solvent and the temperature,

$B'm$ is a term to account for the so-called "salting-out" effect, M is the molecular weight of the solvent, d is the density of the solvent, $\mathring{a}$ is an ion-size parameter (often used as the closest distance of approach of the two ions).

Substituting Eqn. 4.10 into Eqn. 4.8 and rearranging, we obtain

$$E_m^\circ = E + 2k \log m - \frac{2kA\sqrt{md}}{1 + \mathring{a}B\sqrt{md}} - 2k \log (1 + 0.002\, Mm) + 2kbm \tag{4.11}$$

where

$$k = \frac{2.303RT}{F}$$

Letting the first four terms on the right-hand side of Eqn. 4.11 $= E_m^{\circ\prime}$, and rearranging gives

$$E_m^{\circ\prime} = E_m^\circ - 2kbm \tag{4.12}$$

then a plot of $E_m^{\circ\prime}$ against m should give a straight line of slope $-2kB'$ and intercept E_m°. With the value of E_m° known, the activity coefficients can be calculated from Eqn. 4.8 for each concentration.

For the case of hydriodic acid, a cell 4-III buffered to high pH values can be used, that is,

$$\text{Pt; H}_2 \mid \underset{(m_1)}{\text{HA}}, \underset{(m_2)}{\text{NaA}}, \underset{(m_3)}{\text{KI}} \mid \text{AgI} \mid \text{Ag} \tag{4-III}$$

where HA is a weak acid and NaA its sodium salt.

From the law of mass action

$$K_a = \frac{a_H a_A}{a_{HA}} \tag{4.13}$$

and Eqn. 4.10, the following equation may be derived:

$$E_{HI} - E_{m,HI}^\circ + k \log \frac{m_I m_{HA}}{m_A} = -k \log \frac{(\gamma_I \gamma_{HA})}{\gamma_A} - k \log K_a \tag{4.14}$$

where K_a is the dissociation constant of HA and γ_{HA} is the activity coefficient of the undissociated acid. For the chloride system, a similar equation can be derived,

$$E_{HCl} - E_{m,HCl}^\circ + k \log \frac{m_{Cl} m_{HA}}{m_A} = -k \log \frac{(\gamma_{Cl} \gamma_{HA})}{\gamma_A} - k \log K_a \tag{4.15}$$

Consider a pair of cells in which $m_1 = m_2 = m_3 = m$; m_{HA} and m_A differ slightly from m because of the hydrolysis of the A^- ion, but they will have the same values in each cell. Subtraction of Eqns. 4.14 and 4.15 gives

$$E_{m,HI}^\circ = E_{m,HCl}^\circ + E_{HI} - E_{HCl} + k \log \frac{\gamma_I \gamma_{HA} \gamma_A}{\gamma_{Cl} \gamma_{HA} \gamma_A} \tag{4.16}$$

At low concentrations, the activity coefficient term may be taken as zero, and in this event Eqn. 4.16 reduces to

$$E^{\circ}_{m,\mathrm{HI}} = E^{\circ}_{m,\mathrm{HCl}} + E_{\mathrm{HI}} - E_{\mathrm{HCl}} \tag{4.17}$$

Therefore, using the literature value for $E^{\circ}_{m,\mathrm{HCl}}$, the value of $E^{\circ}_{m,\mathrm{HI}}$ may be computed.

Cells 4-I–4-III discussed above are examples of cells without transference and if suitable reference electrodes are available their use is preferred as no problems involving liquid-junction potentials are present. There are numerous other examples of cells without transference, but the general approach follows that presented above.

In cases where it is not possible to use a cell of the type 4-I–4-III, some form of concentration cell can be employed. A combination of two simple cells of type 4-I connected in reverse produces a concentration cell without transference, for example,

$$\mathrm{M} \mid \mathrm{M}^+, \mathrm{X}^-, \mathrm{S}_1 \mid \mathrm{AgX} \mid \mathrm{Ag}\cdots\mathrm{Ag} \mid \mathrm{AgX} \mid \mathrm{X}^-, \mathrm{M}^+, \mathrm{S}_2 \mid \mathrm{M} \qquad \text{(4-IV)}$$

where M represents a metal such as lithium, X^- is a halide ion, and S_1 and S_2 are two organic solvents. M | M^+ is often an amalgam electrode.

The emf of cell 4-IV is given by

$$\Delta E = E_2^{\circ} - E_1^{\circ} - \frac{2RT}{F} \ln \frac{(a_{\pm})_2}{(a_{\pm})_1} \tag{4.18}$$

Alternatively one may use a cell containing a liquid junction such as

$$\mathrm{M} \mid \mathrm{M}^+, \mathrm{X}^-, \mathrm{S}_1 \parallel \mathrm{X}^-, \mathrm{M}^+, \mathrm{S}_2 \mid \mathrm{M} \qquad \text{(4-V)}$$

where the double line represents the liquid junction. For the case where the two solvents are identical, the emf of cell (4-V) is given by

$$E = \frac{-RT}{F} \ln \frac{(a_{\pm})_2}{(a_{\pm})_1} + E_{\mathrm{j}} \tag{4.19}$$

where the liquid junction potential, E_{j}, is given by

$$E_{\mathrm{j}} = (1 - 2t_-) \frac{RT}{F} \ln \frac{(a_{\pm})_2}{(a_{\pm})_1} \tag{4.20}$$

where t_- is the transference number of the anion.

Further details on these and other cells with their various limitations can be found in any general physical chemistry textbook.

4.2.1 Reference Electrodes in Nonaqueous Solvents The selection and use of reference electrodes in a variety of nonaqueous solvents has been treated extensively in two authoritative publications (2,3). The practical problems encountered in the use of each reference electrode are most often peculiar to the particular solvent. At this stage it will be sufficient to summarize some of the salient features.

If at all possible, it is convenient to use the primary reference electrode, that is, the hydrogen electrode, in order to determine the standard electrode potential of other electrode systems. The hydrogen electrode has been used successfully in a variety of organic solvents such as acetone, propylene carbonate, tetrahydrofuran, and ethanol. The catalytic activity of the platinized platinum surface can lead to decomposition of the solvent in some cases with a resulting erratic behavior of the potential and irreproducibility.

Silver–silver halide electrodes have found extensive use in the protic solvents, such as the alcohols, and behave very reversibly. However, in aprotic solvents, such as acetonitrile, the formation of anionic complexes with the halide ion precludes their use. The solubility of the silver halide in the aprotic solvent means that the concentration of silver chloride in the solvent will be about the same as that of the added chloride. This factor will provide a significant contribution to the liquid-junction potential in concentration cells.

Both calomel and other mercurous halides are found to disproportionate in many aprotic solvents according to the reaction

$$Hg_2X_2 + X^- \rightleftharpoons HgX_3^- + Hg$$

and thereby give rise to unreliable potentials.

One of the most successful reference electrodes that have been used in polar aprotic solvents is the silver–silver ion electrode, except in cases where the solvent is oxidized by Ag^+. Silver nitrate is normally the source of the silver ion.

Amalgam electrodes such as $Li(Hg) \mid Li^+$ and $Cd(Hg) \mid Cd^{2+}$ have been used fairly widely in solvents such as dimethylsulfoxide, propylene carbonate, and dimethylformamide. In addition, redox systems such as ferrocene–ferrocinium ion and iron(II)–(III) *o*-phenanthroline complexes have found several applications in acetonitrile, for instance.

Reference electrodes that have been used in such solvents as liquid ammonia, liquid sulfur dioxide, pyridine, and anhydrous hydrofluoric acid have been described by Hills (3).

4.2.2 Applications of Emf Data Several approaches have been used in the analysis of emf data for electrochemical cells. The standard electrode potentials are important and fundamental quantities in their own right, and some investigators have focused their attention on establishing a reliable value for this quantity in the nonaqueous solvent of interest.

If the objective is to examine the departure of the system from ideal conditions, the activity coefficients are of concern. On the other hand, specific solute–solvent interactions are determined from quantities derived from emf data; that is, standard free energies, enthalpies, and entropies.

The solvation of ions in nonaqueous solvents is most commonly approached by comparisons of the free energies of transfer of an electrolyte from one solvent to another. In many cases, the transfer considered is that from water to the nonaqueous solvent using the relation

$$\Delta G_t^\circ = nF({}_wE^\circ - {}_sE^\circ) \tag{4.21}$$

where $\Delta G_t^\circ \equiv ({}_sG^\circ - {}_wG^\circ)$ is the standard free energy of transfer per mole of electrolyte and ${}_wE^\circ$ and ${}_sE^\circ$ are the standard potentials obtained in water and the nonaqueous solvent, respectively. (See Eqn. 5.5.) If the emf's are measured over a range of temperature, the corresponding enthalpies of transfer (ΔH_t°) and entropies of transfer (ΔS_t°) can be evaluated.

4.2.3 Electrode Potentials and Related Quantities in Selected Nonaqueous Solvents The purpose of this section is to indicate the approaches to and the type of information that can be gained from emf studies and will be illustrated by discussions of investigations in several types of solvents. Several comprehensive texts (3–6) have reviewed and tabulated a large amount of the information available in this field. As in other chapters, it is convenient to divide the studies into those involving protic solvents and those involving dipolar aprotic solvents.

Protic Solvents A large percentage of the emf studies in protic solvents have involved the alcohols, particularly methanol and ethanol and their aqueous mixtures. The experimental difficulties encountered in obtaining satisfactory and reproducible emf values in nonaqueous solvents are illustrated by the large discrepancies occurring for the standard electrode potential for the cell without liquid junction, Pt; H_2 | HCl | AgCl–Ag in ethanol. Strehlow has tabulated values obtained by several investigators and these values are shown in Table 4.1. A critical examination of the various studies tends to

Table 4.1 The Standard Electrode Potentials E_m° of the Silver–Silver Chloride Electrode in Ethanol at 25°C as Determined by Different Investigators

E_m° (V)	Investigator	Year
−0.0442	Harned and Fleysher	1925
−0.0365	Lucasse	1926
−0.0883	Woodcock and Hartley	1928
+0.02190	Mukherjee	1954
−0.08138	Taniguchi and Janz	1957
−0.079	LaBas and Day[a]	1960
−0.0723	Teze and Schaal	1961

[a] Value taken from a diagram.

Reprinted with permission from *The Chemistry of Non-Aqueous Solvents,* Vol. I, J. J. Lagowski, Ed., Academic Press, New York (1966), Chap. 4.

indicate that the value reported by Janz and Taniguchi (7) is the most reliable, as they took into account the effect of ion association of HCl in anhydrous ethanol. The large discrepancies occurring in Table 4.1 are probably due to contamination of the ethanol with water, and it is a continuing problem in all emf studies in nonaqueous solvents to obtain reliable data, especially at low concentrations. The extrapolation procedure used by Janz to obtain the standard electrode potential accounted for ion association by employing the relation

$$(E^{\circ\prime} - E_{\alpha}) = E + 2k \log \alpha C y_{\alpha} = E_{C}^{\circ} \tag{4.22}$$

where α is the degree of dissociation and y_{α} is the activity coefficient at the ionic concentration αC, and $k = 2.303\ RT/F$. The quantities α and y_{α} were obtained by simultaneous solution of the thermodynamic dissociation constant equation and the Gronwall, LaMer, and Sandved extended equation for the activity coefficient

$$\alpha = \frac{1}{2}\left[-\frac{K}{y_{\alpha}^{2}C} + \left(\frac{K^{2}}{y_{\alpha}^{4}C^{2}} + \frac{4K}{y_{\alpha}^{2}C}\right)^{1/2}\right] \tag{4.23}$$

where K is the dissociation constant for HCl in anhydrous ethanol and is normally evaluated from conductivity data. A first-order approximation of y_{α} was obtained by calculating $y_{\pm}$ (the mean ionic activity coefficient at concentration C) from the Gronwall, LaMer, and Sandved extended terms of the Debye–Hückel theory, using a particular value of $\mathring{a}$. Once the value of y_{α} has been approximated, a value of α may be determined from Eqn. 4.23 and the value gained for α may be used to recalculate y_{α}, using the same equation, but replacing C with αC. The method of successive approximations was then used to determine precise values of y_{α} and α. A plot of Eqn. 4.22 was then used to obtain a value of E_{C}°. Extrapolations based on several values of $\mathring{a}$, the ion-size parameter, were made and are shown in Fig. 4.1. As is seen in the figure, the extrapolation based on $\mathring{a} = 3.99$ Å was in accord with the theoretical predictions of a straight line with zero slope up to a concentration of 0.008 m. Pronounced curvatures were obtained for extrapolations based on other values of $\mathring{a}$. Comparisons of values of $\mathring{a}$ used by other investigators are discussed in this paper (7). The value obtained for the standard potential on the molal scale was $E_{m}^{\circ} = -0.08138$ V.

The mean molal activity coefficients, $\gamma_{\pm}$, were calculated from the thermodynamic equation:

$$\log \gamma_{\pm} = \frac{E_{m}^{\circ} - E}{2k} - \log m \tag{4.24}$$

and are given in Table 4.2. Values for the degree of dissociation of HCl in ethanol at 25°C are given in Table 4.3. The relatively large departure from ideality of the activity of hydrogen chloride in ethanol was attributed to the presence of ion association (ion-pair formation). The fairly detailed analysis by Janz described here highlights the approach used in many other studies.

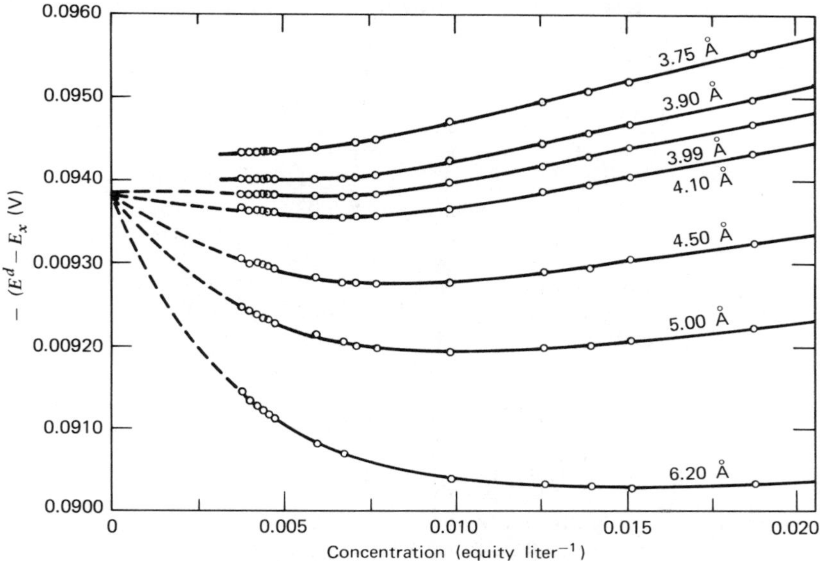

Figure 4.1 Extrapolation ($E^{\circ\prime} - E_\alpha$) for E°_C. Reprinted with permission from *J. Phys. Chem.*, **61,** 688 (1957). Copyright 1957. American Chemical Society.

For example, a very similar study by Nunez and Day (8) on the investigation of the thermodynamic properties of HBr in anhydrous ethanol used the cell Pt; H_2(1 atm) | HBr | AgBr | Ag to obtain the standard electrode potential ($E^\circ_m = -0.1816$) and mean molal activity coefficients. In this case, a value of $\mathring{a}$ = 5.0 Å gave the only curve approaching a zero slope.

As indicated earlier, a good deal of attention has been focused on emf studies in mixed solvents, particularly alcohol–water mixtures. An example of this type of investigation is illustrated by the study of Dill, Itzkowitz, and Popovych (9), who measured the standard potentials of the potassium elec-

Table 4.2 Mean Molal Activity Coefficients of HCl in Ethanol at 25°C

m (equiv. kg^{-1})	$\gamma_\pm$	m (equiv. kg^{-1})	$\gamma_\pm$
0.005	0.5871	0.03	0.3552
0.007	0.5397	0.05	0.3000
0.01	0.4903	0.07	0.2632
0.02	0.4018	0.1	0.2316

Reprinted with permission from *J. Phys. Chem.*, **61,** 688 (1957). Copyright 1957. American Chemical Society.

Table 4.3 Degree of Dissociation of HCl in Ethanol at 25°C

m (equiv. kg^{-1})	α	m (equiv. kg^{-1})	α
0.001	0.9595	0.01	0.8332
0.002	0.9315	0.015	0.8030
0.005	0.8800	0.02	0.7812
0.007	0.8582	0.025	0.7634

trode, E°_K, at 25°C in ethanol–water mixtures using the cell

$$\mathrm{K(Hg)\mid KCl\mid AgCl(s)\mid Ag} \tag{4-VI}$$

with a dropping amalgam electrode.

The potential E_{VI} of cell 4-VI at 25°C is given by

$$E_{VI} = E^\circ_{VI} + 0.05916 \log a_{K(Hg)} - 0.05916 \log a_K a_{Cl} \tag{4.25}$$

where $E^\circ_{VI} = E^\circ_{AgCl} - E^\circ_{K(Hg)}$ referred to E°_H in the given solvent and $a_{K(Hg)}$ is the activity of potassium in the amalgam. Activity coefficients for KCl in the various solvent mixtures were evaluated by rearranging Eqn. 4.25 in the form:

$$E_{VI} = E^{\circ\prime} - 0.05916 \log (a_K a_{Cl}) \tag{4.26}$$

where $E^{\circ\prime} = E^\circ_{VI} + 0.05916 \log a_{K(Hg)}$. Then, the mean molar activity coefficient $y_\pm$ of KCl is given by

$$\log y_\pm = \frac{E^{\circ\prime} - E_{VI}}{0.1183} - \log (\alpha C_{KCl}) \tag{4.27}$$

where α is the degree of dissociation of KCl. The quantity $E^{\circ\prime}$ was evaluated by using an expression involving the use of the Debye–Hückel theory:

$$E_{VI} + 0.1183 \log (\alpha C_{KCl}) = E^{\circ\prime} + 0.1183 A I^{1/2} \tag{4.28}$$

and plotting the left-hand side of the equation vs $I^{1/2}$, where I is the ionic strength.

The standard potential of the potassium electrode, E°_K was obtained by adding the standard potentials of the cells

$$\mathrm{K(Hg)\mid KI\ (ethylamine)\mid K(s)} \tag{4-VII}$$

$$\mathrm{Ag\mid AgCl\mid KCl\ (given\ solvent)\mid K(Hg)} \tag{4-VIII}$$

$$\mathrm{Pt;\ H_2(g)\mid HCl\ (given\ solvent)\mid AgCl\mid Ag} \tag{4-IX}$$

Table 4.4 Standard Potentials and Medium Effects at 25°C

Wt % ethanol	$10^3 a_{K(Hg)}$	E°_{VI}	${}_sE^\circ_K$	$(\log {}_m\gamma_K + \log {}_m\gamma_{Cl})$	$(\log {}_m\gamma_K - \log {}_m\gamma_H)$
100.0	1.136	1.8187	−2.865	6.121	0.987
92.3	1.328	1.8721	−2.757	5.242	2.803
80.3	1.909	1.9696	−2.799	3.626	2.099
60.2	2.282	2.0464	−2.830	2.377	1.574
40.0	2.206	2.1075	−2.8687	1.388	0.918
20.3	1.377	2.1558	−2.9021	0.6034	0.353
15.0	1.325	2.1714	−2.9136	0.3448	0.159
0.0	—	2.1931	−2.9230	0.0000	0.000

where $E^\circ_{VII} = E - 0.05916 \log a_{K(Hg)}$. Using tables of activities of potassium in amalgams as a function of mole fraction and a measured potential for cell 4-VII, a value for E°_{VII} can be obtained. As cell 4-VIII is the reverse of cell 4-VI, $E^\circ_{VIII} = -E^\circ_{VI}$. The standard potentials of cell 4-IX were interpolated from literature values. Finally, the values of E°_{VI} were obtained from $E^{\circ\prime}$ discussed earlier.

The activities of the amalgams and the standard potentials E°_{VI} and E°_K are given in Table 4.4. In addition, the medium effect, ${}_m\gamma_i$ (discussed more fully in Chapter 5), was evaluated for each solvent composition. The values given in Table 4.4 are represented by the Eqns.

$$\log {}_m\gamma_K + \log {}_m\gamma_{Cl} = \frac{{}_wE^\circ_{VI} - {}_sE^\circ_{VI}}{0.05916} \tag{4.29}$$

and

$$\log {}_m\gamma_K - \log {}_m\gamma_H = \frac{{}_sE^\circ_K - {}_wE^\circ_K}{0.05916} \tag{4.30}$$

where the subscripts s and w indicate nonaqueous solvent and water, respectively. The significance of these results is discussed in Chapter 5.

As an example of a study involving an electrolyte other than a 1:1 electrolyte, Hefley and Amis (10) carried out emf studies of cadmium chloride in ethanol and ethanol–water mixtures. The cell used was

$$\text{Cd} - \text{Cd}_x\text{Hg}(11\%) \mid \text{CdCl}_2(m) \mid \text{AgCl–Ag} \tag{4-X}$$

and measurements were undertaken at 25, 30, and 35°C.

The emf of cell 4-X is given by

$$E = E^\circ - k \log 3m^3\gamma_\pm{}^3 \tag{4.31}$$

where $\gamma_\pm$ is the mean ionic activity coefficient. Substituting for

$$\log \gamma_\pm = -A\sqrt{m} + B'm \tag{4.32}$$

where A is the Debye–Hückel limiting slope, one obtains

$$E + k \log 4m^3 - 3kA\sqrt{m} = E^\circ - 3kB'm = E_H \tag{4.33}$$

However, it is recognized from electrical conductivity data that cadmium chloride is incompletely dissociated in ethanol, and allowance has to be made for this effect.

The dissociation mechanisms considered by Hefley and Amis were

$$CdCl_2 \longrightarrow CdCl^+ + Cl^- \tag{4.34}$$

$$CdCl^+ \rightleftharpoons Cd^{2+} + Cl^- \tag{4.35}$$

If $\gamma_{Cd} = \gamma_{aM} = \gamma_{aX}$, then the equilibrium constant, K, for Eqn. 4.35 is

$$\begin{aligned} K &= \frac{(\alpha m\gamma_{aM})(1 + \alpha)\gamma_{aX} m}{m(1 - \alpha)\gamma_{aX}} \\ &= \frac{m\gamma_{aM}\alpha(1 + \alpha)}{1 - \alpha} \end{aligned} \tag{4.36}$$

and

$$\alpha = \frac{1}{2}\left\{-\left(1 + \frac{K}{m\gamma_{aM}}\right) + \left[\left(1 + \frac{K}{m\gamma_{aM}}\right)^2 + \frac{4K}{m\gamma_{aM}}\right]^{1/2}\right\} \tag{4.37}$$

In the above expressions, α is the degree of dissociation corresponding to Eqn. 4.35, and γ_{aM} and γ_{aX} are the activity coefficients of Cd^{2+} and $CdCl^+$, respectively.

Iterative procedures were used to determine K, α, and γ_{aM}. Several algebraic manipulations were involved in arriving at the final working equation from which E° could be obtained and is given as follows:

$$\begin{aligned} E^\circ = E &+ \frac{3 \times 2.303RT}{2F} \log m + \frac{3 \times 2.303RT}{2F} \log \gamma_a \\ &- \frac{3 \times 2.303RT}{2F} A\sqrt{m} + \frac{3\sqrt{2} \times 2.303RT}{2F} A\sqrt{I} \\ &+ \frac{2.303RT}{2F} \log \alpha\,(1 + \alpha)^2 \end{aligned} \tag{4.38}$$

where γ_a is the real mean activity coefficient of cadmium chloride. Satisfactory values of the standard potential were determined over the entire solvent, temperature, and concentration ranges by assuming the one method of ionization and simultaneously varying the equilibrium constant at a fixed distance of closest approach of the ions as the solvent composition was varied. The standard potentials were then used to calculate the mean ionic activity coefficients of cadmium chloride in the various solvents, using the form of the Nernst equation applicable to 2:1 electrolytes. The general behavior found was that deviations from the Debye–Hückel theory occurred at lower concentrations as the dielectric constant of the solvent was decreased. In addition, the thermodynamic functions, ΔG°, ΔH°, and ΔS° were calculated for the cell process.

Several studies have been undertaken in methanol and methanol–water mixtures with particular reference to the structure of the mixtures and ion-solvation in terms of acid–base properties of the solvent mixture. Thermodynamic transfer functions have also been evaluated and their single-ion values generated by various methods (see Chapter 5).

Feakins et al. (11) and DeLigny et al. (12) have made emf measurements on cells without transference, such as

$$\mathrm{Pt;\ H_2(1\ atm)\ |\ HX\ |\ AgX\ |\ Ag} \qquad \mathrm{X = Cl,\ or\ Br}$$

or buffered cells as discussed earlier, in a whole range of methanol–water mixtures and pure methanol. DeLigny used the cell

$$\begin{array}{l} \mathrm{Ag\ |\ AgCl\ |\ NaCl}(m),\ \mathrm{LiH\text{-}succ}(m)\ |\ \mathrm{Pt,\ H_2}\ |\ \mathrm{LiH\text{-}succ}(m)\ |\ \mathrm{AgX\ |\ Ag} \\ \qquad\qquad \mathrm{H_2\text{-}succ}(m) \qquad\qquad\qquad\qquad \mathrm{H_2\text{-}succ}(m) \\ \qquad\qquad\quad (\mathrm{A}) \qquad\qquad\qquad\qquad\qquad\quad \mathrm{NaX}(m) \\ \qquad\qquad\qquad\qquad\qquad\qquad\qquad\qquad\qquad (\mathrm{B}) \end{array} \tag{4-XI}$$

where X = Br or I, LiH-succ = lithium hydrogen succinate, and H_2-succ = succinic acid.

The emf of cell 4-XI is given by

$$E_{\mathrm{XI}} = E^\circ_{\mathrm{AgX}} - E^\circ_{\mathrm{AgCl}} - \frac{RT}{F}\ln\frac{\gamma_{\mathrm{X}}}{\gamma_{\mathrm{a}}} + \frac{RT}{F}\ln\frac{\gamma_{\mathrm{H\text{-}succ}}(\mathrm{B})}{\gamma_{\mathrm{H\text{-}succ}}(\mathrm{A})} \tag{4.39}$$

where γ_X is the activity coefficient of X and γ_a is the activity coefficient of Cl^-. The investigations showed departures in the standard potentials for E°_{AgBr} and E°_{AgI}, which were thought to be due to the effect of light on the halide electrodes.

Other investigations in some of the higher alcohols, such as propanol and butanol, are highlighted in Refs. 5 and 6, but essentially the electrochemical cells employed and the derived quantities are similar to those discussed above for ethanol and methanol.

One other study that is worth some discussion is that by Kundu et al. (13), who investigated the standard potentials of Ag | AgBr in propylene glycol and Ag | AgI in ethylene and propylene glycols at temperatures of 5–45°C. The cell employed was

$$\begin{array}{l} \mathrm{Pt;\ H_2(1\ atm)\ |\ HOAc,\ NaOAc,\ NaX\ |\ AgX\text{–}Ag} \\ \qquad\qquad\qquad\qquad (m_1) \qquad\quad (m_2) \end{array} \tag{4-XII}$$

The dissociation constants of HOAc at the different temperatures were obtained from independent measurements in the cell

$$\text{Pt; } H_2 \text{ (1 atm)} \mid \underset{(m_1)}{\text{HOAc}}, \underset{(m_2)}{\text{NaOAc}}, \text{NaCl} \mid \text{AgCl} \mid \text{Ag} \qquad \text{(4-XIII)}$$

where the standard potential of the Ag | AgCl electrode in the particular glycol at the different temperatures was known.

The dissociation constant was obtained from the equation

$$E_X - E^\circ_{m,\mathrm{Ag|AgX}} + \log \frac{m_{\mathrm{HOAc}} m_X}{m_{\mathrm{OAc}}} = \mathrm{p}K_a - \log \frac{\gamma_{\mathrm{HOAc}}\gamma_X}{\gamma_{\mathrm{OAc}}} \qquad (4.40)$$

where E_X is the observed emf and the activity coefficient term vanishes at zero ionic strength.

If we let the left-hand side of Eqn. 4.40 equal $\mathrm{p}K_a'$, then we can rewrite the equation as

$$\mathrm{p}K_a' = \mathrm{p}K_a - \log \frac{\gamma_{\mathrm{HOAc}}\,\gamma_X}{\gamma_{\mathrm{OAc}}} = \mathrm{p}K_a + f(I) \qquad (4.41)$$

when I is the ionic strength. A plot of $\mathrm{p}K_a'$ vs I will yield a value for $\mathrm{p}K_a$ when extrapolated to zero ionic strength. For propylene glycol, the expression for $\mathrm{p}K_a$ as a function of temperature was

$$\mathrm{p}K_a = \frac{827.82}{T} + 4.817 + 0.00581T \qquad (4.42)$$

The standard electrode potentials were obtained by rearrangement of Eqn. 4.40 to obtain

$$E^{\circ\prime}_X = E_X - k(\mathrm{p}K_a) + k \log \frac{m_1 m_3}{m_2}$$

$$= E^\circ_{m,\mathrm{Ag|AgX}} - k \log \frac{\gamma_{\mathrm{HOAc}}\gamma_X}{\gamma_{\mathrm{OAc}}} \qquad (4.43)$$

$$= E^\circ_{m,\mathrm{Ag|AgX}} + \mathrm{f}(I) \qquad (4.44)$$

The value of $E^\circ_{m,\mathrm{Ag|AgX}}$ was obtained from the intercept of a plot of $E^{\circ\prime}_X$ vs I.

The emf data were used to generate values of the standard free energies, $\Delta G^\circ_{\mathrm{soln.}}(\mathrm{HX})$, heats of formation, $\Delta H^\circ_{\mathrm{soln.}}(\mathrm{HX})$, and partial molal entropies, $S^\circ_{\mathrm{soln.}}(\mathrm{HX})$ of hydrogen halides in glycols and water. The values are given in Table 4.5. The standard molal free energy of formation, $\Delta G^\circ_{\mathrm{soln.}}(\mathrm{HX})$ of solvated HX (X = Cl, Br, I) assumed to be completely ionized is given by

$$\Delta G^\circ_{\mathrm{soln.}}(\mathrm{HX}) = \Delta G^\circ_{\mathrm{AgX(s)}} - nFE^\circ \qquad (4.45)$$

where $\Delta G^\circ_{\mathrm{AgX}}$ is calculated from data given by Latimer (see General References).

In propylene glycol, it was observed that the free energies, entropies, and enthalpies of the solvated hydrogen halides tend to be less negative as the

Table 4.5 Standard Free Energies, $\Delta G^\circ_{soln.}(HX)$, Heats of Formation, $\Delta H^\circ_{soln.}(HX)$, and Partial Molal Entropies, $\bar{S}^\circ_{soln.}(HX)$ of Hydrogen Halides in Glycols and Water

	$\Delta G^\circ_{soln.}(HX)$ (kcal mol^{-1})	$\Delta H^\circ_{soln.}(HX)$ (kcal mol^{-1})	$\bar{S}^\circ_{soln.}(HX)$ (cal mol^{-1} deg^{-1})
		Ethylene Glycol	
HCl	−26.77	−26.28	1.6
HBr	−20.61	−19.32	4.3
HI	−9.08	−6.55	8.5
		Propylene Glycol	
HCl	−25.48	−27.89	−8.1
HBr	−19.17	−20.12	−3.2
HI	−7.86	−7.62	+0.8
		Water	
HCl	−31.34	−27.32	13.5
HBr	−24.57	−18.67	19.8
HI	−12.34	−4.78	25.4

halides are in the order of Cl^-, Br^-, I^-, whereas, in ethylene glycol, a similar trend is observed except for the entropy data, which become more positive.

Nayak et al. (14) measured the emf of the cells

$$\text{Pt; } H_2 \mid HCl \mid AgCl \mid Ag$$

and

$$\text{Pt; } H_2 \mid HCl \mid Hg_2Cl_2 \mid Hg$$

in formamide. The purpose of employing both cells was to establish a cross-check on results obtained for the mean molal activity coefficients for HCl in formamide. Standard potentials for both the silver–silver chloride electrode and the calomel electrode were obtained by the Hitchcock extrapolation method (15), which has been referred to earlier in this chapter, and which is a popular method for treating emf data. One of the main conclusions of this study was that both the silver–silver chloride and the calomel electrodes behave satisfactorily in formamide. Emf measurements were also made over a temperature range of 25–55°C and the standard thermodynamic values, ΔG°, ΔH°, and ΔS° were calculated for the cell reaction.

The mean molal activity coefficients exhibited the usual trend of decreasing with concentration at all temperatures. Plots of log $\gamma_{\pm}$ vs concentration yielded smooth curves of the expected shape at all temperatures and the limiting slopes were in good agreement with the theoretical slopes predicted by the Debye–Hückel limiting law. Because of the higher dielectric constant of formamide, these slopes are expected to be less negative than in the case of water and consequently the mean molal activity coefficient of HCl in formamide decreases much less with increasing molality than it does in aqueous solution. A typical plot of log $\gamma_{\pm}$ vs concentration is shown in Fig. 4.2. The trend in the values of the activity coefficients in *N*-methylacetamide is opposite to that in formamide. A comparison of the activity coefficients of HCl at selected molalities in water, formamide, and *N*-methylacetamide is given in Table 4.6. It was also observed that the $E°$ and $-\Delta G°$ values for the cell reaction appeared in the same order as the basicity of the three solvents, that is, water being the most basic, followed by NMA and then formamide.

For the cell reaction

$$\tfrac{1}{2}H_2(g) + AgCl(s) \rightleftharpoons Ag(s) + HCl\ (\text{solvated}) \qquad (4.46)$$

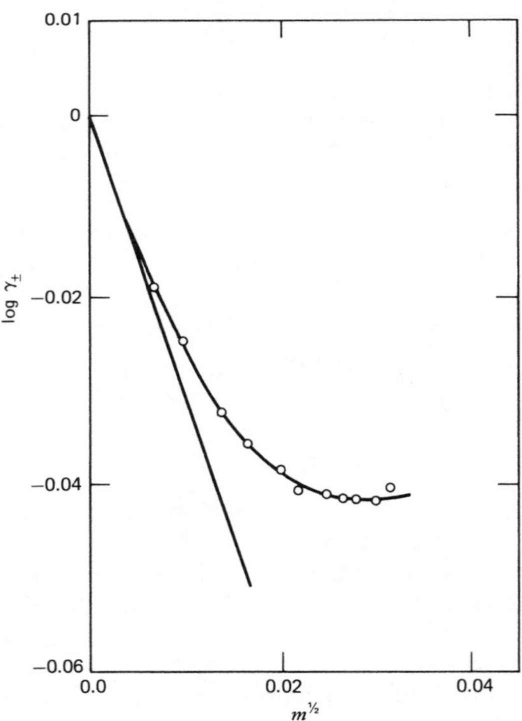

Figure 4.2 Variation of log $\gamma_{\pm}$ vs $m^{1/2}$ at 35°C. Reprinted with permission from *Electrochimica Acta,* **16,** 1757 (1971). Copyright 1971. Pergamon Press Inc., Oxford, U.K.

Table 4.6 Mean Molal Activity Coefficients of HCl in Different Solvents

m	$\gamma_\pm$ in Water at 40°C	$\gamma_\pm$ in Formamide at 40°C	$\gamma_\pm$ in NMA at 40°C
0.01	0.901	0.943	0.971
0.05	0.825	0.909	0.952
0.1	0.789	0.909	0.950

any change in $\Delta G°$ is largely due to the variation of free energy of solvation of HCl, which is largely due to the differences in the solvation energy of the proton in the various solvents.

As was the case for alcohols, the thermodynamic properties of cadmium chloride in formamide are of interest, particularly from the viewpoint of complex formation. One such study by Broadbank et al. (16) used the cell

$$\text{Cd(Hg) 2 phase amalgam} \mid CdCl_2 \mid AgCl \mid Ag \qquad \text{(4-XIV)}$$

The emf measurements were corrected to the values corresponding to pure cadmium electrodes by adding the factor 0.0505 V in each case.

The cell reaction is given by

$$Cd + 2AgCl = Cd^{2+} + 2Cl^- + 2Ag \qquad (4.47)$$

and the emf of the cell is given by

$$E = E° - \frac{RT}{2F} \ln m_{Cd}\, m_{Cl}^2\, \gamma_\pm^3 \qquad (4.48)$$

where $\gamma_\pm$ is the mean ionic activity coefficient of cadmium chloride. It was assumed that the cadmium chloride was incompletely dissociated in solution as was previously demonstrated for aqueous solutions with the dissociation scheme being represented by

$$CdCl_2 \longrightarrow CdCl^+ + Cl^- \qquad (4.49)$$

$$CdCl^+ \rightleftharpoons Cd^{2+} + Cl^- \qquad (4.50)$$

A trial-and-error analysis was used to calculate the degree of dissociation α of the $CdCl^+$ ions and the ionic strength I, using different values for the dissociation constant K of the equilibrium 4.50.

On the introduction of the expression

$$\log \gamma_\pm = \frac{-2AI^{1/2}}{1 + I^{1/2}} + BI \qquad (4.51)$$

into Eqn. 4.48 the following equation was obtained:

$$E' = E + \frac{2.303RT}{2F}\left[\frac{\log \alpha(1 + \alpha)^2 m^3 - 6AI^{1/2}}{1 + I^{1/2}}\right] \\ = E^\circ - \frac{2.303RT}{2F} 3BI \tag{4.52}$$

If the correct value of K is selected, the plot of E' vs I should be a straight line; however, the analysis based on the dissociation scheme given by Eqns. 4.49 and 4.50 failed to produce a straight line.

Based on evidence from polarographic measurements, a dissociation mechanism using a coordination number of 4 for cadmium halide complexes was proposed:

$$2CdCl_2 \longrightarrow Cd^{2+} + CdCl_4^{2-} \tag{4.53}$$

$$CdCl_4^{2-} \rightleftharpoons Cd^{2+} + 4Cl^- \tag{4.54}$$

The dissociation constant K' corresponding to Eqn. 4.54 is given by:

$$K' = \frac{16m^4\alpha^4(1 + \alpha)\gamma_{Cl}^4}{1 - \alpha} \tag{4.55}$$

where m is the initial concentration of $CdCl_2$ and α is the degree of dissociation of $CdCl_4^{2-}$, and the activity coefficients of Cd^{2+} and $CdCl_4^{2-}$ were assumed to be equal. This particular dissociation scheme produced a better straight-line plot and suggests that in $CdCl_2$ solutions in formamide the ions Cd^{2+} and $CdCl_4^{2-}$ predominate.

The molal standard potential of the $Cd^{2+} \mid Cd$ electrode was found to be -0.412 V at 298.15 K. The solubility product of cadmium chloride (2.7×10^{-7}) in formamide was obtained by combining the value for the standard potential of the $Cd^{2+} \mid Cd$ electrode with the standard potential of the $Cd \mid CdCl_2(s), Cl^-$ electrode, where $CdCl_2(s)$ denotes a solid solvate of composition $CdCl_2{\cdot}2\tfrac{1}{2}HCONH_2$. The emf results were also used to calculate the ΔG° and ΔH° for the interaction of anhydrous $CdCl_2$ with formamide to form the solid solvate $CdCl_2{\cdot}2\tfrac{1}{2}HCONH_2$.

Another area of interest has been the study of the dissociation constants of carboxylic acids in formamide, from the viewpoint of examining the effect of several factors upon the equilibrium constant of the proton transfer accompanying the acid dissociation. Mandel and Decroly (17) measured the dissociation constants of both formic and acetic acids at several temperatures between 15 and 45°C using the buffered cell

$$\text{Pt; } H_2 \mid \underset{(m_1)}{HA},\ \underset{(m_2)}{NaA},\ \underset{(m_3)}{NaCl} \mid AgCl \mid Ag \tag{4-XV}$$

where HA represents the acid and NaA the sodium salt of the acid.

The dissociation constants were obtained by a linear extrapolation to zero ionic strength of the auxiliary function $-\log K_a'$, given by the following relation:

$$-\log K_a' = \frac{(E - E^\circ)}{2.303RT} + \log m_3 \frac{(m_1 - m_H)}{(m_2 + m_H)}$$
$$= -\log K_a - \log \frac{\gamma_{Cl}\gamma_{HA}}{\gamma_A} \tag{4.56}$$

In Eqn. 4.56, E is the observed potential of the buffered cell; E°, the standard potential of the Ag | AgCl electrode; K_a, the dissociation constant of the acid HA. The activity coefficient term can be neglected as it tends to zero at zero ionic strength.

The values of the dissociation constants for the two acids as a function of temperature are represented graphically in Fig. 4.3. The variation with temperature was fitted to a parabolic curve of the type:

$$\log(K_a)_t - \log(K_a)_\theta = -p(t - \theta)^2 \tag{4.57}$$

where θ was the temperature at the minimum of the curve and p was a constant. It was observed that both acids were slightly stronger in water than in formamide.

A solvent with an even higher dielectric constant than that of formamide is N-methylformamide ($\epsilon = 182.4$) and the thermodynamic properties of several alkali-metal halides in this solvent have been investigated by

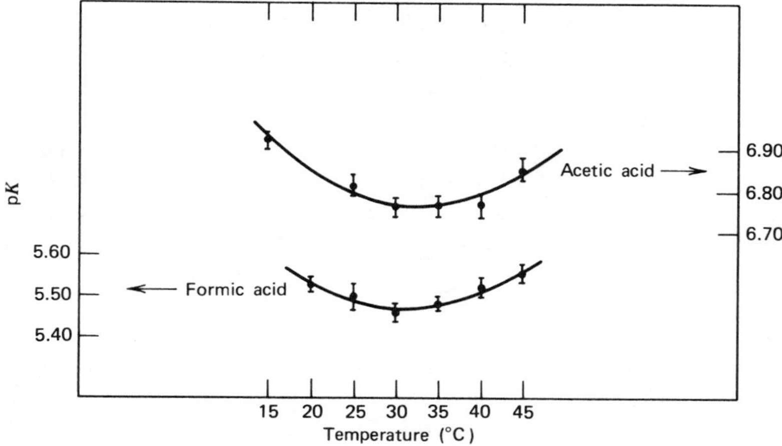

Figure 4.3 Variation of the dissociation constants of formic acid and acetic acid in formamide as a function of temperature. Reprinted with permission from *Trans. Faraday Soc.*, **56,** 29 (1960).

Criss (18). Emf measurements were carried out on cells of the type:

$$M_y(Hg) \mid MX \mid AgX \mid Ag \qquad (4\text{-XVI})$$

where M = Li, Na, K, or Cs and X = Cl^- or Br^-, and using literature data for the cell M | MX | M_y(Hg), the standard potentials for cells of the type:

$$M \mid MX \mid AgX \mid Ag \qquad (4\text{-XVII})$$

were evaluated. The usual type of information was derived from the emf measurements, that is, activity coefficients and standard free energies of formation. It was observed in this study that all the amalgam electrodes underwent abrupt decreases in emf at low concentrations (~0.01 *m*), probably as a result of decomposition of the amalgam, causing an increase in the metal-ion concentration around the electrode surface.

The standard potentials for cells of type 4-XVII were obtained by the usual extrapolation procedures and are summarized in Table 4.7.

The derived activity coefficient data indicate that any ion association in these systems is very slight, although the results were in conflict with heats of solution data by the same author, where the departures of the limiting slopes from the theoretical values were fairly large.

The free energies of formation for MX in solution were obtained from the $E°$ values and combined with heat data to obtain the standard partial molal entropies for each electrolyte.

Dawson et al. (19) have determined the standard potential of the Ag | AgCl electrode in *N*-methylacetamide using the cell

$$Pt;\ H_2 \mid HCl \mid AgCl \mid Ag$$

at temperatures between 35 and 70°C and found that the activity coefficients of HCl decreased with increasing molality much less than in aqueous solutions, indicating that solutions of HCl in *N*-methylacetamide are more nearly ideal than solutions of HCl in water.

Some of the older literature included emf studies on carboxylic acids, such as acetic acid and formic acid, but more recent interest has focused on the study of acetic acid–water mixtures, with particular reference to ion solvation. For example, Feakins et al. (20) determined the standard emf's for the cells

$$Pt;\ H_2 \mid HX \mid AgX \mid Ag$$

in 10, 20, 40, and 60% (w/w) solutions of acetic acid and water, where X = Br or I. The unknown standard electrode potentials were determined by a modification of Owen's method (21), in which the working equation is

$$E_{HX} - E_{HCl} = E^{\circ}_{m,\,HX} - E^{\circ}_{m,\,HCl} + \lambda m \qquad (4.58)$$

where the values for E_{HCl} and $E^{\circ}_{m,\,HCl}$ had been previously determined. Values of $E^{\circ}_{m,\,HX} - E^{\circ}_{m,\,HCl}$ were obtained by plotting the left-hand side

Table 4.7 Standard Potentials for the Cell M|MX|AgX|Ag in *N*-Methylformamide at 25°C

Cell	E° (V)
Li|LiCl|AgCl|Ag	3.123_7
Na|NaCl|AgCl|Ag	2.806_7
K|KCl|AgCl|Ag	3.021_2
Cs|CsCl|AgCl|Ag	2.987
Na|NaBr|AgBr|Ag	2.713_5

Reprinted with permission from *J. Phys. Chem.*, **70,** 1496 (1966). Copyright 1966. American Chemical Society.

of Eqn. 4.58 vs m and extrapolating to $m = 0$. The standard electrode potentials were used to generate molar free energies of transfer, ΔG_t°, from water to the mixtures for HCl, HBr, and HI. The free energies were split into ion values by a plot involving the reciprocal of the anionic radii.

$$\Delta G_t^\circ = \Delta G_t^\circ(H^+) + ar_a^{-1} \tag{4.59}$$

The results of this analysis are given in Table 4.8, together with a comparison of the results obtained for the same ions in methanol–water and dioxane–water mixtures. It is seen that the proton is in higher free-energy states in the acetic acid–water mixtures than in the comparable methanol–water or dioxane–water mixtures. The results were correlated in terms of an acid–base concept, that is, the specific interaction between the proton and the basic center of the solvating molecule (the oxygen atoms), or between the anions and the hydroxylic hydrogen atoms on the solvent molecule. Further discussion of these concepts is given in Chapter 5.

Heston and Hall (22) determined the activity of hydrogen chloride in glacial acetic acid using the cell

Pt/chloranil (satd.), hydrochloranil (satd.) HCl | AgCl | Ag (4-XVIII)

and the E° value was obtained by evaluating $E^{\circ\prime}$ from the Gronwall, La Mer, Sandved extension of the Debye–Hückel theory using the equation:

$$\begin{aligned} -\log f = {} & 19.63 \left(\frac{Z^2}{a}\right) \frac{x}{1+x} \\ & - 320.7 \left(\frac{Z^2}{a}\right)^3 10^3[\tfrac{1}{2}X_3(x) - 2Y_3(x)] \\ & - 26{,}200 \left(\frac{Z^2}{a}\right)^5 10^5[\tfrac{1}{2}X_5(x) - 4Y_5(x)] \end{aligned} \tag{4.60}$$

Table 4.8 Free Energies of Transfer, from Water to Aqueous Organic Solvents, of Individual Ions and Values of a (Eqn. 4.59)

Organic solvent (% w/w)	ΔG_t° (cal g ion^{-1})				a (cal Å)
	H^+	Cl^-	Br^-	I^-	
Acetic acid					
10	−190	440	410	380	800
20	−260	820	780	690	1490
40	−120	1480	1410	1230	2700
60	210	2320	2300	1930	4280
Methanol					
10	−570	750	700	630	1360
20.22	−920	1270	1140	1060	2270
33.4	−2500	3000	2800	2500	5460
43.12	−3400	4100	3900	3400	7500
Dioxane					
20	−2400	2800	2600	2300	5050
45	−5700	7000	6600	5800	12700

Reprinted with permission from *J. Chem. Soc.*, **A,** 1211 (1966).

where $x = \kappa a$ and for acetic acid $\kappa = 1.734 \times 10^8\sqrt{C}$. The standard potential was obtained from the equation:

$$E^\circ = E^{\circ\prime} - 0.1183 \log f \qquad (4.61)$$

It was found that the activities were in good agreement with previous vapor pressure measurements. The results for acetic acid were correlated with activity measurements in other solvents, such as benzene, nitrobenzene, ethanol, and methanol, in order to compare the basic strengths of the different solvents. Plots of log a_2 (the activity of acetic acid) versus log X_2 (the mole fraction of HCl) were performed, with the positions of the curves indicating the relative basicities of the solvents. Solvents that had little tendency to accept protons gave curves that lay at high values of log a_2, with the solution obeying Henry's law.

Some early work on activity coefficients of sulfuric acid in anhydrous acetic acid was carried out by Hutchison (23) using the cell

$$\text{Pt; } H_2 \mid H_2SO_4 \mid Hg_2SO_4 \mid Hg \qquad (4\text{-XIX})$$

Hutchison considered sulfuric acid as a ternary electrolyte dissociating into $2H^+$ and SO_4^{2-}. La Mer (24) interpreted the Hutchison data in terms of the sulfuric acid acting as a binary electrolyte and dissociating into H^+ and HSO_4^-, and found that on this basis the emf data could be accounted for using the Gronwall, La Mer, and Sandved extension of the Debye–Hückel theory.

Finally, for examples of emf studies involving alcohol-water mixtures, the work of Feakins et al. (23), Amis (24), or Roy (25) represent a selection of fairly widely investigated systems using the Pt; H_2 | HX | AgX | Ag cell. In each case, the approach is basically the same, namely, a determination of the standard electrode potential using one of several extrapolation procedures, the evaluation of activity coefficients or medium effects, calculation of free energies or enthalpies of transfer of the electrolyte from water to the solvent mixture and the determination of single-ion values for the transfer process.

A point worthy of mention in the study by Roy (25) is that a curve-fitting technique was used to obtain the value for $E°$, as the Gronwall–La Mer–Sandved extension produced curvatures in the plots at the lower concentrations. The equation used was

$$E + \frac{2RT}{F} \ln m = E° + A_1 m^{1/2} + A_2 m + A_3 m^{3/2} + \cdots \tag{4.62}$$

which is analogous in form to Eqn. 4.75.

Aprotic Solvents A fairly substantial literature exists of emf studies in aprotic solvents (2,5,6). The intent here is to focus attention on some of the more commonly used solvents of practical interest.

Studies of the thermodynamic properties of lithium halides in the dipolar aprotic solvent dimethylsulfoxide have received attention with particular application of the use of lithium electrodes in organic battery systems. For example, LiCl has been investigated by Butler (26) and by Tobias (27). Both investigators determined activity coefficients of LiCl in DMSO using the cell

$$\text{Li(Hg)} \mid \text{LiCl} \mid \text{TlCl(s)} \mid \text{Tl (Hg)} \tag{4-XX}$$

and found that the results obtained in the two studies from the emf method were in excellent agreement and also agreed with an evaluation of activity coefficients from cryoscopic data. The standard potential of the cell was evaluated from a mean value of the quantity, $E°'$, where

$$E°' = E + 0.1183\left(\log m - \frac{1.115\sqrt{m}}{1 + \sqrt{m}} + 0.282m\right) + 0.511(t - 25) \tag{4.63}$$

where m is the molal concentration of LiCl and t is the temperature in °C.

For each value of the observed emf, a value of the activity coefficient $\gamma_\pm$ was calculated using the expression.

$$\log \gamma_{\pm} = \frac{E^{\circ} - E}{0.1183} - \log m \tag{4.64}$$

The results obtained for the mean ionic activity coefficients could be represented (up to $\sim m = 0.15$) by an equation of the Guggenheim form:

$$\log \gamma_{\pm} = -\frac{1.115\sqrt{m}}{1 + \sqrt{m}} + 0.282\, m \tag{4.65}$$

or by the extended Debye–Hückel equation.

Salomon (28) used the cell

$$\text{Li} \mid \text{LiX} \mid \text{TlX, Tl(Hg)} \qquad \text{(4-XXI)}$$

to determine the activity coefficients at several temperatures of lithium bromide and lithium iodide in DMSO. The observed emf's of cell XXI were converted to the emf's of cell 4-XXII for solid thallium by the relation

$$E_{\text{XXII}} = E_{\text{XXI}} - E_{\text{XXIII}} \tag{4.66}$$

where E_{XXII} and E_{XXIII} are the emf's of the cells

$$\text{Li} \mid \text{LiX} \mid \text{TlX} \mid \text{Tl} \qquad \text{(4-XXII)}$$

$$\text{Tl} \mid \text{Tl}^{+} \mid \text{Tl(Hg)} \qquad \text{(4-XXIII)}$$

The standard potential E°_{XXII} was evaluated using the Guggenheim approximation for $\gamma_{\pm}$, that is,

$$\ln \gamma_{\pm} = -\frac{Am^{1/2}}{1 + m^{1/2}} + 2\beta m \tag{4.67}$$

where A is the Debye–Hückel constant and β is an empirical constant. Free energies, enthalpies, and entropies of cell 4-XXII were also calculated.

A plot of $\ln \gamma_{\pm}$ vs $m^{1/2}$ is shown in Fig. 4.4, where it is observed that at about $m = 0.4$, the activity coefficient for LiBr begins to increase, compared to the case for LiCl in DMSO, where no evidence of curvature occurred up to concentrations of 0.9 M. The curvature is even more marked for LiI. According to the theory proposed by Stokes and Robinson (29), the curvature in plots such as these is due to decreasing solvent activity, which implies that the chloride ion is to all intents and purposes unsolvated in DMSO, whereas the bromide ion is solvated to a fairly large extent; and the iodide ion is solvated to a greater extent by the DMSO, although considerably less than the cations. Information relative to the free energies, enthalpies, and entropies of transfer of the lithium halides from water to DMSO is discussed in Chapter 5.

Propylene carbonate is another aprotic solvent that has received attention due to its possible application in nonaqueous batteries and several thermodynamic studies have been undertaken in this solvent. Salomon (30)

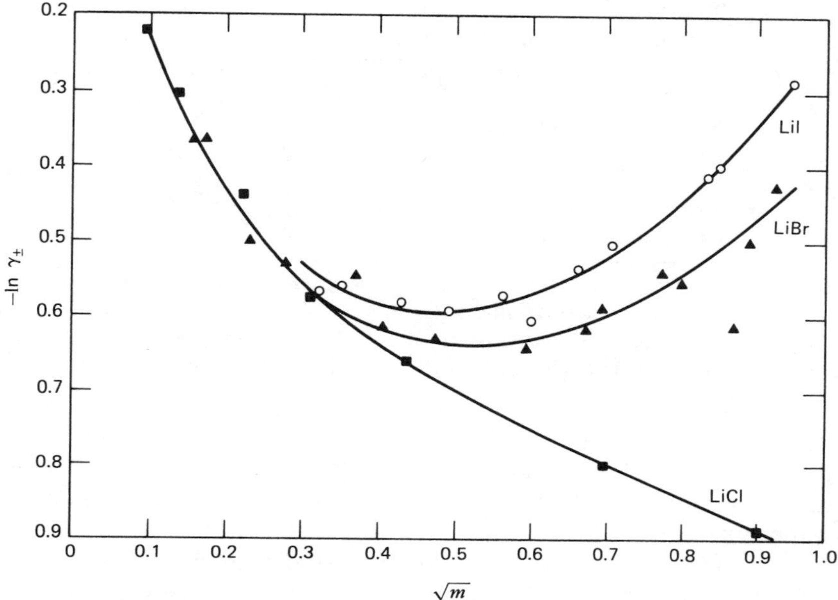

Figure 4.4 Plot of ln $\gamma_\pm$ vs $m^{1/2}$ for lithium halides in DMSO. (○) LiI, (▲) LiBr, (■) LiCl. Reprinted with permission of the publisher, The Electrochemical Society, Inc., from *J. Electrochem. Soc.*, **117,** 325 (1970).

investigated the cell

$$\text{Li} \mid \text{Li}^+, \text{X}^- \mid \text{TlX}, \text{Tl(Hg)} \qquad \text{(4-XXIV)}$$

in propylene carbonate (X = Cl or Br) over a temperature range of 15–45°C. Both activity coefficients and the thermodynamic quantities $\Delta G°$, $\Delta H°$, and $\Delta S°$ were obtained and a comparison of these thermodynamic quantities was made in propylene carbonate, water, *N*-methylformamide and methanol. Measured emf's for cell 4-XXIV were converted to emf's for the cell

$$\text{Li} \mid \text{LiX} \mid \text{TlX}, \text{Tl(s)} \qquad \text{(4-XXV)}$$

by the method discussed previously for DMSO. The energetics for the transfer of LiCl from water to various organic solvents are presented in Table 4.9. The large value for ΔG_t°(LiCl) for the transfer from water to propylene carbonate is attributed to the effect of solvent structure on the anion transfer. This effect is discussed in terms of the relatively unstructured nature of propylene carbonate with a localized negative dipole centered at the carbonyl group and a diffuse positive charge spread over the remainder of the molecule. The cation is attracted to the carbonyl group, while the anions are solvated by the positive end of the solvent dipole. The large value for the

Table 4.9 Energetics of LiCl Transfer from Water to Various Organic Solvents at 25°C[a]

Solvent	ΔG_t°	ΔH_t°	ΔS_t°
H_2O–20% dioxane	0.055	−0.585	−2.15
DMSO	4.75	−3.00	−26.01
PC	14.74	6.98	−26.01
NMF	3.20	−4.22	−24.89
CH_3OH	4.800	−2.600	−24.15

[a] All units refer to molarity scale. ΔG and ΔH are in kcal mol^{-1} and ΔS in eu.

Reprinted with permission from *J. Phys. Chem.*, **73**, 3299 (1969). Copyright 1969 American Chemical Society.

free energy of transfer is also reflected in the positive enthalpy of transfer as shown in Table 4.9. The failure of the Born equation is also demonstrated with these data.

One of the problems associated with emf studies in propylene carbonate is the selection of a suitable reference electrode. One careful study that focuses on this problem is that by Jasinski (31), in which the Li | Li^+ reference electrode is evaluated. Three forms of the electrode were evaluated in solutions containing 1 *M* lithium perchlorate. The following observations were made. Electodeposited lithium was susceptible to poisoning by electrolyte impurities. A lithium powder electrode resulted in too large an ohmic resistance. A lithium ribbon electrode which was scraped clean before use was found to give the most satisfactory behavior.

Propylene carbonate is also an interesting solvent for the study of solubility and complex formation. Butler et al. (32) investigated such equilibria potentiometrically for silver chloride in propylene carbonate–water mixtures. It was found that the principal species in solution was $AgCl_2^-$ and that the overall complex formation constants and solubility product increase with increasing water concentration.

Emf studies in acetonitrile have also presented problems in selection of satisfactory reference electrodes. Silver chloride has been found to slowly form a series of anionic and polynuclear complexes, which give rise to drifting potentials. Calomel type electrodes are also unsuitable because the large formation constants for mercury(II) halides in this solvent cause extensive mercury(I) disproportionation. Kratochvil (33) has investigated the silver–silver nitrate (0.01 *M*) couple in acetonitrile. The electrode was found to behave reversibly and its potential was unaffected by the common impurities found in acetonitrile. Kolthoff and Thomas (34) determined the standard potentials of ferrocene–ferricinium picrate and tris(*o*-phenanthroline) iron(II)–(III) perchlorates using a platinum indicator electrode and a silver–0.01 *M* silver nitrate reference electrode. The purpose of the study was to estimate liquid-junction potentials between acetonitrile solutions and the aqueous saturated calomel electrode.

Dimethylformamide is a nonaqueous solvent that has been of particular interest for the study of electrochemical reactions and for coordination chemistry in general. Butler (35,36) has investigated the thermodynamics of lithium chloride in DMF (35) and the solubility and complex formation of silver chloride in anhydrous DMF (36). Lithium chloride was investigated using the cell

$$\text{Tl (1.01 mol \% in Hg)} \mid \text{TlCl(s)} \mid \text{LiCl} \mid \text{Li (1.06 mol \% in Hg)} \quad \text{(4-XXVI)}$$

for the temperature range of 13–46°C. The emf data were used to calculate standard potentials, free energies, enthalpies, and entropies for the reactions

$$Li^+ + Cl^- + Tl(s) \rightarrow Li(s) + TlCl(s)$$

$$Li^+ + Tl(s) \rightarrow Li(s) + Tl^+$$

as well as activity coefficients for LiCl in DMF. It was observed that the effect on the measurements of water impurity in the range 0.002–0.013 m was negligible and there was no evidence for any strong ion pairing between Li^+ and Cl^-. Transfer functions, ΔG_t°, ΔH_t°, and $T\Delta S_t^\circ$ were determined in the usual way. In the study involving silver chloride, only mononuclear complexes AgCl and $AgCl_2^-$ were observed.

Inorganic Solvents The following discussion is intended to indicate the general scope of equilibrium emf measurements in selected inorganic solvents. Liquid hydrogen fluoride has been the subject of several investigations. For example, Koerber et al. (37) evaluated the standard free energies of formation of mercurous, cupric and lead fluorides using the cell

$$M, MF_2(s) \mid HF\ (NaF) \mid Hg_2F_2(s) \mid Hg \qquad \text{(4-XXVII)}$$

where M = Cd, Cu or Pb, at 0 and 10°C. In addition, the cell

$$Hg, Hg_2F_2(s) \mid NaF \parallel NaF, H_3OF, AgF \mid Ag \qquad \text{(4-XXVIII)}$$

was used to obtain activity coefficients of silver fluoride in liquid HF.

Clifford (38) was interested in contrasting the behavior of HF and H_2O. The following single-compartment cells were studied:

$$Cu(s), CuF_2(s) \mid TlF \mid TlF_3(s) \mid Pt \qquad \text{(4-XXIX)}$$

$$Ag(s) \mid AgF, TlF \mid TlF_3(s) \mid Pt \qquad \text{(4-XXX)}$$

$$Ag(s) \mid AgF \mid AgF_2(s) \mid Pt \qquad \text{(4-XXXI)}$$

Values of E° and ΔG° at 0°C were obtained. The trend in the potentials in HF was found to be very similar to the corresponding potentials in water. The thallium and silver couples were compared to the copper couple:

$$Cu(s) + TlF_3(s) \rightarrow CuF_2(s) + TlF(HF)$$

Thermodynamic ionization constants $K_{b\,(apparent)}$ and $K_{b\,(molar)}$ were calculated for the reactions

$$AgF \rightleftarrows Ag^+ + F^-$$

$$TlF \rightleftarrows Tl^+ + F^-$$

in HF at 0°C. Apparent and molal activity coefficients were also calculated. The results indicated that the fluorides in HF appeared to be considerably more highly dissociated than the corresponding hydroxides in water, reflecting the greater solvating power of HF for anions, compared to that of water.

Liquid ammonia is of interest in studying amino complexes of various metal salts, using emf techniques. Garner et al. (39) used the cell

$$Zn(Hg) \mid ZnCl_2{\cdot}6NH_3(s),\ NH_4Cl(NH_3),\ CdCl_2{\cdot}6NH_3(s) \mid Cd(Hg) \qquad \text{(4-XXXII)}$$

to evaluate the thermodynamic constants of the amminecadmium chlorides and cadmium chloride itself. From the emf and the temperature coefficient of the cell, together with other data, the standard free energies, enthalpies, and entropies of the solid amminecadmium chlorides and cadmium chloride were calculated and are given in Table 4.10. It was also established that the cadmium amalgam–hexamminecadmium chloride half-cell behaved as a suitable reversible reference electrode.

Table 4.10 The Thermodynamic Constants of the Amminecadmium Chlorides and Cadmium Chloride at 25°C

Substance	$\Delta G°$ (cal)	$\Delta H°$ (cal)	$S°$ (cal deg)[a]
$CdCl_2 \cdot 10NH_3(s)$	−134,633	−318,741	145.1
$CdCl_2 \cdot 6NH_3(s)$	−125,653	−242,781	90.8
$CdCl_2 \cdot 4NH_3(s)$	−115,533	−201,321	56.4
$CdCl_2 \cdot 2NH_3(s)$	−105,363	−146,781	65.7
$CdCl_2 \cdot NH_3(s)$	−95,772	−125,041	36.7
$CdCl_2$ (s)	−81,162	−90,591	33.5

[a] $S°$ is the standard virtual entropy.

Reprinted with permission from *J. Amer. Chem. Soc.*, **57**, 2055 (1935). Published 1935. American Chemical Society.

Summary Equilibrium emf measurements using galvanic cells provide a variety of information concerning both the energetics of electrode reactions and the structure of electrolyte solutions. One of the primary concerns is the establishment of the standard electrode potential $E°$, from which the Gibbs free energy for the cell reaction can be calculated. Various extrapolation procedures, which in most cases are based on some form of the Debye–Hückel equation for the mean ionic activity coefficient, can be used to gain a precise value of the $E°$. The determination of $E°$ as a function of temperature leads to the thermodynamic functions $\Delta H°$ and $\Delta S°$ for the cell reaction. Electrode potentials in conjunction with the standard electrode potential are needed to evaluate experimental mean ionic activity coefficients for electrolytes, the values of which give an indication of the departure from ideal solution behavior.

Measurements of electrode potentials are also very useful for the determination of the dissociation constants of weak acids and for studying complex formation of various electrolytes in solution. Solubility product constants can also be obtained using electrochemical cells.

Some of the problems involved in the design and use of reference electrodes in both protic and aprotic nonaqueous solvents are discussed in this chapter.

In mixed solvent systems, the information obtained from emf measurements is often used to compute transfer functions, such as the free energy of transfer of electrolytes from one solvent to another. With the aid of extrathermodynamic assumptions it is possible to generate transfer functions for single ions, which elucidate further the effect of the medium on ion–solvent interactions.

4.3 VAPOR PRESSURE STUDIES

Vapor pressure measurements on solutions of nonaqueous electrolytes are useful for determining both solvent and electrolyte activities. However, experimental difficulties inherent in the application of the vapor pressure technique to nonaqueous solutions have led to rather limited data. In fact the most widely studied solvent has been liquid ammonia.

For an aqueous solution, the activity of the water is calculated from the equation

$$a_1 = \frac{P_1}{P_1^0} \tag{4.68}$$

where P_1 is the vapor pressure of the water in the solution and P_1^0 is the vapor pressure of the pure water. An analogous equation could be written for a nonaqueous solution. The activity of the electrolyte is calculated from the activity of the solvent using the Gibbs–Duhem equation

$$S\,dT - V\,dP + \sum X_i\,d\bar{G}_i = 0 \tag{4.69}$$

which for constant temperature and pressure simplifies to

$$d \ln a_3 = -(X_1/X_3)\,d \ln a_1 \tag{4.70}$$

where subscripts 1 and 3 refer to pure solvent and electrolyte, respectively, and X is the mole fraction. Details of this method are given by Randall and White (40).

In the case of a mixed solvent system, it is necessary to know both the vapor pressure and the vapor composition in order to calculate the activities of both components of the solvent. Assuming the applicability of Dalton's law,

$$p_i = X_i P \tag{4.71}$$

and knowing X_i, the mole fraction of the ith component in the vapor, the partial pressure is obtained and the activity of component i in solution is

$$a_i = p_i/p_i^0 \tag{4.72}$$

The addition of a solute to a two-component solvent at constant X_1 is given by

$$X_1 \, d \ln a_1 + X_2 \, d \ln a_2 = -m_3 M_{12} \, d \ln a_3/1000 \tag{4.73}$$

when subscript 2 refers to the other solvent component, and M_{12} (the average molecular weight of the binary solvent) $= X_1 M_1 + X_2 M_2$.

A treatment discussed by Grunwald and Baccarella (41) yields the following equation for a 1:1 electrolyte.

$$(1000/M_{12})(X_1 \, d \ln \alpha_1 + X_2 \, d \ln \alpha_2) = -2dm - 2md \ln \gamma_{\pm} \tag{4.74}$$

where $\alpha_1 = p_1/p_1^*$ and $\alpha_2 = p_1/p_2^*$; p_1 and p_2 are the partial pressures of the two solvents over the solution containing the solute and p_1^* and p_2^* are the partial pressures of the solvents over the solvent of composition X_1, in the absence of solute.

An application of this approach is that of Campbell (42), who calculated the mean molal activity coefficients of lithium and sodium chlorates in water and water–dioxane mixtures.

A power series for $\ln \gamma_{\pm}$ due to Scatchard (43) is given as:

$$\ln \gamma_{\pm} = -Am^{1/2} + Bm + Cm^{3/2} + Dm^2 + Em^{5/2} + \cdots \tag{4.75}$$

where A is the Debye–Hückel limiting slope and $B, C, D, E, \ldots$ are adjustable constants. Substituting Eqn. 4.75 in Eqn. 4.74 gives

$$(-1000/M_{12})(X_1 \ln \alpha_1 + X_2 \ln \alpha_2)$$
$$= 2m - \tfrac{2}{3}Sm^{3/2} + Bm^2 + \tfrac{6}{5}Cm^{5/2} + \tfrac{4}{3}Dm^3 + \tfrac{10}{7}Em^{7/2} + \cdots \tag{4.76}$$

In Eqn. 4.76, the terms on the left-hand side and the first two terms on the right-hand side involve only solvent properties and experimental data and the remaining adjustable constants can be evaluated by a least-squares analysis. Once the constants are obtained, they are used in Eqn. 4.75 to evaluate $\gamma_{\pm}$.

The values of $\gamma_{\pm}$ obtained from this analysis were tested with two theoretical approaches, namely, the Stokes and Robinson method (44) and the Glueckauf method (45).

The relevant equations are, respectively,

$$\log \gamma_{\pm} = -\frac{A \mid Z_1Z_2 \mid I^{1/2}}{1 + B\mathring{a}I^{1/2}} - \frac{h}{\nu}\log a_{\text{A}} - \log\left[1 + 0.001W_{\text{A}}(\nu - h)m\right] \qquad (4.77)$$

where h is the solvation number and $\mathring{a}$, the distance of closest appoach and

$$\ln \gamma_{\pm} = -\frac{A \mid Z_1Z_2 \mid I^{1/2}}{1 + B\mathring{a}I^{1/2}} + 0.018mr(r + h - \nu) + \frac{h - \nu}{\nu}\ln(1 + 0.018mr) - \frac{h}{\nu}\ln(1 - 0.018mh) \qquad (4.78)$$

Good agreement for the water solutions between the experimental activity coefficients and the theoretical predicted values was obtained over the entire concentration range (4.0–9.4 m) for sodium chlorate, but in the case of lithium chlorate, the theory begins to break down above 2 m, indicating the greater hydration of the lithium ion. The hydration numbers obtained from the Glueckauf equation were lower than those obtained from the Stokes and Robinson equation. For the dioxane–water mixtures, the Grunwald and Bacarella method (41) was used to analyze the data and thus obtain activity coefficients. A graph of the activity coefficients of sodium and lithium chlorate in 44.5% dioxane is shown in Fig. 4.5. The higher degree of solvation that occurs with the lithium ion is demonstrated by the curve for $LiClO_3$ lying above the curve for $NaClO_3$. As the dioxane content increases with an accompanying decrease in dielectric constant, the activity coefficients of both electrolytes are greatly reduced as a result of an increasing tendency towards ion pairing. Figures 4.6 and 4.7 show the behavior of the activities of water and dioxane as a function of the concentration of sodium chlorate and lithium chlorate, respectively. For sodium chlorate, the water activity decreases with the addition of salt, while the dioxane activity increases. This type of behavior is to be expected if the sodium chlorate is preferentially solvated by water. As more water molecules attach themselves to the sodium ion, the relative content of the dioxane in the solvent mixture increases and therefore the partial pressure and activity of the dioxane will increase. A different type of behavior occurs in the case of lithium chlorate (Fig. 4.7). The decrease in water activity with concentration of solute is more pronounced, while the dioxane activity at first shows a slight increase, followed by a significant decrease. The larger decrease in water activity is a result of the higher hydration of the lithium ion.

Kushchenko (46) measured the vapor pressures of sodium iodide–acetonitrile solutions from 5 to 50°C and found that, except at low concentrations of sodium iodide, the activity of the solvent remained almost constant with change of temperature.

The activity of p-xylene in solutions of tetra-n-pentyl ammonium thio-

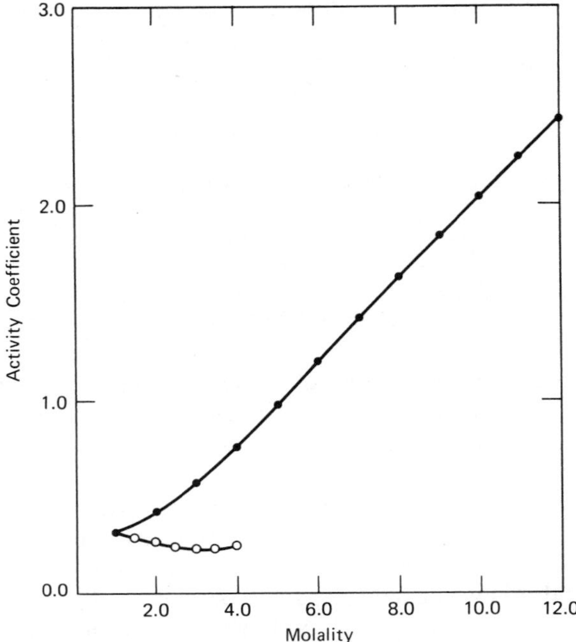

Figure 4.5 Mean molal activity coefficients vs molality for sodium chlorate and for lithium chlorate in the solvent 44.5% dioxane–55.5% water at 25°C. ○ Sodium chlorate ● lithium chlorate. Reproduced by permission of the National Research Council of Canada from the *Canadian Journal of Chemistry*, **47**, 2671 (1969).

cyanate was measured by O'Malley et al. (47) over the temperature range 50–90°C. The concentration range under study covered dilute solutions and ultraconcentrated solutions. All the data for the solvent and the solution were reproduced by the equation:

$$\ln P = A + \frac{B}{T} + \frac{C}{T^2} \tag{4.79}$$

The activity of *p*-xylene, defined as the pressure above the solution divided by the pressure of the pure solvent, was found to be linearly dependent upon the concentration of electrolyte at mole fractions of electrolyte greater than about 0.3.

Liquid ammonia has been the most popular solvent for vapor pressure studies, with the attention being focused on ion association and ion solvation. For example, Linhard (48) used the differential static method to measure the vapor pressure of several alkali-metal halides and nitrates, as well as alkaline-earth metal nitrates in liquid ammonia. A high degree of solvation of the ions was observed, with the solvation order $Cl^- < Br^- < I^-$ and $Cs^+ < Rb^+ < K^+ < Na^+$. Strontium and barium nitrates yielded exceedingly high vapor pressure depressions, which was attributed to a high lattice energy. Hunt and Larson (49) observed large deviations from Raoult's law for ammonium halides and ammonium nitrate in liquid ammonia, an indication of significant association.

Daniels (50) determined the vapor pressure of concentrated solutions of lithium nitrate in liquid ammonia. Straight-line plots of log P vs $1/T$ were observed, indicating conformity with the Claussius–Clapeyron equation. For the case of sodium thiocyanate (51), large negative deviations from Raoult's law were observed. The deviation from nonideality was larger for the systems in which the concentration of ammonia was low. Also, the nonideal behavior decreased with an increase of temperature. However, at high con-

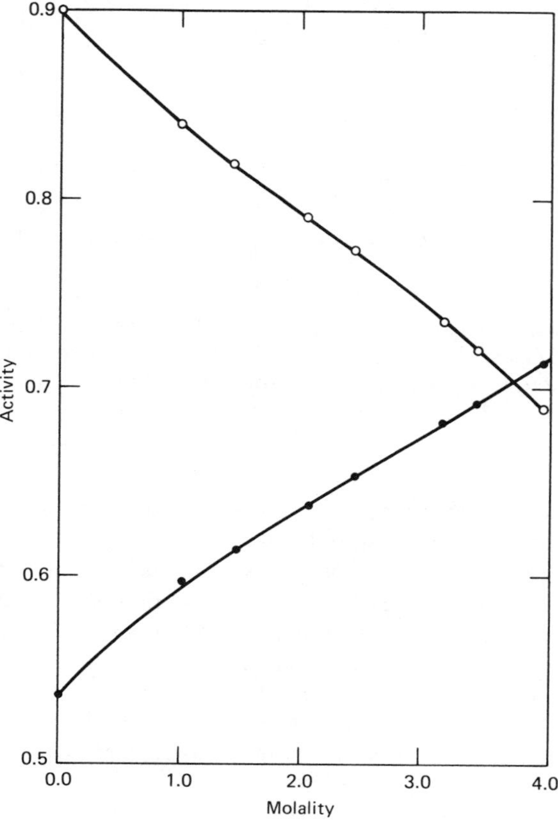

Figure 4.6 Water and dioxane activity vs molality for the system: Sodium chlorate–44.5% dioxane–55.5% water, at 25°C, as determined from vapor pressure and vapor composition. (○) Water activity, (•) dioxane activity. Reproduced by permission of the National Research Council of Canada from the *Canadian Journal of Chemistry*, **47**, 2671 (1969).

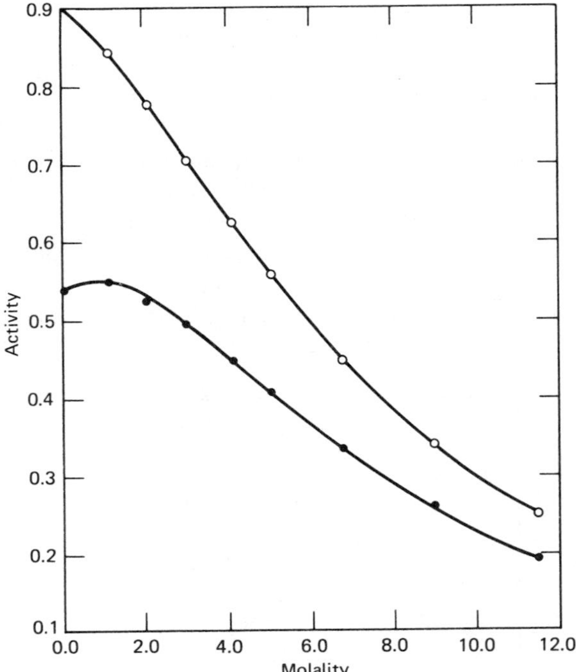

Figure 4.7 Water and dioxane activity vs molality for the system: Lithium chlorate–44.5% dioxane–55.5% water, at 25°C, as determined from vapor pressure and vapor composition. (○) Water activity, (●) dioxane activity. Reproduced by permission of the National Research Council of Canada from the *Canadian Journal of Chemistry*, **47,** 2671 (1969).

centrations and low temperatures, the deviation first increased and then decreased. The change in nonideality with temperature occurred in a concentration range closely corresponding to the region at lower temperatures where the adduct $NaSCN \cdot 3.5NH_3$ precipitates out. All the solutions studied were concentrated, in which the formation of ion pairs and ion clusters is favored. The ion pairs and higher aggregates are less effective than the bare charged ions in causing deviations, because such species have less polarizing effect on the ammonia molecules. At the higher temperature, some of the clusters are broken down and therefore the nonideality of the solution increases.

A limited number of studies have been undertaken in mixed solvent systems where selective solvation of ions by one of the components has been the feature of interest. One example of this approach is the investigation of the vapor pressure of the lithium perchlorate–methanol–water system (52). Vapor pressures were measured at 40°C and 1 *M* concentration. The addition of lithium perchlorate reduced the total vapor pressure and the partial vapor pressure of the methanol relative to the methanol–water system, whereas no change was observed for the partial vapor pressure of the water. In another experiment, an equimolar mixture of methanol and water was selected and the variation of vapor pressure with concentration of lithium perchlorate was examined. The total and partial vapor pressures all decreased with increasing lithium perchlorate concentration. The activities of the methanol and the water both decreased, while the activity coefficients of both components passed through a maximum at a concentration of lithium perchlorate of 4 *M*. The results are explained in terms of selective solvation of the lithium perchlorate by the methanol molecules, with a corresponding reduction in the vapor pressure of the methanol. For cases where the alcohol content is low, the salt is mainly hydrated and the vapor pressure of the water decreases, whereas for systems of high alcohol content, the solvation is due mainly to methanol molecules and therefore the partial vapor pressure of the water will increase, relative to its value in the binary solvent system. For the equimolar mixture, as the concentration of lithium perchlorate increases, the vapor pressure of both components decreases, but the change in the activities of the alcohol and the water show preferential solvation by the methanol.

A similar type of investigation was undertaken by Gross (53) for the system alkali-metal nitrate–water-hydrogen peroxide, in which the partial pressures of hydrogen peroxide and water were determined over the complete composition range at 50°C. For lithium nitrate, the partial pressure of the water component is lowered over the entire composition range, while that for hydrogen peroxide decreases only up to about 0.9 mole fraction hydrogen peroxide. The negative deviations were attributed in part to the specific interactions of the hydrogen peroxide and the water, causing always a negative change in the total pressure. The slight increase in the partial pressure of the hydrogen peroxide at about 0.9 mole fraction is an indication that there is very little specific interaction between the lithium nitrate and the H_2O_2. For the case of sodium nitrate, the deviations from ideality are less negative, indicating that the solvation is less extensive than for the $LiNO_3$, as a result of the smaller charge density for Na^+. On the other hand, potassium nitrate exhibits different behavior. A strong interaction occurs between the solute and the H_2O_2, and in systems with a high water content, the hydrogen peroxide–water clusters are broken to allow H_2O_2 molecules to interact with each other. This results in an increase in the pressure of the water, in accord with experimental observation. However, as the H_2O_2 content is increased, the bonding between the two solvent molecules increases, and therefore the partial pressure of the water decreases. Similar trends were observed for $RbNO_3$. The general conclusion was that the smaller ions were preferentially solvated by the water, while the larger ions were preferentially solvated by the hydrogen peroxide.

Finally, mention should be made of vapor pressure measurements in mixed solvents, which are used to determine the activity coefficient of the individual solvents. For example, Ryder et al. (54) measured the total vapor pressures for the system CCl_4–$TiCl_4$ at 30, 40, and 50°C over the entire composition range, where the total pressure P_t, is given by

$$P_t = \gamma_1 X_1 P_1^\circ + \gamma_2 X_2 P_2^\circ \tag{4.80}$$

where γ and X represent activity coefficient and mole fraction, respectively, and P_1° and P_2° are the vapor pressures of the pure solvents 1 and 2. The Van Laar equation was used to represent log γ_1 where

$$\log \gamma_1 = \frac{AX_2^2}{[(AX_1/B) + X_2]^2} \tag{4.81}$$

and

$$\log \gamma_2 = \frac{BX_1^2}{[X_1 + (B/A)X_2]^2} \tag{4.82}$$

where A and B are Van Laar constants. The uncertainties in the activity coefficients obtained were 3% for compositions of 50%, but increased as the composition decreased for a particular component.

Summary Vapor pressure measurements in nonaqueous solvents have been limited to selected solvents with the largest effort directed to the study of solutions in liquid ammonia. Solute and solvent activities are derived from vapor pressure data, using equations based on the Gibbs–Duhem relationship. In some of the studies, mean ionic activity coefficients have been determined and the experimental results tested with theoretical approaches, such as the Stokes and Robinson method or the Glueckauf method.

The activities and activity coefficients obtained from vapor pressure measurements in mixed solvents are normally discussed in terms of selective solvation by one of the solvents or extent of solvation of a particular ion. Solvation numbers can be obtained by use of appropriate equations. Deviations of mean ionic activity coefficients from ideal behavior yield evidence for ion association.

In most cases, the data in liquid ammonia were tested with Raoult's law and deviations treated in terms of ion pairs or ion clusters. In addition, the orders of solvation for series of ions have been established in liquid ammonia, with the sodium ion being more strongly solvated than the cesium ion and the iodide ion more strongly solvated than the chloride ion.

4.4 CRYOSCOPY

Cryoscopic measurements in nonaqueous electrolyte solutions are valuable for studying deviations from ideality. However, the disadvantage is that measurements have to be made close to the melting point of the solvent.

The osmotic coefficient (ϕ) of the solvent can be obtained directly and the activity coefficient (γ) of the solute is obtained using the Gibbs–Duhem equation. The relationship between these two quantities is

$$-\ln \gamma = 1 - \phi + \int_{o}^{m_i} (1 - \phi)\, d \ln m + \int_{m_i}^{m} (1 - \phi)\, d \ln m \quad (4.83)$$

where m_i is the lowest molality at which the osmotic coefficient can be obtained experimentally.

To introduce the definition of the osmotic coefficient, consider the equilibrium between a liquid mixture of two substances (nonreacting) and the pure solid phase of one of them. The relationship between the activity of species 1 and the depression of the freezing point, $\theta = T_o - T$, is

$$-\ln a_1(T) = \int_{T_o}^{T} -\frac{\Delta H_f}{RT^2}\, dT \quad (4.84)$$

where ΔH_f is the molal heat of fusion of the pure substance.

For solutions that are nonideal

$$-\ln a_1(T) = \phi M_1 \sum_i m_i \quad (4.85)$$

where ϕ is the practical osmotic coefficient of the solvent, M_1 is the molecular weight of the solvent, and $\sum_i m_i$ is the sum of the molalities of the solutes. Combining Eqns. 4.84 and 4.85 and assuming ΔH_f is a constant over the temperature range considered

$$\phi \sum_i m_i = \frac{\Delta H_f}{RM_1T_o^2} \frac{T_o\theta}{(T_o - \theta)} \quad (4.86)$$

where $M_1RT_o^2/\Delta H_f$ is the molal cryoscopic constant, K_f of the solvent.

For cases where ΔH_f is not constant, it can be shown that

$$K_f \phi \sum m_i = \theta + \left[\frac{1}{T_o} - \frac{\Delta C_f^\circ}{2\Delta H_f}\right] \theta^2 + \frac{\theta^3}{T_o^2} \quad (4.87)$$

where ΔC_f° is the change in ΔC_p on fusion of the pure solvent and ΔC_p is the difference in the partial molal heat capacity of the solvent in the solution and the molal heat capacity of the solid solvent.

Cryoscopic studies have been limited to a few solvents, including acetic acid, benzene, dimethylsulfoxide, ethylene carbonate, ethylenediamine, formamide, formic acid, *N*-methylacetamide (NMA), and sulfolane. Trends in osmotic and activity coefficients have been compared for a variety of electrolytes in the solvents enumerated above and the behavior discussed in terms of specific ion–ion and ion–solvent interactions.

A very careful cryoscopic study has been made by Wood et al. (55,56) in *N*-methylacetamide. The electrolytes investigated consisted of a selection of alkali-metal halides and nitrates (55) and tetraalkylammonium halides and some alkali-metal formates, acetates, and propionates. All measurements were conducted over a concentration range of 0.1–0.8 m. Activity coefficients were calculated from freezing-point measurements and were represented by an extended form of the Guggenheim equation:

$$\log \gamma_{\pm} = -\frac{Am^{1/2}}{1 + m^{1/2}} + Bm + Cm^{3/2} \tag{4.88}$$

where A is the Debye–Hückel constant. Integration of Eqn. 4.88 yields an equation for the osmotic coefficient

$$\phi = 1 - \frac{2.303Am^{1/2}}{3}\sigma(m^{1/2}) + \frac{2.303Bm}{2} + 2.303\frac{5}{6}Cm^{3/2} \tag{4.89}$$

where

$$\sigma(m^{1/2}) = \frac{3}{m^{3/2}}\left[1 + m^{1/2} - \frac{1}{1 + m^{1/2}} - 2\ln(1 + m^{1/2})\right] \tag{4.90}$$

Figure 4.8 shows the variation of osmotic coefficients for the alkali-metal halides and nitrates in NMA. At low concentrations, the osmotic coefficients are much higher than in water, due to the high dielectric constant of NMA (178 at 30.5°C), which gives a Debye–Hückel slope that is only one quarter as large as that for water. At the highest concentrations, the order of the osmotic coefficients follows the trend NaI > LiBr > LiCl > $LiNO_3$ > NaBr > NaCl > KI > KBr > CsI > CsBr > CsCl > $NaNO_3$, an order which is

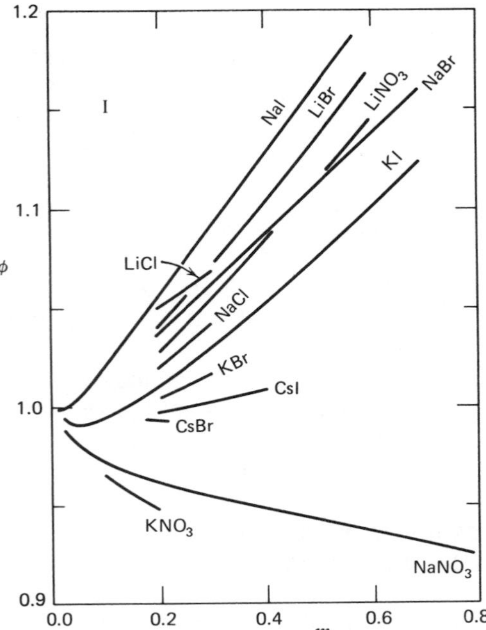

Figure 4.8 Osmotic coefficients in N-methylacetamide vs m. Reprinted with permission from *J. Phys. Chem.*, **75**, 2313 (1971). Copyright 1971. American Chemical Society.

consistent with $Li^+ > Na^+ > K^+ > Cs^+$ and $I^- > Br^- > Cl^- > NO_3^-$. It is important to note that the order observed in NMA is similar to that obtained in water, indicating that similar factors operate.

One has to consider what factors affect the values of the osmotic coefficients. At moderate concentrations, the osmotic coefficients will depend mostly on the interactions of the two oppositely charged ions, although there is also likely to be some contribution from ions of like charges. Also, the interactions between the ions are influenced by the solvation of the ions and the relative ease with which ions of opposite charge displace solvent molecules from the coordination sphere of an ion. It was proposed that the competition of both solvent and oppositely charged ions for an ion is an important effect of the solvent.

The suggestion was that if both ions were small or both ions were large, they could come into contact with each other more easily. This postulate can be explained by comparing the interaction energy of a small ion with a coordinated solvent molecule and with an oppositely charged ion.

For the case of an oppositely charged ion that is large relative to the solvent molecule, the interaction energy between the two ions is less than the interaction energy of the ion with the solvent molecule, and therefore it will be difficult for a large ion to enter the coordination sphere of a small ion. For the case of two small ions or two large ions, the ion interaction energies are greater than the ion–solvent interaction energies and therefore ion–ion contact occurs. Such a model explains the cation and anion order of the osmotic coefficients observed in NMA, assuming that the cations are strongly solvated and the anions are weakly solvated. As the cation gets smaller and the anion gets larger, it is more difficult for the solvent displacement to occur, ion–ion repulsion increases, and the osmotic coefficient increases.

The similarity of the trends of the activity coefficients observed for water and NMA indicates that the structural properties of water and NMA may be similar. All the evidence indicates that no structure breaking occurs in solutions of alkali-metal halides in NMA, compared to aqueous solutions, in which structure breaking and making account for the trends in activity coefficients. Viscosity measurements indicate that alkali metal halides are strong structure makers in NMA, where each ion acts as a crosslinking agent for the chains of hydrogen-bonded solvent molecules that occur in the pure solvent. The conclusion drawn is that net structure breaking cannot account for the order of the osmotic coefficients, and it is possible that osmotic coefficients in both NMA and water are not sensitive to small changes in the structure of the solvent.

A similar ordering of osmotic coefficients occurred for the acetates and propionates of lithium, sodium, and potassium, the formates of lithium and sodium, and the chlorides, bromides, and iodides of some *n*-tetraalkylammonium ions in NMA (56). The behavior is shown in Figs. 4.9 and 4.10. At low concentrations, it is seen that the order of the osmotic coefficients for

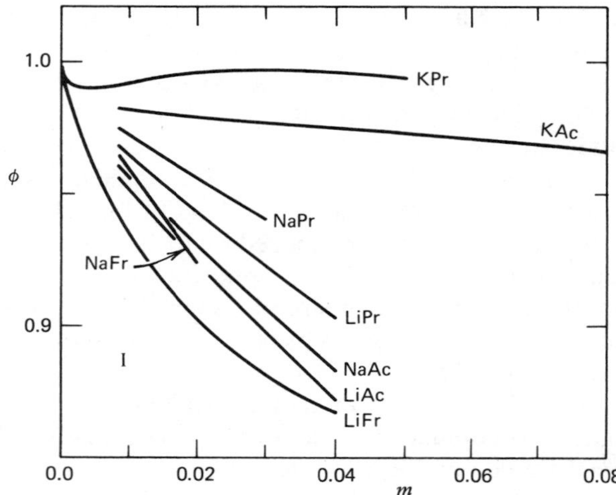

Figure 4.9 Osmotic coefficients of alkali metal formates (Fr), acetates (Ac), and propionates (Pr) in NMA as a function of molality. Reprinted with permission from *J. Phys. Chem.*, **75,** 2319 (1971). Copyright 1971. American Chemical Society.

the *n*-tetraalkylammonium halides follows the sequence, $Cl^- > Br^- > I^-$, the same as observed in aqueous solutions. For the smaller ions, the order follows:

$$Bu_4N^+ > Pr_4N^+ > Et_4N^+ > Me_4N^+ \qquad \text{in both water and NMA.}$$

As the anion increases in size it is seen that the order reverses in water and becomes less pronounced in NMA.

Because the trends in the two solvents are similar, it is evident that the water-structure-enhancing properties of the larger tetraalkylammonium ions are not responsible for these trends, but they can be explained by the penetration of the larger, weakly solvated anions into the hydrocarbon portion of the cation.

Dimethylsulfoxide has been used by several investigators to determine activity coefficients of selected electrolytes and in some cases comparisons have been made between the activity coefficients obtained from cryoscopy and those obtained by emf measurements. In such comparative studies, the agreement is not good due to heat transfer effects in the cryoscopic method. Lithium chloride has been the most widely studied electrolyte, but reference to Garnsey and Prue (57) shows serious discrepancies between the activity coefficients obtained. In most studies, any anomalies in the activity coefficients were attributed to ion-pair formation.

The Garnsey and Prue study (58) involved the evaluation of osmotic coefficients for several alkali-metal salts (mostly perchlorates) in DMSO and sulfolane. The osmotic coefficients were calculated from the relation:

$$\phi = \frac{T°\theta}{2\,\lambda_c m(T° - \theta)} \tag{4.91}$$

where $T°$ is the freezing point of the pure solvent, m is the molality of the solution, λ_c is the cryoscopic constant, and θ is the freezing-point depression. The values of ϕ in DMSO were all greater than those predicted by the Debye–Hückel limiting law for the entire concentration range (m = 0.01–0.25). Also, with increasing concentration it was observed that the values for $LiClO_4$ were greater than those for $KClO_4$ and the divergence increased with concentration. This effect suggests that the cation–anion interaction is greater for the $KClO_4$ and that the strong solvating power of DMSO, like that of water, inhibits the interaction of cations with large anions as the size of the cation decreases.

In contrast, in sulfolane, the osmotic coefficients for $LiClO_4$ and $NaClO_4$ lie below the Debye–Hückel limiting law values, while the values for $KClO_4$ and $RbClO_4$ are close to those predicted by the limiting law up to a concentration of m = 0.01, but show increasingly positive deviations with increasing concentration. Also, the osmotic coefficients in sulfolane were found to decrease with decreasing size of cation, which is the opposite sequence to that found for water and DMSO. The direct correlation with cation size

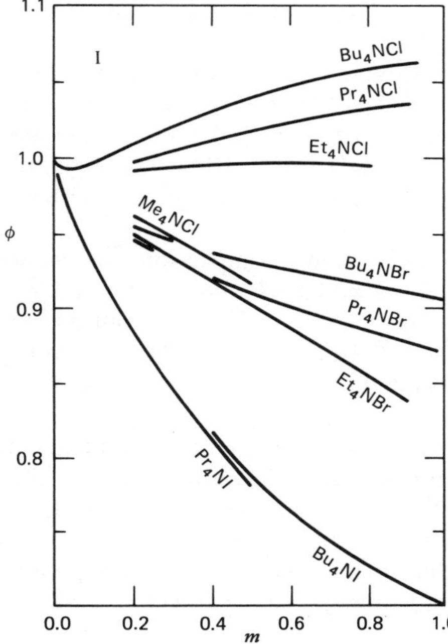

Figure 4.10 Osmotic coefficients of tetraalkylammonium halides in NMA vs molality. Reprinted with permission from *J. Phys. Chem.*, **75,** 2319 (1971). Copyright 1971. American Chemical Society.

observed in this case suggests that sulfolane has a low solvating power. This study illustrates that even though the two solvents have similar dielectric constants, they possess very different powers of ionic solvation, giving rise to the opposite sequences of osmotic coefficients relative to cation radius for the alkali-metal perchlorates and incomplete dissociation of the perchlorates in sulfolane.

It is of interest to compare the behavior of nonelectrolytes and electrolytes in a nonaqueous solvent and their behavior in different solvents. Bonner et al. (59) have measured the osmotic and activity coefficients of six nonelectrolytes and ten electrolytes in DMSO and compared the behavior in DMSO and ethylene carbonate. The nonelectrolytes considered were carbon tetrachloride, urea, 1,3-dimethylurea, benzene, and ethylene glycol, while the electrolytes studied were LiCl, NaCl, KCl, NH_4Cl, and CsCl.

Considering first the nonelectrolytes, the effect of hydrogen bonding is observed between DMSO and the hydroxyl hydrogens on the ethylene glycol and *meso*-erythritol by the negative deviations from Raoult's law. A similar type of behavior is observed for these particular nonelectrolytes in aqueous solution, while in an aprotic solvent such as ethylene carbonate, the osmotic coefficient of ethylene glycol decreases rapidly as the concentration is increased. Apparently, this contrasting behavior can be accounted for in terms of the size of the different solvent molecules, as DMSO and water have similar sizes and are different in size compared to ethylene carbonate. Also, in DMSO, the solvation due to solvent–solute hydrogen bonding exerts a greater influence than any effects due to solvent structure breaking or self-association of the solvent molecules.

In addition, it was observed that positive deviations from Raoult's law were observed for solutions of urea and 1,3-dimethylurea in DMSO, ethylene carbonate, and water. Also, the osmotic coefficients of both solutes decreased in the order $H_2O > DMSO \gg EnCO_3$. This is further evidence pointing to the fact that solute–solvent interactions may be greater with the smaller solvent molecules, H_2O and DMSO, than with $EnCO_3$.

For the case of the electrolyte solutions, again a comparison was undertaken in the solvents DMSO and ethylene carbonate. It should be noted that the dielectric constant of $EnCO_3$ is about twice that of DMSO, but its molecular size is also much larger, both effects to be considered when estimating the relative importance of solvation and dielectric constant on solution properties.

As an example, sodium iodide yielded larger osmotic coefficients in DMSO than in $EnCO_3$, indicating the importance of solvation by the smaller solvent molecules. It is observed that the larger ions such as K^+ and Et_4N^+ were less solvated and also the osmotic coefficients of the iodides were significantly larger than those of the chlorides in DMSO, a result of greater solvation of the more polarizable iodide ion and to a lower ion pairing between cations and the iodide ion.

The osmotic and activity coefficients of the salts studied decreased in the sequence

$$Li^+ > Na^+ > K^+ > Cs^+ > NH_4^+ > (C_2H_5)_4N^+$$

a trend that reflects the relative solvation of the cations. In addition, it was observed that the osmotic coefficients of ethylene–bis(trimethylammonium) iodide were quite low and similar in magnitude in water, DMSO, and $EnCO_3$, indicating that this large bolaform electrolyte is relatively unsolvated in any solvent and that appreciable ion pairing occurs. The general conclusion in this particular study was that solvation, which is larger for the smaller solvent molecules, is more important in determining solution properties than is dielectric constant.

Another study involving a comparison of osmotic and activity coefficients of solutes in structured solvents, such as ethylene carbonate (layered structure), *N*-methylacetamide (polymeric chains), and water (three-dimensional network), is that by Bonner (60). Measurements were undertaken with nonelectrolytes (carbon tetrachloride, dibenzyl, *o*-dichlorobenzene, ethylene glycol, urea, and 1,3-dimethylurea) and electrolytes (NaI, KI, CsI, NH_4I, ZnI_2, $ZnCl_2$, and $(C_3H_7)_4NI$).

In ethylene carbonate, it was found that the osmotic coefficients of the nonelectrolytes decreased fairly rapidly with increasing concentration and then leveled off, suggesting disruption of the solvent structure by the nonelectrolyte. The fact that the decrease in osmotic coefficient of CCl_4 in NMA is more marked indicates that the structure of NMA is more easily broken down than that of ethylene carbonate. A further observation is that in ethylene carbonate, the osmotic coefficients of carbon tetrachloride at any concentration are greater than those of dibenzyl. On the other hand, the compound phenanthrene, with a similar molecular weight to dibenzyl, has higher osmotic coefficients than those of CCl_4 in NMA. This effect is apparently caused by the solvation of aromatic solutes in NMA by the formation of solvent complexes.

Comparison of the behavior of urea and 1,3-dimethylurea (DMU) shows that the osmotic coefficients of urea are larger than those of DMU. This difference is attributed to the fact that urea is sufficiently solvated in water and NMA to prevent excessive association of the solute from occurring.

For the electrolytes studied, the osmotic coefficients of the alkali-metal iodides decrease in the order Li > Na > K > Rb > Cs at any fixed concentration, an order attributed to the solvating properties of the ions. Osmotic coefficients of sodium and potassium iodides at any concentration in the three solvents were found to be in the order H_2O > ethylene carbonate > NMA. In the case of H_2O > ethylene carbonate, this effect is due to the greater solvating ability of the smaller water molecule, whereas the smaller coefficients in NMA are a reflection of the case of breaking NMA chains upon the addition of a solute.

In contrast, it is found that the osmotic coefficients of tetrapropylammonium iodide are higher in ethylene carbonate than they are in water, probably a result of structure-enforced ion pairing in which the ions are forced together to minimize the interaction with the surrounding water. A similar phenomenon would not be operative in ethylene-carbonate solutions.

Several cryoscopic studies, especially some of the earlier ones, were undertaken to obtain molecular weights and the extent of association of electrolytes in nonaqueous solvents (i.e., existence of dimers, trimers, etc.). Acetic acid, a solvent of low dielectric constant, was a popular choice for these types of studies. For example, Eichelberger (61) examined the salts NH_4NO_3 and H_2SO_4 in anhydrous acetic acid. The association behavior was studied in terms of the Lewis and Randall j function, where

$$j = 1 - \frac{\Delta T}{\nu m K_F} \tag{4.92}$$

where ΔT is the freezing point lowering, ν the number of ions, m is the concentration, and K_F is the molal freezing-point constant. The following criteria were used for association: $j = 0$ (complete dissociation); $j = 1$ (complete association), and $j = 0.5$ (no association or dissociation). The data showed that ammonium nitrate in the concentration range 0.005–0.05 m was associated as ion pairs, whereas at low concentrations, sulfuric acid was appreciably dissociated. However, at concentrations of about 0.1 m, sulfuric acid behaved as an associated electrolyte in anhydrous acetic acid. Also it was observed that both electrolytes deviated widely from the predictions of the Debye–Hückel limiting law, as demonstrated by plots of j vs $\sqrt{m}$.

Cryoscopic studies using cobalt and nickel acetates (62) in acetic acid indicated that both salts existed mainly as dimeric species, such as $2Co(C_2H_3O_2)_2 \cdot HC_2H_3O_2$.

Kraus et al. (63, 64) used cryoscopy to study the behavior of both carboxylic acids and quaternary ammonium salts in benzene. For the organic acids, dissociation constants were derived for the monomer–dimer equilibrium using the relation

$$\frac{2m\alpha^2}{1 - \alpha} = K \tag{4.93}$$

where K is the dissociation constant of the dimer, m is the stoichiometric concentration of the acid in moles per 1000 g of benzene, and α is the fraction of material existing as the monomer.

$$\alpha = 2r - 1 \tag{4.94}$$

where $r = \Delta T/\Delta T_o$, the ratio of the observed freezing-point depression to that of the ideal substance at the same concentration.

Rearranging Eqn. 4.93 we obtain

$$2m\alpha = \frac{K}{\alpha - K} \tag{4.95}$$

so that a plot of $m\alpha$ vs $1/\alpha$ yields the dissociation constant K. The values thus obtained are given in Table 4.11. It is seen that, with the exception of hydrocinnamic acid, the order of the dissociation constants of the dimer in benzene is the same as that of their ionic dissociation constants in water, $_wK_a$.

In the study involving the determination of the molecular weights of the quaternary ammonium salts, an association number, n (the ratio of the ob served molecular weight to formal weight) was evaluated. The trends in the behavior of the association number as a function of salt concentration are shown in Fig. 4.11. It is observed that all of the salts investigated exhibited marked association in benzene solutions, with the degree of association being dependent on both structural and constitutional factors. Of the salts measured, the octadecyltri-*n*-butylammonium formate exhibited the highest degree of association.

Depending upon the extent of association, it was found convenient to divide the salts into five types:

A Salts with large unsymmetrical cations and large anions.
B Salts with large unsymmetrical cations and small anions.
C Salts of large symmetrical cations and large anions.
D Salts of large symmetrical cations and small anions.
E Highly associated salts.

The association of ion dipoles is essentially governed by two factors, namely, the dipole moment of the ion pair and the size and symmetry of the ions. For example, tetraisoamylammonium picrate with a dipole moment of 18×10^{-18} D shows an association of 30% at 0.001 N, while triisoamylammonium picrate, with a dipole moment of 13×10^{-18} D shows an association of 1% at 0.001 m and 7% at 0.01 m. On the other hand, tri-*n*-butylammonium

Table 4.11 Values of Dissociation Constants of Organic Acids in Benzene and Water (63,64)

Acid	$K_{C_6H_6} \times 10^4$ (5.4°C)	$_wK_a \times 10^5$ (25°C)
Phenylpropiolic	27	590
o-Bromobenzoic	12	140
Hydrocinnamic	10	2.19
Benzoic	6.4	6.27
Cinnamic (trans)	2.2	3.65

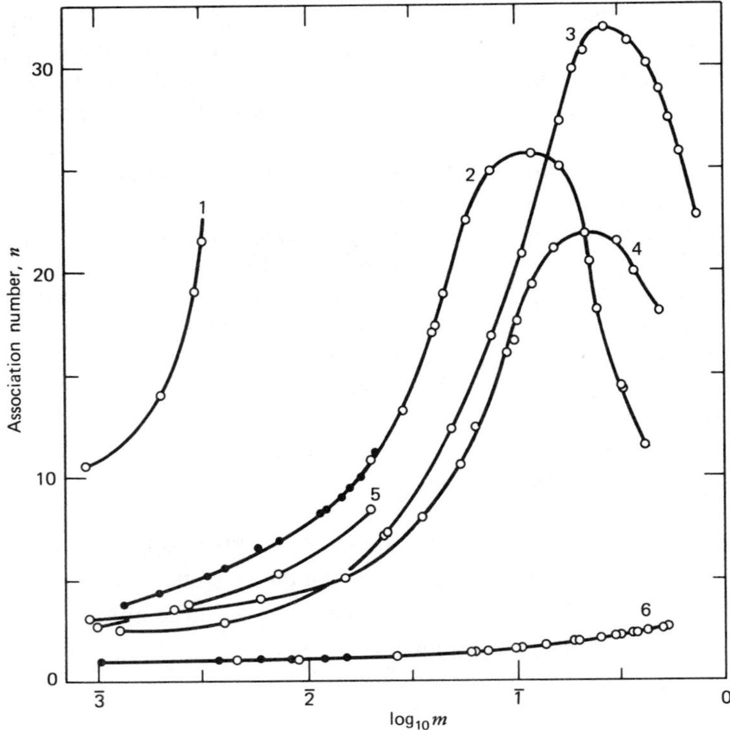

Figure 4.11 Association numbers of salts in benzene at the freezing point: 1, octadecyltributylammonium formate; 2, tetraisoamylammonium thiocyanate; 3, tetra-*n*-butylammonium thiocyanate; 4, *n*-amyltri-*n*-butylammonium iodide; 5, equimolar mixture of 2 and 3; 6, trisoamylammonium picrate: (○) Reference 64; (•) Barton and Kraus (63). Reprinted with permission from *J. Amer. Chem. Soc.*, **73,** 4557 (1951). Copyright 1951. American Chemical Society.

iodide with a dipole moment of only 8×10^{-18} D shows an association of 12% at 0.001 *m* and 40% at 0.01 *m*.

It was observed that the association numbers of class D salts increased rapidly with the concentration until a maximum was reached between 0.12 and 0.28 *m*. The existence of the maxima indicates a competition between kinetic interactions and association interactions, with the kinetic factor overcoming the association effect as the concentration increases.

Summary Freezing-point measurements or cryoscopic studies have been used to investigate two distinct areas. Early studies were directed to the determination of molecular weights of solvents such as acetic acid and in particular to examine the formation of dimers and higher aggregates. More precise data are used to determine both osmotic coefficients (ϕ) of solvents and activity coefficients (γ) of solutes using the Gibbs–Duhem equation. Cryoscopic studies of nonaqueous electrolyte solutions have been limited to a few selected solvents, and the trends in the osmotic and activity coefficients have been compared for a variety of electrolytes. The relative orders for the cation and anion coefficients can normally be rationalized in terms of specific ion–ion or ion–solvent interactions. Correlations with cation size and therefore solvating power of the cation are often useful. For example, in the solvent *N*-methylacetamide (NMA) the cation and anion order of the osmotic coefficients can be explained by assuming that the cations are strongly solvated, while the anions are weakly solvated. As the cation gets smaller and the anion gets larger, it is more difficult for the solvent displacement to occur, ion–ion repulsion increases, and the osmotic coefficient increases. Trends in activity coefficients in NMA and water can be accounted for in terms of structure breaking and structure making. No structure breaking occurs in solutions of alkali-metal halides in NMA, compared to aqueous solutions, in which structure breaking and making account for the trends in activity coefficients. The size of the solvent molecule is another factor. Solute–solvent interactions are greater with the smaller solvent molecules, for example, water and DMSO, compared to ethylene carbonate.

4.5 SOLUBILITY STUDIES

The relative solubilities of electrolytes in protic and dipolar aprotic solvents are influenced by the extent of solvation of the ions, solvent–solvent interactions, and other specific effects, such as volume energy. Solubility products, and dissociation constants of salts in nonaqueous solvents can give insights into ionic solvation. It is normally found that solubilities of salts do not lie in the order of the dielectric constants of the solvents, but depend more on specific interactions. This phenomenon was elaborated upon in Chapter 2. For the case when the solute is in equilibrium with its ions, $\Delta G = 0$ and the standard free energy of solution is given by

$$\Delta G^\circ_{\text{soln.}} = -RT \ln K_{\text{sp}}$$

where (4.96)

$$K_{\text{sp}} = a_{\text{M}}^{\nu_+} a_{\text{X}}^{\nu_-}$$

where ν_+ represents the number of cations M and ν_- represents the number of anions X produced by the dissociation of the electrolyte.

Solubility data are normally obtained over a range of temperatures and use can be made of such data to obtain the standard enthalpy of solution using the relation:

$$\left(\frac{\partial \ln K_{\text{sp}}}{\partial T}\right)_P = -\frac{1}{R}\left[\frac{\partial(\Delta G^\circ)/T}{\partial T}\right]_P = \frac{\Delta H^\circ_{\text{soln.}}}{RT^2} \tag{4.97}$$

It is assumed that $\Delta H^\circ_{soln.}$ is constant over a moderate temperature range and thus the integrated form of Eqn. 4.97 becomes

$$\log K_{sp} = -\frac{\Delta H^\circ_{soln.}}{R \ln 10}\left(\frac{1}{T}\right) + C \tag{4.98}$$

where C is the integration constant. $\Delta H^\circ_{soln.}$ is obtained from the slope of a plot of log K_{sp} vs T^{-1}.

One of the most extensive studies of solubility in anhydrous solvents is that by Parker et al. (65). The solubility products of silver, cesium, and potassium salts and the dissociation constants of AgX_2^- and I_3^- were determined in the protic solvents water, methanol, and formamide and in the dipolar aprotic solvents DMF, dimethylacetamide, dimethylsulfoxide, acetonitrile, and hexamethylphosphoramide. The solubility products of the silver salts and of the alkali-metal salts are given in Tables 4.12 and 4.13, respectively. Solubilities were obtained in this study using a variety of techniques, including potentiometric titrations, spectrophotometry, gravimetric analysis, and atomic absorption measurements.

Correlations were made between the equilibrium constants for each class of solvent, i.e., protic and aprotic, and a fairly linear relation was found in each case. For example, in the protic solvents, water, methanol, and formamide equations of the form:

$$\log K_s(H_2O) = \log K_s\,(CH_3OH) + 2.6 \tag{4.99}$$

were obtained for the solubility products of the silver salts. Deviations from these linear correlations were attributed to different extents of hydrogen bonding and ion polarizability. For example, silver salts of large polarizable anions, like AgSCN or $AgBPh_4$, are weak hydrogen-bond acceptors and tend to be more soluble in methanol than predicted by Eqn. 4.99, whereas silver salts of small anions, such as AgCl, which are strong hydrogen-bond acceptors, tend to be less soluble than predicted by Eqn. 4.99. For the case of the dipolar aprotic solvents, it was found that, with few exceptions, the solubility products of silver salts in one dipolar aprotic solvent could be predicted from the solubility product in another dipolar aprotic solvent with a fair degree of reliability using equations similar to 4.99. However, no correlations existed between solubility products of silver salts in protic and dipolar aprotic solvents.

A striking feature is that of the influence of anion solvation on the solubility of salts. For instance it is observed that potassium chloride is much less soluble (log $K_s = -5.4$) than potassium perchlorate (log $K_s = -0.1$) in DMF, whereas in methanol the trend is reversed (KCl; log $K_s = -2.5$; $KClO_4$; log $K_s = -4.5$).

An examination of the solubilities of the alkali-metal salts (Table 4.13) shows that within each class of solvent, the free energies of transfer of salts of the same cation are more or less independent of the anion, whereas free energies of transfer from a protic solvent to a dipolar aprotic solvent are

Table 4.12 Solubility Products of Silver Salts at 25°C

				$pK_s = -\log[Ag^+][X^-]$				
AgX	H_2O	CH_3OH	$HCONH_2$	DMF	DMAC	DMSO	CH_3CN	HMPT
AgCl	9.8	13.1	9.4	14.5	14.3	10.4	12.9	11.9
AgBr	12.3	15.2	11.4	15.0	14.5	10.6	12.9	12.3
AgI	16.0	18.3	14.5	15.8	14.7	11.4	14.2	—
AgN_3	8.6	11.2	7.7	11.0	10.8	6.5	9.6	8.5
AgSCN	11.9	13.9	9.9	11.5	10.5	7.1	10.0	7.4
AgOAc	2.4	6.1	—	10.2	9.7	4.4	7.4	—
AgOTs	1.3	3.2	—	1.3	—	—	—	—
$AgB(C_6H_5)_4$	13.4	13.2	10.3	6.7	5.9	4.6	7.2	4.7

Table 4.13 Solubility Products of Alkali-Metal Salts at 25°C

	$-\log K_s = pK_s$							
Salt	H_2O	CH_3OH	$HCONH_2$	DMF	DMSO	CH_3CN	HMPT	$CH_3CONHCH_3$
KCl	−0.90	2.5	—	5.4	—	7.20	—	−1.84
KBr	—	1.7	—	2.4	0.60	5.60	4.0	−0.74
KI	—	0.2	—	−0.5	—	2.00	—	+0.26
$KClO_4$	1.66	4.5	—	0.1	−0.80	—	—	−0.76
$KBPh_4$	7.53	5.0	—	—	—	—	—	—
$KOC_6H_4NO_2$-*p*	—	0.2	—	0.4	—	—	—	—
KPi	3.36	4.2	—	−0.2	—	4.55	—	—
$NaOC_6H_4NO_2$-*p*	0.91	−0.2	—	−0.58	—	5.42	—	—
NaN_3	—	0.9	—	1.9	0.64	—	—	—
NaCl	—	1.5	—	—	—	9.0	4.0	−0.94
CsCl	—	1.7	0.53	4.9	—	6.8	—	—
CsBr	—	2.2	0.29	3.3	—	4.4	—	—
CsI	—	1.9	0.23	1.7	—	3.0	—	—
CsPi	—	4.2	—	0.5	—	—	—	—
NEt_4Pi	—	2.0	—	0.0	—	—	—	—

strongly dependent on the anion, accounting for the different behavior of KCl and $KClO_4$ above in terms of hydrogen-bond acceptor abilities (see also Fig. 2.9).

Fairly extensive solubility measurements have been made in the protic solvent methanol and correlations with ionic size established. For example, Harner et al. (66) examined the solubility behavior of salts consisting of the cations Li^+, Na^+, K^+, Ca^{+2}, and Ba^{+2} with the anions F^-, Cl^-, Br^-, I^-, NO_3^-, CO_3^{-2}, and SO_4^{-2} in both methanol and water.

For the alkali-metal ions the salts of all the strong acids, except the sulfates, were soluble in methanol, whereas the salts of the weak acids were relatively insoluble, except for the fluorides and carbonates. Also, the solubilities increased with anion size, except in the case of KF. However, for the alkaline-earth ions the solubilities decreased with increasing anion size for F^-, Cl^-, and NO_3^- but increased with cation size for CO_3^{-2} and SO_4^{-2}. The solubilities in methanol are comparable, although smaller than the corresponding solubilities in water, except in the case of the alkali sulfates and some of the nitrates. Also, the solubilities of the alkaline-earth sulfates increased with cation size in methanol, but decreased with cation size in water.

Popovych and Friedman (67) used the solubility approach to study medium effects in water and methanol with the tetraphenylborates and picrates of potassium, triisoamylbutylammonium, and tetrabutylammonium. Ultraviolet spectroscopy was used to determine the solubilities of the tetraphenylborates, while visible spectra were used for the picrates. For the saturated methanolic solutions, association constants obtained from conductance data were used to calculate ionic concentrations, $(C\alpha)$ and activities. Conventional mean ionic coefficients, $y_\pm$ (molar scale) were obtained from the relation

$$-\log y_\pm^2 = \frac{3.803(C\alpha)^{1/2}}{1 + 0.5099\mathring{a}(C\alpha)^{1/2}} \tag{4.100}$$

where $\mathring{a}$ is the ion-size parameter.

The justification for using the Debye–Hückel approach is based on the assumption that solvation effects for the large ions used in this study are known to be negligible and it is the solvation effect that gives rise to deviations from the Debye–Hückel law. The medium effects will be discussed in Chapter 5, and therefore only the results obtained from $\Delta G^\circ = 2RT \ln {}_my_\pm$ will be given here and they are presented in Table 4.14. The negative sign reflects the fact that the distribution of ions favors the nonaqueous phase, compared to the positive medium effects for most electrolytes.

The effect of solvent–solvent interactions and the contribution of volume energies is the significant feature in the study of Deno and Berkheimer (68), who investigated the solubilities of organic salts, R_4NClO_4 in water, ethanol, benzene, and their aqueous mixtures. It was observed that with increasing size of R, the solubility in benzene and ethanol relative to water was found to increase, indicating that factors other than ion–solvent electrostatic forces

Table 4.14 Medium Effects for Transfer of Electrolytes from Water to Methanol

Electrolyte	Medium effect $\log {}_m y_{\pm}^2$
$KBPh_4$	−2.300
KPi	+1.101
(TAB)Pi	−5.398
Bu_4NPi	−4.438
$(TAB)BPh_4$	−8.799
Bu_4NBPh_4	−7.839

Reprinted with permission from *J. Phys. Chem.*, **70,** 1671 (1966). Copyright 1966. American Chemical Society.

must contribute to the solubility process because these forces are always greater in water.

Instead, the solubility trends are discussed in terms of the energy required to accommodate a solvent molecule in a "hole," the volume energy. This energy arises from the product of the volume of the hole times the internal pressure of the solvent, which is high for a molecule such as water, due to the intermolecular hydrogen bonding. The main forces in most organic liquids are those arising from the London forces, and therefore the energy required to form a hole is smaller.

As the size of R increases, the volume energy increasingly favors the greater solubility in benzene relative to water. At $(C_4H_9)_4NClO_4$, the volume energy effect equals the ion–solvent electrostatic effect, giving rise to equal solubilities in water and benzene. However, as the size of R increases beyond butyl, with a corresponding decrease in the electrostatic effect, the ratio of the solubility in benzene to the solubility in water appears to increase without limit. The practical application of the solubility behavior discussed above is that "large" salts can be extracted from water by organic solvents.

Addison et al. (69) investigated the solution properties of anhydrous copper nitrate, silver nitrate and zinc nitrate in ethyl acetate, acetonitrile and nitromethane. The results are given in Table 4.15. It was observed that the solubilities in acetonitrile were considerably higher than in nitromethane, reflecting the complexing powers of the acetonitrile molecule.

Free energies and entropies of solvation can be determined from solubility data and this approach was used by Criss and Luksha (70) for some alkali-metal halides in *N,N*-dimethylformamide. The solubilities, standard free energies of solution, together with the standard free energies of for-

Table 4.15 Solubilities of $Cu(NO_3)_2$, $AgNO_3$, and $Zn(NO_3)_2$ in Various Solvents

	Solubility g/100 g solvent		
Solvent	$Cu(NO_3)_2$	$AgNO_3$	$Zn(NO_3)_2$
H_2O	150 (25°C)	228	128 (25°C)
EtOAc	151 (25°C)	2.7	>136
$MeNO_2$	5.1	Negligible	0.45
MeCN	33.7	112	71

Reprinted with permission from *J. Chem. Soc.*, 4308 (1960).

mation are given in Table 4.16. The free energies of formation of the salts in DMF were obtained from a combination of the standard free energies of solution and the free energies of formation of the crystalline salts.

The standard free energies of solvation were obtained by using calculated values of the free energies of formation of pairs of alkali-halide gaseous ions. A modified Born equation (Latimer, Pitzer, Slansky approach, see Chapter 9) was used to develop a set of single-ion free energies of solvation in DMF and these values are given in Table 4.17. A plot of the free energy of solvation vs $(r + \delta)^{-1}$ for the alkali-metal ions and the halide ions showed a linear relationship as predicted by the modified Born equation. The entropies of solvation obtained by a combination of the free energy and enthalpy data are also given in Table 4.17. It was observed that the entropies obtained in DMF were much more negative than those obtained in water or in NMF, which is the result of the greater degree of internal hydrogen bonding in the latter two solvents. When an ion is placed in DMF, the entropy increase

Table 4.16 Solubilities, Standard Free Energies of Solution and Standard Free Energies of Formation of Electrolytes in *N,N*-Dimethylformamide

Salt	Solubility (mol kg^{-1})	$\Delta G^\circ_{soln.}$ (kcal mol^{-1})	$\Delta G^\circ_{f(DMF)}$ (kcal mol^{-1})	$\Delta G^\circ_{solv.}$ (kcal mol^{-1})
LiF	5.34×10^{-5}	11.69	−127.9	−219.8
NaF	4.68×10^{-5}	11.84	−117.5	−193.1
NaCl	6.08×10^{-3}	6.38	−85.41	−166.8
KCl	2.28×10^{-3}	7.41	−90.18	−149.2
CsCl	3.06×10^{-3}	7.10	−89.5	−135.8
CsBr	2.62×10^{-2}	5.00	−86.6	−132.6
AgBr	9.55×10^{-4}	8.37	−14.56	—
AgI	4.01×10^{-4}	9.36	−6.49	—

Reprinted with permission from *J. Phys. Chem.*, **72**, 2966 (1968).

Table 4.17 Ionic Free Energies and Entropies of Solvation in DMF

Ion	$\Delta G^\circ_{solv.}$ (kcal mol^{-1})	$\Delta S^\circ_{solv.}$ (cal deg^{-1} mol^{-1})
Li^+	−129.2	−45.6
Na^+	−102.5	−39.6
K^+	−84.9	−36.1
Cs^+	−71.5	−29.9
F^-	−90.6	—
Cl^-	−64.3	−32.2
Br^-	−61.1	−28.6
I^-	−56.4	−25.9

Reprinted with permission from *J. Phys. Chem.*, **72,** 2966 (1968).

from structure breaking is small, while the entropy decrease from orientation of the solvent molecules is fairly large.

Studies of the solubility of silver halides and the stability of silver-halide complexes in several nonaqueous solvents were undertaken by Luehrs et al. (71). The solvents investigated were acetonitrile, DMSO, nitroethane, acetone, and methanol. The solubility product constants for the silver halides are given in Table 4.18. It is evident that the solubility of each of the silver halides parallels the relative solvating ability of the solvents for silver ion, as indicated by the formal reduction potentials according to nitroethane < acetone < methanol < acetonitrile < DMSO (0.61, 0.40, 0.34, 0.13, 0.03 V). This also strongly suggests that cation solvation plays the dominant role in determining solubility.

In each solvent the overall formation constants of the AgX_2^- complexes increase from $AgCl_2^-$ to AgI_2^-. The tendency for the reaction $AgX(s) + X^- \rightleftharpoons AgX_2^-$ to occur is much less pronounced in methanol than in other solvents. This is attributed to the ability of methanol to solvate

Table 4.18 Solubility Products of Silver Halides in Nonaqueous Solvents at 23°C

Solvent	AgCl	AgBr	AgI
DMSO	$10^{-10.4}$	$10^{-10.6}$	$10^{-12.0}$
Acetonitrile	$10^{-12.4}$	$10^{-13.2}$	$10^{-14.2}$
Methanol	$10^{-13.0}$	$10^{-15.2}$	$10^{-18.2}$
Acetone	$10^{-16.4}$	$10^{-18.7}$	$10^{-20.9}$
Nitroethane	$10^{-21.1}$	$10^{-21.8}$	$10^{-22.6}$

Reprinted with permission from *Inorg. Chem.*, **5,** 201 (1966). Copyright 1966. American Chemical Society.

halide ions through a hydrogen-bonding mechanism, thus making it less available for reaction with solid silver halide.

Summary Solubilities of electrolytes in nonaqueous solvents can be obtained using a variety of techniques, including potentiometric titrations, spectrophotometry, gravimetric analysis, and atomic absorption. The information attainable from solubility data includes solubility product and dissociation constants of salts, standard enthalpies of solution, and free energies and entropies of solvation. This information is useful as an insight into ionic solvation.

Correlations can be made between the equilibrium constants for protic and aprotic solvents. Deviations from these linear correlations can be attributed to different extents of hydrogen bonding and ion polarizability. For example, silver salts of large polarizable anions like AgSCN or $AgBPh_4$ are weak hydrogen-bond acceptors and tend to be more soluble in a solvent such as methanol than predicted by the correlation equation, whereas silver salts of small anions, such as AgCl, which are strong hydrogen-bond acceptors, tend to be less soluble than predicted. For dipolar aprotic solvents, it is often possible to predict the solubility products of silver salts in a certain solvent if the solubility product in another solvent is known.

Anion solvation can also play an important role in determining solubility behavior. For example, this is illustrated by the appreciable variation in the free energies of transfer of several salts of the same cation from a protic solvent to a dipolar aprotic solvent.

The contribution of volume energies is significant in accounting for solubilities of organic salts such as R_4NClO_4 in various solvents. For example, with increasing size of R, the solubility in benzene and ethanol relative to water is found to increase. The solubility trends can be discussed in terms of the energy required to accommodate a solvent molecule in a hole, the volume energy. This volume energy is high in water due to the intermolecular hydrogen bonding. The main forces in most organic solvents are those due to London forces, and therefore the energy required to form a hole is smaller. An important practical application of this trend is that "large" salts can be extracted from water by organic solvents. Finally, the complexing power of a solvent such as acetonitrile can strongly influence solubilities.

4.6 HEATS OF SOLUTION—CALORIMETRY

Both heats of solution and dilution have been measured in nonaqueous solvents with varying degrees of precision. In addition, several investigations have focused on aqueous-organic mixtures in order to obtain heats of transfer. For comparison purposes, the standard enthalpy of solution $\Delta H^\circ_{\text{soln.}}$ is required, and normally the procedure is to extrapolate heats of solution data obtained at finite concentrations to infinite dilution values, using some form of the Debye–Hückel theory for this extrapolation. The working equation is of the form

$$\Delta H_{\text{soln.}} = \frac{2}{3} A_{\text{H}} \frac{\nu}{2} \left| Z_{+}Z_{-} \right| I^{1/2} + \Delta H^{\circ}_{\text{soln.}} \tag{4.101}$$

where

$$A_{\text{H}} = -3 \ln 10RT^2 A \left[\frac{\alpha}{3} + \frac{1}{T} + \left(\frac{\partial \ln \epsilon_r}{\partial T} \right)_P \right],$$

α is the temperature coefficient of expansion, and A is a Debye–Hückel term containing two temperature dependent terms, ϵ_r and d_o, where ϵ_r is the dielectric constant and d_o the density of the pure solvent. The value of $\Delta H^{\circ}_{\text{soln.}}$ in Eqn. 4.101 is obtained from a plot of $\Delta H_{\text{soln.}}$ vs $I^{1/2}$, where I is the ionic strength. Alternative methods for obtaining $\Delta H^{\circ}_{\text{soln.}}$ and a more thorough treatment of the method discussed briefly above are given by Criss (72). Heats of solution measurements in nonaqueous solvents have originated from a limited number of laboratories, and it is the intent here to indicate the approach used by several key investigators to attempt to gain an understanding of the treatment of the data and their interpretation in terms of solution properties.

Criss and co-workers (73) have been interested in heats of solution in *N*-methylformamide, *N,N*-dimethylformamide and methanol. Normally, the electrolytes studied were alkali-metal halides or quaternary-ammonium halides. One of the features in these series of studies was to compare the experimental limiting slopes with the corresponding theoretical slopes predicted by the Debye–Hückel theory. For example, for LiCl, NaCl, NaBr, and NaI in *N*-methylformamide, the limiting slopes from the heat data vary from 7 to 280 times the theoretical values, whereas CsCl and KCl gave slopes that were similar to those predicted. Positive slopes such as those found for these systems generally arise from incomplete dissociation of the electrolyte and the existence of ion pairing in these systems was supported by electrical conductance measurements. Similar trends existed for the case of heats of solution in *N,N*-dimethylformamide, where, in addition to the alkali-metal halides, both $MgCl_2$ and $GdCl_3$ were also included. It was generally observed that salts having cations with high charge densities had limiting slopes many times greater than predicted by theory, again suggesting ion association. Criss also examined the Latimer, Pitzer, and Slansky modified Born equation for the free energy of solvation of an ion ΔG_i, that is,

$$\Delta G_i = -e_i^2 (1 - 1/\epsilon)/2(r_i + \delta) \tag{4.102}$$

where e_i and r_i are the charge and radius of the ion, respectively, ϵ is the dielectric constant of the solvent, and δ is a parameter depending on the sign of the charge on the ion and the solvent (see Chapter 9 and Ref. 58).

Table 4.19 Standard Heats of Solution of Tetraalkylammonium Iodides in DMF and Water at 25°C

Electrolyte	$\Delta H^\circ_{\text{soln.}}$ (kcal mol^{-1}) DMF	H_2O
$(CH_3)_4NI$	4.01	9.99
$(C_2H_5)_4NI$	3.42	6.74
$(C_3H_7)_4NI$	2.00	2.77

Reprinted with permission from *J. Phys. Chem.*, **73,** 174 (1969). Copyright 1969. American Chemical Society.

Differentiation of Eqn. 4.102 and use of the Gibbs–Helmholtz equation yields the enthalpy of solvation ΔH_i as

$$\Delta H_i = \Delta G_i \left[1 - \frac{T(\partial\epsilon/\partial T)}{\epsilon(\epsilon - 1)} + \frac{T}{r_i + \delta}\left(\frac{\partial\delta}{\partial T}\right)\right] \tag{4.103}$$

so that ΔH_i is a linear function of $(r_i + \delta)^{-1}$ if it is assumed that $\partial\delta/\partial T$ is negligible. For the systems studied in *N,N*-dimethyformamide, a plot of solvation enthalpies for the alkali-metal bromides vs $(r + \delta)^{-1}$ gave a straight line with $\delta = 0.8$ Å.

Criss also measured the standard heats of solution of several tetraalkylammonium iodides in water and in *N,N*-dimethylformamide at 25°C. The standard heats of solution in the two solvents are given in Table 4.19. These standard heats of solution were used to generate standard enthalpies of transfer of the salts from water to DMF and these are given in Table 4.20 (see also Table 5.3). For purposes of comparison, the enthalpies of transfer are also given for the alkali-metal iodides and, in addition, results are also given for transfers to other solvents. Two points arise from this study. First, the modified Born equation (Latimer, Pitzer, and Slansky) does not account adequately for heats of transfer data. Second, the large change observed in ΔH°_t in going from Cs^+ to $(CH_3)_4N^+$ is not explained by the water structure alone, but is the result of special interactions of the tetraalkylammonium ions with solvents in general.

As an example of the very sparse measurements of heat capacities, Criss (73) determined the heat capacities of solution, ΔC°_p, for tetraalkylammonium bromides in water and in methanol at several temperatures. The heat capacities were evaluated using the integral enthalpy of solution method as follows.

At any given temperature, the heat of solution can be expressed by

$$\Delta H_s = n_1\bar{H}_1 + n_2\bar{H}_2 - n_1\bar{H}^\circ_1 - n_2H^\circ_2 \tag{4.104}$$

Table 4.20 Standard Enthalpies of Transfer of Electrolytes from Water to Various Nonaqueous Solvents

	$\Delta H_{t(\text{water}\rightarrow\text{nonaq. solv.})}$ (kcal mol^{-1})			
Salt	DMF	Propylene carbonate	DMSO	CH_3OH
LiI	−3.97	−0.10		
NaI	−11.35	−3.22	−9.67	−6.21
KI	−12.90	−6.02	−11.36	
CsI	−12.22	−7.18	−10.30	−3.82
$(CH_3)_4NI$	−5.98	−4.74		−0.39
$(C_2H_5)_4NI$	−3.32	−0.59	−1.97	+1.25
$(C_3H_7)_4NI$	−0.769			

Reprinted with permission from *J. Phys. Chem.*, **73,** 174 (1969). Copyright 1969. American Chemical Society.

where $\bar{H}_1$ and $\bar{H}_2$ are the partial molal enthalpies of the solvent and the solute, respectively, and $\bar{H}_1^\circ$ and H_2° are the molal enthalpies of the pure solvent and pure solid solute, respectively, and n_1 and n_2 are the respective numbers of moles of solvent and solute. For the case of an infinitely dilute solution, $\bar{H}_1 = \bar{H}_1^\circ$, $\bar{H}_2 = \bar{H}_2^\circ$ and if 1 mole of solute is considered, $n_2 = 1$ then

$$\Delta H_s^\circ = \bar{H}_2^\circ - H_2^\circ \tag{4.105}$$

where $\bar{H}_2^\circ$ is the partial molal enthalpy of the solute at infinite dilution. For small temperature differences

$$\frac{d\Delta H_s^\circ}{dT} = \frac{d\bar{H}_2^\circ}{dT} - \frac{dH_2^\circ}{dT} = \Delta C_{p_2}^\circ = \bar{C}_{p_2}^\circ - C_{p_2}^\circ \tag{4.106}$$

or

$$\bar{C}_{p_2}^\circ = C_{p_2}^\circ + \Delta C_{p_2}^\circ \tag{4.107}$$

where $\Delta C_{p_2}^\circ$ is the change in heat capacity for the solution process and $C_{p_2}^\circ$ is the heat capacity of the pure solute.

In this particular study, several interesting features arise. First, the temperature coefficient for the aqueous tetramethylammonium bromide is negative, while for tetrabutylammonium bromide the coefficient is positive. This effect is shown in Fig. 4.12, where the temperature dependence of the heat capacities of solution is shown and is of course opposite in sign to the enthalpies of solution. The tetramethyl salt shows a maximum in the heat capacity around 65°C, while the very positive ΔC_p° values for the tetrabutyl salt decrease monotonically over the entire temperature range. In comparison, the values of $C_{p_2}^\circ$ for the tetramethyl salt in water and methanol were found to have different behaviors. In contrast to the characteristic maximum observed for $C^{\circ}_{p_2}$ in aqueous solutions, the $C^{\circ}_{p_2}$ values in methanol decrease monotonically. Apparently the disruption of the solvent structure by the tetraalkyl ion is more pronounced in water solutions than in methanolic solutions, which are less structured. A similar effect is shown for sodium perchlorate in Fig. 4.13.

Propylene carbonate and DMSO have been the subject of a number of heats of solution studies by Friedman and co-workers (74). In one such study heats of solution were measured for some alkali-metal trifluoroacetates, tetraphenylborates, iodides, and perchlorates in water and propylene carbonate. The heats of solution in the two solvents were used to compute standard heats of transfer from water to propylene carbonate as shown in Table 4.21. Using the entries in the last column of Table 4.21, the standard enthalpies of transfer of the alkali-metal ions relative to the Na^+ have been computed and these are shown in Table 4.22. The process can be represented as $M^+_{(aq)} + Na^+_{(PC)} \rightarrow M^+_{(PC)} + Na^+_{(aq)}$, where M is an alkali metal and PC is propylene carbonate. The values in the second column of Table 4.22 are calculated from the heats of formation of gaseous and aqueous ions from tables. These values are then combined with the heats of transfer to obtain the relative solvation enthalpies in propylene carbonate given in the last column of Table 4.22. Comparison of values in water and propylene carbonate shows that the range of enthalpies of solvation is smaller for the ions in propylene carbonate than in water, implying that the magnitudes of the solvation energies are smaller as well. In the case of lithium trifluoracetate, it was observed that the heat of solution depended on the concentration of the salt, implying ion association of the type $Li_2(CF_3CO_2)_2$, the existence of

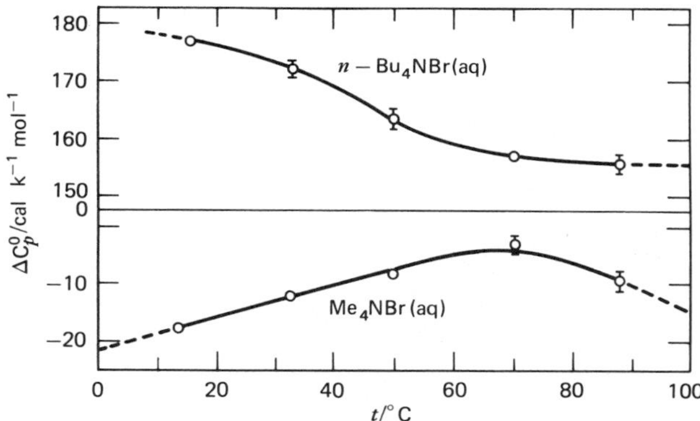

Figure 4.12 Temperature dependence of the heat capacity of solution ΔC°_{p} of tetramethylammonium bromide and tetra-*n*-butylammonium bromide in water. Reprinted with permission from *J. Chem. Thermo.*, **4**, 321 (1972). Copyright 1972. American Chemical Society. Academic Press, New York.

Table 4.21 Standard Heats of Transfer of Alkali-Metal Ions from Water to Propylene Carbonate

Metal ion	$\Delta H_t^\circ(MX)$ (kcal mol^{-1})	$\Delta H_t^\circ(MX) - \Delta H^\circ(NaX)$ (kcal mol^{-1})
	Trifluoroacetates	
Li^+	8.69	3.34
Na^+	5.35	0
K^+	2.52	−2.83
Rb^+	1.89	−3.46
Cs^+	1.52	−3.83
	Tetraphenylborates	
Na^+	−5.93	0
K^+	−8.67	−2.74
Rb^+	−9.32	−3.39
Cs^+	−10.02	−4.09
	Perchlorates	
Li^+	−3.16	3.21
Na^+	−6.37	0
	Iodides	
Li^+	−0.10	3.12
Na^+	−3.22	0

Table 4.22 Heats of Transfer of Alkali-Metal Ions Relative to Na^+(kcal mol^{-1})

Ion	$\Delta H^\circ_{t\,aq \to PC}$	$\Delta H^\circ_{t\,aq \to gas}$	$\Delta H^\circ_{t\,PC \to gas}$
Li^+	3.17	26.28	23.11
Na^+	0	0	0
K^+	−2.80	−20.18	−17.38
Rb^+	−3.43	−26.09	−22.66
Cs^+	−3.96	−30.61	−26.65

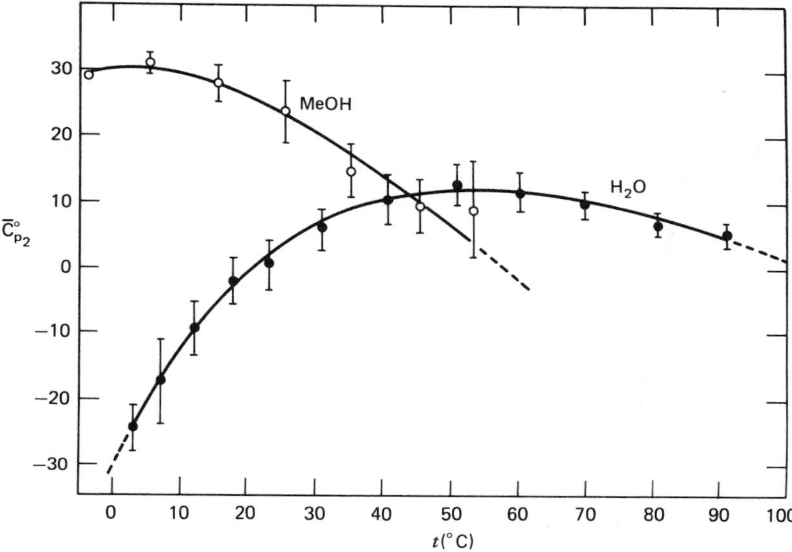

Figure 4.13 Partial molal heat apacities (cal deg^{-1} mol^{-1}) of sodium perchlorate in water and anhydrous methanol as a function of temperature. Reprinted with permission from *J. Chem. Eng. Data,* **17,** 222 (1972). Copyright 1972. American Chemical Society.

which was supported by evidence from electrical conductivity measurements, where $\Lambda/\Lambda_o = 0.6$ in a 0.001 *M* solution.

The formation of ion pairs and possible higher aggregates accounts for the poor solvation of the trifluoroacetate ion. In another study by Friedman et al. (74), this investigation was extended to include the heats of solution of some tetraalkylammonium salts in water and propylene carbonate, and thereby include solvation enthalpies for halide ions. The results for $\Delta H°$ relative to that for Na^+ were Me_4N^+, −1.45; Et_4N^+, 2.65; Cl^-, 3.83; Br^-, 0.85; and I^-, −3.22 (all in kcal mol^{-1}).

A further study by Friedman et al. (74) incorporates a large collection of heats of solution data in water, propylene carbonate, DMSO, and D_2O. The bulk of the data were for quaternary ammonium and tetraphenylborates of the alkali metals. Single-ion enthalpies of transfer for the ions from PC to DMSO were determined. It was found that the enthalpy of transfer of the R_4N^+ ions to PC from DMSO consisted of additive contributions from the CH_2 groups. The single-ion enthalpies of transfer were obtained using the convention that $(Ph_4As^+)_{PC \leftarrow DMSO} = (BPh_4^-)_{PC \leftarrow DMSO}$.

Single-ion enthalpies of transfer from water to propylene carbonate were also reported. Figure 4.14 shows the dependence on ionic radius of the ionic enthalpies of transfer to propylene carbonate from DMSO. It is clear from the figure that the enthalpy of transfer of the tetraalkylammonium ions is

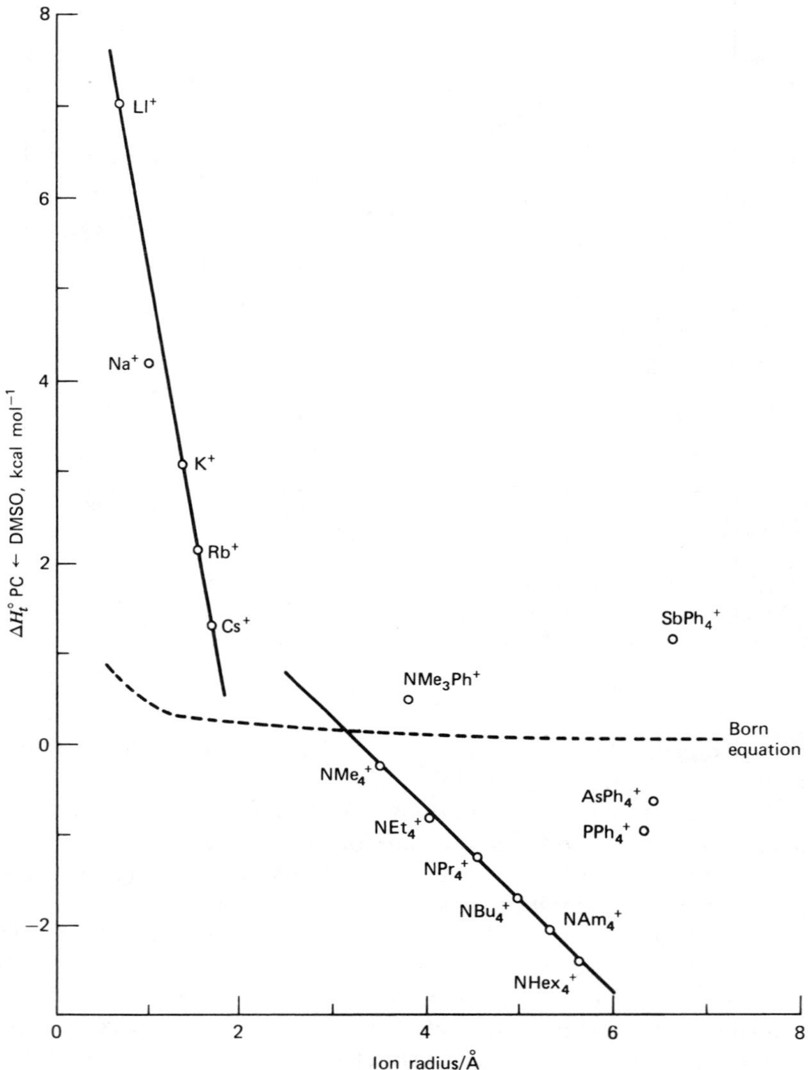

Figure 4.14 Ionic enthalpies of transfer to propylene carbonate from DMSO. Reprinted with permission from *J. Phys. Chem.*, **73**, 3934 (1969). Copyright 1969. American Chemical Society.

negative and linear in the ionic radius and also linear in the carbon number, i.e., there are additive contributions from methylene groups mentioned above. However, the corresponding additivity principle does not work for the aryl-substituted ions. The trends shown for the alkali-metal ions in Fig. 4.14 appear to be dominated by the interaction of the ions as Lewis acids with the solvents as Lewis bases, as DMSO is more basic than PC and Li^+ is more acidic than Na^+, etc. In contrast, the enthalpy of transfer of the R_4N^+ ions to PC from water was found to be dominated by an effect which was not consistent with additive contributions from methylene groups, but actually the value passed through a maximum as a function of chain length at the tetraamylammonium ion (Fig. 4.15).

This effect was discussed in terms of structural changes ("icebergs") in the water in the neighborhood of the ions in aqueous solution. In terms of solvation effects, Me_4N^+ was a net structure breaker, structure making and breaking mutually cancel for Et_4N^+, and structure-making effects dominate for Pr_4N^+, Bu_4N^+, and Am_4N^+.

An illustration of the solvation enthalpies of several nonelectrolytes in water, propylene carbonate, and DMSO is also provided by Friedman et al. (74). The nonelectrolytes studied consisted mostly of alcohols, hydrocarbons, and some aryl halides. From the standard enthalpies of solution, enthalpies of transfer to various media from water were calculated. A plot of

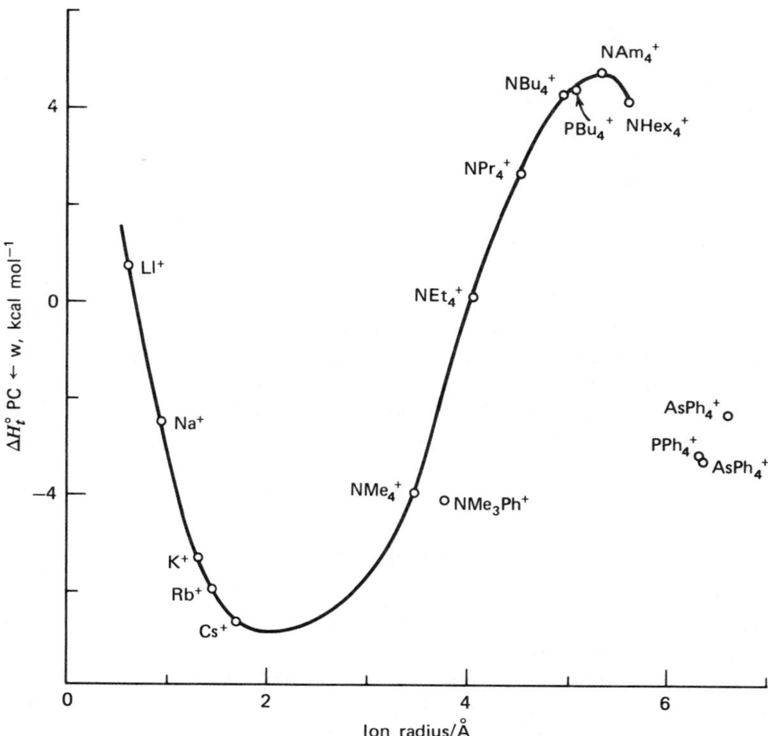

Figure 4.15 Ionic enthalpies of transfer to propylene carbonate from water. Reprinted with permission from *J. Phys. Chem.*, **73**, 3934 (1969). Copyright 1969. American Chemical Society.

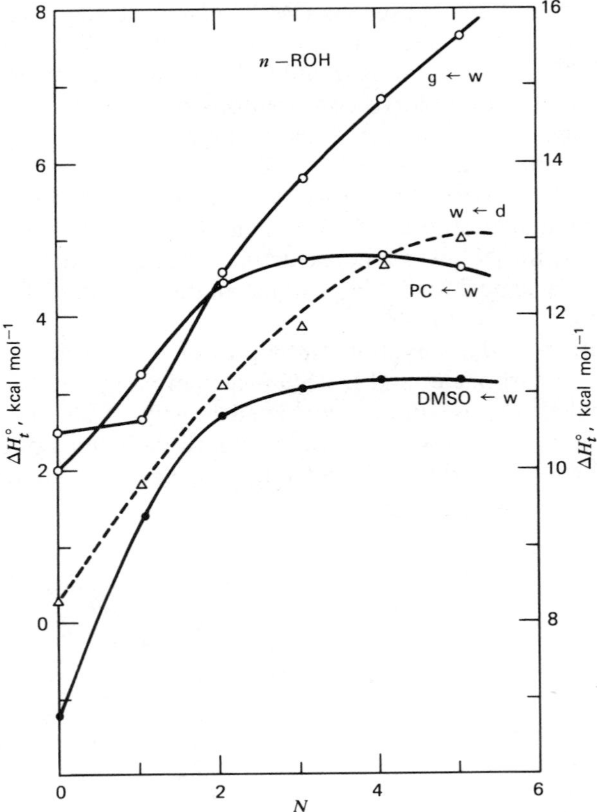

Figure 4.16 Enthalpies of transfer of normal alcohols to various media from water. The scale for the transfer to the gas phase from water is on the right; for the others the scale is on the left. The scale for the transfer to H_2O from D_2O (W ← d) is multiplied tenfold. Reprinted with permission from *J. Phys. Chem.*, **73,** 1572 (1969). Copyright 1969. American Chemical Society.

this type for the alcohols as a function of carbon number is shown in Fig. 4.16. It is observed that the shapes of all the curves are similar, suggesting that a significant factor contributing to the shape may be the structural (iceberg) effect in water, an effect that is considered to be larger in D_2O than in H_2O and absent in DMSO, and PC.

This structural effect was discussed on a more quantitative basis by considering the various parts contributing to the total solvation process. The treatment was based on the following equation for the partial molar enthalpy of an infinitely dilute solute species X in a solvent:

$$\mathrm{H}(x;a) = \mathrm{H}(x;g) + \mathrm{CAV}(x;a) + \mathrm{VDW}(x;a) + \mathrm{HB}(x;a) + \mathrm{STR}(x;a) \tag{4.108}$$

Here, $H(x;g)$ is the enthalpy of X in the gas phase at infinite dilution, CAV is the enthalpy increase in the process of making a cavity in the solvent to accommodate the solute molecule, VDW includes the van der Waals interaction of the solute and the solvent and polarization–dipole and dipole–dipole interactions, HB is the enthalpy of formation of solute–solvent hydrogen bonds, and STR is the enthalpy associated with the structural change produced in the solvent by the solute or by the cavity (see also Section 2.2.2).

It was then assumed that for any straight-chain primary alcohol $C_NH_{2N+1}OH$

$$\mathrm{CAV} + \mathrm{VDW} + \mathrm{HB} = (\mathrm{HO}\cdots\mathrm{H})_a + N(\mathrm{CH_2})_2 \qquad (4.109)$$

where $(\mathrm{HO}\cdots\mathrm{H})_a$ and $(CH_2)_a$ are coefficients that are independent of the species of the primary alcohols and where $(CH_2)_a$ encompasses mainly the effect of a CH_2 group on CAV (larger volume) and on VDW (larger polarizability).

It was found that this equation was consistent with the linear dependence for the enthalpy of transfer to the gas from PC on N as shown in Fig. 4.17.

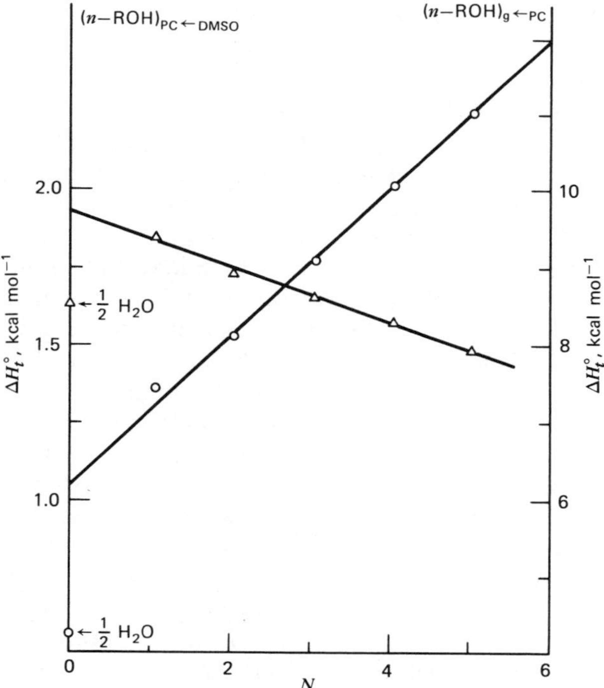

Figure 4.17 Enthalpies of transfer of alcohols (○) to the gas from propylene carbonate (scale on right); (△) to propylene carbonate from DMSO (scale on left). Reprinted with permission from *J. Phys. Chem.*, **73**, 1572 (1969). Copyright 1969. American Chemical Society.

This trend supports the idea of group additivity of solvation energies. The plot in Fig. 4.17 can be used to determine the coefficients $(HO\cdots H)_{PC}$ and $N(CH_2)_{PC}$ from the relation

$$(C_NH_{2N+1}OH)_{g\leftarrow PC} = -(HO\cdots H)_{PC} - N(CH_2)_{PC} \qquad (4.110)$$

Also in Fig. 4.17, a similar plot is given for the transfer from DMSO to PC.

This linear dependence of enthalpies of transfer of alcohols upon N is derived in the absence of structural effects. The structural term in aqueous solutions was arrived at by comparing the transfers PC ← W and d ← W (D_2O from H_2O) and deriving a reduced enthalpy of transfer given by

$$-H^{red} \equiv (C_NH_{2N+1}OH)_{PC\leftarrow w} + y^{-1}(C_NH_{2N+1}OH)_{d\leftarrow w} = j + Nj' \qquad (4.111)$$

where y is assumed to be independent of N and

where $j \equiv (HO\cdots H)_{PC} - (1 - z/y)(HO\cdots H)_w$

$$j' \equiv (CH_2)_{PC} - (1 - z'/y)(CH_2)_w$$

where $STR_d(N) \equiv (1 + y)\, STR_w(N)$

$$(HO\cdots H)_d \equiv (1 + z)(HO\cdots H)_w$$

$$(CH_2)_d \equiv (1 + z')(CH_2)_w$$

Equation 4.111 is plotted for various values of y in Fig. 4.18. The linearity of the plots in Fig. 4.18 indicates that the curvature of each plot in Fig. 4.16 indeed has the same origin, the structural effect in water. The structural effect in the enthalpy of hydration of the normal alcohols ranged from about -2.5 kcal mol^{-1} for methanol to about -8.5 kcal mol^{-1} for amyl alcohol, and the plot of the structural effect as a function of chain length tends to level off near amyl alcohol. It was found that this simple analysis was less successful for the aromatic compounds included in this study.

Somsen et al. (75) have undertaken the most comprehensive study of heats of solution of alkali-metal halides in the amide solvents, formamide, *N,N*-dimethylformamide, and dimethylacetamide. The approaches used in this series of studies were basically the same and the results for formamide will illustrate them. The heats of solution of some 18 alkali-metal halides were measured in formamide at 25°C, using an adiabatic calorimeter. The resulting enthalpies of solution at infinite dilution are given in Table 4.23. Somsen (75) analyzed the enthalpy of solution data using a model according to Van Eck which treated the total enthalpy of solvation of an ion ($\Delta H_{solv.}$) as consisting of three terms: ΔH_c, the enthalpy resulting from the charge of the ion, ΔH_{nc}, the enthalpy contribution which does not depend on the charge and ΔH_h, the enthalpy required to make a hole in the solvent in order to accommodate the ion, that is,

$$\Delta H_{soln.} = \Delta H_c + \Delta H_{nc} + \Delta H_h \qquad (4.112)$$

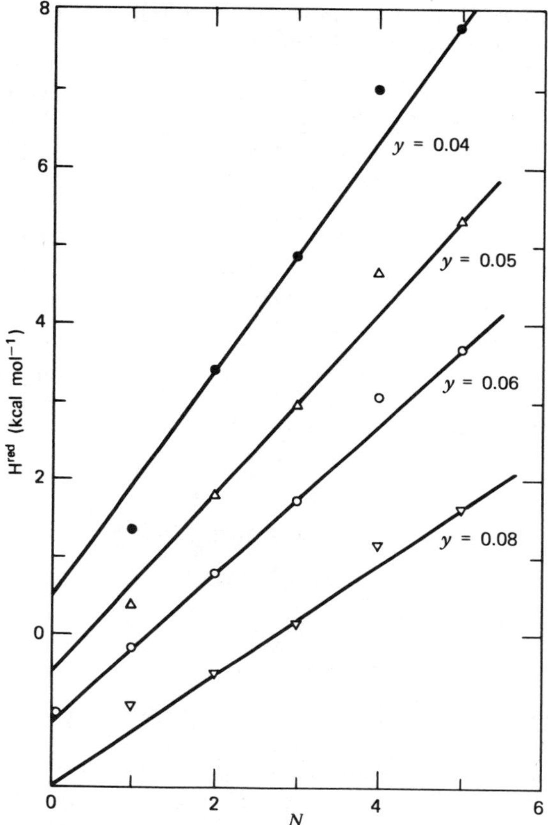

Figure 4.18 Reduced enthalpies of transfer of normal alcohols. Reprinted with permission from *J. Phys. Chem.*, **73,** 1572 (1969). Copyright 1969. American Chemical Society.

By considering the enthalpies of solvation of a salt to be built up additively from the individual ionic values, the enthalpy of solvation of a salt MX in water may be expressed as

$$\begin{aligned}\Delta H_{\text{solv.}}(\text{MX}) &= \Delta H_{\text{solv.}}(\text{M}^+) + \Delta H_{\text{solv.}}(\text{X}^-) \\ &= -(I_{\text{M}} + E_{\text{M}}) + \Delta H_{\text{nc}} + \Delta H_{\text{h}} + \Delta H_{\text{solv.}}(\text{X}^-) \qquad (4.113)\end{aligned}$$

and in formamide

$$\Delta H^*_{\text{solv.}}(\text{MX}) = -(I_{\text{M}} + E_{\text{M}}) + \Delta H^*_{\text{nc}} + \Delta H^*_{\text{h}} + \Delta H^*_{\text{solv.}}(\text{X}^-) \qquad (4.114)$$

where I_{m} and E_{m} are the ionization potentials and electron affinities of the metal atoms, respectively.

The enthalpy of solvation of a salt is related to the lattice enthalpy and

Table 4.23 Enthalpies of Solution at Infinite Dilution of Alkali-Metal Halides in Formamide at 25°C

Salt	$\Delta H^\circ_{soln.}$ (kJ mol^{-1})	Salt	$\Delta H^\circ_{soln.}$ (kJ mol^{-1})
LiCl	−39.43	KI	−4.250
LiBr	−56.04	RbF	−22.04
LiI	−76.39	RbCl	+2.971
NaCl	−8.80	RbBr	+3.14
NaBr	−18.44	RbI	+0.96
NaI	−31.07	CsF	−31.78
KF	−13.32	CsCl	+3.98
KCl	+3.429	CsBr	+7.58
KBr	+0.973	CsI	+9.29

Reprinted with permission from *Rec. Trav. Chim.*, **84,** 985 (1965).

the enthalpy of solution by (see also Eqns. 2.9–2.13):

$$-\Delta H_{solv.}(MX) = \Delta H_{cryst.}(MX) - \Delta H^\circ_{soln.}(MX) \qquad (4.115)$$

Combination of Eqns. 4.113–4.115 gives

$$\Delta H^\circ_{soln.}(H_2O) - \Delta H^\circ_{soln.}(HCONH_2)$$
$$= \Delta H_{nc} + \Delta H_h - \Delta H^*_{nc} - \Delta H^*_h + \Delta H_{solv.}(X^-) - \Delta H^*_{solv.}(X^-) \qquad (4.116)$$

For salts with the same anion, the right-hand side of Eqn. 4.116 has a constant value, provided that $\Delta H_{nc} + \Delta H_h$ is constant for the different ions. The differences in enthalpies of solution for the alkali halides in formamide and water are given in Table 4.24. With the exception of Li, the differences were found to be constant and independent of the cation. The deviation produced by lithium was attributed to its different coordination ability.

Benoit (76) considered solvation in several dipolar aprotic solvents and evaluated ionic enthalpies of transfer from heats of solution measurements based on the assumption that $\Delta H_t(AsPh_4^+) = \Delta H_t(BPh_4^-)$. Enthalpies of solution were determined for a series of alkali perchlorates and tetraethylammonium salts in sulfolane (TMS) with a few measurements in propylene carbonate, dimethylformamide, dimethylsulfoxide, and methanol to complement data already existing. Ionic enthalpies of transfer were calculated using propylene carbonate as a reference solvent because it was felt that the peculiar structure of water may lead to some distortion of the values, and also propylene carbonate is less basic than the other solvents that were considered. The resulting cationic and anionic enthalpies of transfer from PC to the various solvents are presented in Figs. 4.19 and 4.20. Values were deduced from the data according to

$$\Delta H_t(PC \rightarrow S) = \Delta H_t(H_2O \rightarrow S) - \Delta H_t(H_2O \rightarrow PC) \qquad (4.117)$$

It was shown that the Born equation was inadequate in explaining the observed behavior; in other words, coulombic interactions alone were insufficient to account for the behavior of the ionic enthalpies of transfer.

Cation solvation was related to solvent basicity, with the order DMSO >DMF > TMS > PC found for the solvating strength of the aprotic solvents towards Na^+, K^+, and Rb^+. Water, and to some extent methanol, do not solvate the larger cations as well as do the aprotic solvents. However, NEt_4^+ may be specifically solvated by both protic solvents, possibly because it enhances their structure. Relative to anionic solvation, it was found that there was no simple ionic radii relationship existing to account for the behavior. The order of solvating strengths of solvents toward anions is DMF > DMSO > TMS > PC. The order of solvation of anions is $Cl^- > Br^- > I^- > ClO_4^- > BPh_4^-$ and is common to the aprotic solvents, water and methanol.

Fuchs and co-workers (77) have correlated heats of solution data and single-ion enthalpies of transfer with the phenomenon that certain dipolar aprotic solvents such as DMSO accelerate the rates of certain nucleophilic displacement reactions. In one study the enthalpies of solution were measured for LiCl, LiBr, LiI, KBr, and KI in DMSO, while in the other study the enthalpies of solution of some 16 organic and inorganic salts were obtained in methanol, DMSO, *N,N*-dimethylformamide, *N*-ethylacetamide, and *N*-methylpyrrolidinone. Single-ion enthalpies of transfer from methanol to other solvents were calculated on the basis of three different extrathermodynamic assumptions:

$$\Delta H_{\text{solv.}}[(C_4H_9)_4N^+] = \Delta H_{\text{solv.}}[^-B(C_4H_9)_4]$$

$$\Delta H_{\text{solv.}}[(C_5H_{11})_4N^+] = \Delta H_{\text{solv.}}[^-B(C_5H_{11})_4]$$

$$\Delta H_{\text{solv.}}[(C_6H_5)_4\,As^+] = \Delta H_{\text{solv.}}[^-B(C_6H_5)_4]$$

Good agreement was found in all cases between single-ion enthalpies of

Table 4.24 Differences in Enthalpies of Solution of Alkali Halides in Formamide and Water (kcal mol^{-1})

Cation \ Anion	F^-	Cl^-	Br^-	I^-
Li^+	—	+0.54	+1.66	+3.13
Na^+	—	+3.03	+4.26	+5.62
K^+	−1.07	+3.29	+4.56	+5.92
Rb^+	−1.06	+3.30	+4.48	+5.97
Cs^+	−1.00	+3.15	+4.39	+5.7

Reprinted with permission from *Rec. Trav. Chim.*, **84,** 985 (1965).

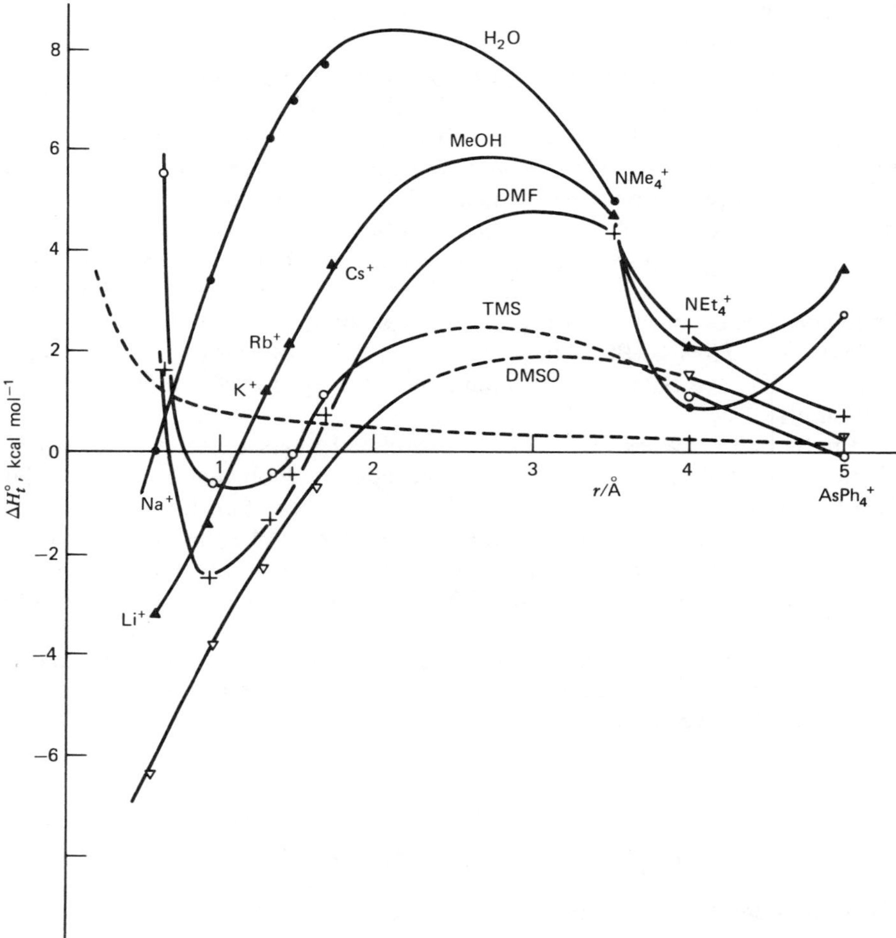

Figure 4.19 Cationic enthalpies of transfer from PC to several solvents. (○) TMS, (▽) DMSO, (+) DMF, (▲) MeOH, (•) water. Dashed lines represent the calculated Born term for transfer from PC to TMS. Reprinted with permission from *J. Amer. Chem. Soc.*, **91,** 6221 (1969). Copyright 1969. American Chemical Society.

transfer calculated from the first two assumptions, but values based on the tetraphenylarsonium tetraphenylborate assumption showed deviations.

For the case of DMSO, the enthalpies of solution and solvation of some alkali halides are given in Table 4.25. The enthalpies of solvation were calculated from

$$\Delta H_{\text{solv.}} = \Delta H_{\text{soln.}} - \Delta H_{\text{latt.}} \tag{4.118}$$

Table 4.25 Enthalpies of Solution and Solvation of Some Alkali Halides in DMSO and H_2O at 25°C

Halide	$\Delta H_{soln.}$ (DMSO)	$\Delta H_{soln.}$ (H_2O)	Lattice energy	$\Delta H_{solv.}$ (DMSO)	$\Delta H_{solv.}$ (H_2O)
LiCl	−10.9	−8.85	203.4	−214.3	−212.2
LiBr	−17.1	−11.67	191.3	−208.4	−203.0
LiI	−24.2	−15.13	177.0	−201.2	−192.1
KBr	−2.7	4.75	162.1	−164.8	−157.4
KI	−6.5	4.86	151.1	−157.6	−146.2

Reprinted with permission from *J. Amer. Chem. Soc.*, **90,** 6698 (1968). Copyright 1968. American Chemical Society.

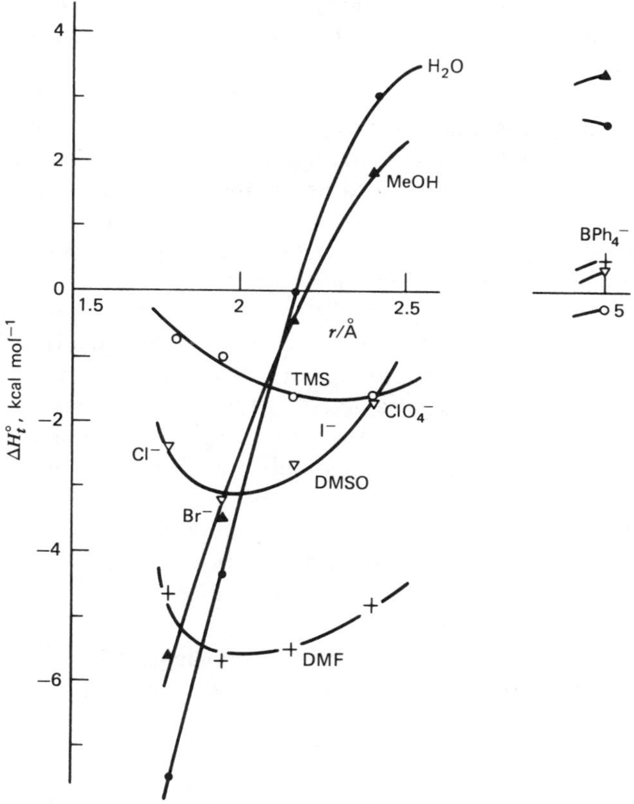

Figure 4.20 Anionic enthalpies of transfer from PC to several solvents (○) TMS, (▽) DMSO, (+) DMF, (▲) MeOH, (●) water. Reprinted with permission from *J. Amer. Chem. Soc.*, **91,** 6221 (1969). Copyright 1969. American Chemical Society.

Kinetic data for the displacement of tosylate from *n*-propyl tosylate by Cl^-, Br^-, and I^- are given in Table 4.26. The rates of displacement on *n*-propyl tosylate by halides in aqueous DMSO are in the order $I^- > Br^- > Cl^-$, while in DMSO, the opposite order is observed. This reversed nucleophilic order is attributed to the smaller differences in halide solvation in DMSO than in water. This solvation effect is seen if one considers the lithium and potassium salts separately, when it is evident that in both cases the iodide ion is more weakly solvated than is the bromide and the chloride ion is the most strongly solvated. The rate enhancement in DMSO is due to the fact that an unsolvated nucleophile should be more reactive than a solvated one, from which solvent molecules must be removed before the reaction can occur.

Nonaqueous–aqueous solvent mixtures have received some attention relative to heats of solution studies. For example, Stern et al. (78) have measured the enthalpies of transfer of hydrochloric acid and perchloric acid from water to aqueous ethylene glycol and aqueous acetic acid. The enthalpies of transfer of HCl and $HClO_4$ from water to aqueous ethylene glycol as a function of solvent composition are shown in Fig. 4.21, while a similar plot for the enthalpies of transfer from water to aqueous acetic acid is shown in Fig. 4.22. The two similar S-shaped curves in Fig. 4.21 indicate complex structural changes. In the water-rich region, both curves show small enthalpy maxima, a region often referred to as having an enhanced structure. The second region of decreasing enthalpies is due to the relative structure-breaking efficiency of the ions, with the perchlorate anion producing the largest structure-breaking effect. In the glycol-rich region, the enthalpies rise sharply, and the rise appears to be independent of the anion. This region is interpreted as one in which the nonelectrolyte begins penetrating the inner hydration structure of the ions, with complete exchange of solvating species between mixed solvent and pure glycol at the observed minimum.

The plots in Fig. 4.22 are very similar. The left-hand side corresponds to transfer from pure water to mixed solvent where both HCl and $HClO_4$ are predominantly ionized, and it appears that the mixed solvent is less ordered than the pure water, as in the analogous second region of the ethylene glycol–water system. At concentrations above $X_3 = 0.75$, the acetic acid lowers the dielectric constant to such an extent that attraction between

Table 4.26 Rates of the *n*-Propyl Tosylate–Halide Ion Reaction in DMSO and DMSO–Water (70:30) at 50°C

	Chloride $k_2 \times 10^3$	Bromide $k_2 \times 10^3$	Iodide $k_2 \times 10^3$
DMSO	7.93	4.88	1.53
DMSO–H_2O (70:30)	0.250	0.318	1.663

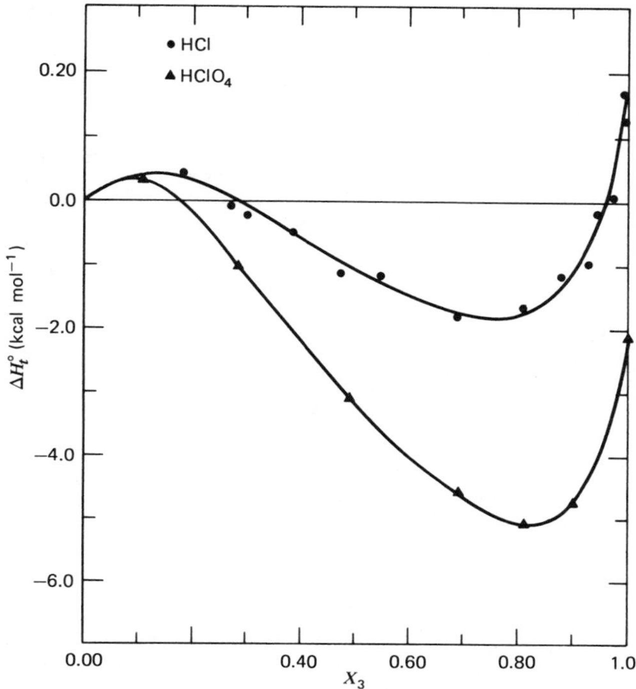

Figure 4.21 Enthalpies of transfer of HCl and $HClO_4$ from water to aqueous ethylene glycol. Reprinted with permission from *J. Phys* Chem., **73**, 928 (1969).

the proton and the anion may result in extensive ion-pair formation. The increase in the enthalpy of transfer of HCl in this region is about three times that of $HClO_4$, probably as a result of the relative order of acid strengths [pK_{HCl} = 8.55; pK_{HClO_4} z 4.87] in HOAc and the difference of solvated association complexes.

Feakins (79) has carried out studies of ion solvation in dioxane–water mixtures using heats of solution of alkali-metal halides. A comparison of the enthalpies of transfer ΔH_t° against mole fraction for LiCl and NaCl in dioxane–water and methanol–water and for NMe_4Br in dioxane–water is shown in Fig. 4.23. At low mole fractions (X_2) of dioxane, ΔH_t° is uniformly negative for the alkali-metal halides and HCl, in contrast to their behavior in alcohol–water mixtures, in which ΔH_t° is positive and passes through a maximum. While a maximum in ΔH_t° for electrolytes with small ions indicates a region of maximum solvent structure, the absence of such a maximum, as in the dioxane–water system, does not necessarily imply that there is no structural enhancement. If the other factors affecting ΔH_t° should cause a strong downward trend in it with increasing X_2, the structural effect might be swamped.

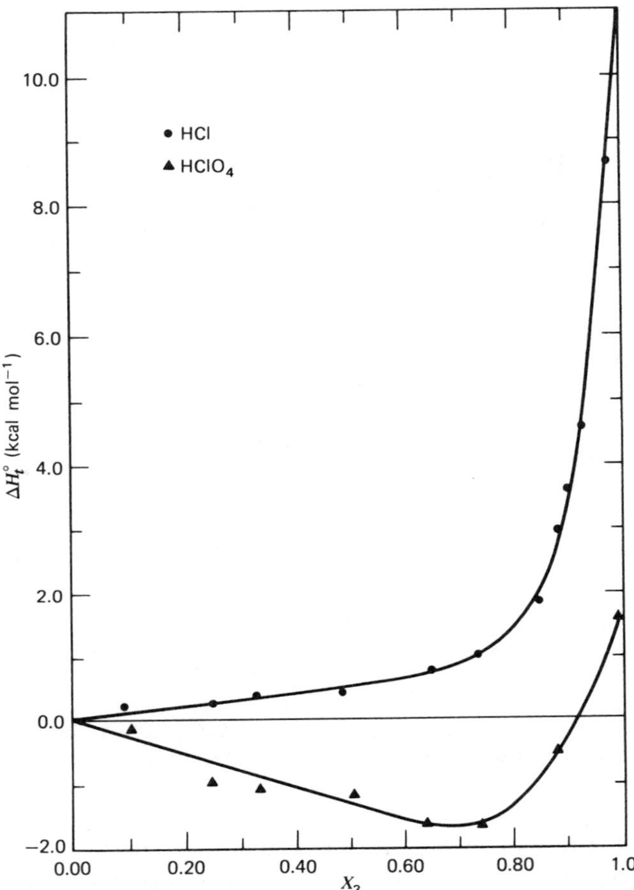

Figure 4.22 Enthalpies of transfer of HCl and $HClO_4$ from water to aqueous acetic acid. Reprinted with permission from *J. Phys. Chem.*, **73,** 928 (1969).

Table 4.27 Enthalpies of Transfer of Electrolytes from Water to Acetonitrile (kcal mol^{-1}) at 25°C

Solute	ΔH_t°	Solute	ΔH_t°
NaI	−5.08	Me_4NClO_4	−7.71
KI	−7.19	Et_4NClO_4	−4.24
RbI	−7.23	Pr_4NClO_4	−1.65
Me_4NI	−5.49	Bu_4NClO_4	0.00
Et_4NI	−2.02	MeNCl	+1.90
Pr_4NI	+0.62	Et_4NCl	+4.43
$NaClO_4$	−7.50	Me_4NBr	−1.71

Reprinted with permission from *J. Chem. Soc. Faraday I,* 1375 (1973).

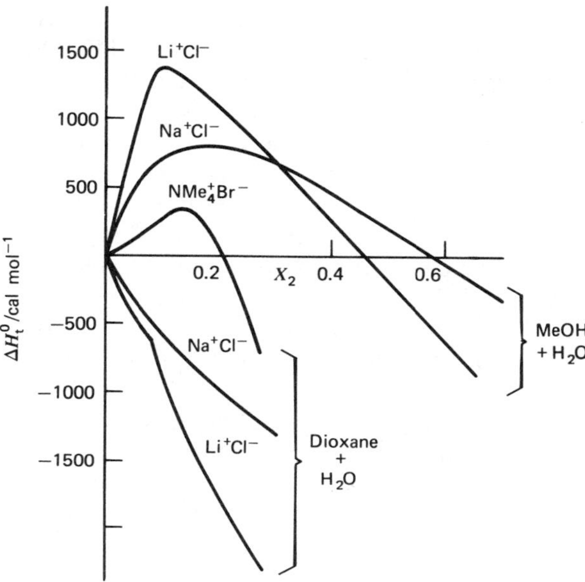

Figure 4.23 Plot of ΔH°_t vs X_2 for LiCl and NaCl in dioxane + water and methanol + water mixtures; also $NMe_4^+Br^-$ in dioxane + water. Reprinted with permission from *J. Chem. Soc. Faraday I.* 314 (1976).

Abraham (80) has produced a fairly extensive set of data for heats of solution of 1:1 electrolytes in acetonitrile and used the data to calculate heats of transfer of the electrolytes from water to acetonitrile. These enthalpies of transfer are given in Table 4.27. The enthalpies of transfer were combined with data for free energies of transfer to yield entropies of transfer. Single-ion transfers were also determined.

Summary Heats of solution obtained from calorimetric studies are principally used for the determination of enthalpies of solvation of electrolytes and for the evaluation of enthalpies of transfer of electrolytes from one solvent to another. Enthalpies of transfer of single ions may be obtained by using assumptions such as the equal sharing of the heats between two large ions. The standard enthalpy of solution, ΔH°_{soln}, is obtained by extrapolating heats of solution data measured at finite concentrations to infinite dilution, using some form of the Debye–Hückel theory.

One approach in the analysis of heats of solution data is to compare experimental limiting slopes with the corresponding theoretical slopes predicted by the Debye–Hückel theory. Positive deviations generally arise in those cases where the dissociation is incomplete due to ion pairing, which often occurs in nonaqueous solvents for cations having high charge densities.

An example of this effect occurs for the alkali-metal halides in amide solvents and methanol. The Latimer, Pitzer, and Slansky-modified Born equation is often used to test enthalpy of solvation data, but with varying degrees of success.

The total solvation process can be viewed from several standpoints. For example, the overall process can be thought of as the combined effects of the enthalpy of the species in the gas phase at infinite dilution, the enthalpy increase in the process of making a cavity in the solvent to accommodate the solute molecule, van der Waals interaction of the solute and the solvent, polarization–dipole and dipole–dipole interactions, enthalpy of formation of solute–solvent hydrogen bonds, and enthalpy associated with structural change produced in the solvent by the solute. In the Van Eck model, the total enthalpy of solvation of an ion ($\Delta H_{\text{solv.}}$) is considered to consist of three terms, viz., ΔH_c, the enthalpy resulting from the ion; ΔH_{nc}, the enthalpy contribution which does not depend on the charge, and ΔH_h, the enthalpy required to make a hole in the solvent in order to accommodate the ion.

In many of the calorimetric studies, the focus is on the standard enthalpies of transfer of electrolytes or single ions from one solvent to another. Various approaches have been used to examine the trends in enthalpies of transfer of single ions. In many cases, correlations between ΔH_t° and ionic radii are attempted, while in the case when quaternary ammonium salts are used as electrolytes, correlations between ΔH_t° and the number of carbon atoms in the chain are examined. In propylene carbonate–water mixtures, the trends shown for the alkali-metal ions appear to be dominated by the interaction of the ions as Lewis acids with the solvents as Lewis bases. For a series of solvents, the order of basicity toward the alkali-metal ions is DMSO > DMF > TMS > PC. Also single-ion transfers can be discussed in terms of structural changes in the water in the neighborhood of the ions. The structure-making and -breaking effects of the ions may either mutually cancel, or one of them may be the dominating factor.

The large change in ΔH_t° in going from Cs^+ to $(CH_3)_4N^+$ is not explained by the water structure alone, but is the result of special interactions of the tetraalkylammonium ions with solvents in general.

Using heats of formation of gaseous and aqueous ions, single-ion enthalpies of transfer can be obtained and thus relative solvation enthalpies can be generated. It has been found that the range of solvation enthalpies of ions is smaller in propylene carbonate than in water. Poor solvation of the trifluoroacetate ion in some nonaqueous solvents is due to the formation of ion pairs and possible higher aggregates.

Ionic enthalpies of transfer can be determined, using the assumption that, for example, $\Delta H_t(AsPh_4^+) = \Delta H_t(BPh_4^-)$. In many aqueous–nonaqueous mixtures, plots of enthalpies of transfer vs composition of the solvent mixture often exhibit pronounced maxima or minima, which reflects either an enhanced structural region or a structure-breaking zone. This is particularly true for the alcohol–water mixtures, but in dioxane–water mixtures no such maxima or minima are found.

4.7 DENSITY STUDIES: MOLAL VOLUMES

The determination of partial molal volumes of electrolyte solutions provides additional evidence for the structural properties of electrolyte solutions. Much of the precise data available are for aqueous solutions, but some attention has been directed to nonaqueous electrolyte solutions.

The molal volume of the solute is given by

$$\bar{V}_2 = \phi_v + \frac{m^{1/2}}{2}\left(\frac{\partial \phi_v}{\partial m^{1/2}}\right) \tag{4.119}$$

where the partial molal volume ϕ_v is given by

$$\phi_v = \frac{V - n_1 V_1^\circ}{n_2} \tag{4.120}$$

where V_1° is the molal volume of the solvent.

The standard partial molal volume $\bar{V}_2^\circ$ is given by

$$\bar{V}_2^\circ = \phi_v^\circ \tag{4.121}$$

An excellent review article by Millero (81) presents a very thorough coverage of the molal volumes of electrolytes in both aqueous and nonaqueous solutions. Summary tables of nonaqueous data form part of this review and it is recommended that the reader consult this article. Examples are given that illustrate the application of partial molal volumes to study ion–solvent and solvent–solvent interactions and how the concentration dependence of the partial molal volume is used to study ion–ion interactions.

Gopal et al. (82) measured the partial molal volumes of some uni-univalent electrolytes in formamide and some tetraalklammonium iodides in N-methylacetamide. Both of these solvents were selected because of their high dielectric constants with a resulting absence of any possible ion association. Density data were used first to calculate the apparent molal volumes ϕ_v given by

$$\phi_v = \frac{1000(d_o - d)}{mdd_o} + \frac{M}{d} \tag{4.122}$$

where d is the density of the solution, d_o that of the pure nonaqueous solvent, m is the molality, and M is the molecular weight of the solute.

Using the Masson equation:

$$\phi_v = \phi_v^\circ + S\sqrt{C} \tag{4.123}$$

where C is the molar concentration and ϕ_v° is the partial molal volume at infinite dilution, and S is a constant, plots of ϕ_v vs $C^{1/2}$ were used to gain

Table 4.28 Partial Molal Volumes at Infinite Dilution of Some Electrolytes in *N*-Methylacetamide at Various Temperatures

	ϕ_v°						
Salt	35°C	40°C	45°C	50°C	60°C	70°C	80°C
Et_4NI	175.4	177.1	178.2	179.9	181.9	182.8	183.7
Pr_4NI	247.5	249.2	251.0	252.5	254.8	256.0	256.9
Bu_4NI	322.9	324.6	325.8	327.0	329.0	330.6	331.9
Pen_4NI	391.1	393.4	395.6	397.8	402.1	406.2	409.9
Hex_4NI	461.2	463.1	464.7	466.4	468.7	471.1	473.4
Hep_4NI	527.3	530.0	532.4	535.4	539.4	543.5	547.6
KI	45.9	46.2	46.5	46.7	46.4	45.4	44.6
KCl	20.5	20.2	20.0	19.7	19.2	18.4	17.2

the value of ϕ_v° by extrapolation. In formamide, it was found that the values of the partial molal volumes of the ions were higher than in water, but were found to be additive, that is,

$$\phi_v^\circ\ (\text{KCl}) - \phi_v^\circ\ ((\text{NaCl}) = 10.90\ \text{ml}$$

$$\phi_v^\circ\ (\text{KBr}) - \phi_v^\circ\ (\text{NaBr}) = 10.90\ \text{ml}$$

In the study using *N*-methylacetamide, the measurements were carried out at different temperatures to examine the rate of change of partial molal volumes with temperature. The plots of apparent molal volumes vs $C^{1/2}$ gave straight lines in all the cases studied, in accord with the Masson equation. A positive slope S was obtained for Et_4NI and Pr_4NI, but the slope was negative for Bu_4NI, Pen_4NI, Hex_4NI, and Hep_4NI. The values of the partial molal volumes at infinite dilution, ϕ_v° are given in Table 4.28. The values of ϕ_v° obtained for the lower-molecular-weight quaternary ammonium iodides in NMA were somewhat smaller than the corresponding values in water, suggesting the possibility of a stronger R_4N^+–NMA interaction with a consequent larger electrostriction in NMA than in water. However, the ion–solvent interactions in NMA and water are very different.

It was also observed that the values of ϕ_v° for a particular R_4NI increased with increasing temperature, with some curvature occurring in the plots for the iodides of lower molecular weight. For small electrolytes such as KI and LiCl, the plots of ϕ_v° vs T passed through a maximum. However $d\phi_v^\circ/dT$ decreased with rising temperature for the smaller R_4N iodides, while it was almost temperature independent for the larger iodides. The curvature in the ϕ_v° vs T plots only occurred for those systems that gave a positive slope for the ϕ_v° vs $C^{1/2}$ curves. This type of behavior is characteristic of the contraction of the solvent around an ion, or electrostriction. On heating, some of the solvent may be released from the loose solvation layers so that the volume would increase a little more rapidly than that of the pure solvent and hence $d\phi_v^\circ/dT$ would start out being positive.

However, as the temperature increases, the value of $d\phi_v^\circ/dT$ would decrease, because the loosely bound solvent molecules around the ions would slowly pass into solution, so that, at the higher temperatures, the pure solvent would expand more rapidly than the solution containing ions with the firmly bound first solvation layer. For the larger R_4N^+ ions with their lower surface charge density, electrostatic solvation and structure breaking of NMA are negligible. However, one effect that is possible is the penetration of the space between the coiled alkyl chains attached to the nitrogen atom by the solvent molecules, mainly due to the directing influence of the positively charged nitrogen atom. At the higher temperatures, this penetration effect would be less, due to higher thermal energy, and hence ϕ_v° would increase. In addition, the coiled chains would open out at the higher temperatures, causing the volume of the system to increase even in the absence of penetration.

Millero (83) determined apparent molal volumes of benzene, pyridine, benzoic acid, NaCl, KCl, NaBr, KBr, $NaNO_3$, and sodium benzoate in *N*-methylpropionamide from 15 to 40°C. The molal volumes $\bar{V}^\circ$ and molal expansivities $\bar{E}^\circ$ at infinite dilution are given in Table 4.29. It was observed that the apparent molal volume and apparent molal expansivity of benzene in NMP were independent of concentration and approximately equal to the values in pure benzene, indicating that no or very little solute–solvent interaction occurs between benzene and NMP. On the other hand, the solute–solvent interactions between pyridine and NMP are large, shown by the

Table 4.29 Molal Volume $\bar{V}^\circ$ and Molal Expansivity $\bar{E}^\circ$ at Infinite Dilution for Various Solutes in NMP at 25°C

Solute	$\bar{V}^\circ$ (ml mol^{-1})	$\bar{E}^\circ$ (ml mol^{-1} deg^{-1})
Benzene	88.1	0.09
Pyridine	31.3	0.18
$NaNO_3$	39.6	−0.04
NaCl	30.7	−0.06
KCl	35.5	−0.07
NaBr	35.8	−0.05
KBr	40.9	−0.04
NaOBz	103.6	—
HOBz	100.2	—

Reprinted with permission from *J. Phys. Chem.*, **72,** 3209 (1968).

Table 4.30 Molal Volume of Ions, $\bar{V}°$ (ion), in NMP and H_2O and $\Delta\bar{V}°$ (trans) from H_2O to NMP at 25°C

Ion	Crystal radius (Å)	$\bar{V}°$ (ion) ($ml\ mol^{-1}$) NMP	H_2O	$\bar{V}°$ (trans) $H_2O \rightarrow NMP$ ($ml\ mol^{-1}$)
H^+		3.4	−4.5	7.9
Na^+	0.95	6.0	−5.7	11.9
K^+	1.33	11.1	4.5	6.6
Cl^-	1.81	24.7	22.3	2.4
Br^-	1.95	29.8	29.2	0.6
NO_3^-	2.03	33.6	33.6	0
OBz^-		97.7		

Reprinted with permission from *J. Phys. Chem.*, **72**, 3209 (1968). Copyright 1968. American Chemical Society.
120

fact that the molar volume of pyridine in NMP is smaller than the molar volume of pure pyridine. Other evidence (viscosity data) indicates that pyridine breaks down the structure of NMP by forming terminal hydrogen bonds that decrease the volume. The apparent molal expansivity is positive, since at higher temperatures there is less structure in NMP to break down.

It is also of interest to note that the additivity of $\bar{V}°$ for ions in NMP can be shown by comparing the differences between salt pairs, for example, $\bar{V}°(KCl)-\bar{V}°(NaCl)$, $\bar{V}°(KBr)-\bar{V}°(NaBr)$, $\bar{V}°(NaBr)-\bar{V}°(NaCl)$, and $\bar{V}°(KBr)-\bar{V}°(KCl)$. This analysis produced values of $\bar{V}°(K-Na)$ = 4.8, 5.1 $ml\ mol^{-1}$ and $\bar{V}°(Br-Cl)$ = 5.1, 5.4 $ml\ mol^{-1}$. The present study indicated an approach for yielding single ion values for $\bar{V}°$. This approach used the initial assumption that $\bar{V}°(H^+)$ = O or $\bar{V}°(HOBz) = \bar{V}°(OBz^-)$. The values of $\bar{V}°$(ion) were then plotted vs r^3 (the crystal radius). $\bar{V}°(H^+)$ was then adjusted until the best fit for both cations and anions was obtained. In Table 4.30, single-ion values for $\bar{V}°$(ion) in NMP are given and compared to values obtained in water. Volumes of transfer of ions from water to NMP are also given in Table 4.30. These volumes of transfer can be attributed to the differences in electrostriction or solute–solvent interactions, since the intrinsic volume of an ion in NMP and H_2O should be nearly equal.

For most of the electrolytes (Table 4.29), the molal expansivities are negative, compared to positive values in aqueous solutions. The positive values are related to the ability of the salt to change the structure of the water over and above the normal electrostriction effect.

In a further study, Millero (84) discusses a more rigorous interpretation of partial molal volumes of ions in various solvents (methanol, NMP, water), using the Frank and Wen model for ion–solvent interactions. The partial molal volume of ions was assumed to be due to three major contributions:

$$\bar{V}°(\text{ion}) = \bar{V}°(\text{int.}) + \bar{V}°(\text{elec.}) + \bar{V}°(\text{disord.}) \qquad (4.124)$$

where $\bar{V}°$(int.) is the intrinsic partial molal volume, equal to the crystal volume; $\bar{V}°$(elec.) is the electrostriction partial molal volume (decrease in volume due to ion–solvent interactions) and $\bar{V}°$(disord.) is the disordered partial molal volume due to void space effects. The results of fitting the $\bar{V}°$(ions) in the various solvents to various semiempirical equations indicates that $\bar{V}°$(disord.) is related to the structure of the solvent. The disordered

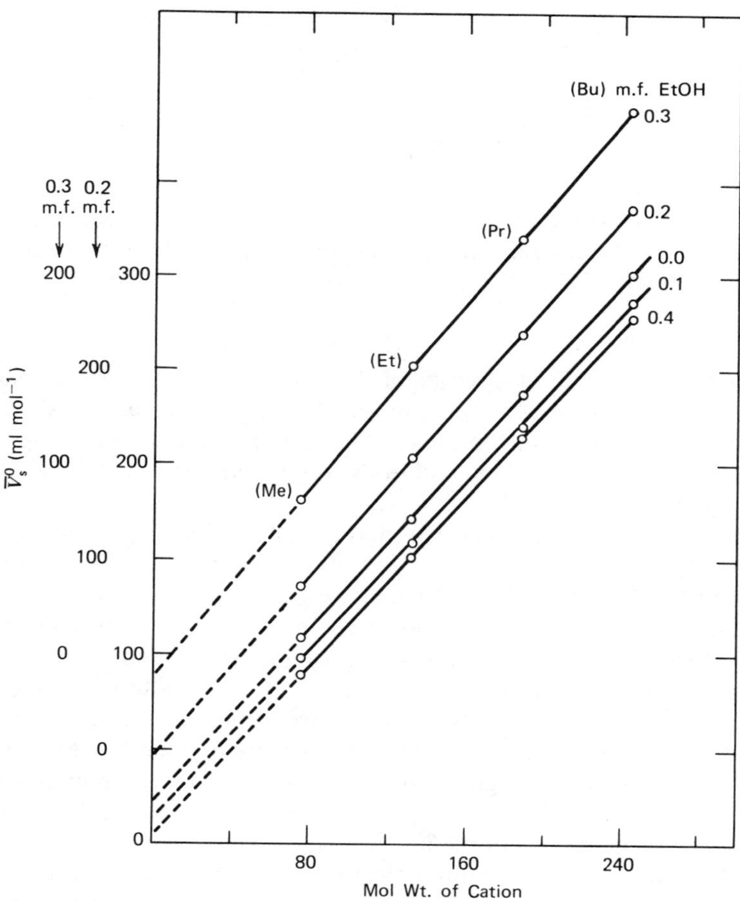

Figure 4.24 Partial molal volumes of tetraalkylammonium chlorides as a function of molecular weight of cation in various ethanol–water mixtures at 50.25°C. m.f. = mole fraction of ethanol. Reprinted by permission of the National Research Council of Canada from the *Canadian Journal of Chemistry,* **46,** 2333 (1968).

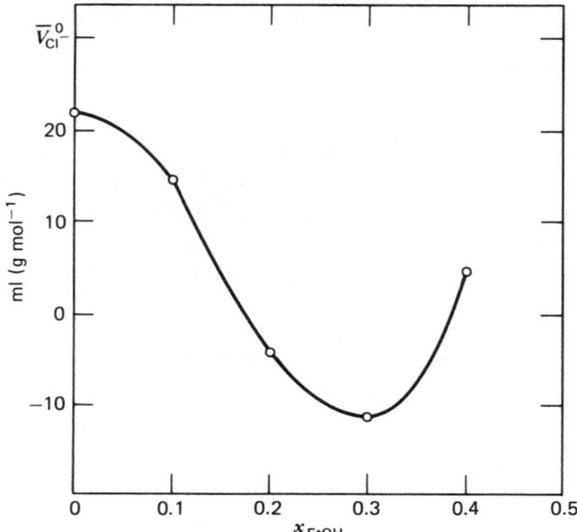

Figure 4.25 Ionic partial molal volumes of chloride ion as a function of solvent composition at 50.25°C. Reprinted by permission of the National Research Council of Canada from the *Canadian Journal of Chemistry,* **46,** 2333 (1968).

effect could be thought of in terms of the void space caused by the solvated ion, rather than improper packing in the electrostricted region. Also, it was observed that $\overline{V}°$(elec.) of the ions in the various solvents was proportional to $1/r$ (crystallographic radius) and in the predicted order, $\overline{V}°$(elec.) of methanol > water > NMP.

Finally, an example of a study of partial molal volumes in a mixed solvent system is that by Lee and Hyne (85). They determined the partial molal volumes of Me_4NCl, Et_4NCl, Pr_4NCl, and Bu_4NCl in a series of ethanol–water mixtures. The purpose of the study was to examine the structure-making and structure-breaking abilities of tetraalkylammonium ions in a mixed solvent system. First, it was noted that plots of ϕ vs $C^{1/2}$ had two distinct linear regions. In the case of Me_4NCl, the sharp transition occurred at about $C^{1/2} = 0.35$. Single-ion partial molal volumes were obtained by extrapolation of $\overline{V}^{\circ}_{R_4NX}$ vs molecular weight of the cations R_4N^+ to yield anionic partial molal volumes. A plot such as this is shown in Fig. 4.24. The values of $\overline{V}°(Cl^-)$ were obtained from the intercepts. The ionic partial molal volumes of the chloride ion as a function of solvent composition are shown in Fig. 4.25. The minimum in this plot was suggested as probably due to the maximum in the electrostriction effect.

The ionic partial molal volumes of the cations were determined using the relation

$$\overline{V}^{\circ}_{R_4N^+} = \overline{V}^{\circ}_{R_4NCl} - \overline{V}^{\circ}_{Cl^-} \tag{4.125}$$

The single-ion values for the cations are plotted in Fig. 4.26 as a function of solvent composition. All the cations show a minimum at 0.1 mole fraction and a maximum at 0.3 mole fraction. The similarity in all the curves suggests that the same effects are operating in all the cases. To analyze the trends in Figs. 4.25 and 4.26, one must realize that they represent the combined effects of cation and anion. The minimum at $X = 0.3$ in the anion plot (Fig. 4.25) is due to the electrostriction effect of the chloride ion and this effect is opposed by the hydrophobic and size effect of the ammonium cation, which is characterized by a maximum at the same solvent composition (Fig. 4.26).

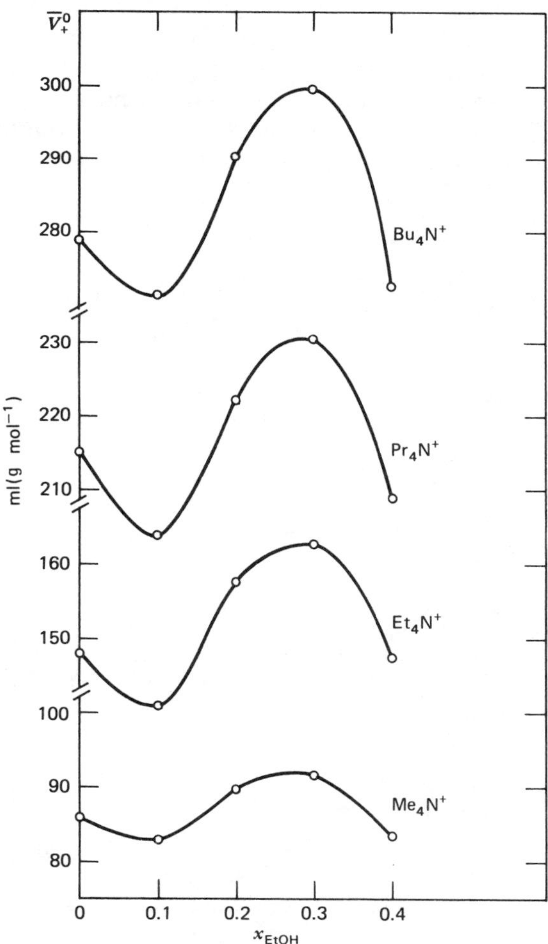

Figure 4.26 Ionic partial molal volumes of the tetraalkylammonium cations as a function of solvent composition at 50.25°C. Reprinted by permission of the National Research Council of Canada from the *Canadian Journal of Chemistry*, **46**, 2333 (1968).

The resulting balance between these two effects is demonstrated by a plot of the partial molal volumes of the total electrolyte as a function of solvent composition, as shown in Fig. 4.27. The maximum at 0.3 mole fraction observed for the Me_4N^+ ion has now become a minimum, indicating that the electrostriction effect of the chloride ion more than compensates for the small hydrophobic and size effect of the tetramethylammonium cation. However, as the cation effect becomes larger, the Cl^- electrostriction is exactly compensated and eventually outweighed by the hydrophobic and size effects in tetrapropyl- and tetrabutylammonium salts. The behavior of the partial molal volumes of the quaternary ammonium salts is also discussed in terms of the structure of the binary solvent system, for example, the enhanced structure of water at $\sim X = 0.1$, giving rise to the minimum in the plot, due to the existence of a maximum number of preformed "holes" into which the solute molecule may fit. At $X = 0.3$, there are no such preformed holes, as the solvent structure has been broken down and the addition of the solute requires a maximum expansion of the solvent in order to accomodate it,

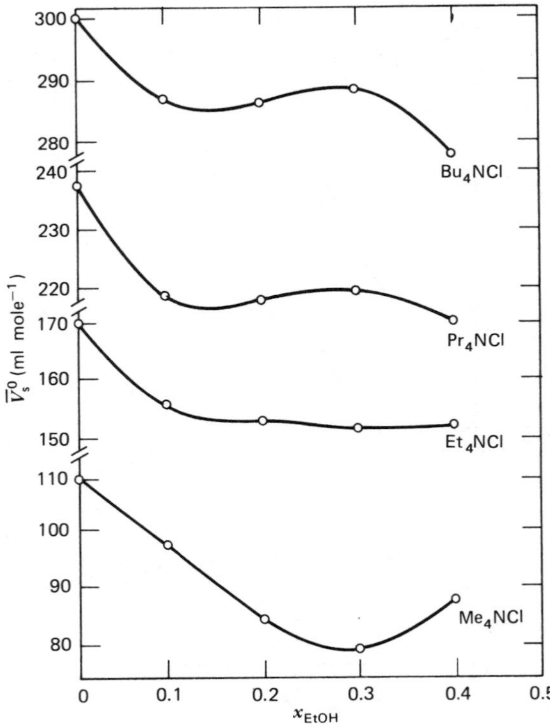

Figure 4.27 Partial molal volumes of tetraalkylammonium chlorides as a function of solvent composition at 50.25°C. Reprinted by permission of the National Research Council of Canada from the *Canadian Journal of Chemistry*, **46**, 2333 (1968).

thereby giving a maximum in the partial molal volume of the quaternary ammonium salts (Fig. 4.26).

Summary The determination of partial molal volumes of electrolyte solutions is useful as a guide to the structural properties of such solutions, in particularly in the study of ion–solvent and solvent–solvent interactions. The concentration dependence of the partial molal volume is used to study ion–ion interactions. The Masson equation, which relates the apparent molal volume, the partial molal volume and the concentration, is often used for the analysis of the data. The relative magnitudes of the partial molal volumes provide information about the strengths of the interactions between ions and solvent molecules. Measurements of partial molal volumes as a function of temperature can produce an extremum (maximum or minimum), which is an indication of electrostriction or the contraction of solvent around an ion.

An interesting additivity relationship for molal volumes of ions in certain nonaqueous solvents is sometimes observed by comparing the differences between salt pairs such as KCl with NaCl and KBr with NaBr. Single-ion values for $\overline{V}°$ can be generated by assuming either that $\overline{V}°(H^+) = 0$ or $\overline{V}°(HOBz) = \overline{V}°(OBz^-)$. Volumes of transfer of ions can also be obtained and can be attributed to the differences in the electrostriction or solute–solvent interactions. A more rigorous interpretation of partial molal volume data used the Frank and Wen model for ion–solvent interactions. In this treatment, the partial molal volume of ions is assumed to be due to three major contributions, namely, the intrinsic partial molal volume (equal to the crystal volume), the electrostriction partial molal volume (the decrease in volume due to ion–solvent interactions) and the disordered partial molal volume due to void space effects.

LITERATURE CITED

1 Ives, D. J. G., and G. J. Janz, "General and Theoretical Introduction," in *Reference Electrodes: Theory and Practice*, D. J. G. Ives and G. J. Janz, Eds., Academic Press, New York, 1961, Chap. 1.

2 Butler, J. N., "Reference Electrodes in Aprotic Organic Solvents," in *Advances in Electrochemistry and Electrochemical Engineering*, P. Delahey, Ed., Vol. VII, Interscience, New York, 1970, pp. 77–175.

3 Hills, G. J., "Reference Electrodes in Non-Aqueous Solvents," in *Reference Electrodes*, D. J. G. Ives and G. J. Janz, Eds., Academic Press, New York, 1961, Chap. 10.

4 Strehlow, H., "Electrode Potentials in Non-Aqueous Solvents," in *The Chemistry of Non-Aqueous Solvents*, Vol. I, J. J. Lagowski, Ed., Academic Press, New York, 1966, Chap. 4.

5 Janz, G. J., and R. P. T. Tomkins, *Nonaqueous Electrolytes Handbook*, Vol. II, Academic Press, New York, 1973.

6 Salomon, M., "Thermodynamic Measurements Part 2 Electrochemical Measurements," in *Physical Chemistry of Organic Solvent Systems*, A. K. Covington and T. Dickinson, Eds., Plenum Press, New York, 1973, Chap. 2.

7 Taniguchi, H., and G. J. Janz, *J. Phys. Chem.*, **61,** 688 (1957).
8 Nunez, L. J., and M. C. Day, *J. Phys. Chem.*, **65,** 164 (1961).
9 Dill, A. J., L. M. Itzkowitz, and O. Popovych, *J. Phys. Chem.*, **72,** 4580 (1968).
10 Hefley, J. D., and E. S. Amis, *J. Phys. Chem.*, **69,** 2082 (1965).
11 Feakins, D., and P. Watson, *J. Chem. Soc.*, 4686, 4734 (1963). Feakins, D., K. G. Lawrence, and R. P. T. Tomkins, *J. Chem. Soc.*, 753 (1967).
12 Alfenaar, M., C. L. De Ligny, and A. G. Remijnse, *Rec. Trav. Chim.*, **86,** 555 (1967).
13 Kundu, K. K., D. Jana, and M. N. Das, *J. Phys. Chem.*, **74,** 2625 (1970).
14 Nayak, B., and D. K. Sahu, *Electrochim. Acta*, **16,** 1757 (1971).
15 Hitchcock, D., *J. Amer. Chem. Soc.*, **50,** 2076 (1928).
16 Andrews, A. W., D. A. Armitage, R. W. C. Broadbank, K. W. Morcom, and B. L. Muju, *Trans. Faraday Soc.*, **67,** 128 (1971).
17 Mandel, M., and P. Decroly, *Trans. Faraday Soc.*, **56,** 29 (1960).
18 Luksha, E., and C. M. Criss, *J. Phys. Chem.*, **70,** 1496 (1966).
19 Dawson, L. R., W. H. Zuber, Jr., and H. C. Eckstrom, *J. Phys. Chem.*, **69,** 1335 (1965).
20 Bennetto, H. P., D. Feakins, and D. J. Turner, *J. Chem. Soc. A*, 1211 (1966).
21 Owen, B. B., *J. Amer. Chem. Soc.*, **57,** 1526 (1935).
22 Heston, B. C., and N. F. Hall, *J. Amer. Chem. Soc.*, **56,** 1462 (1934).
23 Andrews, A. L., H. P. Bennetto, D. Feakins, K. G. Lawrence, and R. P. T. Tomkins, *J. Chem. Soc. A*, 1486 (1968).
24 McIntyre, J. M., and E. S. Amis, *J. Chem. Eng. Data*, **13,** 371 (1968).
25 Roy, R. N., and A. Bothwell, *J. Chem. Eng. Data*, **15,** 548 (1970).
26 Holdeck, G., D. R. Cogley, and J. N. Butler, *J. Electrochem. Soc.*, **116,** 952 (1969).
27 Smyrl, W. H., and C. W. Tobias, *J. Electrochem. Soc.*, **115,** 33 (1968).
28 Salomon, M., *J. Electrochem. Soc.*, **116,** 1392 (1969); **117,** 325 (1970).
29 Robinson, R. A., and R. H. Stokes, *Electrolyte Solutions*, 2nd rev. ed., Butterworths, London, 1968.
30 Salomon, M., *J. Phys. Chem.*, **73,** 3299 (1969).
31 Burrows, B., and R. Jasinski, *J. Electrochem. Soc.*, **115,** 365 (1968).
32 Butler, J. N., D. R. Cogley, and W. Zurosky, *J. Electrochem. Soc.*, **115,** 445 (1968).
33 Kratochvil, B., E. Lorah, and C. Garber, *Anal. Chem.*, **41,** 1793 (1969).
34 Kolthoff, I. M., and F. G. Thomas, *J. Phys. Chem.*, **69,** 3049 (1965).
35 Butler, J. N., and J. C. Synnott, *J. Amer. Chem. Soc.*, **92,** 2602 (1970).
36 Butler, J. N., *J. Phys. Chem.*, **72,** 3288 (1968).
37 Koerber, G. G., and T. De Vries, *J. Amer. Chem. Soc.*, **74,** 5008, (1952).
38 Clifford, A. F., W. D. Pardieck, and M. W. Wedley, *J. Phys. Chem.*, **70,** 3241 (1966).
39 Garner, C. S., E. W. Green, and D. M. Yost, *J. Amer. Chem.Soc.*, **57,** 2055 (1935).
40 Randall, M., and A. M. White, *J. Amer. Chem. Soc.*, **48,** 2514 (1926).
41 Grunwald, E., and A. L. Bacarella, *J. Amer. Chem. Soc.*, **80,** 3840 (1958).
42 Campbell, A. N., and B. G. Oliver, *Can. J. Chem.*, **47,** 2671 (1969).
43 Scatchard, G., and S. S. Prentiss, *J. Amer. Chem. Soc.*, **56,** 1486 (1934).
44 Stokes, R. H., and R. A. Robinson, *J. Amer. Chem. Soc.*, **70,** 1870 (1948).
45 Glueckauf, E., *Trans. Faraday Soc.*, **51,** 1235 (1955).
46 Kushchenko, V. V., and K. P. Mishchenko, *Zhur. Prikl. Khim.*, **41,** 620 (1968).
47 O'Malley, J. A., C. Owens, C. Schmid, D. Quimby, and C. M. King, *J. Phys. Chem.*, **72,** 3584 (1968).

48 Linhard, M., *Z. Physik. Chem.*, **A175,** 438 (1936).

49 Hunt, H., and W. E. Larsen, *J. Phys. Chem.*, **38,** 801 (1934).

50 Blytas, G. C., D. J. Kertesz, and F. Daniels, *J. Amer. Chem. Soc.*, **84,** 1083 (1962).

51 Blytas, G. C., and F. Daniels, *J. Amer. Chem. Soc.*, **84,** 1075 (1962).

52 Sergeeva, V. F., and E. S. Moiseeva, *Russ. J. Gen. Chem.*, **32,** 2370 (1962).

53 Everhard, M. E., P. M. Gross, and J. W. Turner, *J. Phys. Chem.*, **66,** 923 (1962).

54 Ryder, G. A., M. R. Karmal, and L. N. Canjar, *J. Chem. Eng. Data*, **6,** 594 (1961).

55 Wood, R. H., R. K. Wicker, and R. W. Kreis, *J. Phys. Chem.*, **75,** 2313 (1971).

56 Kreis, R. W., and R. H. Wood, *J. Phys. Chem.*, **75,** 2319 (1971),

57 Garnsey, R., and J. E. Prue, "Thermodynamic Measurement, Part 3 Precision Cryoscopy," in *Physical Chemistry of Organic Solvent Systems*, A. K. Covington and T. Dickinson, Eds., Plenum Press, New York, 1973, Chap. 2.

58 Garnsey, R., and J. E. Prue, *Trans. Faraday Soc.,*, **64,** 1206 (1968).

59 Kim, S. J., O. D. Bonner, and D. S. Shin, *J. Chem. Thermodyn.*, **3,** 411 (1971).

60 Bonner, O. D., S. J. Kim, and A. L. Torres, *J. Phys. Chem.*, **73,** 1968 (1969).

61 Eichelberger, W. C., *J. Amer. Chem. Soc.*, **56,** 799 (1934).

62 Toppmeyer, W. P., and A. W. Davidson, *Inorg. Chem.*, **2,** 823 (1963).

63 Barton, B. C., and C. A. Kraus, *J. Amer. Chem. Soc.*, **75,** 4561 (1951).

64 Copenhafer, D. T., and C. A. Kraus, *J. Amer. Chem. Soc.*, **73,** 4557 (1951).

65 Alexander, R., E. C. F. Ko, Y. C. Mac, and A. J. Parker, *J. Amer. Chem. Soc.*, **89,** 3703 (1967).

66 Harner, R. E., J. B. Sydnor, and E. S. Gilreath, *J. Chem. Eng. Data*, **8,** 411 (1963).

67 Popovych, O., and R. M. Friedman, *J. Phys. Chem.*, **70,** 1671 (1966).

68 Deno, N. C., and H. E. Berkheimer, *J. Org. Chem.*, **28,** 2143 (1963).

69 Addison, C. C., B. J. Hathaway, N. Logan, and A. Walker, *J. Chem. Soc.*, 4308 (1960).

70 Criss, C. M., and E. Luksha, *J. Phys. Chem.*, **72,** 2966 (1968).

71 Luehrs, D. C., R. T. Iwamoto, and J. Kleinberg, *Inorg. Chem.*, **5,** 201 (1966).

72 Criss, C. M. "Thermodynamic Measurements, Part 1 Solubility, Calorimetry, Volume Measurements and Viscosities," in *Physical Chemistry of Organic Solvent Systems*, A. K. Covington and T. Dickinson, Eds., Plenum Press, New York, 1973, Chap. 2.

73 See, e.g., Held, R. P., and C. M. Criss, *J. Phys. Chem.*, **69,** 2611 (1965); *J. Phys. Chem.*, **71,** 2487 (1967); Bhatnagar, O. N., and C. M. Criss, *J. Phys. Chem.*, **73,** 174 (1969); Mastroianni, M. J., and C. M. Criss, *J. Chem. Thermo.*, **4,** 321 (1972); *J. Chem. Eng. Data*, **17,** 222 (1972).

74 See, e.g., Wu, Y. C., and H. L. Friedman, *J. Phys. Chem.*, **70,** 2020 (1966); *J. Phys. Chem.*, **70,** 501 (1966); Krishnan, C. V., and H. L. Friedman, *J. Phys. Chem.*, **73,** 3934 (1969); *J. Phys. Chem.*, **73,** 1572 (1969).

75 See, e.g., Somsen, G., and J. Coops, *Rec. Trav. Chim.*, **84,** 985 (1965); Weeda, L., and G. Somsen, *Rec. Trav. Chim.*, **86,** 893 (1967); *Rec. Trav. Chim.*, **86,** 263 (1967); *Rec. Trav. Chim.*, **85,** 159 (1966).

76 Choux, G., and R. L. Benoit, *J. Amer. Chem. Soc.*, **91,** 6221 (1969).

77 Fuchs, R., J. L. Bear, and R. F. Rodewald, *J. Amer. Chem. Soc.*, **91,** 5797 (1969); Rodewald, R. F., K. Mahendran, J. L. Bear, and R. Fuchs, *J. Amer. Chem. Soc.*, **90,** 6698 (1968).

78 Stern, J. H., and J. Nobilione, *J. Phys. Chem.*, **72,** 1064 (1968); *J. Phys. Chem.*, **72,** 3937 (1968); *J. Phys. Chem.*, **73,** 928 (1969).

79 Feakins, D., and C. T. Allan, *J. Chem. Soc. Faraday, I*, 314 (1976).

80 Abraham, M., *J. Chem. Soc., Faraday, I*, 1375 (1973).

81 Millero, F. J., *Chem. Rev.*, **71,** 147 (1971).

82 Gopal, R., and R. K. Srivastava, *J. Phys. Chem.*, **66,** 2704 (1962); Gopal, R., and M. A. Siddiqi, *J. Phys. Chem.*, **73,** 3390 (1969).

83 Millero, F. J., *J. Phys. Chem.*, **72,** 3209 (1968).

84 Millero, F. J., *J. Phys. Chem.*, **73,** 2417 (1969).

85 Lee, I., and J. B. Hyne, *Can. J. Chem.*, **46,** 2333 (1968).

GENERAL REFERENCES

Bates, R. G., "Electrode Potentials," in *Treatise on Analytical Chemistry*, Part 1, Vol. I, 2nd ed., I. M. Kolthoff and P. J. Elving, Eds., Wiley-Interscience, New York, 1978.

Butler, J. N., "Reference Electrodes in Aprotic Organic Solvents," in *Advances in Electrochemistry and Electrochemical Engineering*, Vol. VII, P. Delahey, Ed., Wiley-Interscience, New York, 1970, pp. 77–175.

Covington, A. K., and T. Dickinson, Eds., *Physical Chemistry of Organic Solvent Systems*, Plenum Press, New York, 1973.

Ives, D. J. G., and G. J. Janz, Eds., *Reference Electrodes*, Academic Press, New York, 1961.

Janz, G. J., and R. P. T. Tomkins, *Nonaqueous Electrolytes Handbook*, Vol. II, Academic Press, New York, 1973.

Latimer, W. M., *Oxidation Potentials*, 2nd ed., Prentice-Hall, Englewood-Cliffs, N.J., 1952.

Lee, T. S., revised by Popovych, O., "Chemical Equilibrium and the Thermodynamics of Reactions" in *Treatise on Analytical Chemistry*, Part 1, Vol. I, 2nd ed., I. M. Kolthoff and P. J. Elving, Eds., Wiley-Interscience, New York, 1978.

Strehlow, H., "Electrode Potentials in Non-Aqueous Solvents," in *The Chemistry of Non-aqueous Solvents*, Vol. I, J. J. Lagowski, Ed., Academic Press, New York, 1966.

CHAPTER FIVE

Correlation of Properties in Different Solvents

It is now generally recognized that the transfer of a solute from one solvent to another can lead to enormous changes in both equilibrium and rate constants of chemical reactions. Perhaps the most familiar manifestation of solvent effects on equilibria is the variation of solubility as a function of the solvent. For example, AgCl is virtually insoluble in water but very soluble in such liquids as ammonia and dimethylthioformamide, while the reverse is true of $BaCl_2$, which is very soluble in water but insoluble in liquid ammonia. Analytical chemists have long used nonaqueous solvents as a means of shifting precipitation, solvent-extraction, and acid–base equilibria to effect separations and identifications that would have been impossible in aqueous media. Thus it is not uncommon to add organic solvents to aqueous solutions to decrease the solubility of precipitates in gravimetric analysis. Metal chelates are generally insoluble in water but are extracted into chloroform or carbon tetrachloride, in which they are soluble. Phenol cannot be titrated successfully as an acid in aqueous solution, but a change to a medium that is either sufficiently aprotic (e.g., acetone) or basic (e.g., ethylenediamine) produces a favorable shift in the acid–base equilibria, so that a quantitative titration with a base becomes feasible.

A change in the solvent can also bring about large shifts in oxidation–reduction equilibria. For example, the equilibrium constant for the disproportionation of cuprous ion, $2Cu^+ \rightleftarrows Cu + Cu^{2+}$, changes from 10^{+6} in water to 10^{-11} in a 0.5 mole fraction acetonitrile–water mixture, to 10^{-20} in anhydrous acetonitrile (1). When an oxidation–reduction equilibrium that experiences such major shifts determines the voltage of a galvanic cell, the large solvent effects on the equilibrium constant are reflected in the corresponding electrode potentials as well. Thus, it was noted by Parker (2) that the emf of silver–potassium galvanic cells could be varied by up to 2.1 V by using appropriate solvent combinations in the two half-cells.

Physical–organic chemists have discovered that a change from water and alcohols to dipolar aprotic solvents can increase some reaction rates by a factor of anywhere from 10^2 to 10^7, depending on the reactants involved. For example, the S_N2 reaction of methyl iodide with fluoride ion is about 10^7 times faster in dipolar aprotic solvents than in water or alcohols (3).

What determines the relationship between properties such as equilibrium constants, rate constants, and electrode potentials in different solvent media? Does a pH value in methanol have the same meaning as in water? Is it possible to interpret the potential of a galvanic cell in which one electrode is inserted into an aqueous, and the other, into a nonaqueous medium? Can a salt bridge minimize a liquid-junction potential at the boundary of electrolyte solutions in different solvents? These and many related questions have perplexed experimentalists and theoretical chemists for decades. The need to correlate free-energy functions, such as the Gibbs free energies, enthalpies, entropies, equilibrium constants, pH values, and emf data, determined in different solvents, has been acquiring increasing importance and sometimes urgency in view of the ever-expanding application of nonaqueous solvents in all branches of chemistry and the resulting accumulation of data on the physicochemical properties of nonaqueous solutions. In particular much effort has been devoted to the establishment of solvent-independent ion-activity scales, such as a universal pH scale, and of a solvent-independent emf series.

To the uninitiated, it might seem strange that a change in the nature of the solvent should present correlation problems in the first place. After all, calories, volts, grams, moles, and liters have the same meaning in ethanol as in water and a 0.1 *M* KCl in aqueous solution represents the same analytical concentration as 0.1 *M* KCl in ethanolic solution. The reason why correlation is a problem for quantities related to free energy is that they are functions of the *activity*, and not of the concentration. Activities, in turn, depend on the *difference* between the Gibbs free energies of a species in the given state and in some standard reference state (Eqn. 5.1). Generally, the reference state is chosen in the solvent in which the property is determined and will have a different energy value in each solvent medium. Differences between the free energies, enthalpies, and entropies of a solute in two solvents are known as the thermodynamic *transfer functions* for that solute (see below). Thus one incentive for determining the transfer properties for electrolytes and ions is to be able to predict values of equilibrium and rate constants, to express activities and electrode potentials in different media on a single, solvent-independent scale, or to answer the other related questions posed in the beginning of this section. Beyond these immediate applications, however, values of the free energies, enthalpies and entropies of transfer provide a wealth of information about the energetics and mechanisms of solvent–solute and solvent–solvent interactions and about the structure of the solvents and their solutions.

5.1 THERMODYNAMIC TRANSFER FUNCTIONS

5.1.1 Definitions and General Applications The relationships between free energy, activity, and concentration were discussed in Chapter 4 (Section 4.1 and Eqns. 4.1–4.3). For convenience, we shall repeat here the two key relationships:

$$\bar{G}_i = \bar{G}_i^\circ + RT \ln a_i \tag{5.1}$$

and

$$a_i = m_i \gamma_i \tag{5.2}$$

The familiar salt-effect activity coefficient γ, which in the case of electrolytes is the mean ionic activity coefficient, is a measure of deviations from ideality resulting from solute–solute interactions only; the solvent remains so far the same both in the reference and in the measuring state. A change in the nature of the solvent medium, however, will generally require a change in the numerical value of the $\bar{G}_i^\circ$ in Eqn. 5.1. For example, when applied to hydrogen ions, Eqn. 5.1 would be rewritten in the form of Eqn. 5.3a for water and 5.3b for ethanol:

$$_{\mathrm{w}}\bar{G}_{\mathrm{H}} = {}_{\mathrm{w}}\bar{G}_{\mathrm{H}}^\circ + RT \ln a_{\mathrm{H}} \tag{5.3a}$$

$$_{\mathrm{e}}\bar{G}_{\mathrm{H}} = {}_{\mathrm{e}}\bar{G}_{\mathrm{H}}^\circ + RT \ln a_{\mathrm{H}}^* \tag{5.3b}$$

where subscripts w and e indicate that the free energy is referred to standard states in water and ethanol, respectively, while the nonaqueous activity is designated by an asterisk. In the above equations, the activities and molalities are related as usual:

$$a_{\mathrm{H}} = m_{\mathrm{H}}\, {}_{\mathrm{w}}\gamma_{\mathrm{H}} \tag{5.4a}$$

$$a_{\mathrm{H}}^* = m_{\mathrm{H}}\, {}_{\mathrm{s}}\gamma_{\mathrm{H}} \tag{5.4b}$$

where the activity coefficients ${}_{w}\gamma_{H}$ and ${}_{s}\gamma_{H}$ become unity at infinite dilution in water and in ethanol, respectively.

It is clear now why numerical values of activities and other free-energy functions cannot be directly compared for different solvent media. Consider, for example, the meaning of pa_H ($pa_H \equiv -\log a_H$) from Eqn. 5.3a and 5.3b. Since $pa_H = (\bar{G}^\circ_H - \bar{G}_H)/2.3RT$, it is obvious that an identical numerical value of pa_H in ethanol and in water will not represent the same free energy for the hydrogen ion in the two media because the values of ${}_{e}\bar{G}^\circ_H$ and of ${}_{w}\bar{G}^\circ_H$, which are the solvation energies of the proton, are not likely to be equal. Another way of looking at the problem is to focus on the nature of the activity coefficients in water and in ethanol. Since they are referred to different standard states, infinite dilution in water and in ethanol, respectively, the corresponding free energies are measured from different points of origin.

To express activities in two solvents on the same scale, we must refer them to a single standard state. This amounts to evaluating the difference between the standard free energies in the two reference states. For example, to correlate the pa_H scales in ethanol and in water, it is necessary to evaluate the quantity $({}_{e}\bar{G}^\circ_H - {}_{w}\bar{G}^\circ_H)$. In the general case, the quantity required for the correlation of free-energy properties of species i in two solvents is the difference between the partial molal free energies of that solute in its aqueous and nonaqueous standard states:

$$\Delta\bar{G}^\circ_t(i) = {}_{s}\bar{G}^\circ_i - {}_{w}\bar{G}^\circ_i = RT \ln {}_{m}\gamma_i \tag{5.5}$$

This difference is known as the *standard free energy of transfer*, $\Delta\bar{G}^\circ_t(i)$, or simply the *transfer free energy* for the solute i from water to the given nonaqueous solvent. The corresponding activity coefficient ${}_{m}\gamma_i$ is known as the *transfer activity coefficient*, or *medium effect** of solute i between the two solvents. Since ${}_{m}\gamma_i$ is a property of two ideal solutions (cf. discussion of Eqn. 4.2), it is determined by the nature of the solute and the two solvents, but at a given temperature and pressure it is otherwise a constant, independent of the concentration or of other substances present in solution. Thus, it is a measure of the difference between the solvent–solute interactions in the two (ideal) reference states, with the solute–solute interactions (salt or concentration effects) eliminated. For uncharged solute species and electroneutral combinations of ions, values of $\Delta\bar{G}^\circ_t$ can be determined experimentally from a variety of thermodynamic data, as will be shown in the following section. In the case of individual ions, the estimation of $\Delta\bar{G}^\circ_t$ requires extrathermodynamic assumptions, which we discuss in Section 5.2.

The operational meaning of the transfer activity coefficient (medium effect) can be derived by considering again Eqns. 5.3. Normally, the activity

* Some of the other synonyms found in the literature are "primary medium effect," "medium activity coefficient," "solvent activity coefficient," "distribution coefficient," and "degenerate activity coefficient."

of hydrogen ions in ethanol would be referred to the ethanolic standard state (Eqns. 5.3b and 5.4b). However, it is equally permissible to refer the hydrogen-ion activity in ethanol to the aqueous standard state, that is, to apply Eqns. 5.3a and 5.4a to an ethanolic solution. When Eqns. 5.4a and 5.4b are applied to *the same* solution *in ethanol*, the activity coefficient ${}_{s}\gamma_{H}$ will still become unity at infinite dilution in ethanol, but the activity coefficient ${}_{w}\gamma_{H}$ will become unity only at infinite dilution in water. In ethanol, as the molality approaches zero, ${}_{w}\gamma_{H}$ will approach ${}_{m}\gamma_{H}$, the medium effect. Consequently, when referred to the aqueous standard state, the activity coefficient of the hydrogen ion in ethanol is a product of the salt effect, ${}_{s}\gamma_{H}$, and the medium effect, ${}_{m}\gamma_{H}$:

$$ {}_{w}\gamma_{H} = {}_{s}\gamma_{H}\,{}_{m}\gamma_{H} \qquad (5.6)$$

Of course, for aqueous solutions, ${}_{m}\gamma_{H}$ has the value of unity and the overall activity coefficient is equal to the salt effect, ${}_{s}\gamma_{H}$, which for electrolytes is identical with the mean ionic activity coefficient, $\gamma_{\pm}$.

From Eqns. 5.4 and 5.6, it is clear that the medium effect ${}_{m}\gamma_{i}$ can be used as a conversion factor from a nonaqueous to an aqueous activity scale. Thus, in the case of hydrogen-ion activity we have

$$a_{H} = a_{H}^{*}\,{}_{m}\gamma_{H} \qquad (5.7)$$

and

$$\mathrm{p}a_{H} = \mathrm{p}a_{H}^{*} - \log {}_{m}\gamma_{H} \qquad (5.7a)$$

to correlate $\mathrm{p}a_{H}$ scales in water and ethanol. It is now obvious why the evaluation of the medium effects for the proton is the key to a correlation of $\mathrm{p}a_{H}$ scales in different solvents.

For a simple general equilibrium

$$\mathrm{A} + \mathrm{B} \rightleftarrows \mathrm{X} + \mathrm{Y} \qquad (5.8)$$

the equilibrium constant in water, ${}_{w}K$, and in a nonaqueous solvent, S, ${}_{s}K$, can be related by means of transfer activity coefficients, or medium effects, by applying Eqn. 5.7 in a general way,

$${}_{w}K = {}_{s}K\,\frac{{}_{m}\gamma_{X}\,{}_{m}\gamma_{Y}}{{}_{m}\gamma_{A}\,{}_{m}\gamma_{B}} \qquad (5.9)$$

Similarly, when Eqn. 5.9 is applied to a simple dissociation of a monobasic acid, $HA \rightleftarrows H^{+} + A^{-}$, governed by the constant K_{HA}, the correlation becomes

$${}_{w}K_{HA} = {}_{s}K_{HA}\,\frac{{}_{m}\gamma_{H}\,{}_{m}\gamma_{A}}{{}_{m}\gamma_{HA}} \qquad (5.10)$$

and

$$\Delta \mathrm{p}K_{HA} \equiv \mathrm{p}({}_{s}K_{HA}) - \mathrm{p}({}_{w}K_{HA}) = \log\left(\frac{{}_{m}\gamma_{H}\,{}_{m}\gamma_{A}}{{}_{m}\gamma_{HA}}\right) \qquad (5.10a)$$

When the stoichiometric coefficients of the reactants or products in Eqn. 5.8 differ from unity, the corresponding medium effects are raised to the powers of those coefficients. Thus for the generalized equation

$$bB + cC \rightleftarrows xX + yY \tag{5.11}$$

the relationship between equilibrium constants becomes

$$_{w}K = {}_{s}K \frac{_{m}\gamma_{X}^{x}\ _{m}\gamma_{Y}^{y}}{_{m}\gamma_{B}^{b}\ _{m}\gamma_{C}^{c}} \tag{5.12}$$

Similarly, the rate constants in water, $_{w}k$, and in solvent, S, $_{s}k$, for a reaction that goes through a transition state $AB^{\ddagger}$ in its rate-determining step:

$$A + B \rightleftarrows AB^{\ddagger} \tag{5.13}$$

are related via medium effects:

$$_{w}k = {}_{s}k \frac{_{m}\gamma_{AB^{\ddagger}}}{_{m}\gamma_{A}\ _{m}\gamma_{B}} \tag{5.14}$$

It is noteworthy that if the same reference state were used for activities in all solvents, the numerical values of equilibrium and rate constants would be identical in all solvents, since then all $_{m}\gamma$'s would equal unity.

Standard electrode potentials in water, $_{w}E^{\circ}$, and in a nonaqueous solvent, $_{s}E^{\circ}$, are also related through the appropriate medium effects. This arises from the fact that for any oxidation–reduction process, $\Delta G^{\circ} = -nF\,\Delta E^{\circ}$, where F is the Faraday and n is the number of electrons.† For example, in the case of the galvanic cell which serves to define the standard (reduction) potential of the silver–silver ion electrode

$$\text{Pt; } H_2(g) \mid H^+ \parallel Ag^+ \mid Ag(s) \tag{5-I}$$

$$p_{H_2} = 1 \quad a_{H} = 1 \quad a_{Ag} = 1$$

for which the overall cell reaction is $Ag^+ + \frac{1}{2}H_2(g) \rightleftarrows H^+ + Ag(s)$, it follows that

$$\frac{_{s}E^{\circ}_{Ag} - {}_{w}E^{\circ}_{Ag}}{2.3RT/nF} = \log {}_{m}\gamma_{Ag} - \log {}_{m}\gamma_{H} \tag{5.15}$$

In Eqn. 5.15, $_{s}E^{\circ}_{Ag}$ and $_{w}E^{\circ}_{Ag}$ are the standard potentials of cell 5-I in the nonaqueous solvent and in water, respectively.

So far, we have been limiting the definition of transfer free energies and activity coefficients, or medium effects, to the process of transfer from water to a nonaqueous solvent. Although this is the most common usage, it is sometimes of interest to deal with transfers between two nonaqueous media. When a reference solvent other than water is chosen, that convention must be made perfectly clear in notation. For example, if the transfer is defined

† The relationship between solvation energy and E° is derived in Chapter 9.

as taking place from acetonitrile (reference) to ethanol, the symbol for the corresponding transfer activity coefficient could be ${}^{AN}\gamma^{e}$. In this case, the ${}_{w}\bar{G}_i^\circ$ in Eqn. 5.5 would be replaced by ${}_{AN}\bar{G}_i^\circ$ and ${}_{s}\bar{G}_i^\circ$, by ${}_{e}\bar{G}_i^\circ$, where e denotes ethanol. Unless otherwise stated, the transfer processes in this text are assumed to take place from aqueous reference states.

While for most purposes the transfer functions of interest are the free energies and the activity coefficients, additional information on solvation phenomena can be obtained from a consideration of the enthalpies $\Delta\bar{H}_t^\circ$ and the entropies $\Delta\bar{S}_t^\circ$ of transfer. The usual relationship between these thermodynamic functions holds for their transfer properties as well:

$$\Delta\bar{G}_t^\circ = \Delta\bar{H}_t^\circ - T\,\Delta\bar{S}_t^\circ \tag{5.16}$$

5.1.2 Determination of Thermodynamic Transfer Functions We have shown in the preceding section how knowledge of the numerical values for the transfer activity coefficients would enable us to correlate equilibrium constants, rate constants, electrode potentials, and activity scales in different solvent media. The question remaining now is how to determine the transfer activity coefficients required for such correlations. The problem of evaluating transfer free energies, transfer activity coefficients, and other transfer functions is analogous to that of determining salt-effect activity coefficients in a single solvent. For uncharged solutes and for electroneutral combinations of ions, activity coefficients and other thermodynamic functions can be evaluated from experimental data without approximations and assumptions. In the case of transfer activity coefficients, electroneutral combinations are such quantities as ($\log {}_{m}\gamma_{H} + \log {}_{m}\gamma_{Cl}$) or ($\log {}_{m}\gamma_{Ag} - \log_{m}\gamma_{H}$). For single ions, however, values of the transfer functions can be estimated only from extrathermodynamic models and assumptions. In this section we will consider the common methods for the determination of the thermodynamic transfer functions for uncharged solutes and electroneutral combinations of ions. In Section 5.2, we will deal with the splitting of the thermodynamic transfer functions into individual ionic contributions and with the application of the single-ion transfer functions thus obtained.

Transfer activity coefficients are commonly determined from the solubilities or the electrode potentials in the two solvents compared; occasionally, distribution and vapor-pressure methods are employed.

Solubility The partial molal free energies of a solute species in a saturated solution and in the solid in equilibrium with that solution are equal. Thus when the saturated solutions in two solvents are in equilibrium with the same solid phase (i.e., in the absence of crystal solvates) their free energies are also equal. By analogy with Eqn. 5.3 we can write for a saturated aqueous solution of solute i:

$$\bar{G}_i = {}_{w}\bar{G}_i^\circ + RT \ln {}_{w}a_i(\text{satd.}) \tag{5.17a}$$

and for a nonaqueous solution saturated with the same solute:

$$\bar{G}_i = {}_s\bar{G}_i^\circ + RT \ln {}_s a_i(\text{satd.}) \tag{5.17b}$$

Since $\bar{G}_i$ is the same in both equations, we obtain

$$\Delta\bar{G}_t^\circ(i) \equiv {}_s\bar{G}_i^\circ - {}_w\bar{G}_i^\circ = RT \ln \frac{{}_w a_i(\text{satd.})}{{}_s a_i(\text{satd.})} = RT \ln {}_m\gamma_i \tag{5.18}$$

When the solid forms a hydrate i·y·H_2O and a solvate i·x·s, Eqn. 5.18 (5) can be modified as follows:

$${}_s\bar{G}_i^\circ - {}_w\bar{G}_i^\circ - RT\frac{y}{x} \ln \frac{p_{H_2O}p_{i\cdot x\cdot s}}{p_s p_{i\cdot y\cdot H_2O}} = RT \ln {}_m\gamma_i \tag{5.19}$$

where the p's denote the partial pressures of the pure solvents and of the solvates, as indicated. As a rule, however, one simply avoids those solute–solvent combinations that show evidence of solvate formation.

It follows from Eqn. (5.18) that the transfer activity coefficient of an uncharged solute species is given by the ratio of its solubilities in the reference solvent and in the solvent to which the transfer takes place, where both solubilities are expressed as activities. For an electrolyte, the activity in saturated solution is equivalent to the solubility (ion-activity) product, K_{sp}. It is common practice to calculate ${}_m\gamma$ values in logarithmic form as

$$\log {}_m\gamma(\text{electrolyte}) = p({}_sK_{sp}) - p({}_wK_{sp}) \tag{5.20}$$

For any symmetrical (1:1, 2:2, etc.) electrolyte

$$\log {}_m\gamma(\text{electrolyte}) \equiv \log {}_m\gamma_\pm^2 = (\log {}_m\gamma_+ + \log {}_m\gamma_-) \tag{5.21}$$

Solubility products for a number of electrolytes in different solvents can be found in Tables 4.12, 4.13, and 4.18.

Because transfer activity coefficients are properties of ideal solutions, their values for individual ions can be added and subtracted (in logarithmic form) in a manner analogous to the handling of ionic limiting equivalent conductances. We can calculate the quantity ($\log {}_m\gamma_K - \log {}_m\gamma_{Na}$) for example, from the difference ($\log {}_m\gamma_{KCl} - \log {}_m\gamma_{NaCl}$). The additivity rule also makes it possible to calculate indirectly the transfer activity coefficients of those electrolytes for which accurate K_{sp} values cannot be obtained in one or both of the solvents. Thus values of $\log {}_m\gamma$ for tetraphenylarsonium tetraphenylborate, which is only sparingly soluble in water, are usually obtained from those of other electrolytes, for example,

$$\log {}_m\gamma_{Ph_4AsBPh_4} = \log {}_m\gamma_{Ph_4AsPi} + \log {}_m\gamma_{KBPh_4} - \log {}_m\gamma_{KPi} \tag{5.22}$$

(Pi^- = picrate).

It is clear that once the value of $\log {}_m\gamma$ is known, the corresponding transfer free energy, $\Delta\bar{G}_t^\circ$, is calculated simply from Eqn. 5.5. A useful relationship in this connection is $\Delta\bar{G}_t^\circ = 1.364 \log {}_m\gamma$ (in kcal mol^{-1} at 25°C).

When transfer free energies rather than activity coefficients are desired, an alternative calculation route can be adopted via the free energy changes for the solution, $\Delta\bar{G}^\circ_{\text{soln.}}$. Since $\Delta\bar{G}^\circ_{\text{soln.}} = -RT \ln K_{sp}$ (Eqn. 2.13), it follows that

$$\Delta\bar{G}^\circ_t = {}_s(\Delta\bar{G}^\circ_{\text{soln.}}) - {}_w(\Delta\bar{G}^\circ_{\text{soln.}}) \tag{5.23}$$

which is, of course, equivalent to Eqn. 5.18 or 5.19.

As Eqn. 5.18 suggests, the transfer activity coefficient between a pair of immiscible solvents should be equal to the distribution ratio of the solute species between the two solvents, that is, ${}_m\gamma = {}_wa/{}_sa$. The trouble with determining transfer activity coefficients by distribution methods lies in the fact that even nominally immiscible solvents become saturated with each other at equilibrium, so that we are never dealing with a transfer between the pure solvents. Moreover, it is well known that even traces of water can alter drastically the solvation equilibria in some nonaqueous solvents.

Electromotive Force It was mentioned earlier in this chapter that because $\Delta G^\circ = -nF\,\Delta E^\circ$, values of the $\Delta\bar{G}^\circ_t$ or of the ${}_m\gamma$ for electroneutral combinations of ions can be obtained from the standard potentials of galvanic cells reversible to these ions in the two solvent media. Equation 5.15 illustrated this method on the example of evaluating the quantity ($\log {}_m\gamma_{Ag} - \log {}_m\gamma_H$). A very popular method for the calculation of the quantity ($\log {}_m\gamma_H + \log {}_m\gamma_{Cl}$) makes use of the E°'s for the hydrogen–silver–silver chloride cells:

$$\text{Pt(s); } H_2\text{(g)} \mid \text{HCl} \mid \text{AgCl(s); Ag(s)} \tag{5-II}$$

When the standard potential of cell 5-II is ${}_wE^\circ_{AgCl}$ in water and ${}_sE^\circ_{AgCl}$ in the nonaqueous solvent, it follows at 25°C that

$$\log {}_m\gamma_{HCl} \equiv (\log {}_m\gamma_H + \log {}_m\gamma_{Cl}) = \frac{{}_wE^\circ_{AgCl} - {}_sE^\circ_{AgCl}}{0.05916} \tag{5.24}$$

Transfer activity coefficients for HCl are of fundamental importance because the values of $\log {}_m\gamma_H$ and $\log {}_m\gamma_{Cl}$ derived from them are required for a correlation of pa_H scales and of emf series in different solvents. In Table 5.1 we collected some of the values of $\log {}_m\gamma_{HCl}$ [which is of course identical with ($\log {}_m\gamma_H + \log {}_m\gamma_{Cl}$)] calculated from Eqn. 5.24 for a number of solvents. In Table 4.4 the quantities ($\log {}_m\gamma_K + \log {}_m\gamma_{Cl}$) and ($\log {}_m\gamma_K - \log {}_m\gamma_H$) for ethanol–water solvents, similarly calculated from emf data, are listed.

Vapor Pressure On rare occasions, transfer activity coefficients for volatile solutes, such as HCl, have been determined by the isopiestic vapor-pressure method. This method is based on the same principle as the solubility method, namely, that two phases that are in equilibrium with the same third phase have equal partial molal free energies. It follows that, when the partial

Table 5.1 Transfer Activity Coefficients for HCl in Different Solvents at 25°C (Molal Scale, Water Is the Reference Solvent)[a]

Solvent	$\log {}_m\gamma_{HCl}$	Solvent	$\log {}_m\gamma_{HCl}$
Wt % Ethanol in Water		Wt % Methanol in Water	
10.0	0.13	43.3	0.48
20.0	0.25	64.0	0.78
30.0	0.37	84.2	1.53
40.0	0.47	94.2	2.34
50.0	0.62	100	3.93
60.0	0.79	Wt % Dioxane in Water	
70.0	1.12		
80.0	1.57		
90.0	2.21	20	0.32
100	5.13	45	0.99
n-Propyl alcohol	5.48	70	2.64
Isopropyl alcohol	5.82	Acetone	2.86
Acetic acid	9.54	Formamide	0.31

[a] Calculated from Eqn. 5.24 using data compiled by Strehlow (4). Table 16.3 in Ref. (5).

Reprinted with permission from *Treatise on Analytical Chemistry,* Part I, Vol. I, 2nd ed., I. M. Kolthoff and P. J. Elving, Eds., Wiley-Interscience, New York, 1978, p. 724. Table courtesy of John Wiley & Sons, Inc.

pressures of a solute above an aqueous and a nonaqueous solution are equal (hence the term isopiestic), the transfer activity coefficient for that solute is given by the ratio of its activities in water and in the nonaqueous solvent:

$${}_m\gamma = \frac{{}_wa}{{}_sa} \tag{5.25}$$

When the partial pressures above the aqueous solution ${}_wp$ and the nonaqueous solution ${}_sp$ are not equal, Eqn. 5.25 is modified to

$${}_m\gamma = \frac{{}_sp\,{}_wa}{{}_wp\,{}_sa} \tag{5.26}$$

Calorimetry Enthalpies of transfer, $\Delta\bar{H}^\circ_t$, are usually calculated from experimental values for the enthalpies of solution in water and in the nonaqueous solvent:

$$\Delta\bar{H}^\circ_t = {}_s(\Delta\bar{H}^\circ_{soln.}) - {}_w(\Delta\bar{H}^\circ_{soln.}) \tag{5.27}$$

Enthalpies of solution are most commonly obtained directly from calorimetric measurements of the heats of solution in the case of moderately soluble substances or estimated from the heats of precipitation in the case of slightly soluble substances. Corrections to infinite dilution are made whenever the data show a concentration dependence in the range of the measurements. Less commonly, values of $\Delta\bar{H}^\circ_{\text{soln.}}$ are derived from the temperature dependence of the emf, E, of galvanic cells, utilizing the Gibbs–Helmholtz relationship:

$$\Delta H = -nF\left[E - T\left(\frac{\partial E}{\partial T}\right)_P\right] \tag{5.28}$$

Entropies of solution are obtained by difference from the equation $\Delta\bar{G}^\circ_{\text{soln.}} = \Delta\bar{H}^\circ_{\text{soln.}} - T\,\Delta\bar{S}^\circ_{\text{soln.}}$ and then the values of $\Delta\bar{S}^\circ_{\text{t}}$ are calculated as for the other functions from the difference:

$$\Delta\bar{S}^\circ_{\text{t}} = {}_{\text{s}}(\Delta\bar{S}^\circ_{\text{soln.}}) - {}_{\text{w}}(\Delta\bar{S}^\circ_{\text{soln.}}) \tag{5.29}$$

5.1.3 Discussion of Representative Data Cox, Hedwig, Parker, and Watts (6, 7) published a selective compilation of the available free energies, enthalpies, and entropies of solution for a number of 1 : 1 electrolytes in water and in several nonaqueous solvents. From these data they computed the corresponding transfer functions from water to the nonaqueous media, accompanying the results with an excellent discussion. In Tables 5.2–5.4 some of their results on the transfer functions are reproduced and in the text that follows, the interpretations presented by these authors are used as a basis for discussion.

Additional compilations of the enthalpies of transfer for a number of electrolytes can be found in Chapter 4, Tables 4.20, 4.21, 4.25, and 4.27 as well as in Figs. 4.21–4.23. Also, the thermodynamic functions for the transfer of LiCl from water to a number of organic solvents are given in Table 4.9.

In view of what we know from Chapter 2, the data in Table 5.2 offer no surprises. Simple electrolytes are generally most soluble in water and, to a lesser extent, in formamide, which means that their free energies of solution in these solvents (not shown in the Tables) will be the lowest (most negative). Consequently, for most electrolytes, the free energies for the transfer from water to nonaqueous solvents are positive. One exception is the transfer of silver salts to media where the Ag^+ ion is specifically solvated; there we observe negative values of $\Delta\bar{G}^\circ_{\text{t}}$ in some instances. Another exception are the salts of large polarizable organic ions, such as Ph_4As^+ or BPh_4^-, which are more soluble in nonaqueous solvents than in water. Their $\Delta\bar{G}^\circ_{\text{t}}$ values in nonaqueous solvents are negative. Of the liquids shown in Table 5.2, the poorest solvents for most electrolytes are propylene carbonate and acetonitrile.

More surprising are the results in Table 5.3. While the free energies for the transfer of most electrolytes from water to nonaqueous solvents are positive, their *enthalpies* for the corresponding transfers are usually negative. At the root of this phenomenon lies the difference between the degree of structure in liquid water and in nonaqueous liquids. When an ion is

Table 5.2 Free Energies of Transfer of 1 : 1 Electrolytes from Water to Nonaqueous Solvents at 25°C[a]

	Solvent						
Electrolyte	Methanol	Formamide	DMSO[b]	DMF[c]	PC[d]	Sulfolane	Aceto-nitrile
AgCl	4.9	−0.4	1.4	6.8	13.9	11.7	4.7
AgBr	4.3	−1.0	−1.7	4.0	11.2	8.6	2.1
AgI	3.5	−1.9	−4.8	0.1	7.4	4.0	−2.1
AgN_3	4.3	−0.8	−2.3	4.3	10.7	8.6	1.8
AgSCN	3.1	−0.3	−6.1	−0.3	6.0	3.0	−2.2
AgOAc	5.4	—	3.0	10.6	—	—	8.2
$AgBPh_4$	−3.8	−9.4	−17.0	−13.2	−6.0	−9.5	−13.2
LiCl	4.0	1.0	4.8	5.7	14.7	15.5	17.2
LiBr	3.8	—	2.4	1.8	13.0	—	—
LiI	2.2	—	0.8	—	10.2	—	—
Me_4NI	3.5	—	—	3.0	—	—	3.8
Et_4NI	1.8	—	—	2.5	—	—	2.4
NaF	—	4.0	—	11.2	17.2	—	—
NaCl	5.0	1.4	6.0	8.5	12.7	12.1	13.4
NaBr	5.0	—	—	4.7	—	8.5	11.0
KCl	5.6	2.2	—	8.6	10.6	—	12.0
KBr	4.8	0.7	3.2	4.8	—	6.8	9.1
KI	2.2	0.7	—	—	5.6	5.6	5.5
$KClO_4$	3.7	—	—	−0.4	—	—	3.0
$KBPh_4$	−3.1	−6.1	—	—	—	—	−6.1
CsCl	6.0	2.0	—	9.3	6.1	10.2	10.1
CsI	4.0	−0.6	—	2.4	1.7	—	4.3
RbCl	6.2	2.0	—	—	8.3	10.5	11.9
RbBr	5.3	—	5.0	6.0	6.3	—	9.2
$RbClO_4$	3.9	—	—	−0.6	—	—	−2.6
Ph_4AsBPh_4	−11.3	−11.5	−17.7	−18.2	−17.2	−16.4	−15.6
Ph_4AsI	−3.8	−4.2	−3.8	−4.3	−4.6	−3.8	−3.2

[a] Data in this Table were selected from Ref. (7), Table 1. Molar scale. The values are in kcal mol^{-1}.
[b] Dimethylsulfoxide.
[c] *N,N*-Dimethylformamide.
[d] Propylene carbonate.

introduced into water, enthalpy is gained due to ion–water interactions in the first hydration sphere, but that gain is more than offset by the enthalpy loss resulting from the breaking of the solvent–solvent hydrogen bonds in the extensive three-dimensional water structure in order to accommodate the ion. The net effect is that the dissolution of most electrolytes in water is enthalpically disfavored. In less structured liquids, however, any enthalpy loss due to the breaking of solvent–solvent bonds is more than compensated for by the enthalpy gain from the ion–solvent interactions in the solvation sphere. One consequence of the exothermic nature of the enthalpies for the transfer from water to nonaqueous solvents is the fact that as the temperature is raised, the solubilities of most simple electrolytes in water increase relative to their solubilities in nonaqueous solvents.

Exceptions to the generally exothermic trend in the transfer enthalpies are the endothermic transfers of several electrolytes from water to propylene carbonate and the endothermic transfer from water to nonaqueous solvents of most tetraalkylammonium salts. Presumably, in the case of the large "hydrophobic" ions, there is an enthalpy gain when water reinforces its structure in the vicinity of such solutes (see Chapter 2).

It is very interesting to compare the free energies and the enthalpies for the transfer of LiCl, NaCl, and KCl from water to methanol–water solvents, shown in Fig. 5.1. As the methanol content of the mixture increases, ΔG_t° shows an almost continuous rise, which can be rationalized on the basis of electrostatic models for the solvation of ions, in which the free energy of

Table 5.3 Enthalpies of Transfer of 1:1 Electrolytes from Water to Nonaqueous Solvents at 25°C[a]

	Solvent						
Electrolyte	Methanol	Formamide	DMSO[b]	DMF[c]	PC[d]	Sulfolane	Acetonitrile
AgCl	−3.5	−4.6	−8.8	—	3.7	—	—
AgBr	−3.0	—	−11.1	−8.4	1.0	0.1	−10.9
AgI	−5.7	—	−14.4	—	−2.8	−5.7	−14.3
AgN_3	−4.9	—	−13.7	−8.9	0.9	−0.5	−10.5
LiCl	−3.2	−0.6	−2.0	−2.9	7.0	—	—
LiBr	—	−1.6	−5.5	−6.8	4.9	—	—
LiI	—	−3.1	−9.1	−11.0	−0.1	—	—
NaCl	−2.8	−3.0	—	—	5.2	—	—
NaBr	−4.0	−4.3	−6.1	−7.3	3.0	—	—
NaI	−5.3	−5.6	−9.7	−11.4	−2.6	−5.6	−4.8
$NaClO_4$	−5.9	−5.0	−11.2	−13.3	−6.4	−8.7	−7.5
$NaBPh_4$	−5.3	−4.0	−9.4	−12.5	−5.9	−6.1	−5.6
KBr	−3.5	−5.0	−7.5	−8.7	−0.8	—	—
KI	−5.0	−5.8	−11.4	−13.0	−5.9	—	−7.2
$KBPh_4$	—	—	−11.1	−14.1	−8.7	—	−7.9
RbI	—	−5.8	—	−12.6	−1.2	—	−7.2
CsI	−3.8	−5.9	−10.8	−12.3	−7.0	—	—
NMe_4I	−0.4	—	−6.8	−7.2	−4.8	—	−5.5
NEt_4I	1.6	—	−1.9	−3.2	−0.7	−1.3	−2.0
NBu_4I	4.8	—	3.5	1.0	4.1	—	2.7
Ph_4AsI	0.4	−1.9	−4.9	−5.3	−2.6	−4.4	−4.1

[a] Data in this Table were selected from Ref. 7, Table 2. Molar scale. The values are in kcal mol^{-1}. Sulfolane values are at 30°C.
[b] Dimethylsulfoxide.
[c] *N,N*-Dimethylformamide.
[d] Propylene carbonate.

Table 5.4 Entropies of Transfer for 1:1 Electrolytes from Water to Nonaqueous Solvents at 25°C[a]

	Solvent					
Electrolyte	Methanol	Formamide	Sulfolane	DMF[b]	PC[c]	Acetonitrile
LiCl	7.2	1.6	6.8	8.6	7.6	—
LiBr	—	—	7.9	8.6	8.1	—
LiI	—	—	8.3	—	10.3	—
NaCl	7.8	4.4	—	—	7.5	—
NaBr	9.0	—	—	12.0	—	—
KCl	8.1	5.5	—	—	9.1	—
KBr	8.3	5.7	10.7	13.5	—	—
KI	7.2	6.5	—	—	11.5	12.7
RbCl	8.1	5.4	—	—	7.6	—
CsCl	7.4	5.4	—	—	5.8	—
CsI	7.8	5.3	—	14.7	8.7	—
AgCl	8.4	4.2	10.6	—	10.2	—
AgBr	7.3	—	10.3	12.4	10.2	13.0
AgN_3	9.2	—	11.4	13.2	9.8	12.3
AgI	9.2	—	11.1	—	10.2	12.2
NEt_4I	0.2	—	3.9	5.7	—	4.4
Ph_4AsI	−4.2	−2.3	1.1	−1.0	−2.0	0.9
NMe_4I	+3.9	—	—	10.2	—	9.3

[a] Data in this Table were selected from the data in Ref. 7, Table 4. Molar scale. The values are $-298\ \Delta\bar{S}_t^\circ$ in kcal mol^{-1}.
[b] *N,N*-Dimethylformamide.
[c] Propylene carbonate.

solvation increases with decreasing dielectric constant of the medium (see Eqn. 2.16, Table 2.3, and Eqn. 5.33a). The transfer enthalpy, however, rises at first reaching a maximum in the vicinity of a mole fraction of methanol of 0.1–0.2, and then decreases continuously with increasing methanol content. The enthalpy maximum occurs in a region where the water structure has been increased by the introduction of *small* amounts of the organic component, methanol (cf. discussion of Fig. 5.3). Obviously, the transfer of ions from water to a solvent that is even more strongly hydrogen bonded is enthalpically disfavored. Furthermore, it is most disfavored for LiCl (highest $\Delta\bar{H}_t^\circ$) perhaps because the strong ionic field of the small Li^+ ion would tend to cause the most destruction of the existing solvent structure. Addition of larger quantities of methanol to water causes a gradual breakdown of the water structure, and therefore the enthalpy of transfer keeps decreasing for the same reasons as in the transfer of ions from water to pure liquids of low structure (see above).

For most electrolytes, the entropies of transfer from water to nonaqueous solvents are negative (Table 5.4). This is a reflection of the fact that most electrolytes gain entropy when transferred from a crystalline solid to aqueous solution, but they lose entropy when dissolved in a nonaqueous liquid. Thus, the preferential solubility of most electrolytes in water is a result of the favorable entropy change. Significantly, the entropy losses are much smaller for the transfer from water to the highly structured liquids, formamide and *N*-methylformamide, than to other nonaqueous solvents. Also significant is the observation that for salts containing the large tetraalkylammonium and the tetraaryl ions, the entropy losses upon transfer from water to nonaqueous solvents are considerably smaller than for other electrolytes. Evidently the entropies of solution and their changes are governed by the relative degree of order in the pure liquid and in the given solution. In highly structured liquids (water, formamides), the entropy loss due to formation of a primary solvation sphere around an ion is more than offset by the effect of disrupting the structure of the pure liquid and by the formation of a structure-broken region, which results from the competing influences of the ionic field and the normal bonding forces in the liquid (see Section 2.2.3). In weakly structured solvents, however, there is no such compensating mechanism for the entropy losses resulting from ionic solvation.

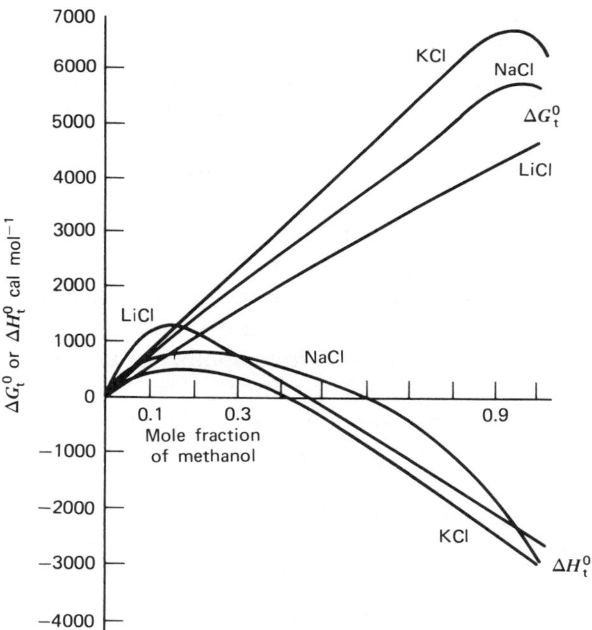

Figure 5.1 Free energies and enthalpies for the transfer of LiCl, NaCl, and KCl from water to methanol–water solvents (molar scale, 25°C). Reprinted with permission from *Physico-Chemical Processes in Mixed Aqueous Solvents,* F. Franks, Ed., American Elsevier Publishing Company, Inc., New York, 1967, p. 87. By courtesy of F. Franks and the publishers.

Large hydrophobic ions, such as R_4N^+ and Ph_4As^+, dissolve in water with a loss of entropy because water forms highly structured surfaces at the interfaces of such solutes. Therefore, the entropy losses accompanying the transfer of their salts from water to nonaqueous solvents are less than for salts composed of smaller ions. The dissolution of these organic "hydrophobic" solutes in water (sometimes called "hydration of the second kind") occurs because of the previously mentioned gain in the enthalpy resulting from the formation of the hydrogen-bonded structures known as "icebergs" or "cages" (Chapter 2). Additional interpretation of the transfer functions is best deferred to the next section, where the transfer functions for single ions are considered.

5.2 TRANSFER FUNCTIONS FOR SINGLE IONS AND THEIR APPLICATIONS

The usefulness of thermodynamic transfer functions for electrolytes is rather limited. Instead, the key to some of the most important questions in solution chemistry lies in the knowledge of the transfer free energies or transfer activity coefficients for *single* ions.

One example is the task of establishing solvent-independent ion-activity scales. The convenience and reproducibility of measurements using the commercial glass–calomel pH cell has long extended the analytical applications of pH from water to partially and totally nonaqueous media. The expanding use of other ion-selective electrodes promises a similar future for the determination of other ion activities as well. The problem is that the ion activities determined in each solvent are referred to the standard state in that solvent and are therefore thermodynamically unrelated to the activities of the same ion in other solvent media. If we consider that every pure liquid and every mixture of liquids constitutes a distinct solvent medium, we are faced, in effect, with a potentially infinite number of activity scales for each ion. In practical terms it means, for example, that a solution having a pa_H^* value of, say, 3.0 in ethanol probably has a different acidity than a solution of pa_H of 3.0 in water. We have shown earlier (Eqn. 5.7a) that the pa_H^* in any nonaqueous solvent can be expressed on the single aqueous scale, that is, referred to the aqueous standard state, if we know the value of ${}_m\gamma_H$, the transfer activity coefficient for the hydrogen ion, for that particular solvent.

Similarly there is a need to express standard electrode potentials on a single, solvent-independent scale. By convention, the emf series in any solvent is referred to the arbitrary zero of the standard hydrogen electrode (SHE) in the same solvent. Again, this generates as many thermodynamically unrelated emf series, as there are solvents. They are unrelated because, as a rule, absolute values of E_H° will not be equal in different solvents, since the solvation energy of the proton, $\bar{G}_H^\circ$, is likely to show considerable solvent dependence; (remember that $\Delta\bar{G}_H^\circ = -nF\ \Delta E_H^\circ$). However, once we know the value of ${}_m\gamma_H$ for a given solvent, the potential of the SHE in that solvent can be referred to the SHE in water by the relationship

$$ {}_wE^\circ(H,S) = (RT/F) \ln {}_m\gamma_H + {}_wE^\circ(H,H_2O) \tag{5.30} $$

In Eqn. 5.30 (5), ${}_wE^\circ(H,S)$ represents the potential of the SHE in the nonaqueous solvent S, but expressed on the aqueous emf scale, referred to ${}_wE^\circ(H,H_2O) = 0$ V. Once the value of the SHE in the nonaqueous solvent has been "translated" into the aqueous scale, the E°'s of other electrodes in the same medium can then be referred easily to the single zero point of the aqueous SHE.

The third related problem that requires knowledge of the transfer activity coefficients for single ions is the estimation of the liquid-junction potentials, E_j, at the interfaces of different solvent media. When analytical pH measurements, other forms of potentiometry, and polarography became extended to nonaqueous media, chemists found it often convenient to retain aqueous reference half-cells and salt bridges. Furthermore, in nonaqueous galvanic cells, salt bridges often contain a solvent different from that in the measuring or the reference half-cell. The resulting unknown and possibly large liquid-junction potentials at the boundaries of different solvents distort the observed emf and pH readings, sometimes contributing errors of the same order of magnitude as the quantity being determined. However, when the transfer activity coefficient of the relevant single ion is known, it is possible to estimate closely the liquid-junction potentials at the interface of two solvent media and to introduce the necessary correction to the analytical or electrochemical data derived from cells with transference (with liquid junction).

Application of transfer activity coefficients to the estimation of E_j can be illustrated by formulating the potential E of the following cell with liquid junction, which is composed of half-cells that are standard hydrogen electrodes in water and in solvent S:

$$ \underset{E_j}{\text{SHE(water)} \mid \text{SHE(S)}} \tag{5-III} $$

For the above cell, the potential E is given by

$$ E = {}_wE^\circ(H,S) - {}_wE^\circ(H,H_2O) + E_j \tag{5.31} $$

According to Eqn. 5.30, the first two terms on the right-hand side of Eqn. 5.31 are simply $(RT/F) \ln {}_m\gamma_H$, which means that

$$ E = \frac{RT}{F} \ln {}_m\gamma_H + E_j \tag{5.32} $$

Thus, when ${}_m\gamma_H$ is known, E_j for cell 5-III can be calculated. Knowledge of E_j at the interfaces of different solvents is useful in the interpretation of pH measurements because there is evidence that in many solvents the magnitude of E_j between aqueous bridge solutions and dilute nonaqueous buffers is primarily a function of the solvent and is relatively independent of the pH (8).

Aside from their usefulness in solving the above analytical and electrochemical problems, transfer activity coefficients of *single* ions are indispensable to any meaningful interpretation of solvation phenomena. Solvents seldom interact equally with an anion and a cation of a given electrolyte, so that dealing with the "mean" transfer activity coefficient of an electrolyte usually obscures the solvation picture. Specific or preferential solvation of one ion by a solvent is well established. Certainly the most spectacular changes in equilibria and rates of chemical reactions are brought about by the interaction between a solvent and a single ionic species. These are some of the reasons why "single-ion thermodynamics" has developed rapidly in the last few decades, even though to some chemists its legitimacy may still appear to be somewhat controversial.

5.2.1 Estimation of Transfer Functions for Single Ions Over the years, many extrathermodynamic assumptions have been proposed for the estimation of the transfer free energies and transfer activity coefficients of single ions. Several summaries (9–14) and two comprehensive critical reviews (5,15) of the field are available.

At this point it might be worthwhile to consider briefly just what it is that a transfer free energy of an ion represents. From Eqn. 5.5 it follows that the $\Delta\bar{G}_t^\circ$ of any solute is the *difference* between its solvation energies in two solvents, of which one is usually water. The reader will recall from Chapter 2 that the solvation energy of an ion is thought of as a composite of electrostatic, nonelectrostatic, and specific chemical interactions. However, precise mathematical formulations have been attempted only for the electrostatic energy components. The latter comprise the Born charging (Eqn. 2.16), ion–dipole (Eqn. 2.3), ion–quadrupole (Eqn. 2.4), ion-induced dipole (Eqn. 2.5), and dispersion (Eqn. 2.8) energy terms. Numerical values of these energy terms have been computed for several monatomic ions in aqueous solution and their sums were found to correspond closely to experimental values of hydration energies (e.g., Table 2.1). Would it not be possible to compute the nonaqueous solvation energies in an analogous manner and then obtain the $\Delta\bar{G}_t^\circ$ by difference from the hydration energy? (Eqn. 5.5).

One problem with this approach is our insufficient knowledge of the structure of nonaqueous solvents and their solvation shells around ions, including the magnitude of the various electrical and distance parameters required for the calculation. Furthermore, even if an accurate calculation of the *electrostatic* energy terms could be made, this would lead to correct $\Delta\bar{G}_t^\circ$ values only in cases where the differences between nonelectrostatic and chemical interactions in the two solvents could be neglected. Fundamentally, the *ab initio* calculations of the $\Delta\bar{G}_t^\circ$ for ions suffer from the analytically unfavorable situation where a small difference must be computed between two large uncertain values. Furthermore, while the solvation energy of closed-shell monatomic ions is believed to be predominantly electrostatic, its magnitude probably does not vary by much among solvents of moderate dielectric constant, so that the corresponding *transfer* free energies are likely to be determined by other energy components. For example, the Born energy for an ion of $r = 1.5$ Å is -109 kcal mol^{-1} in water and -108 kcal mol^{-1} in acetonitrile. Thus, while the Born energy probably constitutes a major portion of the *solvation* energy of an ion, the *transfer* free energies between water and acetonitrile are more likely to reflect specific chemical interactions, as well as ion–dipole and dispersion energies.

Nevertheless, much literature has been devoted to calculation of transfer free energies of ions from the Born equation. A combination of Eqn. 2.16b from Chapter 2 with Eqn. 5.5 yields the expression for the contribution of the Born energy to the overall transfer free energy of an ion:

$$\Delta\bar{G}_t^\circ(\text{Born}) = \frac{166}{r}\left[\frac{1}{{}_s\epsilon} - \frac{1}{{}_w\epsilon}\right] \quad \text{in kcal mol}^{-1} \tag{5.33a}$$

where r is the ionic radius in Ångstroms and ${}_s\epsilon$ and ${}_w\epsilon$ are the dielectric constants of the nonaqueous solvent and water, respectively. The corresponding transfer activity coefficient (with water as the reference medium) is given at 25°C by

$$\log {}_m\gamma_i = \frac{121.6}{\text{r}}\left[\frac{1}{{}_s\epsilon} - 0.0128\right] \tag{5.33b}$$

Today the uncorrected Born equation is used sometimes as a qualitative guide for predicting the direction of change for ionic solvation energies, but, as will be demonstrated later in this chapter, even such a limited application of the Born equation must be viewed with extreme caution. However, corrected Born equations seem to hold more promise. Thus, Strehlow et al. (4, 16) as well as Coetzee et al. (17, 18) derived values of $\Delta\bar{G}_t^\circ$ for ions by adjusting the ionic radii in the Born equation until experimental and calculated values of the solvation energies or the transfer free energies agreed for electroneutral pairs of ions. This method will be described in greater detail in Chapter 9. Corrections for the dielectric constant in the vicinity of the ions have also been tried with some success.

Historically, the first approach to the estimation of transfer activity coefficients for single ions was probably that of Bjerrum and Larsson (19), who assumed that the liquid-junction potential between hydrogen half-cells in water and in ethanol–water solvents could be eliminated by interposition of a salt bridge of aqueous KCl. If this were true, values of $\log {}_m\gamma_H$ could be obtained directly from the emf of cell 5-III, as shown by Eqn. 5.32. Recently this type of approach has been revived by Parker and his associates, who assume that the E_j is negligible when Ag^+/Ag electrodes in two nonaqueous solvents are connected by a salt bridge of tetraethylammonium picrate in one of the solvents (9, 10).

For a while, extrapolation methods for the estimation of the $\bar{G}^\circ$ and $\Delta\bar{G}_t^\circ$ values for single ions dominated the field. Izmaylov (20–22) plotted experimentally measurable combinations of solvation energies, such as $(\bar{G}_M^\circ - \bar{G}_H^\circ)$, as a function of $1/r_M$, where M was the metal ion being varied and r_M, its radius. Assuming that the solvation of ions is solely electrostatic in nature and hence a function of r^{-n} (see Chapter 2), Izmaylov obtained the values of $\bar{G}_H^\circ$ from the intercepts of such plots. Feakins and his associates (23–25) constructed analogous plots directly for the transfer free energies. DeLigny and Alfenaar (26, 27) adopted the formulation of the transfer free energy of an ion as being the composite of electrostatic and nonelectrostatic contributions. They carried out extrapolations similar to those of their predecessors, except that they first subtracted a nonelectrostatic component of the transfer free energy estimated from a suitable uncharged analog. Subsequently, DeLigny et al. (28–30) drastically modified their extrapolation method by applying to it the ion–quadrupole theory of solvation.

In electrostatic models of ionic solvation, an ion of infinite size has a solvation energy of zero. The solvation energies for large ions of finite size should be small and their transfer free energies, perhaps negligible. On the basis of such considerations, Pleskov (31) in 1947 proposed that the rubidium electrode be adopted in place of the hydrogen electrode as the solvent-independent zero point for electrode potentials. Rubidium was the largest cation for which electrode potentials were known ($r_{Rb^+} = 1.48$ Å) at that time. This proposal was tantamount to assuming that $\log {}_m\gamma_{Rb} = 0$ in all solvents. While today nobody would accept the generalization that the transfer free energy of any solute could be zero, Pleskov's proposal remains important in the historical sense as the first step in the evolution of more plausible assumptions for the estimation of transfer functions for single ions, assumptions based on the solvation of large molecules and ions.

From our discussion of solvation in Chapter 2, it follows that, in the absence of specific chemical interactions, the solvation free energy of an ion can be viewed as a composite of electrostatic and nonelectrostatic contributions. The transfer free energy can be similarly formulated:

$$\Delta\bar{G}_t^\circ(\text{ion}) = \Delta\bar{G}_t^\circ(\text{el}) + \Delta\bar{G}_t^\circ(\text{nonel}) \tag{5.34}$$

While the above formulation may be generally applicable, the assignment of a numerical value to $\Delta\bar{G}_t^\circ(\text{nonel})$ has presented some problems. Ideally, $\Delta\bar{G}_t^\circ(\text{nonel})$ should be identical with the $\Delta\bar{G}_t^\circ$ value for an uncharged molecule of the same size and structure as the ion. However, convincing ion–molecule analogs were nonexistent until large symmetrical organic ions became available. It is the availability of large ion–molecule analogs that has led to a number of approaches in which the transfer free energy of an ion is equated to that of an uncharged molecule which approximates the ion in size and structure. The working assumptions then are that

$$\Delta\bar{G}_t^\circ(\text{ion}) = \Delta\bar{G}_t^\circ(\text{nonel}) \tag{5.35a}$$

and

$$\Delta\bar{G}_t^\circ(\text{nonel}) = \Delta\bar{G}_t^\circ(\text{uncharged analog}) \tag{5.35b}$$

Ion–molecule analogs for which such assumptions might be applicable must be very large, essentially equal in size, with the active atom insulated from the solvent by inert organic groups. The ion should be univalent, characterized by low surface-charge density and an absence of specific interactions with the solvents. When the ion is large, the electrostatic component of its transfer free energy is expected to be small, and its $\Delta\bar{G}_t^\circ$ should be determined mainly by $\Delta\bar{G}_t^\circ$(nonel). The following ion–molecule pairs have found application in the estimation of $\Delta\bar{G}_t^\circ$ values for single ions: ferricinium $[Fe(C_5H_5)_2{}^+]$–ferrocene $[Fe(C_5H_5)_2]$, tetraphenylarsonium (Ph_4As^+)–tetraphenylmethane (Ph_4C), tetraphenylborate $(BPh_4{}^-)$–tetraphenylmethane, Hammett H_o indicator ions (BH^+)–indicator conjugate bases (B), H_- indicator ions (A^-)–indicator conjugate acids (HA), triiodide ion–iodine, 4-nitrofluorobenzene–S_NAr transition-state anion, $4\text{-}NO_2C_6H_4N_3F_{\ddagger}{}^-$, and methyliodide–$S_N2$ transition-state anion, $SCNCH_3I_{\ddagger}{}^-$.

Of the ion–molecule methods, the ferrocene assumption introduced by Strehlow (4) has been used most extensively. The ferrocene–ferricinium electrode was proposed as a solvent-independent reference electrode against which other potentials could be measured on a unified scale. Some of the applications of the ferrocene electrode are discussed in Chapter 9.

All ion–molecule assumptions neglect the electrostatic contribution to the transfer free energy of an ion (see Eqn. 5.34). The error resulting from this omission is eliminated when the ion–molecule assumption is replaced by the "reference electrolyte" assumption. Here, the transfer free energy of a 1 : 1 electrolyte composed of large ions that are very nearly equal in size and structure is apportioned equally between the anion and the cation. Currently the reference electrolyte of choice among the researchers in this field and one from which most of the available data have been derived is tetraphenylarsonium tetraphenylborate (Ph_4AsBPh_4) (e.g., Refs. 9, 10, 14, 32–36). Transfer properties of single ions are determined on the assumption that for the Ph_4As^+ and the $BPh_4{}^-$ ions they can be equated in any pair of solvents, e.g., ${}_m\gamma_{Ph_4As} = {}_m\gamma_{BPh_4}$ or $\Delta\bar{G}_t^\circ(Ph_4As^+) = \Delta\bar{G}_t^\circ(BPh_4{}^-)$. The tetraphenyl*phosph*onium tetraphenylborate assumption was also tested (36, 37) and was found to be equivalent to the $Ph_4As\ BPh_4$ assumption (36).

The key factors that contribute to the credibility of the *assumption* that $\Delta\bar{G}_t^\circ(Ph_4As^+) = \Delta\bar{G}_t^\circ(BPh_4{}^-)$ is the experimentally observable *fact* that $\Delta\bar{G}_t^\circ(Ph_4As^+) = \Delta\bar{G}_t^\circ(Ph_4P^+)$ and that the $\Delta\bar{G}_t^\circ$ values for the tetraphenyl ions are approximated by those of the uncharged tetraphenyl molecules, such as tetraphenylmethane (Ph_4C) and tetraphenylgermane (Ph_4Ge) (Eqns. 5.35).

Earlier evidence in support of the plausibility of the tetraphenylborate assumption was reviewed (5, 15, 35) and was subsequently corroborated by Kim (38), who used calculations based on solvation-energy models. Kim started with the usual formulation of $\Delta\bar{G}_t^\circ(Ph_4As\ BPh_4)$ as a composite of electrostatic and nonelectrostatic contributions (Eqn. 5.34). By determining the molecular volumes of the tetraphenyl molecules and ions he came to the conclusion that Ph_4C and Ph_4Ge were excellent uncharged analogs for the BPh_4^- and the Ph_4As^+ ion, respectively. Consequently, by subtracting the quantity $[\Delta\bar{G}_t^\circ(Ph_4C) + \Delta\bar{G}_t^\circ(Ph_4Ge)]$ from the experimentally obtained $\Delta\bar{G}_t^\circ(Ph_4As\ BPh_4)$, it is possible to evaluate the electrostatic component of the latter. The *difference* between the electrostatic components of the $\Delta\bar{G}_t^\circ$ for the Ph_4As^+ and BPh_4^- *ions*, which have nearly equal radii, is given essentially by the differences in their ion–quadrupole interactions, since all the remaining electrostatic energy terms are insensitive to the sign of ionic charge and hence cancel in the subtraction (Eqn. 2.4). The above differences could be calculated only for the transfers from water to ethanol, methanol, *N*-methylformamide, and acetone, for which the quadrupole moments were known. Thus, for the above transfers, Kim (38) was able to calculate the values of $\Delta\bar{G}_t^\circ(BPh_4^-)$ and $\Delta\bar{G}_t^\circ(Ph_4As^+)$ exactly. In kcal mol^{-1} these values were, respectively, 5.22 and 5.36 for ethanol, 6.13 and 5.77 for methanol, 7.74 and 8.15 for *N*-methylformamide, and 6.85 and 9.90 for acetone. Considering that the estimated error in such calculations is in the range of $\pm(0.2 - 0.3)$ kcal mol^{-1}, the above agreement between the $\Delta\bar{G}_t^\circ$ values of Ph_4As^+ and BPh_4^- is very good for all solvents, except acetone.

For other solvents, for which quadrupole moments were not yet available, Kim recommended to divide the *electrostatic* component of the $\Delta\bar{G}_t^\circ$ equally between the Ph_4As^+ and the BPh_4^- ions. Considering that the transfer free energies of the reference ions are determined primarily by their nonelectrostatic components, the resulting error should be very small.

The tetraphenylborate assumption has been also used to assign the values of $\Delta\bar{H}_t^\circ$ and $\Delta\bar{S}_t^\circ$ for single ions (32).

5.2.2 Interpretation of the Transfer Functions for Single Ions Obtained by the Tetraphenylborate Assumption Once the value for a transfer property between a pair of solvents is known for any single ion, the corresponding transfer properties for other ions can be obtained by difference from the transfer functions for appropriate electroneutral combinations of ions. For example, $\log {}_m\gamma_K = \log {}_m\gamma_{KBPh_4} - \log {}_m\gamma_{BPh_4}$ or $\Delta\bar{G}_t^\circ(I) = \Delta\bar{G}_t^\circ(Ph_4AsI) - \Delta\bar{G}_t^\circ(Ph_4As)$. In this manner, Cox et al. (6, 7) have calculated the values of $\Delta\bar{G}_t^\circ$, $\Delta\bar{H}_t^\circ$, and $\Delta\bar{S}_t^\circ$ for a large number of single ions in several nonaqueous solvents, starting with the corresponding transfer functions for electrolytes, some of which are shown in Tables 5.2–5.4. We list their results for single ions in Table 5.5.†

The transfer functions for ions other than the bulky organic ions display certain regular trends. Thus, for both anions and cations, the transfer from water to nonaqueous solvents is always accompanied by a decrease in the entropy, the values of $298\Delta\bar{S}_t^\circ$ ranging from -2 to -7 kcal mol^{-1}. Since the entropy term contributes to the transfer free energy in the form of $(-298\Delta\bar{S}_t^\circ)$ (at 25°C), a decrease in the entropy of transfer means an increase in the free energy. In the case of cations, this decrease in the entropy is almost invariably opposed by a decrease in the enthalpy of transfer, which can be as large as -13 kcal mol^{-1}. The two functions are not opposed for the transfer of Li^+ to formamide and to propylene carbonate. In the case of anions, the sign of $\Delta\bar{H}_t^\circ$ depends on the ability of the anion to accept hydrogen bonds. Anions which are good hydrogen-bond acceptors and are therefore strongly solvated by water (e.g., F^-, Cl^-) are reluctant to transfer to dipolar aprotic media, which are incapable of stabilizing them by H-bonding. This is why the chloride ion has a large positive enthalpy of transfer in solvents like dimethylformamide and sulfolane. For the transfer of the Cl^- ion from water to methanol or formamide, which are also capable of H-bonding, the enthalpy change is still positive, but much smaller than for the dipolar aprotic solvents. On the other hand, we observe negative (exothermic) enthalpies of transfer for the large anions, which are weak hydrogen-bond acceptors, but are highly polarizable (I^-, ClO_4^-). Once hydrogen bonding between the anion and water is no longer a significant factor in the solvation, anions behave more like cations in the sense that then the contribution of $-298\Delta\bar{S}_t^\circ$ to the transfer free energy is opposed by that of $\Delta\bar{H}_t^\circ$. In other words, large anions prefer dipolar aprotic solvents to water, in the enthalpic sense.

It is interesting to note that the ΔH_t° values for the transfer of the NBu_4^+ ion from water to dipolar aprotic solvents are endothermic to the extent of $+4$ to $+6$ kcal mol^{-1}, while a similarly bulky organic ion Ph_4As^+ has consistently exothermic enthalpies of transfer. The explanation is that while the hydration of both these ions is enthalpically favored because of the previously mentioned increase in the structure of water in the vicinity of hydrophobic solutes, in the case of the more polarizable Ph_4As^+ (and BPh_4^-) ion this effect is overcome by the stronger dispersion interactions between the tetraphenyl ion and the nonaqueous solvent.

Ionic enthalpies of transfer from water to propylene carbonate are depicted as a function of ionic radius in Fig. 4.15.

† The reader will note that in other literature sources, including some of those utilized in Chapter 4, the numerical values for the transfer functions of single ions may differ from those cited here. These discrepancies stem mainly from the differences between the extrathermodynamic assumptions by which the single-ion functions have been estimated. Critical discussions of the extrathermodynamic assumptions are available (5, 15).

Table 5.5 Free Energies, Enthalpies and Entropies for the Transfer of Single Ions from Water to Nonaqueous Solvents at 25°C[a]

	Methanol			Formamide			*N,N*-Dimethylformamide		
Ion	$\Delta\bar{G}_t^\circ$	$\Delta\bar{H}_t^\circ$	$-298\Delta\bar{S}_t^\circ$	$\Delta\bar{G}_t^\circ$	$\Delta\bar{H}_t^\circ$	$-298\Delta\bar{S}_t^\circ$	$\Delta\bar{G}_t^\circ$	$\Delta\bar{H}_t^\circ$	$-298\Delta\bar{S}_t^\circ$
Li^+	0.9	−5.3	6.2	−2.3	−1.3	−1.0	−2.3	−7.7	5.4
Na^+	2.0	−4.9	6.9	−1.9	−3.9	2.0	−2.5	−7.9	5.4
K^+	2.4	−4.4	6.8	−1.5	−4.0	2.5	−2.3	−9.4	7.1
Rb^+	2.4	−3.7	6.1	−1.3	−4.1	2.8	−2.4	−9.0	6.6
Cs^+	2.3	−3.3	5.1	−1.8	−4.1	2.3	−2.2	−8.8	6.6
Ag^+	1.8	−5.0	6.8	−3.7	−5.4	1.7	−4.1	−9.2	5.1
NEt_4^+	0.2	2.2	−2.0	—	—	—	−2.0	−0.2	−1.8
NBu_4^+	−5.2	5.2	−10.4	—	—	—	−6.8	3.6	−10.4
Ph_4As^+	−5.6	−0.4	−5.2	−5.7	−0.1	−5.6	−9.1	−4.7	−4.4
F^-	3.9	3.3	0.6	5.9	5.1	0.8	13.6[b]	—	—
Cl^-	3.0	2.0	1.0	3.3	0.8	2.5	11.0	5.1	5.9
Br^-	2.7	1.1	1.6	2.7	−0.4	3.1	8.5[b]	0.8	6.4
I^-	1.6	−0.5	2.1	1.8	−1.8	3.6	4.5	−3.3	7.8
ClO_4^-	1.4	−0.6	2.0	—	—	—	0.1	−5.4	5.5
N_3^-	2.6	0.1	2.5	—	—	—	8.2	0.3	7.9
SCN^-	1.4	−0.8	2.2	3.4	−2.6	6.0	3.9	−2.4	6.3

	Dimethylsulfoxide			Acetonitrile			Propylene Carbonate			Sulfolane		
	$\Delta\bar{G}_t^\circ$	$\Delta\bar{H}_t^\circ$	$-298\Delta\bar{S}_t^\circ$	$\Delta\bar{G}_t^\circ$	$\Delta\bar{H}_t^\circ$	$-298\Delta\bar{S}_t^\circ$	$\Delta\bar{G}_t^\circ$	$\Delta\bar{H}_t^\circ$	$-298\Delta\bar{S}_t^\circ$	$\Delta\bar{G}_t^\circ$	$\Delta\bar{H}_t^\circ$	$-298\Delta\bar{S}_t^\circ$
Li^+	−3.5	−6.3	2.8	—	—	—	5.7	0.9	4.8	—	—	—
Na^+	−3.3	−6.6	3.3	3.3	−3.1	6.4	3.6	−1.6	5.2	−0.7	−3.6	2.9
K^+	−2.9	−8.3	5.4	1.9	−5.4	7.3	1.4	−5.0	6.4	−1.0	−6.0	5.0
Rb^+	−2.6	−8.0	5.4	1.6	−5.5	7.1	−0.7	−5.6	4.9	−2.1	−6.4	4.3
Cs^+	−3.0	−7.7	4.7	—	—	—	−2.9	−6.2	3.3	−2.4	−5.9	3.5
Ag^+	−8.0	−13.1	5.1	−5.2	−12.6	7.4	3.8	−3.0	6.8	−0.9	−3.2	2.3
NEt_4^+	−3.0	1.0	−4.0	−2.1	−0.3	−1.8	—	—	—	—	—	—
NBu_4^+	—	—	—	−7.9	4.4	−12.3	—	—	—	—	—	—
Ph_4As^+	−8.8	−2.8	−6.0	−7.8	−2.5	−5.3	−8.5	−3.6	−4.9	−8.5	−2.5	−6.0
Cl^-	9.2	4.5	4.7	10.1[b]	—	—	9.0	6.7	2.3	12.6	6.2	6.4
Br^-	6.1	0.8	5.3	7.6	2.0	5.6	7.1	4.2	2.9	9.5	2.9	6.6
I^-	2.2	−3.2	5.4	4.5	−1.7	6.2	4.2	−0.2	4.4	4.9	−2.1	7.0
ClO_4^-	−0.3	−4.6	4.3	—	—	—	—	—	—	—	—	—
N_3^-	6.1	−0.6	6.7	7.3	2.1	5.2	6.9	3.9	3.0	9.5	3.7	5.8

[a] Unless otherwise stated, the data are from Ref. 7, Table 6. Calculated from the data in Tables 5.2–5.4. Molar scale, values in kcal mol^{-1}; Ph_4AsBPh_4 assumption.

[b] Data from I. M. Kolthoff and M. K. Chantooni, Jr., *J. Phys. Chem.*, **76,** 2024 (1972).

So far we have analyzed the interplay between the enthalpies and the entropies of transfer, which elucidates the structural aspects of solvation. However, the function which actually determines how equilibrium and rate properties change with the solvent is the transfer free *energy* and the related transfer activity coefficients.

In Chapter 3, we stressed that the predominance of hydrogen bonding in amphiprotic solvents and the lack of it in dipolar aprotic solvents was at the root of the difference between the solvation phenomena in these two classes of solvents. The relative hydrogen-bonding abilities of water and the nonaqueous liquids are reflected quantitatively in the transfer free energies of the anions. In water and, to a lesser extent, in solvents like methanol and formamide, the solvation of small anions with high charge density, such as F^-, Cl^-, and N_3^-, and of larger anions with localized charge, such as acetate or benzoate, is energetically favored due to formation of hydrogen bonds from the solvent to the anion. The same ions cannot be stabilized in a similar manner by dipolar aprotic solvents, which are incapable of forming hydrogen bonds. For example, solvation of the chloride ion decreases in the order: water > methanol, formamide > propylene carbonate > dimethylsulfoxide, acetonitrile > *N,N*-dimethylformamide > sulfolane. A positive value of $\Delta\bar{G}_t^\circ$ and of log $_m\gamma$ means that the solute is more favorably solvated by the reference medium (here, water) than by the medium to which it is transferred.

As the size of the anions increases, they become weaker H-bond acceptors and their transfer free energies to dipolar aprotic solvents decrease. For example, in DMF the values of $\Delta\bar{G}_t^\circ$ for the series F^-, Cl^-, Br^-, and I^- are 13.6, 11.0, 8.3, and 4.5 kcal mol^{-1}, in that order. The solvation of large polarizable anions (BPh_4^-, I_3^-) is actually favored in dipolar aprotic solvents relative to water because of strong solvent–solute dispersion interactions, so that their transfer free energies are negative. To illustrate the difference between small and large polarizable anions, it is worth noting that the F^- ion is solvated 14 kcal mol^{-1} more favorably in water than in DMF, while the I_3^- ion prefers DMF to water to the extent of 7 kcal mol^{-1}. Similarly, the transfer free energies of the Cl^- ion and the BPh_4^- ion in DMF differ by 20 kcal mol^{-1}. The effect of anion solvation on solubility was illustrated in Table 2.5.

The large increases in the free energy that small anions experience upon transfer from water to dipolar aprotic solvents are responsible for the increased reactivities of these ions in dipolar aprotic media. For example, in acetonitrile, log $_m\gamma_{Cl}$ = 7.3, which means that a solution of Cl^- in acetonitrile has an activity $10^{7.3}$ times greater than a solution of the same chloride-ion concentration in water. When transfer activity coefficients assume such orders of magnitude, it is small wonder that they cause major changes in the rate and equilibrium constants of chemical reactions. Thus large rate increases are observed when reactions of the type

$$Cl^- + CH_3I \rightleftharpoons ClCH_3I^- \longrightarrow \text{Products} \tag{5.36}$$

are transferred from protic to dipolar aprotic solvents. Similarly, the high reactivity of the Cl^- ion in dipolar aprotic media reverses several types of equilibria which in protic solvents lie to the right (39), for example,

$$KCl + ClO_4^- \rightleftharpoons KClO_4 + Cl^- \tag{5.37}$$

$$AgCl_2^- \rightleftharpoons AgCl(s) + Cl^- \tag{5.38}$$

Solvation of small cations is determined generally by the relative basicity of the solvents, also referred to as their donor ability, or donicity. Those dipolar aprotic solvents that contain oxygen atoms with localized negative charges and are therefore relatively basic, solvate cations better than water. Thus the free energies for the transfer of simple cations from water to DMSO and to amides are negative. Dipolar aprotic solvents with weakly basic oxygens (propylene carbonate) or lacking an oxygen atom altogether (acetonitrile) solvate cations rather poorly, as evidenced by the positive values of $\Delta\bar{G}_t^\circ$. An example where donor–acceptor interactions exert a decisive influence is provided by the free energy of transferring the K^+ ion from trifluoroethanol to ethanol. Although these two solvents have very nearly equal dielectric constants and the dipole moment of trifluoroethanol is actually greater than that of ethanol, the $\Delta\bar{G}_t^\circ(K^+)$ value is about -5 kcal mol^{-1} (trifluoroethanol → ethanol), indicating considerably stronger solvation by ethanol. This substantial energy change has been attributed to the greater basicity (donor ability) of ethanol relative to that of trifluoroethanol (2).

By far the largest transfer free energies are observed for those ions which undergo specific interactions with the solvents. The prime example is the acid–base interaction between the proton and basic liquids. Thus it was estimated by Izmaylov (20) that the free energy for the transfer of the proton from water to liquid ammonia is -23 kcal mol^{-1} ($\log {}_m\gamma_H = -17$). When we consider that the corresponding transfer free energies for strong acids are of the order of -25 kcal mol^{-1}, while those for the alkali salts of the same acids are small, it is clear that what dominates the transfer equilibria for the acids is the powerful acid–base interaction between the proton and ammonia.

Equally impressive is the magnitude of the free energy change for the transfer of the Ag^+ ion from water ($\epsilon = 78.5$) to dimethylthioformamide, $HCSN(CH_3)_2$ ($\epsilon = 48$), which amounts to -23.8 kcal mol^{-1} (40). It is instructive to note that the Born energy for the same transfer (Eqn. 5.33a) is $+1$ kcal mol^{-1}. Considering that for a "normal" cation like K^+, the transfer free energy for the same process is $+6.4$ kcal mol^{-1}, the 30 kcal mol^{-1} discrepancy between Ag^+ and K^+ is likely to reflect a powerful specific interaction between the Ag^+ ion and the sulfur atom of dimethylthioformamide (40).

Also illustrative of specific interactions is the magnitude of the free energy for the transfer of the Ag^+ ion from trifluoroethanol (TFE) to hexamethylphosphoramide (HMPA) (39), which amounts to -22 kcal mol^{-1}, so that $\log {}_m\gamma_{Ag} = -16$. TFE and HMPA have similar dielectric constants, so that the large value of $\Delta\bar{G}_t^\circ(Ag^+)$ is a result of both the strong Lewis acid–base interactions between the Ag^+ ion (acid) and the HMPA (base) and the inability of the weakly basic TFE to solvate the Ag^+ ion. The reverse is true for the transfer of the Cl^- ion from TFE to HMPA. The chloride ion is solvated favorably by TFE, which is a strong H-bond donor, but not by the aprotic HMPA, so that $\Delta\bar{G}_t^\circ(Cl^-) = +15.5$ kcal mol^{-1} ($\log {}_m\gamma_{Cl} = 11.4$). Because the contributions of the anion and the cation are opposite in sign for the above transfer process, the transfer activity coefficient of AgCl is relatively small ($\log {}_m\gamma_{AgCl} = 4.6$). It is clear that if one relied exclusively on the information obtainable from thermodynamic transfer functions for AgCl without splitting them into the contributions of the individual ions Ag^+ and Cl^-, the spectacular changes in the solvation of these ions by TFE and HMPA would not be revealed.

Cations having d^{10} electronic configurations, such as Ag^+, Cu^+, and Au^+, are capable of "back-bonding" their *d*-electrons to the molecules of certain solvents, such as the nitriles. The resulting specific solvation is the reason why the $\Delta\bar{G}_t^\circ$ values are negative for the transfer of these ions from water to acetonitrile, which otherwise is a poor solvent of cations. The strong solvation of the Ag^+ ion by acetonitrile has been demonstrated by such independent techniques as polarography, NMR, and Raman spectroscopy (the last is discussed in Chapter 8). Also, the measurements of Hittorf transference numbers in acetonitrile–water mixtures revealed that the Ag^+ ion was preferentially solvated by acetonitrile. Complexes are known to form between Ag^+ and DMF, DMSO, and HMPA, all of which are reflected by the large negative values of the $\Delta\bar{G}_t^\circ(Ag^+)$ in these solvents.

Preferential solvation by acetonitrile is even greater for the Cu^+ ion than for the Ag^+ ion ($\log {}_m\gamma_{Cu^+} = -8.7$, $\log {}_m\gamma_{Ag} = -3.9$). This, coupled with the preferential solvation of the Cu^{2+} ion by water ($\log {}_m\gamma_{Cu^{2+}} = +9.2$ in acetonitrile), has a drastic effect on the redox equilibrium:

$$Ag + Cu^{2+} \rightleftharpoons Ag^+ + Cu^+ \tag{5.39}$$

The equilibrium constant increases by a factor of 10^{22} when the above process is transferred from water to acetonitrile (39). In fact, Parker (41) utilized the above solvent effects to develop an industrial process for the purification of copper (see Chapter 11).

We cited here only a few of the better known examples of specific ion–solvent interactions, as reflected by the values of transfer free energies. It is a general phenomenon, however, that when ions form stable complexes with the solvent molecules, their transfer free energies in these solvents will as a rule be negative. This is the case for the transfer of Ag^+, Cu^{2+}, Zn^{2+}, Cd^{2+}, and Hg^{2+} ions from water to liquid ammonia or hydrazine. According to Izmaylov (22), the values of $\log {}_m\gamma$ in hydrazine for the Ag^+, Zn^{2+}, and Cd^{2+} ions are -13.2, -6.0, and -6.8, in that order. One consequence of the negative transfer free energies is the fact that the potentials of electrodes reversible to such ions will be shifted in the negative direction relative to their potentials in water, when both are expressed on a common emf scale.

5.2.3 Some Applications of the Transfer Activity Coefficients for Single Ions Transfer activity coefficients of the proton are of unique importance in that the values of log ${}_m\gamma_H$ in a given solvent can be used

1 to correlate the pa_H scales in that solvent and in water (Eqn. 5.7a),
2 to refer the emf series in that solvent to the zero point of the standard hydrogen electrode (SHE) in water (Eqn. 5.30)
3 to estimate the liquid-junction potentials at the boundaries of aqueous and nonaqueous solutions in pH cells (Eqn. 5.32).

In Table 5.6, the values of log ${}_m\gamma_H$ are collected for the entire range of ethanol–water solvents, methanol, and a few dipolar aprotic media. Let us illustrate the application of these data to the type of problems just mentioned.

Correlation of Ion-Activity Scales in Different Solvents Consider the correlation of the pa_H scales in water and nonaqueous solvents. For anhydrous ethanol, log ${}_m\gamma_H$ = 1.7, which means that a solution of any given pa_H^* in

Table 5.6 Transfer Activity Coefficients for the Proton in Different Solvents (Molal Scale, 25°C)

Wt. % Ethanol in water	log ${}_m\gamma_H$ [a]
10.0	0.08
20.0	0.06
30.0	−0.06
40.0	−0.32
50.0	−0.62
60.0	−0.80
70.0	−0.86
80.0	−0.67
90.0	−0.46
100	+1.68
Methanol	+1.68[b]
Acetonitrile	+8.0[c]
N,N-Dimethylformamide	−2.5[c]
Dimethylsulfoxide	−3.3[c]

[a] Data from Ref. 36. Estimated as the average from the Ph_4AsBPh_4 and the Ph_4PBPh_4 assumptions.
[b] Data from Ref. 42. Estimated as the average from the Ph_4AsBPh_4 and the Ph_4PBPh_4 assumptions.
[c] Estimated by the Ph_4AsBPh_4 assumption. Data from the compilation by I. M. Kolthoff and M. K. Chantooni, Jr., *J. Phys. Chem.*, **76**, 2024 (1972).

ethanol becomes more acidic (has a lower pa_H) by 1.7 units when expressed on the aqueous pa_H scale (Eqn. 5.7a). The pa_H^* values to which this correlation applies must be referred to the standard state in ethanol. Ordinarily such values are derived from the emf of hydrogen–silver-silver chloride cells without transference measured in ethanol. However, an entirely different calculation applies when the pH in ethanol is measured with glass and calomel electrodes connected by an aqueous KCl bridge and standardized with aqueous buffers. Interpretation of such pH measurements will be discussed later in connection with the estimation of liquid-junction potentials.

Fundamentally, the fact that log ${}_m\gamma_H$ is positive in ethanol means that liquid water is somewhat more basic than liquid ethanol. The reverse is true of DMF and DMSO, where log ${}_m\gamma_H$ is -2.5 and -3.3, respectively. Thus a solution of a given conventional pa_H^* in DMSO would have a higher pa_H (would be more basic) on the aqueous pa_H scale by 3.3 units. In Fig. 5.2 we compare the pa_H^* ranges of several solvents for which values of log ${}_m\gamma_H$

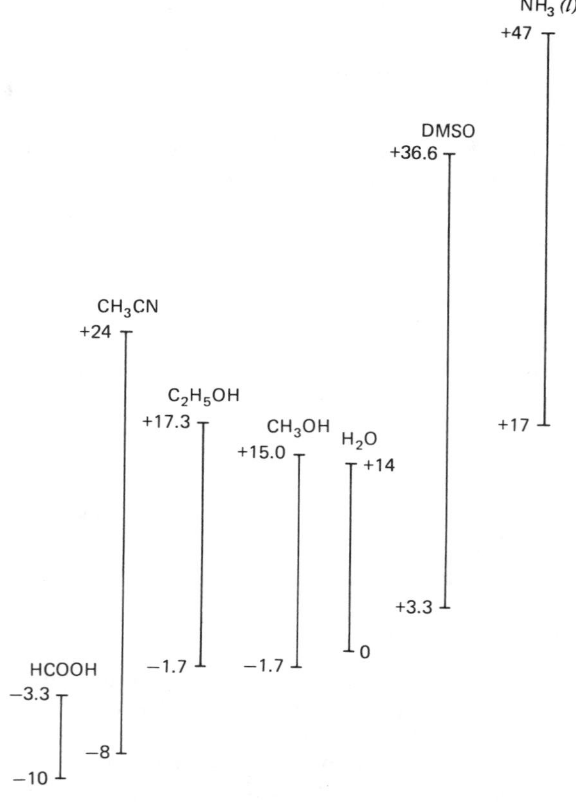

Figure 5.2 A comparison of the pa_H ranges of some liquids on a single aqueous pa_H scale. The estimate of log ${}_m\gamma_H$ for formic acid is 10 (22) and for liquid ammonia, 17 (20).

have been estimated so that they can be expressed on a single aqueous pa_H scale. In each solvent, the conventional pa_H^* scale extends from 0 to pK_s units, where K_s is the autoprotolysis constant. On the aqueous pa_H scale, the same range is equivalent to $-\log {}_m\gamma_H$ to (pK_s − $\log {}_m\gamma_H$) (see Eqn. 5.7a). It is easy to see that the unified pa_H scale makes chemical sense. In liquid ammonia, the pa_H scale extends from +17 to +47 (adopting $pK_s = 30$), which means that a solution of a strong acid in ammonia ($a_H^* = 1$) has a pa_H of +17, being more basic than concentrated alkali solutions in water. The reverse is true of formic acid, where the pa_H scale ranges from −10 to −3.3 units; here, the most basic solutions would still be very acidic in aqueous terms. DMSO starts its pa_H range at +3.3 and is believed to extend to about +37. Strongly basic solutions in that liquid are well known. The most aprotic liquid in Fig. 5.2 is acetonitrile and it does provide a very differentiating range for acidity measurements, extending from −8.0 to +24 pa_H units.

Of course, the activities of ions other than H^+ can be referred to their aqueous standard states in an analogous manner. For example, since $\log {}_m\gamma_{Ag} = -3.9$ in acetonitrile, it follows that a 10^{-3} M solution of the silver ion in acetonitrile would have a pa_{Ag} value on the aqueous scale of $(3.0 + 3.9) = 6.9$. In the absolute sense, the activity of the Ag^+ ions is lowered by acetonitrile because of the powerful complexation by the solvent molecules. Conversely, the activity of a 10^{-5} M solution of chloride ions in acetonitrile, when expressed on the aqueous activity scale, will be increased by a factor of $\log {}_m\gamma_{Cl} = 7.3$, that is, $pa_{Cl} = (5.0 - 7.3) = -2.3$. Of course, this high activity is a consequence of the exceptionally weak solvation of the Cl^- by acetonitrile as compared to its hydration.

Correlation of Emf Series in Different Solvents The correlation of the emf series in water and in other solvents can be achieved by applying Eqn. 5.30 if the relevant $\log {}_m\gamma_H$ value is available. Of course, if the value of $\log {}_m\gamma$ for any single ion is known, $\log {}_m\gamma_H$ can be calculated on its basis from the transfer activity coefficients for a suitable electroneutral combination of ions. When the value of $\log {}_m\gamma_H$ for anhydrous ethanol (1.7) is introduced into Eqn. 5.30, we obtain the value of +0.10 V for the potential of the SHE in ethanol relative to the aqueous SHE. As a result, all conventional electrode potentials in ethanol (i.e., those referred to the arbitrary zero of the SHE in ethanol) now become positive by 0.10 V when referred to the aqueous SHE.

As expected, the potential of the SHE in DMSO ($\log {}_m\gamma_H = -3.3$) and consequently the entire emf series in that solvent is shifted in the negative direction by about 0.19 V when referred to the zero of the aqueous SHE.

Estimation of Liquid-Junction Potentials at the Interfaces of Different Solvents Transfer activity coefficients for the proton also enable us to estimate the liquid-junction potentials which arise at the interfaces of nonaqueous and aqueous solutions in pH cells. Once the liquid-junction potentials are estimated, the interpretation of the pH measurements can be significantly improved.

The familiar pH measurements employing glass and calomel electrode pairs connected by a salt bridge of concentrated aqueous KCl and standardized against aqueous buffers determine so-called "operational pH values." In any medium, the operational pH differs from pa_H by the residual liquid-junction potential, ΔE_j:

$$\text{pH} - pa_H = \frac{\Delta E_j}{0.05916} \quad \text{at } 25°\text{C} \tag{5.40}$$

where pH is the pH-meter reading, pa_H is referred to the aqueous standard state, and ΔE_j is the difference between the liquid-junction potentials encountered in the standardization and in the testing step. The E_j arises at the boundary of the aqueous KCl bridge and the aqueous buffer solutions in the standardization step and between the aqueous KCl and the unknown solution (which may be aqueous or nonaqueous) in the testing step. When both the buffer and the unknown are aqueous, ΔE_j is believed to be generally small, but when the buffer is aqueous and the unknown, nonaqueous, ΔE_j is likely to be appreciable, perhaps of the same order of magnitude as the pH itself. Fortunately, there is evidence that the ΔE_j values between aqueous KCl bridges and dilute nonaqueous buffers are primarily a function of the solvent and are relatively independent of the pH (8, 43–45). Equation 5.40 can be expanded into

$$\text{pH} = pm_H - \log {}_s\gamma_H - \log {}_m\gamma_H + \frac{\Delta E_j}{0.05916} \quad \text{at } 25°\text{C} \tag{5.41}$$

where ${}_s\gamma_H$ is the salt-effect activity coefficient of the hydrogen ion in the nonaqueous medium. Bates et al. (8) have shown that when dilute buffer solutions in ethanol–water and methanol–water solvents are in contact with a salt bridge of 3.5 M aqueous KCl, the quantity $[(\Delta E_j/0.05916) - \log {}_m\gamma_H] \equiv \delta$ is fairly constant for a given solvent medium, independent of the pH. The values of log ${}_m\gamma_H$ for ethanol–water solvents from Table 5.6 are combined with those of the corresponding δ's to calculate the liquid-junction error, ΔE_j, characteristic of these buffer systems.

The results are shown in Table 5.7. Thus, when both δ and log ${}_m\gamma_H$ are available for a given solvent, the pa_H of its solutions, referred to the aqueous standard state, can be evaluated directly from the operational pH readings in the nonaqueous medium (Eqn. 5.41). Values of log ${}_m\gamma$ for ions other than the proton enable us to estimate the liquid-junction potentials in a variety of potentiometric and polarographic cells where an interface exists between two different solvent media (cf. cell 5-III and Eqns. 5.31 and 5.32).

Although we have generally focused here on those transfer activity coefficients that represent major changes in solvation energies, it seems that the values of log ${}_m\gamma_H$ estimated by the tetraphenylborate assumption may be able to reflect subtle structural aspects of solvation as well. Transfer activity

Table 5.7 Liquid-Junction Potentials Between 3.5 *M* Aqueous KCl and Buffer Solutions in Ethanol–Water Solvents at 25°C

Wt % Ethanol in water	δ^a (8)	log ${}_{m}\gamma_{H}$ [b]	ΔE_j pH units	ΔE_j mV
0	0	0	0	0
20	0.02	0.06	0.08	+5
35	0.10	−0.19	−0.09	−5
50	0.21	−0.62	−0.41	−24
65	0.24	−0.83	−0.59	−35
80	0.11	−0.67	−0.56	−33
90	−0.40	−0.46	−0.86	−51
100	−2.91	+1.68	−1.23	−73

[a] $\delta \equiv [(\Delta E_j/0.05916) - \log {}_{m}\gamma_{H}]$ (pH units at 25°C).
[b] Data from Table 5.6, interpolated when necessary.

coefficients for the proton are depicted as a function of solvent composition for the ethanol–water system in Fig. 5.3. As the alcohol is being added to pure water, there is initially a delay in the rate of increase of log ${}_{m}\gamma_{H}$, probably as a result of the reinforcement of the water structure, which typically occurs when small amounts of a hydrophobic substance (here, the alkyl groups of the alcohol) are added to water. Subsequently, most ethanol–water mixtures are characterized by negative values of log ${}_{m}\gamma_{H}$, which pass through a minimum at a composition of about 70 wt % ethanol. This means that most ethanol–water mixtures are more basic than either of the pure liquids, while the minimum corresponds to the solvent mixture of maximum basicity. At very high ethanol contents, the log ${}_{m}\gamma_{H}$ curve rises steeply, presumably signaling the complete conversion of H_3O^+ to $C_2H_5OH_2{}^+$.

Braude and Stern (46) observed a similar variation of the Hammett acidity function, $-H_o$, in aqueous mixtures of ethanol, dioxane, and acetone. In fact, their data for ethanol–water solvents correspond closely to the values of log ${}_{m}\gamma_{H}$ obtained by the tetraphenylborate assumption (14). Braude and Stern reasoned that the basicity of liquid water is relatively low because the basic oxygen atoms in the water molecules are tied up in the extensive intermolecular hydrogen-bonded network. Addition of an organic liquid to water causes a progressive breakdown of the water structure, thus exposing an increasing number of the oxygen atoms for bonding with solute acids and increasing the basicity of the medium. Presumably, maximum breakdown (depolymerization) of the water structure corresponds to the minimum of log ${}_{m}\gamma_{H}$ (and in $-H_o$), which occurs for the solvent mixture of maximum basicity. Beyond the minimum, alcohol–water complexes and eventually the pure alcohol prevail, with an accompanying drop in basicity. As we have seen before, liquid ethanol is less basic than liquid water.

It may be significant that the ΔpK values ($\Delta pK \equiv p_sK - p_wK$) for the primary ammonium acids in ethanol–water and methanol–water solvents

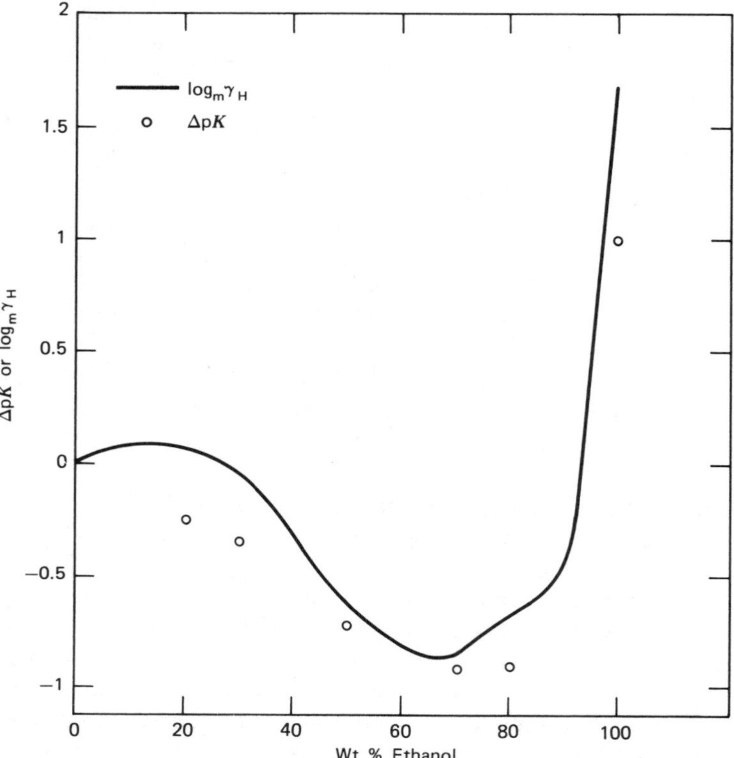

Figure 5.3 Transfer activity coefficients for the proton, log ${}_m\gamma_H$, and ΔpK's of primary ammonium acids in ethanol–water solvents (molal scale, 25°C). Reprinted with permission from *Anal. Chem.*, **46,** 2009 (1974). Copyright 1974. American Chemical Society.

are remarkably constant for each medium, regardless of the size or the nature of the organic residue attached to the ammonium ion (35). Evidently, these changes in pK's reflect some fundamental property of the solutions which is independent of the nature of the acid. Indeed, the near overlap of the ΔpK's and log ${}_m\gamma_H$ function in Fig. 5.3 suggests that the fundamental property is none other than the transfer activity coefficient of the proton (35).

SUMMARY

The changes in the equilibrium and rate constants observed upon the transfer of a chemical reaction from one solvent to another are a function of the transfer free energies, or the transfer activity coefficients (medium effects), of the reactants and the products. Transfer free energies for uncharged substances and electroneutral combinations of ions (including electrolytes) can be determined from solubility, emf, and vapor pressure data. For common electrolytes, the free energies for the transfer from water to nonaqueous media are generally positive. They are negative for electrolytes containing large polarizable organic ions or ions that interact specifically with the molecules of the nonaqueous solvent. Enthalpies and entropies of transfer provide information on the structure of solvents and their solutions. The transfer of common electrolytes from water to nonaqueous media involves a favorable entropy change, but an unfavorable enthalpy change. Tables of the transfer functions for many electrolytes are provided.

Transfer activity coefficients for single ions enable one to express ion-activity scales, such as the pa_H scale, and emf series in a manner independent of the solvent, and to estimate the liquid-junction potentials at the interfaces of electrolyte solutions in different solvent media. Extrathermodynamic assumptions for the estimation of the transfer activity coefficients of single ions are briefly reviewed. Many values of the transfer free energies, enthalpies, and entropies for single ions estimated by the tetraphenylborate assumption are tabulated and interpreted. Transfer activity coefficients of the proton are applied to the interpretation of pa_H scales, emf series, and liquid-junction potentials in selected nonaqueous media.

LITERATURE CITED

1 Waghorne, W. E., Ph.D. Thesis, Australian National University, Canberra, 1973.

2 Parker, A. J., *Electrochim. Acta,* **21,** 671 (1976).

3 Parker, A. J., *Quart. Rev.,* **16,** 163 (1962).

4 Strehlow, H., "Electrode Potentials in Non-Aqueous Solvents," in J. J. Lagowski, Ed., *The Chemistry of Non-Aqueous Solvents,* Vol. I, Academic Press, New York, 1966, Chap. 4.

5 Popovych, O., "Transfer Activity Coefficients (Medium Effects)," in I. M. Kolthoff and P. J. Elving, Eds., *Treatise on Analytical Chemistry,* 2nd ed., Part I, Vol. I, Chap. 12, Wiley, New York, 1978.

6 Cox, B. G., *Annu. Rep. Prog. Chem., Sect. A., Phys. Inorg. Chem.,* **70,** 249 (1974).

7 Cox, B. G., G. R. Hedwig, A. J. Parker, and D. W. Watts, *Aust. J. Chem.,* **27,** 477 (1974).

8 Bates, R. G., M. Paabo, and R. A. Robinson, *J. Phys. Chem.,* **67,** 1833 (1963).

9 Alexander R., and A. J. Parker, *J. Amer. Chem. Soc.,* **89,** 5549 (1967).

10 Alexander, R., A. J. Parker, J. H. Sharp, and W. E. Waghorne, *J. Amer. Chem. Soc.,* **94,** 1148 (1972).

11 Bates, R. G., *Determination of pH,* Wiley, New York, 1964, Chap. 7, 8.

12 Bates, R. G., "Acidity Functions for Amphiprotic Media," in J. J. Lagowski, Ed., *The Chemistry of Non-Aqueous Solvents,* Vol. I, Academic Press, New York, 1966, Chap. 3.

13 Bates, R. G., "Medium Effects and pH in Non-Aqueous Solvents," in J. F. Coetzee and C. D. Ritichie, Eds., *Solute–Solvent Interactions,* Vol. I, Mercel Dekker, New York, 1969, Chap. 2.

14 Popovych, O., and A. J. Dill, *Anal. Chem.,* **41,** 456 (1969).

15 Popovych, O., *Crit. Rev. Anal. Chem.,* **1,** 73 (1970).

16 Koepp, H. M., H. Wendt, and H. Strehlow, *Z. Elektrochem,* **64,** 483 (1960).

17 Coetzee, J. F., and J. J. Campion, *J. Amer. Chem. Soc.,* **89,** 2513, 2517 (1967).

18 Coetzee, J. F., J. M. Simon, and R. J. Bertozzi, *Anal. Chem.,* **41,** 766 (1969).

19 Bjerrum, N., and E. Larsson, *Z. Physik. Chem.,* **127,** 358 (1927).

20 Izmaylov, N. A., *Dokl. Akad. Nauk SSSR,* **126,** 1033 (1959).

21 Izmaylov, N. A., *Zhur. Fiz. Khim.,* **34,** 2414 (1960).

22 Izmaylov, N. A., *Dokl. Akad. Nauk SSSR,* **149,** 884, 1103, 1364 (1963).

23 Andrews, A. L., H. P. Bennetto, D. Feakins, K. G. Lawrence, and R. P. T. Tomkins, *J. Chem. Soc., A.,* 1486 (1968).

24 Bennetto, H. P., D. Feakins, and D. J. Turner, *J. Chem. Soc., A.,* 1211 (1966).

25 Feakins, D., and P. Watson, *J. Chem. Soc.,* 4734 (1963).

26 Alfenaar, M., and C. L. DeLigny, *Rec. Trav. Chim.,* **86,** 929 (1967).

27 DeLigny, C. L., and M. Alfenaar, *Rec. Trav. Chim.,* **84,** 81 (1965).

28 Bax, D., C. L. DeLigny, and M. Alfenaar, *Rec. Trav. Chim.,* **91,** 452, 1225 (1972).

29 Bax, D., C. L. DeLigny, and A. G. Remijnse, *Rec. Trav. Chim.,* **92,** 374 (1973).

30 DeLigny, C. L., H. J. M. Denessen, and M. Alfenaar, *Rec. Trav. Chim.,* **90,** 1265 (1971).

31 Pleskov, V. A., *Usp. Khim.,* **16,** 254 (1947).

32 Cox, B. G., and A. J. Parker, *J. Amer. Chem. Soc.,* **95,** 402, 6879 (1973).

33 Kolthoff, I. M., and M. K. Chantooni, Jr., *J. Phys. Chem.,* **76,** 2024 (1972); *Anal. Chem.,* **44,** 194 (1972); *J. Amer. Chem. Soc.,* **93,** 7104 (1971).

34 Parker, A. J., and R. Alexander, *J. Amer. Chem. Soc.,* **90,** 3313 (1968).

35 Popovych, O., *Anal. Chem.,* **46,** 2009 (1974).

36 Popovych, O., A. Gibofsky, and D. H. Berne, *Anal. Chem.,* **44,** 811 (1972).

37 Grunwald, E., G. Baughman, and G. Kohnstam, *J. Amer. Chem. Soc.,* **82,** 5801 (1960).

38 Kim, J. M., *J. Phys. Chem.,* **82,** 191 (1978).

39 Parker, A. J., *Pure Appl. Chem.,* **25,** 345 (1971).

40 Owensby, D. A., A. J. Parker, and J. W. Diggle, *J. Amer. Chem. Soc.,* **96,** 2682 (1974).

41 Parker, A. J., *Proc. Roy. Austral. Chem. Inst.,* **39,** 163 (1972).

42 Popovych, O., unpublished results (1979).

43 DeLigny, C. L., and M. Rehbach, *Rec. Trav. Chim.,* **79,** 727 (1960).

44 Douheret, G., *Bull. Soc. Chim. France,* 1412 (1967).

45 Gelsema, W. J., C. L. DeLigny, A. G. Remijnse, and H. A. Blijleven, *Rec. Trav. Chim.,* **85,** 647 (1966).

46 Braude, E. A., and E. S. Stern, *J. Chem. Soc.,* 1976 (1948).

CHAPTER SIX

Acid–Base Chemistry

Probably no branch of chemistry has been as closely associated with the growing use of nonaqueous solvents, as acid–base chemistry. In the realm of practical applications, nonaqueous solvents have expanded tremendously the range of acidic and basic strengths that can be measured and differentiated. We recall from Chapter 3 that in water it is impossible to titrate acids and bases having ionic dissociation constants lower than about 10^{-9}. In certain nonaqueous media, on the other hand, it is possible to obtain well-defined titration curves even for such weakly basic substances as amides and aromatic amines and such weakly acidic substances as phenols and enols. Furthermore, we recall that water is a leveling solvent, so that, for example, a mixture of hydrochloric and perchloric acids cannot be resolved when titrated with a base in aqueous solution. However, such resolution can be achieved in methyl isobutyl ketone (Fig. 6.5). On a fundamental level, studies in solutions other than aqueous solutions promoted the adoption of precise and generally applicable definitions of acids and bases, elucidated the role of the solvent in acid–base reactions, and helped provide a proper perspective for the interpretation of a variety of acid–base reactions in solution.

In this chapter we first trace the historical evolution of the Brønsted concept of acidity and devote a separate section to the concept and the chemistry of Lewis acids. Next we discuss the various scales for measuring the strength of acid and base solutions, such as the electrometric pH scales and the acidity functions based on indicators. This is followed by a discussion of acid–base titrations in nonaqueous media and the detailed treatment of acid–base equilibria in acetic acid and acetonitrile.

6.1 CONCEPTS OF ACIDITY

6.1.1 Arrhenius Theory The traditional concept of acids and bases that was generally accepted in chemistry at least through the first quarter of the twentieth century originated in an era when the only solvent of importance in the chemistry of solutions was water. Faraday's discoveries that electrolytic conductance in aqueous solution was due to charged particles (ions) and his characterization of acids, bases, and salts as electrolytes paved the way for the *ionization theory*, or *water theory*, of acids and bases. Proposed by Arrhenius in 1884, this theory defined an acid as a substance that in aqueous solution dissociates into hydrogen ions and a base, as a substance that in aqueous solution dissociates into hydroxyl ions. Neutralization was the reaction of an acid and a base to form a salt and water.

The combination of this definition based on ions with the recognition that electrolytic conductance is due to the presence of ions in solution led to quantitative expressions of acidity in the form of so-called ionization constants determined conductometrically. Ostwald's dilution law, stating that the degree of dissociation of an electrolyte, α, is equal to the ratio of the equivalent conductance at the given concentration, Λ, and at infinite dilution, Λ_o, [$\alpha = \Lambda/\Lambda_o$], was introduced into expressions for the law of mass action and thus used to determine many "ionization constants," $K_d = C\alpha^2(1 - \alpha)^{-1}$, which should be less ambiguously called "ionic dissociation constants." In this manner, values were obtained for the ionic dissociation constants of many acids and bases and an apparently satisfactory interpretation was achieved for the dissociation, hydrolysis, buffer, and indicator equilibria in aqueous solutions.

However, the Arrhenius concept had several limitations: it defined acids and bases only in terms of their solutions and, moreover, in terms of their solutions in one specific solvent, water. Water, as we know, possesses both acid and base properties of its own as well as a high dielectric constant. As a result, the intrinsic acidity and basicity of solutes is modified by water and then manifested in the degree of ionic dissociation. This behavior, however, is not typical of solvents in general. The Arrhenius definition also ran into a problem with bases, some of which do contain hydroxyl ions in their structures (e.g., metal hydroxides), while others (amines) produce hydroxyl ions only upon dissolution in water. The conversion to hydroxyl ions is incomplete, which brought up a general question as to whether only the dissociated fraction of the solute constituted the acid or the base. For example, is pure acetic acid 100% acid by itself, or is it only the 1–2% of it that dissociate into hydronium ions in aqueous solution that should be regarded as acid? Is a covalent liquid like HCl an acid in its pure form, or does it become an acid only upon reaction with water? Furthermore, the Arrhenius concept promoted the erroneous idea that ionic dissociation was indispensable to manifestations of acidity and that, therefore, acid–base reactions could not take place in the majority of nonaqueous solvents and in the absence of solvent.

Gradually, enough data had accumulated to disprove any necessary connection between acid–base reactivity and ionic dissociation. Particularly outstanding in this regard were the contributions of Hantzsch and his associates (1) in the 1920s, who demonstrated that acid–base reactions, including neutralization, indicator-color changes, and catalysis, could take place in aprotic solvents, such as benzene, chloroform, and ether, where no appreciable concentration of ions could be detected. Hantzsch was of the opinion that the extent of an acid–base reaction as manifested by salt formation or the rate of an acid-catalyzed reaction constituted better indices of acidity than electrometric determinations of hydrogen-ion concentration. Thus, chemists were compelled to reorient and redefine their thinking on the subject of acids and bases, and the time became ripe for a new concept of acidity.

6.1.2 The Brønsted Concept The most widely accepted concept of acids and bases, and one that has elucidated the role of the solvent in acid–base reactions, is that proposed by Brønsted in 1923 (2).† An acid is defined as

† Many sources state that the same concept was proposed independently also by Lowry and refer to it as the "Brønsted–Lowry concept," but according to Bell (3), Brønsted deserves sole credit for it.

a proton donor and a base, as a proton acceptor. There is no dependence on solvents or on ionic dissociation.

$$\text{Acid} \rightleftarrows \text{Proton} + \text{Base} \tag{6.1a}$$

$$HA \rightleftarrows H^+ + A^- \tag{6.1b}$$

Thus an acid is a substance that can split off a proton to form its *conjugate base*, while their recombination forms the *conjugate acid* of that base. The intrinsic acidity of acid HA is expressed as the equilibrium constant for reaction (6.1b) and the intrinsic basicity of base A^-, corresponding to the reverse of reaction (6.1b), will be expressed by the reciprocal of that constant:

$$K_a = \frac{(H^+)(A^-)}{(HA)} \qquad K_b = \frac{1}{K_a} \tag{6.2}$$

It should be stressed that the symbol H^+ represents here the unsolvated proton and not the "hydrogen ion," which is a solvated species whose nature usually varies with the medium. Some examples of conjugate acid–base pairs are given below.

Acid	Proton	Conjugate Base
$HClO_4$	H^+	ClO_4^-
H_3O^+	H^+	H_2O
HCO_3^-	H^+	CO_3^{2-}
C_2H_5OH	H^+	$C_2H_5O^-$
NH_4^+	H^+	NH_3

It is clear that a Brønsted acid can be an uncharged molecule, a positive ion, or a negative ion.

Since free protons do not exist in solution in detectable concentrations, reactions of the type expressed by Eqn. 6.1b cannot be realized experimentally, so that the intrinsic acidity or basicity of a substance cannot be measured. This fact is of fundamental importance in understanding acid–base measurements. The only way in which we can determine acid or base strength is by reacting an acid with a reference base and vice versa, by reacting a base with a reference acid:

$$HA \rightleftarrows H^+ + A^- \qquad K_a = \frac{(H^+)(A^-)}{(HA)} \tag{6.3}$$

$$B + H^+ \rightleftarrows BH^+ \qquad K_b = \frac{(BH^+)}{(H^+)(B)} \tag{6.4}$$

$$HA + B \rightleftarrows BH^+ + A^- \qquad K = \frac{(BH^+)(A^-)}{(HA)(B)} = K_aK_b \tag{6.5}$$

Thus the measureable constant for an acid–base reaction, K, is the product of the intrinsic acidity constant of an acid HA and the intrinsic basicity constant of base B. We can determine acid strength only by the extent of an acid–base reaction, or neutralization, in which a proton is transferred from an acid HA to a base B to form the new weaker acid–base pair BH^+ and A^-. For the reaction to proceed from left to right as written, base B must be stronger than the conjugate base of acid HA, i.e., it must have the greater affinity for protons. Every acid–base reaction can be viewed as a sum of two half-reactions, one expressing the intrinsic acidity of acid HA and the other, the intrinsic basicity of base B. Neither half-reaction can be evaluated separately, only their sum. Neither constant K_a, nor K_b can be evaluated individually, only their product. Thus, all acid–base reactions involve the transfer of protons between two conjugate acid–base pairs as represented by Eqn. 6.5.

This is entirely analogous to oxidation–reduction reactions, which instead of proton transfer involve electron transfer. Here also the net reaction consists of two half-reactions, neither of which can be realized independently. A half-reaction $\text{Red}_1 \rightleftarrows \text{Ox}_1 + e$ cannot take place to a measureable extent, unless there is another oxidizing agent Ox_2 with a greater affinity for electrons than Ox_1 that will accept them. We always measure the overall reaction, which is a sum of two half-reactions: $\text{Red}_1 + \text{Ox}_2 \rightleftarrows \text{Ox}_1 + \text{Red}_2$, where Ox and Red stand for the oxidized and the reduced forms of the given species, respectively.

Acidic and basic strengths are expressed quantitatively by the equilibrium constants represented by Eqn. 6.5. In the determination of the strength of an acid, the reference base B is either an indicator, in which case the extent of proton transfer is measured spectrophotometrically, or the solvent, in which case electrometric measurements are employed. The case where the solvent acts as the reference base or acid is both common and important theoretically. Here, the acid–base reaction used to measure the strength of an acid in the solvent S and the necessary equilibrium constants are

$$\text{HA} \rightleftarrows \text{H}^+ + \text{A}^- \qquad K_a(\text{HA}) \tag{6.6a}$$

$$\text{H}^+ + \text{S} \rightleftarrows \text{SH}^+ \qquad K_b(\text{S}) \tag{6.6b}$$

$$\text{HA} + \text{S} \rightleftarrows \text{SH}^+ + \text{A}^- \qquad K = K_a(\text{HA})K_b(\text{S}) \tag{6.6c}$$

Although the intrinsic acidity constants of acids are indeterminable individually, the ratio of the intrinsic strengths of two acids in a given solvent is given by the ratio of their experimentally determined ionic dissociation constants, because the $K_b(\text{S})$ drops out.

Of course, not all solvents have high enough proton affinities to act as reference bases. The classification of solvents based on their proton-transfer behavior is discussed in Chapter 3. We recall that there the solvents were divided into two major categories, amphiprotic and aprotic.

The Brønsted concept of acidity clarified greatly the role of amphiprotic solvents in acid–base reactions. Reactions which used to fall into such apparently diverse categories as ionization of acids and bases, self-dissociation of solvents, hydrolysis, neutralization, dissolution by acids, and displacement were given a unified interpretation. All of these are proton-transfer, or *protolysis*, reactions obeying the fundamental scheme represented by Eqn. 6.5.

Thus the self-dissociation, or self-ionization, of amphiprotic liquids was explained as *autoprotolysis*, or self-protonation, which is an acid–base reaction between two identical solvent molecules, in which one acts as the acid and the other, as the base:

Autoprotolysis of Solvents

$$2H_2O \rightleftarrows H_3O^+ + OH^-$$

$$2C_2H_5OH \rightleftarrows C_2H_5OH_2^+ + OC_2H_5^-$$

$$2HAc \rightleftarrows H_2Ac^+ + Ac^-$$

$$2NH_3 \rightleftarrows NH_4^+ + NH_2^-$$

The "ionization," or ionic dissociation, of acids and bases in solution was recognized as reaction with the solvent involving the transfer of protons:

Ionic Dissociation of Acids and Bases

Solute Solvent

$$HAc + H_2O \rightleftarrows H_3O^+ + Ac^-$$

$$HCl + NH_3 \rightleftarrows NH_4^+ + Cl^-$$

$$CH_3NH_2 + H_2O \rightleftarrows CH_3NH_3^+ + OH^-$$

$$HClO_4 + CH_3OH \rightleftarrows CH_3OH_2^+ + ClO_4^-$$

So-called hydrolysis of salts is in no way different from the proton-transfer reactions of acids and bases exemplified above:

Hydrolysis

$$Ac^- + H_2O \rightleftarrows HAc + OH^-$$

$$CO_3^{2-} + H_2O \rightleftarrows HCO_3^- + OH^-$$

$$NH_4^+ + H_2O \rightleftarrows H_3O^+ + NH_3$$

$$Pb(H_2O)_4^{2+} + H_2O \rightleftarrows H_3O^+ + Pb(OH)(H_2O)_3^+$$

Neutralization

$H_3O^+ + OH^- \rightleftarrows 2H_2O$ (strong acid–strong base in water)

$HCN + NH_3 \rightleftarrows NH_4^+ + CN^-$ (in water)

$KNH_2 + NH_4Cl \rightleftarrows 2NH_3 + KCl$ (in liquid NH_3)

$HClO_4 + C_5H_5N \rightleftarrows C_5H_5NH^+ClO_4^-$ (in glacial HAc)

Note that protolysis leads primarily to *ionization*, that is, the formation of ions from covalent reactants, but the ions dissociate to an appreciable degree only in solvents of high dielectric constant.

Dissolution

$$CuS(s) + 2H_3O^+ \rightleftarrows H_2S + Cu^{2+}$$

$$Ag_2CrO_4(s) + H_3O^+ \rightleftarrows HCrO_4^- + H_2O + 2Ag^+$$

Displacement

$$NH_4^+ + OH^- \rightleftarrows H_2O + NH_3$$

$$HCl + CN^- \rightleftarrows HCN + Cl^-$$

6.1.3 The Lewis Concept The fundamentals of the "electronic theory" of acidity were originally stated in 1923 by G. N. Lewis (4), but were relatively unknown until they were revived later by Lewis and his followers, especially in 1946 by Luder and Zuffanti (5). The Lewis concept extends the definition of acids and bases beyond those of the Brønsted concept both from the theoretical and experimental viewpoints. According to the Lewis theory, a base is defined as an electron-pair donor and an acid, as an electron-pair acceptor. Neutralization is the formation of a coordinate covalent bond between acid and base. Ionization and dissociation may or may not follow the initial formation of an acid–base adduct. Neither protons nor solvents are involved in the definition. From the experimental viewpoint, the Lewis concept is expressed in terms of the four "phenomenological criteria" to which acids and bases should conform:

1 Rapid neutralization (no activation energy).
2 Displacement of weaker acids and bases by their stronger counterparts.
3 Acid–base titrations involving indicator-color changes.
4 Catalytic action.

Lewis bases are the same as Brønsted bases, since species that contain an electron pair that can coordinate with any electron-pair acceptor can also accept a proton, which is a Lewis acid. However, the definition of acids became extended (strictly speaking, *restricted*) to such nonprotonic substances as BF_3, SO_3, $AlCl_3$, $SnCl_4$, Ag^+, and Cu^{2+}. None of these are acids in the Brønsted sense because they contain no hydrogens.

Typical Lewis acid–base reactions would be as follows:

Acid	**Base**		**Acid–Base Adduct**	
SO_3	+ CaO	$\rightleftharpoons$	$CaO \rightarrow SO_3$	$(CaSO_4)$
$SnCl_4$	+ $2Cl^-$	$\rightleftharpoons$	$\left[\begin{matrix} Cl & \\ & \searrow \\ & & SnCl_4 \\ & \nearrow \\ Cl & \end{matrix}\right]^{2-}$	
Cu^{2+}	+ $4NH_3$	$\rightleftharpoons$	$Cu \leftarrow (NH_3)_4{}^{2+}$	
BF_3	+ NH_3	$\rightleftharpoons$	$F_3B \leftarrow NH_3$	

The reaction between Lewis acids and amphiprotic solvents, such as water, leads to the formation of Brønsted acids:

$$SO_3 + H_2O \rightleftharpoons H_2SO_4$$

$$Al^{3+} + 6H_2O \rightleftharpoons Al(H_2O)_6{}^{3+}$$

It is clear that all anhydrides of Brønsted acids are Lewis acids.

While the Lewis concept is useful in providing the basis for a systematic interpretation of coordination equilibria, any idea that it would be in a position to replace the Brønsted concept would seem ill advised. In that connection it may be worthwhile to recount the main criticisms that have been directed against the Lewis concept. One of its weaknesses is that the classical acids, such as HCl and $HClO_4$, have no electron deficiencies and can be included in the Lewis definition only with difficulty. To include protonic acids in the Lewis scheme one has to draw a coordinate covalent bond from the hydrogen of the acid to the base, although a hydrogen atom is not supposed to be associated with more than two electrons at a time. The general criticism is that to include any type of coordination reaction in the definition of an acid–base reaction is to extend the concept to all kinds of unrelated reactions. Furthermore, it leads to the lack of any uniform scale of acid or base strength.

The strengths of acids and bases are dependent on the reaction chosen and their relative strengths (referred to a single substance) are rather specific in nature. According to Bell (3), the Lewis definition does not represent an extension or generalization of the older concepts, but the use of the word "acid" in a fundamentally different sense. This is the main reason why the concept of protonic acids deserves separate treatment.

The greatest justification for a separate treatment of the Brønsted-type acidity lies in its superiority when quantitative relationships are involved. For a given solvent and temperature, the knowledge of one constant for a given conjugate acid–base pair is sufficient for a quantitative determination of the position of its equilibrium in dilute solution (in media of reasonably high dielectric constant). The knowledge of hydrogen-ion concentration (approximated by pH) in conjunction with known equilibrium constants and the analytical concentration enables us to calculate the equilibrium concentrations of all acidic and basic species in the most complex systems. No such quantitative relationships are expected to hold for Lewis acids and bases. Here the *relative* strengths of acids and bases depend on the reference substance chosen. For example, ammonia is a much weaker base than the OH^- ion in their reaction with the proton, but when referred to the Lewis acid Ag^+, the order is reversed, since AgOH is a strong electrolyte, while $Ag(NH_3)_2{}^+$ is a stable complex. In other words, in the Brønsted scheme, the proton is really the reference acid, so that when ammonia reacts with *any* acid, the reaction product is always the same—the ammonium ion. But when Lewis acids react with ammonia, the product will be different in each case. This is the reason why no uniform scale of acidity is possible for Lewis acids. Kolthoff (6) prefers to call Lewis acid–base reactions "coordinations," rather than "neutralizations."

Some Quantitative Aspects Lewis acid–base reactions in their simplest form are governed by the relationship:

$$A + B \rightleftharpoons AB \tag{6.7}$$

$$K = \frac{(AB)}{(A)(B)} \tag{6.8}$$

where A, B, and AB are the acid, the base, and their coordinated adduct, respectively. Of course, adducts can form also at acid–base mole ratios other than unity. There is an interesting formal analogy between the primary Lewis acid–base reactions and the reaction between a Brønsted acid and a base in aprotic media (cf. Eqns. 3.66 and 3.67). In both types of reactions, the primary process is the formation of an adduct AB. The adduct may dissociate further into ions, the extent of any such dissociation depending on the nature of the acid and the base as well as on the dielectric constant of the solvent and its ability to solvate the resulting ions. When such ionic dissociation does occur, the reactions can be followed conductometrically, since electrolytic conductance measures the concentration of free ions and not of the associated species (see Chapter 7).

One example of this type is the conductometric titration of aluminum bromide and aluminum chloride (acids) with dimethyl ether, methanol, and acetone (bases) in nitrobenzene as the solvent (7). The conductance at first decreases to a minimum at the 1:1 acid–base mole ratio, presumably due to the formation of the associated acid–base adduct, and then increases until a maximum is reached at a ratio of 1 mole of acid to 2 moles of base, at which point dissociation into ions such as $[AlBr_2(CH_3)_2O]^+$ and Br^- is believed to be at its maximum (7). In a conductometric titration of titanium tetrachloride (acid) with the chloride ion (base) added in the form of tetraethylammonium chloride in phosphorus oxychloride ($POCl_3$) as the solvent, a conductance maximum occurs at a 1:1 mole ratio of the reactants, corresponding to the formation of a soluble $TiCl_5^-$ complex, followed by a minimum at the 2:1 ratio of base to acid, corresponding to the precipitation of $[(C_2H_5)_4N]_2(TiCl_6)$(8). Similarly, in thionyl chloride ($SOCl_2$), it is possible to titrate tetraethylammonium chloride with stannic chloride conductometrically until the precipitate $[(C_2H_5)_4N]_2(SnCl_6)$ forms at the 2:1 mole ratio of $(C_2H_5)_4NCl$ to $SnCl_4$, giving the first break in the titration curve, and then continue titrating until the precipitate redissolves quantitatively at the 2:2 ratio, producing a second break (9). Iodine monochloride, ICl, acts as a monobasic acid to form a 1:1 product with pyridine in nitrobenzene, a process which can be followed by photometric titration (10). Thermometric titrations were carried out between stannic chloride (a dibasic acid) and dioxane (a diacid base) to a 1:1 endpoint in aprotic solvents (11).

Lewis acids and bases also react with amphoteric aprotic solvents, those which dissociate into solvent cations (lyonium ions) and solvent anions (lyate ions), but do not involve proton transfer. We already discussed the self-dissociation of such solvents in Chapter 3, from which we repeat Eqn. 3.12:

$$nS \rightleftharpoons s^+ + s^- \tag{6.9}$$

The lyonium and lyate ions represent, respectively, the strongest acid and base that can exist in a given amphoteric aprotic solvent. Any solute that increases the lyonium concentration is an acid and any solute that increases the lyate concentration is a base. The analogy with amphiprotic solvents is obvious. Examples of amphoteric solvents are $COCl_2$, $POCl_3$, BrF_3, etc. and their ionization schemes were shown in Eqns. 3.13 in Chapter 3. Lewis acids and bases will react with such solvents as follows (6):

$$2S \rightleftharpoons s^+ + s^- \tag{6.10}$$

(acids)

$$A + s^- \rightleftharpoons As^- \tag{6.11}$$

$$A + 2S \rightleftharpoons As^- + s^+ \tag{6.12}$$

$$2S \rightleftharpoons s^+ + s^- \tag{6.10}$$

(bases)

$$B + s^+ \rightleftharpoons Bs^+ \tag{6.13}$$

$$B + 2S \rightleftharpoons Bs^+ + s^- \tag{6.14}$$

A specific example of reactions 6.10–6.12 is that between aluminum chloride (acid) and the chloride ion (base) in phosgene:

$$COCl_2 \rightleftarrows COCl^+ + Cl^- \qquad (6.15a)$$

$$AlCl_3 + Cl^- \rightleftarrows AlCl_4^- \qquad (6.15b)$$

$$COCl_2 + AlCl_3 \rightleftarrows COCl^+ + AlCl_4^- \qquad (6.15c)$$

The chloride ion was in this case supplied by the ionic dissociation of the solvent. Similarly, acetone acts as a base in thionyl chloride:

$$2SOCl_2 \rightleftarrows 2SOCl^+ + 2Cl^- \qquad (6.16a)$$

$$(CH_3)_2CO + 2SOCl^+ \rightleftarrows (CH_3)_2CO(SOCl)_2^{2+} \qquad (6.16b)$$

$$(CH_3)_2CO + 2SOCl_2 \rightleftarrows (CH_3)_2CO(SOCl)_2^{2+} + 2Cl^- \qquad (6.16c)$$

Equation 6.16c is, of course, a specific example of the process in Eqn. 6.14.

The type of reactions exemplified by Eqns. 6.15 and 6.16 are representative of a large class of Lewis coordination reactions which are governed by the transfer of halide ions between the solute and the solvent (12). Generally, strong acids such as $FeCl_3$, $SnCl_4$, $SbCl_5$, SbF_5, and BF_3 act as the chloride-ion acceptors, while alkylammonium chlorides and alkali-metal chlorides act as the chloride-ion donors, or strong bases. The chloride-transfer reactions seem to be promoted by solvents that themselves dissociate into ions, such as nitrosyl chloride (NOCl), phosphorus oxychloride ($POCl_3$), selenium oxychloride ($SeOCl_2$), arsenic trichloride ($AsCl_3$), etc. In any given solvent, acids can be ranked according to their strength as chloride-ion acceptors.

While the postulated dissociation of oxyhalide solvents, such as $POCl_3 \rightleftarrows POCl_2^+ + Cl^-$, offers an obvious mechanism for the transfer of halide ions to the Lewis acids, the actual process of ionization might be generally different. Thus it was observed that the dissolution of anhydrous $FeCl_3$ in $POCl_3$ produced no $POCl_2^+$ ions; instead the ions were identified as $[Fe(POCl_3)_2Cl_2]^+$ and $[FeCl_4]^-$ (13). Furthermore, solutions of ferric chloride form the tetrachloroferrate(III) ion, $[FeCl_4]^-$, in triethyl phosphate, $(C_2H_5)_3PO$, a solvent that cannot serve as a source of the chloride ion (14). The explanation is that the more general mechanism for the ionization of metal halides involves a coordination between the halide and the solvent. This halide–solvent coordination loosens the metal–halogen bonds, causing halide ions to be liberated, thus forming cationic and anionic species in solution. Obviously the extent of this ionization and dissociation process must depend on the ability of the solvent to coordinate to the halide, to solvate the halide ion and on its dielectric constant. When all three of these solvent properties are appreciably pronounced, the progressive displacement of the halide ions by the solvent will tend to be complete. The coordination model for the reactions between Lewis acids and donor solvents is exemplified here by the following reactions between $FeCl_3$ and solvent S (15):

$$FeCl_3(s) + xS \longrightarrow [(FeCl_3)_2 \cdot xS] \qquad (6.17a)$$

$$[(FeCl_3)_2 \cdot xS] \rightleftharpoons [(FeS_4Cl_2)]^+ + (FeCl_4)^- \qquad (6.17b)$$

$$(FeCl_4)^- + S \rightleftharpoons FeCl_3 \cdot S + Cl^- \qquad (6.17c)$$

$$2[FeCl_3 \cdot S] + 2S \rightleftharpoons [FeS_4Cl_2]^+ + (FeCl_4)^- \qquad (6.17d)$$

$$[FeS_4Cl_2]^+ + S \rightleftharpoons [FeS_5Cl]^{2+} + Cl^- \qquad (6.17e)$$

$$[FeS_5Cl]^+ + Cl^- + S \rightleftharpoons [FeS_6]^{3+} + 2Cl^- \qquad (6.17f)$$

As we noted earlier, the above chloride-displacement reactions will proceed furthest to completion when the solvent coordination ability, its solvation ability for the chloride ion and its dielectric constant are very large. For example, in strong donor solvents of high dielectric constant, such as formamide ($\epsilon = 109.5$), *N*-methylformamide ($\epsilon = 187$), and *N*-methylacetamide ($\epsilon = 37.8$), the principal species present in the above equilibria are $(FeS_6)^{3+}$ and Cl^-. In dimethylsulfoxide ($\epsilon = 49$), the cationic species is $(FeS_5Cl)^{2+}$. In methanol ($\epsilon = 32.6$), both $(FeS_5Cl)^{2+}$ and $(FeS_4Cl_2)^+$ exist in conjunction with Cl^- and $FeCl_4^-$ ions. Weak donor solvents, such as phosphorus oxychloride, trimethyl phosphate, triethyl phosphate, and *N,N*-dimethylacetamide, favor the existence of $FeCl_4^-$ and $(FeS_4Cl_2)^+$ ions in their solutions. For example, when $ZnCl_2$ is added as the chloride-ion donor to a solution of $FeCl_3$ in phosphorus oxychloride or thionyl chloride, the formation of $FeCl_4^-$ is quantitative at zinc–iron mole ratios of both 1:1 and 1:2. Presumably, the solvated zinc ions exist in such solutions in the form of $[ZnS_3Cl]^+$ and $[ZnS_4]^{2+}$ (16). Aluminum chloride in $POCl_3$ solution will donate one or two chloride ions to $FeCl_3$ and will lose all three chloride ions to excess antimony pentachloride, forming the compound $[Al(POCl_3)_6][SbCl_6]_3$ (17).

In solvents of intermediate coordination power and dielectric constant, such as acetonitrile, appreciable concentrations of both the cationic species and of the halometallate ions are likely to be produced. For example, in acetonitrile solutions, the apparent molecular weight of $AlCl_3$ is 201 ± 5 and the major anionic species is $AlCl_4^-$ (18). Therefore the predominant equilibrium is likely to be

$$2nCH_3CN + 3Al_2Cl_6 \rightleftarrows 2[Al_2Cl_5 \cdot nCH_3CN]^+ + 2(AlCl_4)^-$$

No doubt that the well-known preference of acetonitrile to solvate complex anions as opposed to the chloride ion favors the above equilibrium.

We noted earlier that it is difficult to arrange Lewis acids and bases in any uniform order of strength because that order is likely to depend on the nature of the reference substance (base or acid). Ideally, the free energy of adduct formation between an acid and a base in the gas phase should serve as a relative measure of their strength. As such values are not easy to obtain experimentally, the next best choice is to measure the heats of coordination in relatively inert solvents, such as carbon tetrachloride, 1,2-dichloroethane, dichloromethane, or cyclohexane. Presumably, the solvent–solute interactions in these liquids are so weak that the measured heats of adduct formation reflect the same order of relative acid or base strengths as would be observed for these reactions in the gas phase. The reference acids used for the correlation of donor (basic) strengths of organic solvents have been iodine, phenol, and antimony pentachloride. In Table 6.1 we summarize the three basicity scales for a series of carbonyl solvents, referred to the above three acids.

An example where the nature of the reference acid altered the relative order of solvent basicities involves a series of substituted amides (donors). When referred to either iodine or phenol, which are insensitive to steric effects, these amides displayed the same order of basicities, but the order changed when Ni^{2+} ion was used as the reference acid (19).

We mention in Chapter 3 that Gutmann (20) used the Lewis acid–base reaction

$$SbCl_5 + \text{Donor} \rightleftharpoons \text{Adduct} \tag{6.18}$$

as the basis for establishing a scale of solvent donor abilities. The negative enthalpy of reaction 6.18 is called the donor number, DN, which is adopted as the measure of the basicity, or donor ability, of a solvent. In Table 6.2 we reproduce Gutmann's ranking order of solvents according to their donor abilities.

Drago and his associates (21, 22) have interpreted Lewis acid–base strength in terms of a two-parameter equation for the enthalpy of their adduct

Table 6.1 Enthalpies of Formation of Adducts Between Carbonyl Donor Solvents and Lewis Reference Acids

Donor	Lewis acids		
	I_2	Phenol	$SbCl_5$
Acetone $(CH_3)_2CO$	2.5	3.7 ± 0.4	17.03 ± 0.04
Methyl acetate CH_3COOCH_3	2.5 ± 0.1	3.3 ± 0.4	16.38 ± 0.03
Dimethyl carbonate $(CH_3O)_2CO$	—	3.0 ± 0.4	15.17 ± 0.03
N,N-Dimethylacetamide $(CH_3)_2NCH_3CO$	4.0 ± 0.1	6.1 ± 0.4	27.80 ± 0.08
Tetramethylurea $(CH_3N)_2CO$	4.3 ± 0.1	6.0 ± 0.4	29.64 ± 0.03
S-Methylthioacetate CH_3SCH_3CO	3.15 ± 0.12	3.2 ± 0.4	—

Reprinted with permission from *The Chemistry of Non-Aqueous Solvents*, Vol. I, J. J. Lagowski, Ed., Academic Press, New York, 1966, p. 54. Table by courtesy of Academic Press and the editor.

Adducts formed in CCl_4 solutions. The tabulated values are $-\Delta H$ in kcal mol^{-1}.

Table 6.2 Donor Numbers, DN, of Selected Solvents

Solvent	DN	Solvent	DN
1,2-Dichloroethane	~0	Methyl acetate	16.5
Sulfuryl chloride	0.1	*n*-Butyronitrile	16.6
Thionyl chloride	0.4	Acetone	17.0
Acetyl chloride	0.7	Ethyl acetate	17.1
Benzoyl chloride	2.3	Water	18.0
Nitromethane	2.7	Phenylphosphonic dichloride	18.5
Nitrobenzene	4.4	Diethyl ether	19.2
Acetic anhydride	10.5	Tetrahydrofuran	20.0
Benzonitrile	11.9	Diphenylphosphinic chloride	22.4
Selenium oxychloride	12.2	Trimethyl phosphate	23.0
Acetonitrile	14.1	Tributyl phosphate	23.7
Sulfolane	14.8	Dimethylformamide	26.6
Propanediol-1,2-carbonate	15.1	*N,N*-Dimethylacetamide	27.8
Benzyl cyanide	15.1	Dimethylsulfoxide	29.8
Ethylene sulfite	15.3	*N,N*-Diethylformamide	30.9
Isobutyronitrile	15.4	*N,N*-Diethylacetamide	32.2
Propionitrile	16.1	Pyridine	33.1
Ethylene carbonate	16.4	Hexamethylphosphoramide	38.8
Phenylphosphonic difluoride	16.4		

Data from Ref. 10. DN $\equiv -\Delta H$ for the reaction: Donor solvent + $SbCl_5 \rightleftarrows$ Adduct, in kcal mol^{-1}.

formation:

$$-\Delta H = E_A E_B + C_A C_B \tag{6.19}$$

Subscripts A and B refer to acid and base, respectively. The E parameter is a measure of the tendency of an acid or a base to participate in electrostatic interactions, while the C parameter is the measure of their tendency to form covalent bonds. Values of the E and C parameters have been determined from experimental data for the enthalpies of several hundred acid–base reactions. Some of the results are reproduced in Table 6.3. It is significant that on the basis of these average parameters it is possible to predict the enthalpies of acid–base reactions with remarkable accuracy. A few such predictions are compared with the experimental enthalpy values in Table 6.4. One significance of this approach is that it enables us to calculate the enthalpies of some Lewis acid–base reactions which might be inaccessible experimentally. Furthermore, the relative values of E and C parameters are consistent with the structural and electronic properties of the acids and bases. For example, the dipolar and hydrogen-bonding phenol has an E value of 4.33, as compared to $E = 1.00$ for the nonpolar iodine. Not surprisingly, the relative values of the C parameters for these acids are reversed ($C = 1.00$ for iodine and 0.44 for phenol).

Table 6.3 Parameters for Expressing Lewis Acid–Base Strength

Acid	C_A	E_A
Iodine[a]	1.00	1.00
Iodine monochloride	0.830	5.10
Iodine monobromide	1.56	2.41
Thiophenol	0.198	0.987
p-*tert*-Butylphenol	0.387	4.06
p-Methylphenol	0.404	4.18
Phenol	0.442	4.33
p-Fluorophenol	0.446	4.17
p-Chlorophenol	0.478	4.34
tert-Butyl alcohol	0.300	2.04
Trifluoroethanol	0.434	4.00
Pyrrole	0.295	2.54
Isocyanic acid	0.258	3.22
Isothiocyanic acid	0.227	5.30
Boron trifluoride	3.08	7.96
Boron trimethyl	1.70	6.14
Trimethylaluminum	1.43	16.9
Trimethylgallium	0.881	13.3
Trimethylindium	0.694	15.3
Sulfur dioxide	0.808	0.920
Antimony pentachloride	5.13	7.38
Chloroform	0.150	3.31

Base	C_B	E_B
Pyridine	6.40	1.17
Ammonia	3.46	1.36
Methylamine	5.88	1.30
Trimethylamine	11.54	0.808
Triethylamine	11.09	0.991
Acetonitrile	1.34	0.886
Dimethylformamide	2.48	1.23
Dimethylacetamide	2.58	1.32
Acetone	2.33	0.987
Diethyl ether	3.25	0.963
Tetrahydrofuran	4.27	0.978
Dimethylsulfoxide	2.85	1.34
Tetramethylene sulfoxide	3.16	1.38
Tetramethylurea	3.10	1.20
Benzene	0.707	0.486
Hexamethylphosphoramide	1.33	1.73

Selected data from Ref. 22.
[a] Used to define the scale.

Table 6.4 Comparison of Predicted and Experimental Values of Enthalpies for Lewis Acid–Base Reactions

Base	Calculated ΔH (kcal mol^{-1})	Heat measured (kcal mol^{-1})
Pyridine	7.6	7.8
Ammonia	4.8	4.8
Trimethylamine	12.3	12.1
Acetone	3.3	3.3
Diethyl ether	4.2	4.2
Acetonitrile	2.2	1.9
Ethyl acetate	2.7	2.8

Selected data from Ref. 22. Reference acid: iodine.

6.1.4 The Usanovich Concept The electronic concept of acids and bases was carried to its extreme by Usanovich (23). In addition to defining as bases those substances that donate electron *pairs*, he included also substances that donate a *single* electron. By analogy, acids were defined as substances that can accept not only electron pairs, but single electrons as well. This qualifies all reducing agents as bases and all oxidizing agents as acids. In his general concept, Usanovich considered as an acid any substance that forms salts with bases, gives up cations, and combines with anions or electrons. A base is a substance that neutralizes acids, gives up anions or electrons, and combines with cations. Salt formation is the underlying principle of the concept. Typical "Usanovich-only" acid–base reactions would be

$$\tfrac{1}{2}Cl_2 + Na \rightleftharpoons NaCl$$

$$Fe(CN)_2 + 4CN^- \rightleftharpoons Fe(CN)_6^{-4}$$

$$\text{Acid} \qquad \text{Base} \qquad\qquad \text{Salt}$$

While some chemists like this type of concept, we feel that the more general an acid–base definition becomes, the less useful it is likely to be.

Summary In the Arrhenius theory, acids and bases were substances that in aqueous solution dissociated into hydrogen and hydroxyl ions, respectively. This theory was superseded by the Brønsted concept, in which an acid was defined as a proton donor and a base, as a proton acceptor, without reference to solvent or to ionization. In the light of the Brønsted concept, all acid–base reactions are proton-transfer processes, which in amphiprotic media involve the solvent. A more general theory of acidity is that of Lewis, in which a base is defined as an electron-pair donor and an acid, as an electron-pair acceptor, thus extending the definition of an acid to include nonprotonic substances. Lewis acid–base reactions are coordination reactions. Examples of such reactions, including those with amphoteric aprotic solvents, have been discussed. The enthalpy of certain Lewis acid–base reactions has been used as a measure of the relative strength of acids and bases, including the so-called donor numbers of the solvents. However, the relative strengths of Lewis acids and bases depend on the reference substance chosen, which leads to many unrelated scales of acidity. An extreme acid–base concept is that of Usanovich, which expanded the list of acids to include oxidizing agents and the list of bases, to include reducing agents.

6.2 ACIDITY SCALES FOR PROTONIC ACIDS

We must be careful to distinguish between the quantity of an acid (base) in solution and the degree of acidity, or the acid strength, of a solution. Clearly a given volume of a 1 *M* solution of acetic acid in water contains a larger quantity of acid than an equal volume of a 0.1 *M* solution of hydrochloric acid, but the acidity of the latter is much greater. The *quantity* of acid in solution is determined usually from a titration with a base, in which the reaction is forced to completion, indicated by an endpoint, where we observe a sharp change in some property that varies with the progress of the titration. This, however, does not tell us the *degree* of acidity in the original solution. The latter is determined by the extent of reaction between the acid and some reference base, usually under conditions where the reaction is incomplete. If the reference base is the solvent, the acidity will be reflected by the activity of the (solvated) hydrogen ion in solution. If the reference base is an indicator added to the solution, the acidity will be measured by the fraction of that basic indicator that becomes converted to its acid form. Below we will consider in detail both types of acidity measurements. First we consider the case where the extent of acid–solvent reaction is monitored by potentiometric determination of the hydrogen-ion activity—the pa_H measurements.

6.2.1 Measurement and Interpretation of pH For all of its fundamental significance and wide applicability, the measurement and interpretation of pH represents a source of confusion and misinterpretation to chemists at all levels. This unfortunate state can be traced back to the manner in which students have been introduced to pH in the traditional college curricula. In the introductory chemistry courses, pH is defined as the negative logarithm of hydrogen-ion *concentration*. There can be little problem with interpreting pH of this type, which can be derived more or less simply from appropriate analytical concentrations and equilibrium constants. Problems begin to dawn in the intermediate courses, where the concept of activity is gradually introduced and pH is redefined as the negative logarithm of hydrogen-ion *activity*. Fortunately, the two definitions are easily related with the aid of an activity coefficient for the hydrogen ion—a minor correction factor in most dilute solutions, calculable from a Debye–Hückel equation. Unfortunately, this reconciliation is short-lived, as in the physical chemistry courses, the student learns from the writings of thermodynamic purists that activities and activity coefficients of single ions have "no physical significance." While this uncompromising message may haunt some people throughout their graduate years and beyond, most chemists go on measuring pH on commercial pH meters without giving any thought to the apparent discrepancy between theory and practice. Their uncertainty with respect to the meaning of pH may never catch up with them, unless they venture into partially or totally nonaqueous solutions.

It is clear that even before we begin to attack the problem of defining and interpreting pH in *nonaqueous* media, we must acquire a thorough grasp of the definitions, assumptions, and operations involved in the measurement and interpretation of aqueous pH. After all, pH is originally an aqueous scale.

pH in Aqueous Solutions In rigorous nomenclature we distinguish between pH and $\mathrm{p}a_{\mathrm{H}}$. The first is the pH-meter reading and the second is the negative logarithm of hydrogen-ion activity:

$$\mathrm{p}a_{\mathrm{H}} \equiv -\log\,(m_{\mathrm{H}}\gamma_{\mathrm{H}}) \tag{6.20}$$

where m_{H} and γ_{H} are the molality and the (molal) activity coefficient of the hydrogen ion. At 25°C the difference between molality and molarity is practically negligible for aqueous solutions. The activity coefficient is formulated in such a manner that it becomes unity at infinite dilution: $\gamma_{\mathrm{H}} \rightarrow 1$ as $m \rightarrow 0$. Thus $\mathrm{p}a_{\mathrm{H}}$ could be calculated from a known hydrogen-ion concentration and the activity coefficient for the hydrogen ion.

In dilute solutions the activity coefficient can be estimated with the aid of the Debye–Hückel equation in one of its forms, such as

$$\log \gamma = -\frac{Az^2I^{1/2}}{1 + B\mathring{a}I^{1/2}} \tag{6.21}$$

In the above equation, I is the ionic strength, defined as $\frac{1}{2}\Sigma mz^2$, z is the ionic charge, $\mathring{a}$ is the ion-size parameter in Ångstroms, $A = 0.509$, and $B = 0.329$ for aqueous solutions at 25°C. However, estimation of activity coefficients for the hydrogen ion is usually avoided and the pH most frequently is measured for solutions where the hydrogen-ion concentration is unknown. Thus it is highly impractical to "determine pH" by combining hydrogen-ion concentration with a calculated activity coefficient.

For most practical purposes, pH is defined and measured operationally in terms of the electromotive force (emf) of a complete cell composed of a hydrogen electrode and a reference electrode. When the emf readings E_{s} and E_{u} are obtained for the same electrode pair immersed in a standard buffer of known $\mathrm{pH_s}$ and a solution of an unknown having a $\mathrm{pH_u}$, respectively, the relationship between pH_u and the known quantities is

$$pH_u = pH_s + \frac{E_u - E_s}{2.3RT/F} \tag{6.22}$$

where $(2.3RT/F)$ is the familiar factor 0.05916 at 25°C when the emf is in volts. Eqn. 6.22 represents the *operational definition of pH*.

The standard buffer solutions of known pH_s (and their compositions) are provided by the National Bureau of Standards (NBS) in Washington, D.C. Thus the crucial question on the interpretation of pH must be directed to the manner in which these pH numbers have been assigned to the standard buffers by the NBS.

The values of pH_s are assigned to the standard buffers on the basis of emf measurements on cells without transference (without liquid junction between dissimilar solutions). Almost invariably the hydrogen (gas)–silver-silver chloride cell is employed:

$$\underset{\text{buffer}}{\text{Pt(s); H}_2\text{(g)} \mid \text{H}^+, \text{Cl}^- \mid \text{AgCl(s); Ag(s)}} \tag{6-I}$$

The half-cell reactions involved here are:

$$\tfrac{1}{2}H_2(g) = H^+ + e \quad \text{and} \quad AgCl(s) + e = Ag(s) + Cl^- \tag{6.23}$$

and the cell emf is given by:

$$E = E^\circ - 0.05916 \log a_H a_{Cl}$$

or

$$E = E^\circ_{AgCl} + 0.05916\, pa_H - 0.05916 \log m_{Cl} - 0.05916 \log \gamma_{Cl} \tag{6.24}$$

We can solve for pa_H from Eqn. 6.24:

$$pa_H = \frac{E - E^\circ}{0.05916} + \log m_{Cl} + \log \gamma_{Cl} \tag{6.25}$$

The steps involved in arriving at numerical values of pa_H by the NBS procedure are as follows (24):

1 Values of $p(a_H\gamma_{Cl}) = [(E - E^\circ)/0.05916 + \log m_{Cl}]$ are obtained at several low concentrations of added chloride.

2 Values of $p(a_H\gamma_{Cl})^\circ$ are obtained for each buffer by extrapolating the quantities $p(a_H\gamma_{Cl})$ to $m_{Cl} = 0$.

3 $pa_H = p(a_H\gamma_{Cl})^\circ + \log \gamma^\circ_{Cl}$.

4 The activity coefficient for the chloride ion is calculated by the so-called Bates–Guggenheim convention in the form

$$\log \gamma^\circ_{Cl} = \frac{-AI^{1/2}}{1 + 1.5I^{1/2}} \tag{6.26}$$

This is obviously a Debye–Hückel equation (Eqn. 6.21) in which the ion-size parameter $\mathring{a}$ was set equal to 4.5 Å for all ions. Theoretically, $\mathring{a}$ is the mean distance of closest approach between ions in a given solution, but as a practical matter, it is employed as an adjustable parameter to bring about agreement between theoretical and experimental values of mean ionic activity coefficients. It turns out that at an ionic strength of 0.1, the log γ values of a univalent ion calculated under the above convention differ only by 0.01 from those obtained with ion-size values of 2.5 or 6.0.

5 The last step merely identifies the conventional pH_s evaluated by the above procedure with the pa_H:

$$pa_H = pH_s \tag{6.27}$$

Bates (24) is careful to stress that the pa_H scale defined by this procedure is not a thermodynamic, but a conventional one. However, he adds that when the conventional pa_H values obtained in this manner are inserted in equations involving a_H, they do provide equilibrium data that are consistent with those obtained by rigorous thermodynamic methods. The uncertainty in the pH_s values is estimated to be about ± 0.005 units at 25°C.

The hydrogen–silver-silver chloride cell is used in fundamental work, such as the establishment of pH scales, determination of ionization constants, and standard potentials, but it is not convenient for measuring pH on an everyday basis. It requires saturation with hydrogen gas (with which not all solutions are compatible), presence of a known concentration of chloride, calculation of activity coefficients, etc. Therefore, for most practical purposes, pH is measured with an emf cell with transference (with liquid junction). We can represent the common pH cell, composed of a hydrogen–calomel electrode pair, by the following diagram, remembering that the hydrogen *gas* electrode here is usually symbolic only; in practice it is most often replaced by a glass electrode, which functions as if it were a hydrogen gas electrode:

$$\begin{array}{r|l|l} \text{Pt(s); } H_2\text{(g), standard } H^+ & \text{KCl} & \text{KCl, } Hg_2Cl_2\text{(s); Hg(l)} \\ \text{or unknown} & \text{satd.} & \text{satd.} \end{array} \tag{6-II}$$

The overall reaction for the above cell can be written as:

$$\tfrac{1}{2}H_2(g) + \tfrac{1}{2}Hg_2Cl_2(s) \rightleftarrows Hg(l) + H^+ + Cl^-(\text{satd. KCl}) \pm \text{ion transfer} \tag{6.28}$$

The corresponding Nernst equation is

$$E = E^\circ + 0.05916\ pa_H - 0.05916 \log a_{\text{Cl ref.}} + E_j \tag{6.29}$$

Setting $E^{\circ\prime} = E^\circ - 0.05916 \log a_{\text{Cl ref.}}$, that is, collecting within $E^{\circ\prime}$ those terms which, it is hoped, are constant for a given electrode pair, we can solve for pa_H:

$$pa_H = \frac{E - (E^{\circ\prime} + E_j)}{0.05916} \tag{6.30}$$

E_j is the liquid-junction potential arising at the boundary between the KCl salt bridge and the solution the pH of which is being determined. The subscript "ref." attached to chloride-ion activity stresses the fact that it is the chloride in the reference electrode and not in the pH compartment.

From Eqn. 6.30, we can derive the operational definition of pH (Eqn. 6.22) as follows: (1) The quantity $(E^{\circ\prime} + E_j)$ is evaluated via Eqn. 6.30 from the measured emf, E, for a standard buffer of known pH_s, which is assumed to equal pa_H. In effect this is what is done when the pH meter is set to read the pH value of the standard buffer. (2) If we assume that for a given electrode pair the quantity $(E^{\circ\prime} + E_j)$ remains constant when a standard is replaced by an unknown, the operational definition of pH (Eqn. 6.22) will be obtained.

The major assumption when employing the operational definition of pH is that the liquid-junction potential between the salt bridge and the pH solution is the same for the standard and for the unknown, or $E_j(s) = E_j(u)$, so that the E_j's cancel out between the standardization and the testing step. If they do not cancel (and, strictly speaking, they never do, unless the standard and the unknown are identical solutions), then the error in pH introduced by the difference in the liquid-junction potentials is given by

$$pa_H - pH = \frac{E_j(s) - E_j(u)}{0.05916} = \frac{\Delta E_j}{0.05916} \tag{6.31}$$

It should be noted that values of pH_s, having been obtained from cells without liquid junction, contain no compensating E_j errors.

The "residual" (uncancelled) liquid-junction potential, given by the right-hand-side terms of Eqn. 6.31, is the reason why pH-meter readings obtained operationally lie on the same scale as the pH values of the standard buffers only in the limited range between the pH's of approximately 2 and 12 in aqueous solutions. The E_j error is the reason why, for optimum accuracy in pH measurements, the composition of the unknown should be as close as possible to that of the standard. Measured pH will deviate appreciably from the standard pH scale when the composition of the unknown becomes marked by high solute concentration, high acidity or basicity, presence of colloids, suspensions, and nonaqueous solvents.

pH in Nonaqueous Solutions What happens to the pH determined from the emf of a hydrogen–silver-silver chloride cell when, keeping the other variables constant (concentration, temperature), we start mixing in a nonaqueous solvent, such as methanol, at the expense of water? So as not to complicate our discussion more than necessary, let us stay within a high enough dielectric constant to preclude any appreciable ion pairing. Upon addition of methanol, the potential will change, but does that mean that the pH changes as well?

Two parameters in the Nernst equation from which pa_H is derived (Eqn. 6.24) may change with solvent composition. The activity coefficient will change, because the dielectric constant of the solvent alters the constants A and B of the Debye–Hückel equation (Eqn. 6.21). However, in dilute solutions this could be a minor factor. Does the $E°$ change with the solvent? Experimental work shows that it does; in fact the $E°$ of the silver–silver chloride electrode, which is 0.22234 V at 25°C in water, assumes the value of 0.1906 V in 50 wt % methanol mixed with water and even changes the sign (−0.0101 V) in anhydrous methanol.

However, once we take these two changes into account, we can write the same formal expression for the pa_H of a nonaqueous solution as we did for aqueous systems, except that we will designate nonaqueous pa_H^* with an asterisk and the nonaqueous potentials and activity coefficients, with a subscript s on the lower left-hand side of the symbol:

$$pa_H^* = \frac{E - {}_sE°}{0.05916} + \log m_{Cl} + \log {}_s\gamma_{Cl} \tag{6.32}$$

where

$$a_H^* = m_H \, {}_s\gamma_H$$

The new activity coefficients approach unity at infinite dilution as usual, but it is at infinite dilution in the given nonaqueous solvent. In the above manner we can assign pH_s^* numbers to standard buffers in any amphiprotic solvent and thus, in principle, establish in any amphiprotic solvent a pH* scale, modeled after the pH scale in water. Having done so, we could measure pH* operationally in any such solvent (Eqn. 6.22). Indeed, standard pH* buffers are available from the NBS for methanol–water mixtures up to about 70% methanol, and are especially plentiful for the 50–50 methanol–water composition.

Of course, this procedure of defining and measuring pH values leads to as many unrelated pH scales as there are solvents. The scales are unrelated because the activity coefficients of ions in them are referred to infinite dilution in the given solvent. Can anything be done to bring these scales to the same common denominator, to express them quantitatively on a single scale? One approach is to estimate the medium effects or the transfer activity coefficients for the hydrogen ion, ${}_m\gamma_H$, in the solvents for which correlation of pH scales is desirable. We recall that $pa_H = pa_H^* - \log {}_m\gamma_H$ from Eqn. 5.7a, so that when values of $\log {}_m\gamma_H$ are known for a given medium, the pa_H^* numbers in that medium can be expressed on the aqueous pa_H scale. Another, related, approach is to use the *aqueous* value for the $E°$ of the silver–silver chloride electrode when interpreting the measured potential of a hydrogen–silver-silver chloride cell in a *nonaqueous* medium. Equation

6.32 will then assume the form:

$$\mathrm{p}a_{\mathrm{H}} = \frac{E - {}_{\mathrm{w}}E^{\circ}}{0.05916} + \log m_{\mathrm{Cl}} + \log {}_{\mathrm{s}}\gamma_{\mathrm{Cl}} + \log {}_{\mathrm{m}}\gamma_{\mathrm{Cl}} \tag{6.33}$$

In the above expression, the $\mathrm{p}a_{\mathrm{H}}$ is referred to the aqueous standard state, that is, all activity coefficients become unity at infinite dilution *in water*. This is a consequence of using the aqueous standard potential, ${}_{\mathrm{w}}E^{\circ}$, in Eqn. 6.33. As the solution becomes infinitely dilute in the nonaqueous medium, ${}_{\mathrm{s}}\gamma_{\mathrm{Cl}}$ does become unity, but the medium effect ${}_{\mathrm{m}}\gamma_{\mathrm{Cl}}$ persists. It follows that if we knew the numerical value for the medium effect (transfer activity coefficient) for the chloride ion, we could express the $\mathrm{p}a_{\mathrm{H}}$ measured in a nonaqueous solvent on the single aqueous scale. Methods by which medium effects for single ions can be estimated are discussed in Chapter 5.

Nevertheless, a certain amount of interpretation of nonaqueous pH data is possible without the knowledge of medium effects. This was demonstrated in the work of Bates, Paabo, and Robinson (25) in methanol–water mixtures and the full range of ethanol–water solvents. In the above study, certain relationships were derived between pH and $\mathrm{p}\overset{*}{a}_{\mathrm{H}}$, and their practical applications were demonstrated on buffer solutions in the alcohol–water media. Measurements of pH (pH-meter readings) were made with glass–calomel electrodes previously standardized with aqueous buffers and the emf of hydrogen (gas)–silver-silver chloride cells (cell 6-I) was measured in the same solutions. From the latter, values of $\mathrm{p}(a_{\mathrm{H}}\gamma_{\mathrm{Cl}})$ were obtained, since they are equal to the sum of the first two terms on the right-hand side of Eqn. 6.33. The difference between the two measured quantities yields:

$$\mathrm{pH} - \mathrm{p}(a_{\mathrm{H}}\gamma_{\mathrm{Cl}}) = \frac{\Delta E_{\mathrm{j}}}{0.05916} + \log {}_{\mathrm{m}}\gamma_{\mathrm{Cl}} + \log {}_{\mathrm{s}}\gamma_{\mathrm{Cl}} \tag{6.34}$$

After estimating the salt-effect activity coefficient for the chloride ion, ${}_{\mathrm{s}}\gamma_{\mathrm{Cl}}$, from the Debye–Hückel equation, Bates, Paabo, and Robinson isolated the sum of the unknown terms $[(\Delta E_{\mathrm{j}}/0.05916) + \log {}_{\mathrm{m}}\gamma_{\mathrm{Cl}}]$ from which they calculated the related quantity $[(\Delta E_{\mathrm{j}}/0.05916) - \log {}_{\mathrm{m}}\gamma_{\mathrm{H}}] \equiv \delta$.† Recalling again that $\mathrm{p}a_{\mathrm{H}} = \mathrm{p}\overset{*}{a}_{\mathrm{H}} - \log {}_{\mathrm{m}}\gamma_{\mathrm{H}}$, it follows in conjunction with Eqn. 6.31 that

$$\mathrm{p}\overset{*}{a}_{\mathrm{H}} = \mathrm{pH} - \delta \tag{6.35}$$

For each alcohol–water medium studied, δ values turned out to be fairly constant, independent of the pH, so that interpretation of operational pH numbers via Eqn. 6.35 became possible. Of course, such constancy of δ and the applicability of Eqn. 6.35 would have to be demonstrated for each distinctly new solvent medium. The δ values from Ref. 25 are summarized in Table 6.5.

† The calculation makes use of the fact that $\log {}_{\mathrm{m}}\gamma_{\mathrm{H}} = \log {}_{\mathrm{m}}\gamma_{\mathrm{HCl}} - \log {}_{\mathrm{m}}\gamma_{\mathrm{Cl}}$. Values of $\log {}_{\mathrm{m}}\gamma_{\mathrm{HCl}}$ for the alcohol–water systems are known (see Table 5.1).

Table 6.5 Values of the Constant $\delta \equiv [(\Delta E_j/0.05916) - \log {}_m\gamma_H]$ in Methanol–Water and Ethanol–Water Solvents at 25°C, in pH Units

Wt % alcohol	Methanol–water	Ethanol–water
0	0	0
20	0.01	0.02
35	0.06	0.10
50	0.13	0.21
65	0.14	0.24
80	—	0.11
90	—	−0.40
100	—	−2.91

Data from Ref. 25 rounded off to even values of solvent composition in Ref. 24, p. 224.

It can be seen from Table 6.5 that the values of δ are small up to about 80 wt % alcohol. Apparently, the liquid-junction error and the medium effect for the proton cancel each other to a large degree for this range of solvent composition. Further interpretation of the data, however, such as deriving values of pa_H (aqueous scale) or of the E_j error from Eqn. 6.35, require knowledge of the medium effects for the proton, $\log {}_m\gamma_H$. In Chapter 5, we combined the estimates of $\log {}_m\gamma_H$ for ethanol–water solvents (obtained by the tetraphenylborate assumptions) with the values of δ shown in Table 6.5 to calculate the average liquid-junction potential error in the cells from which the δ's were obtained. The results are shown in Table 5.7.

6.2.2 Acidity Functions Based on Indicators In contrast to the electrometric pH scale, which was established for dilute aqueous solutions and only much later extended to a few mixed solvents, the acidity functions based on the spectral changes of acid–base indicators were designed from the beginning to measure the degree of acidity in a wide variety of media, including concentrated solutions of acids and bases in water and nonaqueous solvents. The Hammett acidity function, H_o, provides a quantitative expression for the acidity of a solution based on the extent of protonation of uncharged (hence the subscript zero) indicator bases, which are mainly the substituted anilines (26–29). The governing equilibrium here is the ionic dissociation of the conjugate acid of the uncharged indicator base, HIn^+:

$$HIn^+ \rightleftarrows H^+ + In \qquad (6.36)$$

for which

$$K_{HIn} = \frac{a_H\, a_{In}}{a_{HIn}} \qquad (6.37)$$

or

$$pK_{HIn} = pa_H - \log \frac{m_{In}}{m_{HIn}} - \log \frac{\gamma_{In}}{\gamma_{HIn}} \tag{6.38}$$

In the above equations, m_{HIn} and m_{In} are the concentrations of the acid and the base forms of the indicator, respectively, which are determined spectrophotometrically, while the pK and the activity coefficients are referred to aqueous standard states, regardless of the medium in which the measurements actually take place. The H_o function is then defined in terms of the measureable quantities:

$$H_o \equiv pK_{HIn} + \log \frac{m_{In}}{m_{HIn}} \tag{6.39}$$

which in conjunction with Eqn. 6.38 leads to an equivalent expression:

$$H_o = pa_H + \log \frac{\gamma_{HIn}}{\gamma_{In}} \tag{6.40}$$

Thus H_o becomes identical with pa_H at infinite dilution in water. At finite concentrations in aqueous solutions, the extent to which H_o differs from pa_H will depend on how much the ratio of the salt-effect activity coefficients, $\gamma_{HIn}(\gamma_{In})^{-1}$, deviates from unity. In nonaqueous solutions, an activity coefficient referred to infinite dilution *in water* is a product of the salt effect and the medium effect: $\gamma = {}_s\gamma \; {}_m\gamma$. Therefore, the expression for H_o in a nonaqueous solvent must be expanded using the transfer activity coefficients (medium effects) of the indicator acid and base:

$$H_o = pa_H + \log \frac{{}_s\gamma_{HIn}}{{}_s\gamma_{In}} + \log \frac{{}_m\gamma_{HIn}}{{}_m\gamma_{In}} \tag{6.41}$$

It is clear that before the H_o function can be accepted as a measure of pa_H in any given nonaqueous solvent it must meet two requirements: (1) as in aqueous solutions, the salt-effect ratio, ${}_s\gamma_{HIn}({}_s\gamma_{In})^{-1}$, should not deviate far from unity, and (2) the ratio of the medium effects, ${}_m\gamma_{HIn}({}_m\gamma_{In})^{-1}$, should be close to unity as well. The latter requirement is generally much more difficult to meet, although in Chapter 5 we observed that within the amine–ammonium conjugate acid–base pairs there seems to be substantial cancellation of medium effects for the transfer from water to ethanol–water and methanol–water solvents.

A single indicator cannot cover a wide range of acidities. Therefore, any acidity function requires the use of a "family" of structurally similar indicators that vary in basic strength. If the value of H_o is to be independent of the specific indicator employed, the relative strengths of the indicator acids must also be independent of the medium. This is true when the activity coefficient ratio, $\gamma_{HIn}(\gamma_{In})^{-1}$, is the same for all indicators in a given medium. Obviously, this requirement in the establishment of an H_o function is less stringent than the ideal requirement that the ratio of the indicator activity coefficients be unity. Nevertheless, equality of the $\gamma_{HIn}(\gamma_{In})^{-1}$ ratio can be achieved only within a set of very similar indicators, which in the original Hammett work were substituted primary anilines. Several new acidity functions based on uncharged indicator bases other than the Hammett indicators have been proposed. The results based on these acidity functions generally do not coincide with the H_o values of Hammett. For a modern review of the field of acidity functions, with particular emphasis on the specificity in the behavior of the indicator activity coefficients, see Boyd (30).

While the Hammett H_o function has been employed mainly for the determination of acidities of aqueous solutions of strong acids, the range of measurements extended all the way into some of the pure acid media, that is, into nonaqueous systems. For example, the H_o values were found to be -2.19, -6.3, -7.86, -10.0, and -11.30 for formic acid, nitric acid, methanesulfonic acid, hydrofluoric acid, and 99.5% sulfuric acid, in that order. Very interesting results were obtained when the Hammett H_o function was extended to aqueous mixtures of methanol, ethanol, dioxane, acetone, and tetrahydrofuran. The behavior of the H_o function for 0.1 M HCl solutions in these media is shown in Fig. 6.1. The characteristic minimum in the H_o function when plotted vs solvent composition in aqueous–organic mixtures was also observed for the transfer-activity coefficient of the proton, log ${}_m\gamma_H$, in ethanol–water solvents (Fig. 5.3). There, the values of log ${}_m\gamma_H$ were estimated by the tetraphenylborate assumption. At the end of Chapter 5, we refer to certain evidence that in any given ethanol–water or methanol–water solvent, the medium effects of ammonium acids and their conjugate amine bases do not differ by much. If this should apply to Hammett H_o indicators as well, then it would follow that ${}_m\gamma_{HIn}({}_m\gamma_{In})^{-1} \cong 1$ and the H_o function might represent a fair approximation of pa_H, at least in ethanol–water and methanol–water media. Indeed, when corrections for salt-effect activity coefficients and incomplete dissociation of HCl were made, the values of H_o and log ${}_m\gamma_H$ in ethanol–water solvents showed fair agreement (31).

No such agreement with log ${}_m\gamma_H$ is observed when the acidity function is based on anionic bases. This function, known as H_-, is defined in a manner analogous to H_o on the basis of ionic dissociation equilibria of uncharged acids:

$$HA \rightleftarrows H^+ + A^- \tag{6.42}$$

$$pK_{HA} \rightleftarrows pa_H - \log \frac{m_A}{m_{HA}} - \log \frac{\gamma_A}{\gamma_{HA}} \tag{6.43}$$

leading to the definition:

$$H_- \equiv pK_{HA} + \log \frac{m_A}{m_{HA}} \tag{6.44}$$

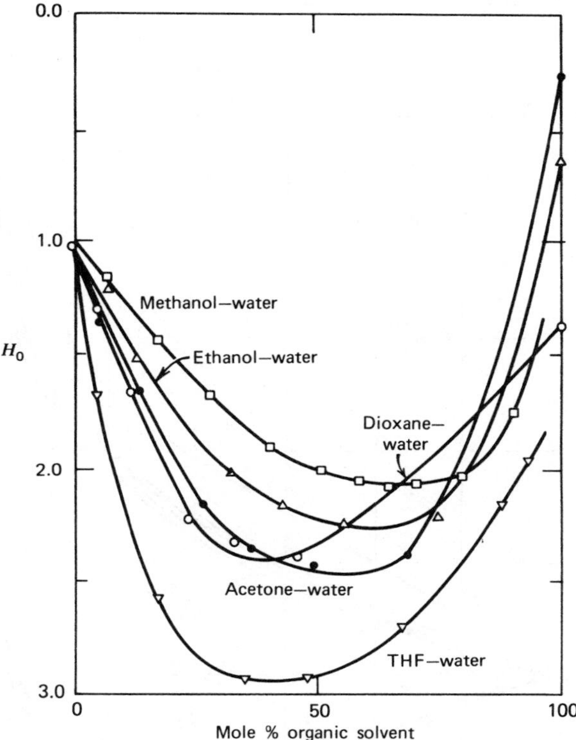

Figure 6.1 Hammett H_o functions (based on anilines) in mixed solvents containing 0.1 *M* HCl. THF is tetrahydrofuran. Reprinted from J. F. Coetzee and C. D. Ritchie, Eds., *Solute-Solvent Interactions*, Vol. I, Marcel Dekker, Inc., New York, 1969, p. 163, by courtesy of Marcel Dekker, Inc.

and

$$H_{-} = pa_{H} - \log \frac{\gamma_{A}}{\gamma_{HA}} \tag{6.45}$$

As in the case of H_o, the acidity function H_- becomes identical with pa_H at infinite dilution in water.

In Fig. 6.2, the behavior of the H_- and H_o functions determined by a series of indicators is compared in ethanol–water solvents containing low concentrations of acid. Compared to the H_o values, the acidities expressed by the H_- function are much higher (by about 5 units in anhydrous ethanol) and show no minimum. A large negative deviation of H_- from pa_H is expected on the basis of the positive medium effects, $\log {}_m\gamma_A$, for the carboxylate ions and the negative medium effects, $\log {}_m\gamma_{HA}$, for the carboxylic acids in nonaqueous media. For example, in methanol, the values of $\log {}_m\gamma$

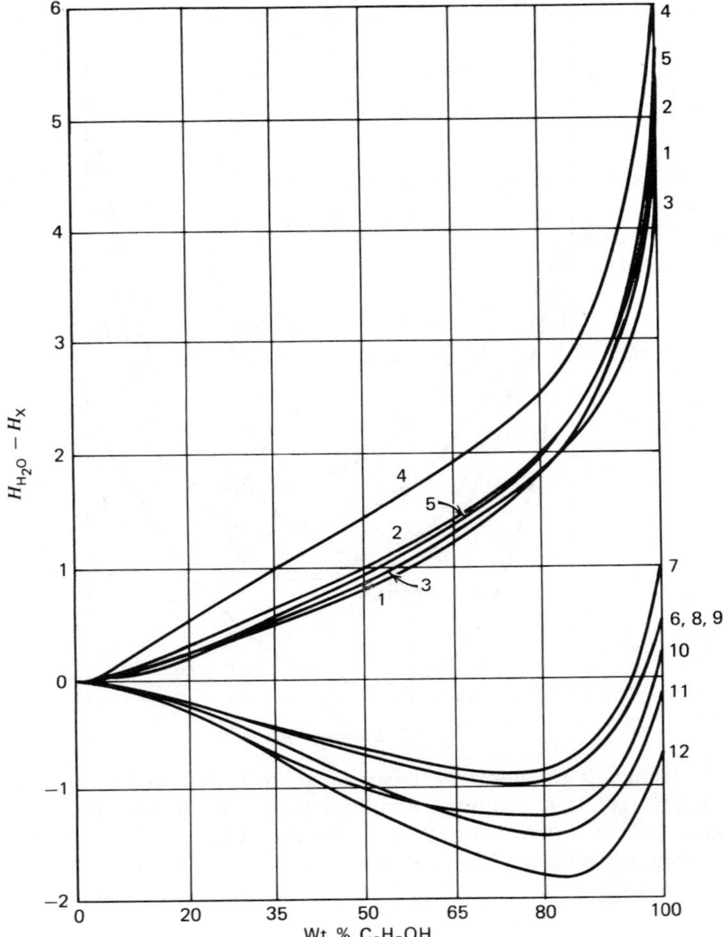

Figure 6.2 Relative values of the Hammett acidity functions H_o and H_- in ethanol–water solvents containing equal (low) concentrations of strong acid. Water is the reference. 1, formic acid; 2, acetic acid; 3, cyanoacetic acid; 4, benzoic acid; 5, salicylic acid; 6, aniline; 7, *p*-toluidine; 8, *m*-toluidine; 9, *o*-toluidine; 10, *p*-nitroaniline; 11, *N*-methylaniline; and 12, *N,N*-dimethylaniline. Reprinted with permission from *The Chemistry of Non-Aqueous Solvents*, Vol. I, J. J. Lagowski, Ed., Academic Press, New York, 1966, p. 154, by courtesy of Academic Press and the editor.

are $+2.7$ and -1.2 for the acetate ion and acetic acid, respectively; the corresponding values are $+1.2$ and -2.0 for the benzoate ion and benzoic acid, respectively (ionic values estimated by the tetraphenylborate assumption). Clearly, the quantities $\log {}_m\gamma_A({}_m\gamma_{HA})^{-1}$, which are tacitly assumed to be zero, are actually of the order of 3–4 log units. It appears that the H_o

function is a much better approximation of the pa_H (aqueous) scale in alcohol–water media than the H_- function.

The H_- function has been more popular in the studies of strongly basic solutions. When the ratio of activity coefficients in Eqn. 6.45 is close to unity, H_- represents the pa_H, so that for dilute aqueous base solutions (up to about 0.1 M OH^-) the following relationship is obeyed:

$$H_- = pK_w + \log (OH^-) \tag{6.46}$$

At higher hydroxyl-ion concentrations, the H_- increases much more rapidly than Eqn. 6.46 would indicate, presumably because the depletion of water molecules in solution results in incomplete hydration of the OH^- ion, leading to an increase in its activity. Analogous behavior is observed in alcohols. In dilute solutions, the relationship between the H_- function and the alkoxide-ion concentration is

$$H_- = B + \log (OR^-) \tag{6.47}$$

where B is a constant, characteristic of the alcohol. For 0.1 M solutions of alkali alkoxides, the increasing order of basicity (as measured by the H_- numbers) was found to be (10): methanol (12.66), ethanol (14.57), isopropyl alcohol (16.95), *t*-pentyl alcohol (18.09), and *t*-butyl alcohol (19.14). However, as the alkoxide solutions become more concentrated, their basicity rises more rapidly than Eqn. 6.47 would predict. Again, it is the reduction in the number of the solvent molecules available for solvation that increases the activity of the OR^- ion.

Interesting is the nature of the constant B in Eqn. 6.47. While in dilute aqueous solutions, $B = pK_w$ (Eqn. 6.46), in an alcohol, B does not simply correspond to the autoprotolysis constant of the solvent. Here, the transfer activity coefficients of three solute species come into play, so that for a solution having a given *conventional* $p\overset{*}{a}_H$ value (i.e., referred to the standard state in the given alcohol) we obtain:

$$H_- = p\overset{*}{a}_H - \log {}_m\gamma_H - \log \frac{{}_m\gamma_A}{{}_m\gamma_{HA}} \tag{6.48}$$

and since

$$pK_s = (ROH_2{}^+)(OR^-) \tag{6.49}$$

where $(ROH_2{}^+) \equiv \overset{*}{a}_H$,

$$H_- = pK_s - \log {}_m\gamma_H - \log \frac{{}_m\gamma_A}{{}_m\gamma_{HA}} + \log (OR^-) \tag{6.50}$$

Consequently,

$$B = pK_s - \log {}_m\gamma_H - \log \frac{{}_m\gamma_A}{{}_m\gamma_{HA}} \tag{6.51}$$

In the above equations, K_s is the autoprotolysis constant of the alcohol ROH and the salt-effect activity coefficients have been omitted. In general, values of log ${}_{m}\gamma_{H}$ and log ${}_{m}\gamma_{A}$ where A^- is a carboxylate anion have been found to be positive, while the values of log ${}_{m}\gamma_{HA}$ for carboxylic acids, were found to be negative for the transfer from water to alcohols (see Chapter 5 and discussion of Eqn. 6.45). This means that values of pK_s should be much more positive than the B values. Indeed, pK_s is reported to be 16.7 in methanol and 19 in ethanol, while the corresponding B values are 13.66 and 15.57, respectively (32).

The H_- function has been employed extensively in the measurement of basicities in aqueous–organic solvent mixtures containing low concentrations of added hydroxide. Such studies have been reported for aqueous mixtures of ethanol, methanol, isopropyl alcohol, *t*-butyl alcohol, *t*-pentyl alcohol, dioxane, hydrazine, ethylenediamine, 2-aminoethanol, pyridine, tetramethylene sulfone, and dimethylsulfoxide (30). Interesting are the results in the last three systems. At constant base concentration, added usually in the form of tetramethylammonium hydroxide, the H_- function increases rapidly as the content of the organic component in the mixture is raised. Presumably, these basic solvents compete effectively with the hydroxyl ion for the protons of the water molecules, thus raising the activity of the OH^- ion. For example, the H_- value of a 0.01 M solution of tetramethylammonium hydroxide in 50 mol % DMSO–water mixture is 17.2 (33), corresponding to a basicity about 10^5 times greater than for a 0.01 M OH^- in water. Considering that the H_- function probably exaggerates the *acidity* of solutions (see discussion of acid solutions), the true basicity of these systems could be even greater.

We observed earlier that acidity functions based on indicators of different charge type, for example, H_o vs H_-, diverge significantly, because the ratios of the transfer activity coefficients (medium effects) for conjugate acid–base pairs of different charge type exhibit very different solvent dependence. Thus the ratio ${}_{m}\gamma_{HIn^+}({}_{m}\gamma_{In})^{-1}$. varies differently with solvent composition than does the ratio ${}_{m}\gamma_{HA}({}_{m}\gamma_{A^-})^{-1}$. Significant in this connection is an old proposal by Schwarzenbach (34) to use the cationic acid indicator 5-pyridiniumglutacondialdehyde perchlorate, which is colorless and is converted to a (red) univalent anion in basic solution. Since the medium effect should be much less for a ratio of two univalent ions as compared to a ratio of an ion and an uncharged molecule, the acidity function based on this indicator might be a better approximation of pa_H than either H_o or H_-.

So far in the discussion of acidity functions we made the tacit assumption that ion pairing presents no problem. This is no longer true if we were to attempt to establish acidity functions in a solvent of low dielectric constant, such as acetic acid ($\epsilon = 6.3$). In a solvent where even the strongest electrolytes have dissociation constants no greater than 10^{-5}, the indicator–acid ion pairs, such as $InH^+ClO_4^-$, would predominate as the acidic species. Under these conditions, values of H_o turn out to be dependent on the nature of the acid and a uniform acidity function is difficult to define (29, 35). The peculiar behavior of indicator equilibria in acetic acid is discussed in the section devoted to the quantitative treatment of acid–base equilibria in that solvent.

It is obvious that in media of low dielectric constant, where free ions exist only in very low concentrations, meaningful scales of acidity should not be based on the activity of ions. Instead, the measure of acidic strength should be the extent of conversion of a reference base to the corresponding acid–base ion pair, for example,

$$\underset{\text{(base)}}{\text{In}} + \underset{\text{(acid)}}{\text{HA}} \rightleftharpoons \underset{\text{(ion pair)}}{\text{InH}^+\text{A}^-} \tag{6.52}$$

Measurements of this type are discussed in Chapter 3 in reference to acid–base reactions in benzene.

Summary Within the Brønsted concept of acidity, the acid strength of a solution can be expressed in terms of the activity of the solvated proton, experimentally approximated either as potentiometric pH or spectrophotometrically as one of the Hammett acidity functions, which are based on indicators. The procedure by which pH values were assigned to standard buffers in water and certain methanol–water mixtures has been described. Operational pH measurements in a nonaqueous medium using glass and calomel electrodes are subject to simple interpretation only if they are based on pH standards in the same medium. When aqueous pH standards are used, a full interpretation of pH in a nonaqueous medium requires knowledge of the transfer activity coefficient for the proton and the liquid-junction potential in the pH cell. However, approximate interpretations are sometimes possible. A solvent-independent pH scale requires knowledge of the transfer activity coefficients for the proton in all the media of interest. Acidity functions H_o and H_- in certain nonaqueous and mixed solvents were discussed and interpreted in terms of transfer activity coefficients.

6.3 ACID–BASE TITRATIONS

Most likely, the original reason why chemists resorted to nonaqueous solvents as titration media for acids and bases was to increase the solubility of organic samples. Initially they may have been surprised that titrations in solvents other than water were feasible in the first place, because the Arrhenius concept limited the manifestations of acidity to aqueous solutions, and then most certainly they were impressed by the fact that the endpoints in nonaqueous solvents were often sharper than in water. Later it was discovered that very weak acids and bases, which could not be titrated in aqueous solutions, could be determined titrimetrically in suitable nonaqueous media. Furthermore, in some nonaqueous solvents, it was possible to obtain separate endpoints for several components of a mixture of acids or bases. Today, nonaqueous acid–base titrimetry is a well-established and widely applied technique, but it took a long time to evolve to the present stage.

As far back as 1910, Folin and Wentworth (37) reported on titrations of fatty acids in aprotic media. However, their work received little attention, probably because the disciples of the strictly aqueous concepts of acidity at that time were not prepared to accommodate such a radical innovation. In the years 1927–1930, we note the pioneering work of Conant, Hall, and Werner (38–42) on the titration of amines in glacial acetic acid, but again there was little followup. Only from about 1950 on, the tempo of the nonaqueous evolution began to accelerate. In 1948, Moss, Elliott, and Hall (43) developed a method for the titration of very weak acids in ethylenediamine. There followed a proliferation of studies, notably those by Fritz (44–53), Pifer and Wollish (54, 55), Bruss and Wyld (56), Coetzee (57–59), and Kolthoff and Bruckenstein (35, 60–71), who, with their many students and associates, have made the major contributions to advancing the art and science of acid–base titrimetry in nonaqueous solvents to its present state. Earlier literature on the subject was reviewed by Fritz (72) and by Kolthoff and Bruckenstein (66). Later, four books appeared dealing with nonaqueous acid–base titrations (73–76), of which the excellent 1973 monograph by Fritz (73) is the most up to date. For current developments in this field, the reader is referred to the biennial Fundamental Reviews in the April issues of *Analytical Chemistry*.

Some of the reasons why water offers a limited range for acid–base titrations (aside from solubility considerations) were pointed out already in Chapter 3. First of all, water is both an acid and a base, so that it tends to reverse the titration equilibria of weak acids and weak bases as follows:

$$\text{Titration of acids:}\quad HA + OH^- \rightleftharpoons A^- + H_2O \tag{6.53}$$

$$\text{Titration of bases:}\quad B + H_3O^+ \rightleftharpoons BH^+ + H_2O \tag{6.54}$$

As a result, only acids and bases having pK's up to about 9 can be titrated in water. Obviously, reaction 6.53 will go to completion only if the solvent has negligible acidic properties, and reaction 6.54 will go to completion only if the solvent has negligible basic properties.

Another drawback of water as a titration medium for acids and bases is its *leveling effect*. In aqueous solution, many acids are converted quantitatively to the H_3O^+ ion and many bases, to the OH^- ion, so that they cannot be distinguished from each other by titration. Aprotic solvents, on the other hand, are *differentiating*, in the sense that when a mixture of acids or bases of different strengths is titrated in them, separate endpoints may be obtained for many of the components. A quantitative index of the leveling or differentiating property of an amphiprotic solvent is its autoprotolysis constant, or the length of its pa_H scale, which are numerically the same. For water, the autoprotolysis constant $pK_w = 14$, so that the pa_H scale in water is arbitrarily thought of as being 14 units long, extending from $pa_H = 0$ to $pa_H = 14$. For ethanol, the autoprotolysis constant $pK_s = 20$, so that its pa_H^* scale is 20 units long. The longer the pa_H scale, (the larger the pK_s value), the more opportunity for the solvent to differentiate acids and bases of various strengths. A leveling solvent, on the other hand, will have a short pa_H scale.

The pa_H of a solution is directly proportional to the potential of a hydrogen electrode. Since all acids and bases are usually titrated potentiometrically, using glass (hydrogen) and calomel (reference) electrodes, it is practical to express the leveling and differentiating properties of solvents in terms of so-called *acidity potential ranges* derived from the emf of the glass–calomel cells. The acidity potential range of a solvent is expressed as the difference between the potential of the glass–calomel electrode pair (in millivolts) in a solution of a strong acid and a strong base (usually both are 10^{-2} *M*) in that solvent. Obviously, the acidity potential range represents the maximum height of a potentiometric titration curve that can be obtained for strong acids and strong bases in the given solvent under the specified conditions of concentration and electrodes. It represents a practical measure of the range of acidic and basic strengths that can be titrated in the given solvent. The longer the acidity potential range, the more differentiating the solvent.

On the basis of the acidity potential ranges it is possible to make semiquantitative predictions about the leveling or differentiation of specific acids or bases in a given solvent. To do this, the relative strengths of acids and bases must be expressed in terms of their half-neutralization potentials, hnp, the potentials at which half of the acid (base) has been neutralized by a strong-base or a strong-acid titrant.† Practical applications of acidity potential ranges in conjunction with the hnp's of acids and bases were introduced by van der Heijde (77, 78). If one considers that a successful titration requires a steep rise of about 200–300 mV between the hnp and the next plateau in the titration curve, it is possible to determine whether the titration curve of a given acid or base would "fit" into the acidity potential range of a given solvent and whether or not leveling would occur.

Figure 6.3 displays the acidity potential ranges of some common titration solvents. The more positive potentials represent acidic solutions and the more negative potentials, basic solutions. Acetic acid has a short potential range situated at the acidic end of the scale. The top of its range corresponds to a solution of the acetate ion. Therefore the tallest titration curves in acetic acid will be fairly short, and solutions with hnp's lying above the upper boundary of its potential range, that is, those more basic than the acetate ion, will be leveled down to that strength. Similarly, a strongly basic solvent like ethylenediamine has a short potential range at the opposite end of the scale. The bottom of its potential range corresponds to a solution of ethy-

† At the hnp, in the absence of complications, $pH \cong pK_a$.

lenediammonium ions, to which all stronger acids are leveled in that solvent. In fact, all acids stronger than about pK_a of 5 in water are leveled in ethylenediamine. Acetone and acetonitrile provide the longest acidity potential ranges for the differentiation of acids and bases. However, it should be noted that the acidity potential ranges or the pK_s values of such nominally aprotic solvents as acetone and acetonitrile are determined as much by small amounts of amphiprotic contaminants (such as water) as by their intrinsic properties, so that in their cases a determination of a practical potential range for each batch of solvent is particularly indicated. Published values of pK_s may not apply to the solvent at hand.

6.3.1 Titration of Bases From the general discussion of solvent properties in Chapter 3, it is clear that to be suitable for the titration of weak bases, a solvent must be as weakly basic as possible (see Eqn. 3.26 and discussion). Furthermore, the titrant should be a strong acid dissolved in a nonbasic solvent.

Glacial acetic acid proved to be an excellent solvent for the titration of weak bases, up to a pK_b of about 12 in water. Perchloric acid dissolved either in acetic acid or in some inert solvent, like dioxane, is the usual titrant. The endpoint in the titrations can be detected either potentiometrically with glass and calomel electrodes or with a visual indicator, such as crystal violet. One drawback of acetic acid is that it levels all bases with pK_b's lower than about 9.3 in water by converting them quantitatively to the acetate ion. As a result, most aliphatic and aromatic amines appear to be of equal strength in acetic acid, so that their mixtures yield a single inflection point in a titration curve. To differentiate amines according to their strengths, one must resort to titrations in solvent media that are either partially or totally aprotic. Pifer, Wollish, and Schmall (55) carried out titrations in a solvent composed of 1 part of acetic acid to 10 parts of chloroform to resolve such mixtures as butylamine and pyridine, potassium and sodium acetates, and potassium acetate, butylamine, and pyridine.

The best differentiating media for the titration of bases are the protophobic dipolar aprotic solvents, such as acetonitrile (45, 79), acetone (46), chloroform (55), methyl isobutyl ketone (56), and sulfolane (59, 80, 81). Occasionally titrations have been reported (79) in such solvents as nitromethane, nitrobenzene, ethylene dichloride, and ethyl acetate. Mixtures of aliphatic (stronger) and aromatic (weaker) amines can be resolved in all of those solvents. This is hardly surprising, since aprotic media do have the longest acidity potential ranges (Fig. 6.3). Sulfolane is so weakly basic that even water (57b), alcohols and ketones can be titrated as bases in it. For titration in aprotic media, the titrant is usually perchloric acid dissolved in dioxane. Anhydrous perchloric acid in sulfolane is also used (57a). However, as the anhydrous acid is difficult to prepare and entails the danger of explosion, the suggestion of using as the titrant $HSbCl_6$, which is completely dissociated in sulfolane, may be worth pursuing (82).

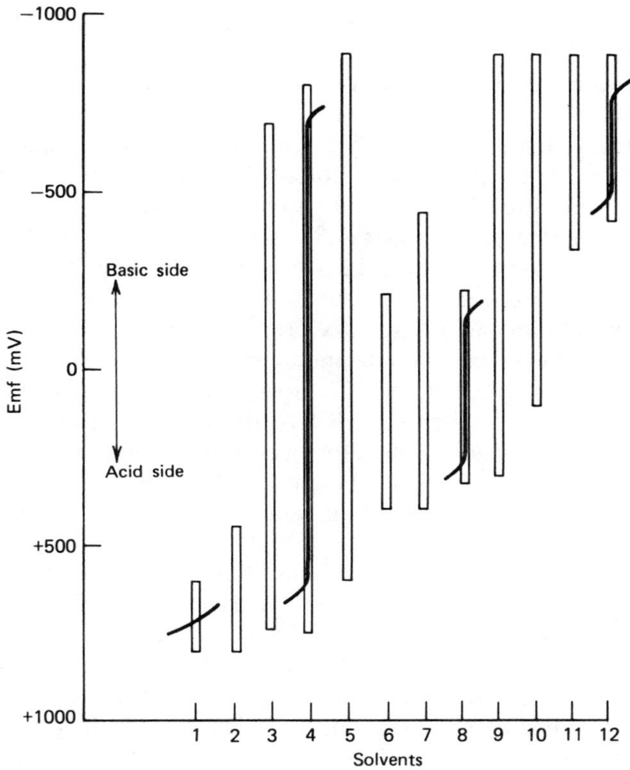

Figure 6.3 Acidity potential ranges of common titration solvents. Potentials are measured with glass and calomel electrodes. The extreme points in the ranges are defined by solutions of 10^{-2} N strong acid or strong base, respectively. Solvents: 1. Trifluoroacetic acid; 2. Acetic acid; 3. Chlorobenzene; 4. Acetone; 5. Acetonitrile; 6. Methanol; 7. Isopropyl alcohol; 8. Water; 9. Dimethylformamide; 10. Pyridine; 11. *n*-Butylamine; 12. Ethylenenediamine. Reprinted with permission from *Anal. Chim. Acta*, **16,** 378 (1957). Copyright 1957. Elsevier Scientific Publishing Co., Amsterdam.

Differentiation between primary, secondary, and tertiary amines can be accomplished with certain reagents that selectively reduce the basicity of a given type of amine. Thus addition of acetic anhydride to a mixture of amines of different type in acetic acid converts the primary and secondary amines to acetamides, which are not titratable in this medium. Tertiary amines can then be determined by difference (83, 84). The differences in the rates of acetylation for various amines are so pronounced that they too can serve as a further aid toward resolution. Furthermore, primary amines in a mixture can be tied up by reaction with salicylaldehyde to form the nontitratable Schiff bases, after which the secondary and tertiary amines, which do not react with salicylaldehyde, can be determined by titration.

Titration of substances so weakly basic that they do not exhibit basic properties in glacial acetic acid or in acetonitrile can be accomplished in acetic anhydride (85, 86). This solvent extends the range of titratable bases to such weakly basic substances as amides, ureas, *N*-heterocyclics, and anions of strong acids. For example, chloride and iodide can be resolved in acetic anhydride by means of a potentiometric titration. Perchloric acid dissolved in a mixture of acetic acid and acetic anhydride is the titrant and the endpoint is detected by a glass–silver-silver chloride electrode system.

One problem with perchloric acid titrant is that commercial perchloric acid contains 28–30% water, corresponding to approximately $HClO_4{\cdot}2H_2O$. Anhydrous perchloric acid titrant is prepared by dissolving the aqueous $HClO_4$ in acetic acid or in sulfolane and then adding enough acetic anhydride to react with all the water. This, of course, adds acetic acid to the titration mixture. Alternatively, when the $HClO_4$ titrant is prepared by simply dissolving the commercial acid in dioxane, a certain amount of water is included in the titration media. It is up to the chemist to decide whether acetic acid or water is the less desirable contaminant in the solutions (66).

6.3.2 Titration of Acids Weak acids are best titrated in solvents that are very weakly acidic (see Eqn. 3.21 and discussion). The titrant should be a strong base dissolved in a nonacidic solvent. Ethylenediamine was the solvent specifically proposed for the titration of very weak acids, such as phenols and enols (43). While ethylenediamine and other very basic solvents (e.g., butylamine) bring out the acidity of weak acids, they level all strong and moderately strong acids. Moreover, they must be used in a dry inert atmosphere because water and carbon dioxide act as acids in them.

A solvent that combines appreciable basicity with a long acidity-potential range and is therefore suitable for the differentiating titrations of weak acids is dimethylsulfoxide (DMSO). Its autoprotolysis constant has been reported (87) to be very low ($pK_s = 33$). An advantage of DMSO over ethylenediamine is its relatively high dielectric constant ($\epsilon = 49$) as compared to that of ethylenediamine ($\epsilon = 13$), which means that complications due to ion-paring equilibria can be avoided in DMSO. Price and Whiting (88) introduced as the titrant the lyate ion of DMSO in the form of its sodium salt, $Na(CH_2SOCH_3)$, which is called dimsyl. The endpoint is detected by means of triphenylmethane indicator, which changes from colorless to red with the excess base. Such weak acids as nitromethane, acetone, alcohols, glycols, phenols, water, amides, thiols, cyclopentadiene, and indene have been successfully titrated in DMSO (88). Nevertheless, such titrations must be carried out under nitrogen. Much attention has been focused therefore on developing titration methods in solvents that can be used in the atmosphere.

One such differentiating solvent for acids is acetone (50). Weakly acidic substances, such as phenols, enols, imides, and sulfonamides as well as carboxylic acids can be titrated in acetone, using tetrabutylammonium hydroxide in benzene–methanol mixtures as the titrant.

Very impressive results on the differentiating titrations of acids (and bases) were reported in methyl isobutyl ketone (MIBK) (56). This solvent has an acidity potential range of about 1450 mV (24 pH units). As shown in Fig. 6.4, in MIBK it is possible to resolve such mixtures of strong acids as $HClO_4$ and H_2SO_4, $HClO_4$ and HNO_3, and the two hydrogens of H_2SO_4. Figure 6.5 illustrates the resolution of five acids, ranging in strength from perchloric acid to phenol. Tetrabutylammonium hydroxide in isopropyl alcohol is the titrant.

Another solvent with a long potential range ($pK_s = 32$?) is acetonitrile. While it is an excellent differentiating medium for the titration of acid or base mixtures, its acid–base equilibria are sometimes complicated by homoconjugation effects (see Section 6.4.2, "Effect of Conjugation on the Shape of Conductometric and Potentiometric Titration Curves"). Other solvents suitable for differentiating titrations of acids are dimethylformamide, pyridine, 2-propanol, and *t*-butyl alcohol (73). The homoconjugation effects are smaller for the more protophilic dipolar aprotic solvents.

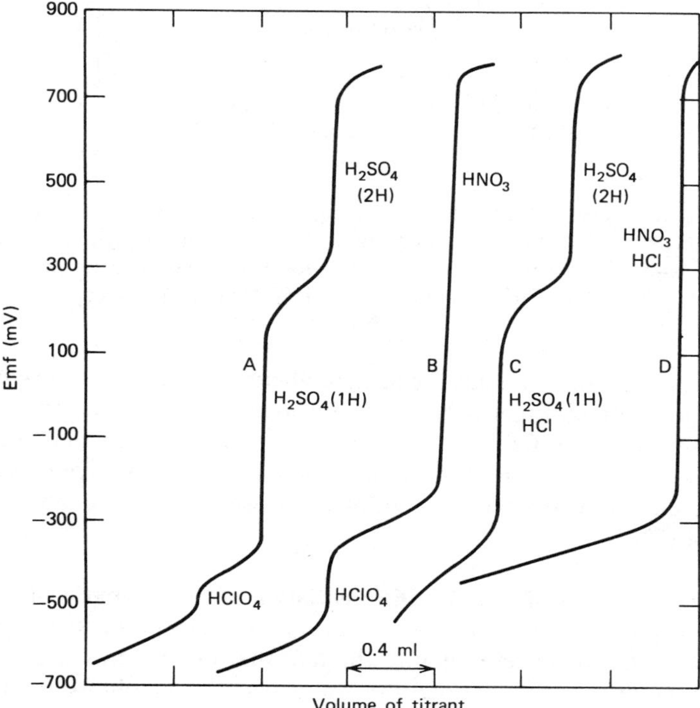

Figure 6.4 Resolution of strong acids in methyl isobutyl ketone; 0.2 *N* tetrabutylammonium hydroxide in isopropyl alcohol is the titrant; glass–calomel electrodes. Reprinted with permission from *Anal. Chem.*, **29**, 232 (1957). Copyright 1957. American Chemical Society.

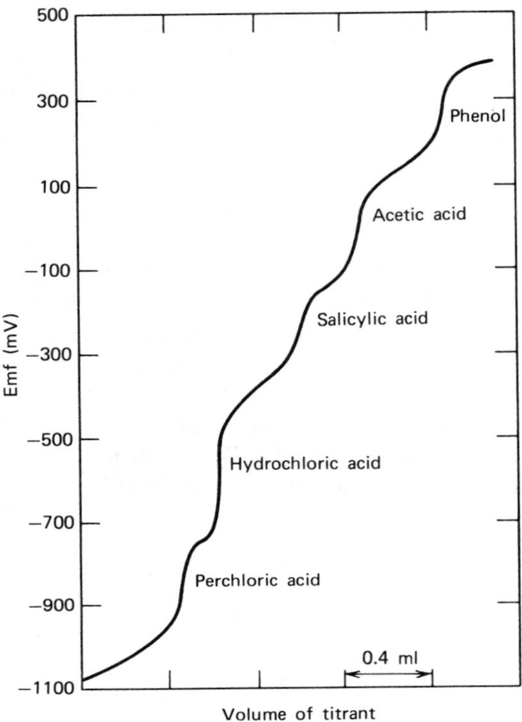

Figure 6.5 Resolution of a mixture of strong, weak and very weak acids in methyl isobutyl ketone; 0.2 *N* tetrabutylammonium hydroxide in isopropyl alcohol is the titrant; glass–platinum (in titrant) electrodes. Reprinted with permission from *Anal. Chem.*, **29**, 232 (1957). Copyright 1957. American Chemical Society.

The usual titrants for acids are tetraalkylammonium hydroxides dissolved in benzene–methanol mixtures or in 2-propanol. Glass electrodes and calomel electrodes modified so that the salt bridge contains methanolic KCl or tetramethylammonium chloride in 2-propanol are employed. A calomel reference electrode specifically modified for use in *t*-butyl alcohol has been described (52).

6.3.3 Theoretical Basis for the Differentiation of the Strengths of Acids and Bases in Dipolar Aprotic Solvents Throughout the preceding section we have repeatedly made reference to the fact that many acids and bases that appear to be of the same strength in water or other amphiprotic media can be differentiated on the basis of their strengths in dipolar aprotic solvents. Such differentiation, or resolution, is achieved when the difference between the p*K*'s of two acids which is small in, say, water, becomes sufficiently large in a dipolar aprotic medium. From Chapter 5 (e.g., Eqn. 5.10a), we recall that the difference between the p*K*'s of an acid in water and in a nonaqueous solvent is governed by the transfer activity coefficients (medium effects) of all the species involved in the dissociation equilibrium:

$$\Delta pK \equiv p_sK - p_wK = \log {}_m\gamma_H + \log \frac{{}_m\gamma_A}{{}_m\gamma_{HA}} \qquad (6.55)$$

The change in the *difference* between the pK's of two acids, HA and HA_1, in water and the nonaqueous solvent is then given by

$$\Delta\Delta pK(HA_1 - HA) = \log \frac{{}_m\gamma_{A_1}}{{}_m\gamma_{HA_1}} - \log \frac{{}_m\gamma_A}{{}_m\gamma_{HA}} \qquad (6.56)$$

Kolthoff and Chantooni (62, 63) have made extensive studies of the factors that affect resolution of acid strengths as formulated in Eqn. 6.56. They noted that, since the resolution is independent of the transfer activity coefficient for the proton, it should be independent of the basicity of the solvent as well. Indeed, when comparisons were made between benzoic acid and benzoic acids substituted in positions other than ortho, the values of $\Delta\Delta pK$ (Eqn. 6.56) were fairly constant (2.2–2.6 log units) for a series of protophilic and protophobic aprotic solvents (63, 89).

In their analysis of Eqn. 6.56 as applied to substituted benzoic acids, Chantooni and Kolthoff (62) made the usual assumptions that the ${}_m\gamma$ of an anion can be thought of as consisting of an electrostatic, ${}_m\gamma$(el), and a nonelectrostatic component and that the latter was equal to the nonelectrostatic component of ${}_m\gamma_{HA}$ for the corresponding acid (cf. Eqns. 5.34 and 5.35b). Then they introduced two additional assumptions: (1) that ${}_m\gamma_{HA}$ can be formulated as a composite of the nonelectrostatic and a hydrogen-bonding contribution, and (2) that the former could be identified with the ${}_m\gamma$ value for the corresponding methyl ester. After determining the ${}_m\gamma$ values for the esters of the benzoic acids studied, they found that the hydrogen-bonding contribution to ${}_m\gamma_{HA}$ was fairly constant, characteristic of the solvent only. Thus, Eqn. 6.56 could be simplified to:

$$\Delta\Delta pK = \log {}_m\gamma_{A_1}(\mathrm{el}) - \log {}_m\gamma_A(\mathrm{el}) \qquad (6.57)$$

According to Eqn. 6.57, the resolution of the strengths of these benzoic acids in dipolar aprotic solvents, relative to water, is determined mainly by the electrostatic components of the medium effects of the acid anions. The values of log ${}_m\gamma_A$(el) were estimated to be 9.2, 9.6, 8.0, and 6.9 for the (unsubstituted) benzoate, 3,4-dimethyl-, 3,4-dichloro-, and 3,5-dinitrobenzoate, respectively (62b). These large medium effects are primarily a function of the H-bonding ability of the anions in water and their ion–dipole and dispersion interactions in the nonaqueous solvent. It is significant that acids whose anions are not hydrogen bonded in amphiprotic solvents, do not exhibit resolution when transferred to aprotic solvents. For example, the ΔpK's of some fluorenes and of fluoradene between methanol and DMSO are determined mainly by the ${}_m\gamma_H$, that is, by the difference in the basicity of the two solvents and lead to no resolution (63).

Dipolar aprotic solvents are particularly effective in increasing the difference between the strengths of pairs of acids of the H_2A and HA^- type. Thus, for succinic acid in water, $(pK_2 - pK_1) = 1.4$, but in acetonitrile, that difference is 11.4. The resulting large resolution is attributed primarily to a much greater solvation in water of the A^{2-} ions as compared to HA^-, which in the case of succinates in acetonitrile is reflected by the medium effects (63): $\log {}_m\gamma_{HA^-} = 6.5$ and $\log {}_m\gamma_{A^{2-}} = 21.9$.

Summary The appreciable acidity and basicity of water imposes limits whereby acids and bases weaker than about $pK \approx 9$ cannot be titrated in aqueous solutions. Furthermore, water levels many strong acids and bases. Proper choice of nonaqueous media, however, expands considerably the limits of detection as well as resolution of acids and bases. Weak bases must be titrated in a weakly basic solvent, such as glacial acetic acid, acetonitrile, acetone, methyl isobutyl ketone, sulfolane, or chloroform. A typical titrant is perchloric acid dissolved in glacial acetic acid or in dioxane. However, glacial acetic acid is a leveling solvent for a wide range of basic strengths. Titration of weak acids requires a weakly acidic solvent, such as acetone, methyl isobutyl ketone, dimethylformamide, pyridine, 2-propanol, *t*-butyl alcohol, dimethylsulfoxide, and ethylenediamine. The last, however, is a leveling medium for most acids. A preferred titrant for acids is tetrabutylammonium hydroxide dissolved in benzene–methanol mixtures or in isopropyl alcohol. The best media for differentiating titrations for mixtures of acids or bases are dipolar aprotic solvents, primarily ketones and acetonitrile. Factors that determine the resolution of acid mixtures are discussed in terms of the transfer activity coefficients of the acid and its anion.

6.4 QUANTITATIVE TREATMENT OF ACID–BASE EQUILIBRIA

In acid–base chemistry, practical applications have often preceded theoretical developments. Thus, acid–base reactions were being carried out in aprotic media long before they could be accounted for by the existing concepts of acidity. Acid–base titrations in nonaqueous solvents, notably in glacial acetic acid, were practiced as an empirical art for about thirty years before they were given a detailed theoretical interpretation. It is primarily the contributions of Kolthoff, Bruckenstein, and their associates that have helped change the status of nonaqueous acid–base titrimetry from an art to a science. In Section 6.4.1 we present a summary of their treatment of acid–base equilibria in acetic acid and, in Section 6.4.2, in acetonitrile. Thus we cover an amphiprotic medium of low dielectric constant and a dipolar aprotic medium.

6.4.1 Equilibria in Acetic Acid Although acid–base titrations in acetic acid date back to the work of Conant, Hall, and Werner in the years 1927–1930

(38–42), their quantitative interpretation had to await the series of papers by Kolthoff and Bruckenstein published some thirty years later (60, 64, 65). Some of their equations are presented in the overview of the chemistry in acetic acid given in Chapter 3. However, for the sake of continuity in reading this section, these equations are repeated here. We recall that an acid or a base in acetic acid undergoes first *ionization* to form an ion pair, followed by *dissociation* of the ion pair into free ions:

$$\text{Acid:}\quad HA + HAc \underset{\text{ionization}}{\rightleftharpoons} H_2Ac^+A^- \underset{\text{dissociation}}{\rightleftharpoons} H_2Ac^+ + A^- \tag{6.58}$$

$$\text{Base:}\quad B + HAc \underset{\text{ionization}}{\rightleftharpoons} BH^+Ac^- \underset{\text{dissociation}}{\rightleftharpoons} BH^+ + Ac^- \tag{6.59}$$

Replacing the symbol H_2Ac^+ for the solvated proton by H^+, we define the corresponding ionization constants, K_i, and dissociation constants, K_d, for acids:

$$K_i^{HA} = \frac{(H^+A^-)}{(HA)} \tag{6.60}$$

$$K_d^{HA} = \frac{(H^+)(A^-)}{(H^+A^-)} \tag{6.61}$$

and for bases:

$$K_i^{B} = \frac{(BH^+Ac^-)}{(B)} \tag{6.62}$$

$$K_d^{B} = \frac{(BH^+)(Ac^-)}{(BH^+Ac^-)} \tag{6.63}$$

For purposes of making most equilibrium calculations, the governing constants are the so-called *overall dissociation constants,* K_{HA} and K_B:

$$K_{HA} = \frac{(H^+)(A^-)}{[(HA) + (H^+A^-)]} = \frac{K_i^{HA}K_d^{HA}}{1 + K_i^{HA}} \tag{6.64}$$

$$K_B = \frac{(BH^+)(Ac^-)}{[(B) + (BH^+Ac^-)]} = \frac{K_i^{B}K_d^{B}}{1 + K_i^{B}} \tag{6.65}$$

Throughout this treatment, activity coefficients have been omitted, so that the quantities in parentheses in the equilibrium expressions are the equilibrium *concentrations*. The total (analytical) concentration of an acid or a base added to the solution is given the symbol C_{HA} and C_B, respectively.

In terms of ionic dissociation, all acids and bases in acetic acid are weak. The strongest acids exist predominantly in the form of H^+X^- ion pairs, and the strongest bases, in the form of BH^+Ac^- ion pairs. Kolthoff and Bruckenstein determined the overall dissociation constants for a number of acids and bases as well as the autoprotolysis constant of acetic acid ($pK_s = 14.45$). Even the strongest acid, $HClO_4$, is a weak electrolyte in acetic acid, having a pK_{HA} of only 4.87. On the basis of these data and the equations just presented, Kolthoff and Bruckenstein developed the following interpretations of acid–base equilibria.

(H^+) in a Pure Solution of an Acid or a Base Since electrolytes in acetic acid are only slightly dissociated, the equilibrium concentration of undissociated acid or base can be equated to the corresponding analytical concentration. Thus, $C_{HA} \cong [(HA) + (H^+A^-)]$ and $C_B \cong [(B) + (BH^+Ac^-)]$. It follows that for acids:

$$(H^+) \cong (K_{HA}C_{HA})^{1/2} \tag{6.66}$$

and for bases:

$$(H^+) \cong \frac{K_s}{(K_B C_B)^{1/2}} \tag{6.67}$$

where K_s is the autoprotolysis constant, $K_s = (H^+)(Ac^-)$. Equations 6.66 and 6.67 are identical to those for a weak acid and a weak base in water. Thus, the pH of all acids and bases in acetic acid changes by 0.5 units upon tenfold dilution.

(H^+) in Mixtures of Base and Its Perchlorate Typically, a base is titrated in acetic acid with perchloric acid:

$$B + HClO_4 \rightleftharpoons BH^+ClO_4^- \tag{6.68}$$

and the quantitative expression for the extent of this neutralization reaction is given by the formation constant, $K_f^{BHClO_4}$:

$$K_f^{BHClO_4} = \frac{(BH^+ClO_4^-)}{(B)(HClO_4)} \tag{6.69}$$

The ion pair $BH^+ClO_4^-$ dissociates slightly into ions ($K_{BHClO_4} \cong 10^{-5}$–$10^{-6}$):

$$K_{BHClO_4} = \frac{(BH^+)(ClO_4^-)}{(BH^+ClO_4^-)} \tag{6.70}$$

Using the overall dissociation constants and the law of electroneutrality, which for these solutions is $(H^+) + (BH^+) = (ClO_4^-) + (Ac^-)$, it is possible to derive an expression for the hydrogen-ion concentration in mixtures of a base and its perchlorate, thus describing the neutralization curves for bases with perchloric acid:

$$(H^+) = \left[\frac{K_s[1 + (K_{BHClO_4}(BH^+ClO_4^-)/K_B C_B)]}{[1 + (K_B C_B/K_s)]}\right]^{1/2} \tag{6.71}$$

When a base is strong enough to be titrated, $(K_B C_B/K_s) \gg 1$, so that Eqn. 6.71 can be simplified to:

$$(H^+) = \frac{K_s[K_{BHClO_4}(BH^+ClO_4^-) + K_B C_B]^{1/2}}{K_B C_B} \tag{6.72}$$

Equation 6.72 formed the basis for the explanation of a number of curious phenomena that could not be accounted for earlier. For example, we are used to the fact that the hydrogen-ion concentration of a buffer solution in water is nearly independent of dilution. However, Hall and Werner (42) found that a tenfold dilution of a partially titrated base in acetic acid caused the pH to decrease by 0.5 units. This phenomenon was not understood at the time, but an inspection of Eqn. 6.72 reveals that in acetic acid the (H^+) should increase as the square root of dilution.

In water, the basicity constant K_B can be estimated from the pH at the 50% neutralization point, since then $K_w(H^+)^{-1} = K_B$. In acetic acid, the 50% neutralization point corresponds to an equimolar mixture of B and $BH^+ClO_4^-$, so that $(BH^+ClO_4^-) = C_B = \frac{1}{2}C$, where C is the total analytical concentration of base B. Introducing C into Eqn. 6.72, we obtain for the hydrogen-ion concentration at the half-neutralization point:

$$(H^+) = \frac{K_s[2(K_{BHClO_4} + K_B)/C]^{1/2}}{K_B} \qquad (6.73)$$

Evidently the pH at the 50% neutralization point in acetic acid varies with the concentration of the base and incorporates the dissociation constant of the base perchlorate as well.

In solutions of pure perchlorate salts, the hydrogen-ion concentration is given by

$$(H^+) = \left[\frac{K_s K_{HClO_4}}{K_B}\right]^{1/2} \qquad (6.74)$$

The above expression implies that the pH at the equivalence point of a titration of an uncharged base with perchloric acid is independent of the concentration of base and acid. It has the same form as that for a salt of a weak acid and a weak base in water. It should be noted, however, that dissociation has been neglected in the derivation of Eqn. 6.74.

The hydrogen-ion concentration for mixtures of a perchlorate salt with perchloric acid, that is, for solutions beyond the equivalence point in a titration of a base with perchloric acid, is given by

$$(H^+) = \frac{K_{HClO_4}C_{HClO_4}}{[K_{HClO_4}C_{HClO_4} + K_{BHClO_4}(BH^+ClO_4^-)]^{1/2}} \qquad (6.75)$$

The above equation was derived from Eqn. 6.72 on the assumption that in excess $HClO_4$, $(K_{HClO_4}C_{HClO_4}) \gg K_s$. It can be seen that for acid-salt mixtures in acetic acid, the hydrogen-ion concentration is inversely proportional to the square root of dilution.

The negligible degree of dissociation of electrolytes in acetic acid has a peculiar consequence in the behavior of acid–base indicators. Whereas in aqueous solutions such indicators respond to hydrogen-ion concentration, in acetic acid the indicator equilibrium is determined by the concentration of the free strong acid (usually perchloric). How this comes about can be shown as follows. An indicator base reacts with perchloric acid to form the corresponding ion pair:

$$I + HClO_4 \rightleftharpoons IH^+ClO_4^- \tag{6.76}$$

$$K_f^{IHClO_4} = \frac{(IH^+ClO_4^-)}{(I)(HClO_4)} \tag{6.77}$$

We can neglect the dissociation of $IH^+ClO_4^-$ into ions because, under the conditions existing near an equivalence point in a titration of a base, that dissociation would be suppressed due to the presence of the base perchlorate. Thus, the ratio of the indicator in its acid and base forms is obtained from Eqn. 6.77 as

$$\frac{(IH^+ClO_4^-)}{(I)} = K_f^{IHClO_4}(HClO_4) \tag{6.78}$$

The value of $(HClO_4)$ in the course of a titration is determined by the formation constant $K_f^{BHClO_4}$ (Eqn. 6.69). Combining it with Eqn. 6.78, we obtain

$$\frac{(IH^+ClO_4^-)}{(I)} = \frac{K_f^{IHClO_4}(BH^+ClO_4^-)}{K_f^{BHClO_4}(B)} \tag{6.79}$$

Apparently the color of a base indicator in acetic acid is independent of hydrogen-ion concentration. Instead, it is governed by the salt: base ratio, which, as long as solvolysis is not appreciable, will not be affected by dilution. In solutions of pure salt $BH^+ClO_4^-$, which correspond to equivalence points in the titration of base B with perchloric acid, $(HClO_4) = (B)$ (this follows from Eqn. 6.68), so that

$$(HClO_4) = \left[\frac{(BH^+ClO_4^-)}{K_f^{BHClO_4}}\right]^{1/2} \tag{6.80}$$

and

$$\frac{(IH^+ClO_4^-)}{(I)} = K_f^{IHClO_4}\left[\frac{(BH^+ClO_4^-)}{K_f^{BHClO_4}}\right]^{1/2} \tag{6.81}$$

Thus the color of an indicator at the equivalence point varies approximately as the square root of the salt concentration, while the pH is independent of the concentration.

6.4.2 Equilibria in Acetonitrile Our discussion of acid–base equilibria in acetonitrile is a brief summary of the treatment presented in the many publications by Kolthoff and Chantooni (62, 63, 67–71) as well as by Coetzee et al. (57–59) and their associates. The fundamentals of acid–base chemistry in acetonitrile are reviewed in Chapter 3 (see Eqns. 3.54–3.65 and discussion). We recall that the process of ionization followed by simple dissociation, which is typical of the behavior of all acids in acetic acid medium, offers a valid description in acetonitrile only for the equilibria of those acids whose anions have delocalized charge and for extremely dilute solutions of other acids. Thus the simple dissociation is known to hold for picric acid and 2,6-dinitrophenol, whose anions do not require hydrogen bonding for their stabilization, but are solvated primarily by the dispersion interactions with acetonitrile molecules. Anions of other acids (carboxylic, inorganic) are solvated in amphiprotic media by accepting hydrogen bonds from the solvent molecules. In aprotic media, where the solvent provides no hydrogen bonds, they are "solvated" by accepting hydrogen bonds either from their conjugate acids HA to form the homoconjugate anions HA_2^-, or from other H-bond donors HR, forming heteroconjugate anions AHR^-. Some consequences of conjugation are discussed in Chapter 3. Here, we will consider its effect on the hydrogen-ion concentration and on the shape of potentiometric and conductometric acid–base titration curves.

Hydrogen-Ion Concentration in Pure Acid Solution The overall dissociation of acids in acetonitrile is determined by a combination of the simple dissociation followed by homoconjugation:†

Simple dissociation:

$$HA \rightleftharpoons H^+ + A^- \qquad K_d^{HA} = \frac{(H^+)(A^-)f^2}{(HA)} \tag{6.82}$$

Homoconjugation:

$$HA + A^- \rightleftharpoons HA_2^- \qquad K_f^{HA_2} = \frac{(HA_2^-)}{(HA)(A^-)} \tag{6.83}$$

Overall dissociation:

$$2HA \rightleftharpoons H^+ + HA_2^- \qquad K_2^{HA} = \frac{(H^+)(HA_2^-)f^2}{(HA)^2} = K_d^{HA}K_f^{HA_2} \tag{6.84}$$

Consequently,

$$(H^+) = (K_2^{HA})^{1/2}(HA) \tag{6.85}$$

In the above equations, H^+ denotes the *solvated* proton, f is the activity coefficient of all univalent ions, and the quantities in parentheses are the equilibrium *concentrations*. Since different processes predominate depending on the concentration and the magnitude of the equilibrium constants, it is more informative to solve for the (H^+) from the law of electroneutrality. In solutions of pure acid:

$$(H^+) = (A^-) + (HA_2^-) \tag{6.86}$$

Substituting into Eqn. 6.86 from the equilibrium constants, we obtain the

† In the original literature, the superscripts and subscripts on the equilibrium constants were reversed, e.g., K_{HA}^{d}. Here, we are being consistent with the symbolism used for equilibria in acetic acid and in liquid ammonia. Also, the ambiguous literature symbol K_2(HA) was replaced by K_2^{HA}. Ionic charges were dropped from subscripts and superscripts.

general expression for the hydrogen-ion concentration:

$$(H^+) = (1/f)\,[K_d^{HA}(HA)\,(1 + K_f^{HA_2}(HA))]^{1/2} \tag{6.87}$$

Under conditions where the acid concentration is low and K_d^{HA} is not very large, $1 \gg K_f^{HA_2}(HA)$, so that when $f \approx 1$, we can write

$$(H^+) = [K_d^{HA}(HA)]^{1/2} \tag{6.88}$$

which is the familiar expression for the hydrogen-ion concentration in a solution of a weak acid undergoing simple dissociation in any solvent. At higher acid concentrations, $K_f^{HA_2}(HA) \gg 1$ and Eqn. 6.87 assumes another limiting form:

$$(H^+) = (1/f)\,(K_d^{HA}K_f^{HA_2})^{1/2}(HA) \tag{6.89}$$

We can see that in the last expression the hydrogen-ion concentration varies directly with the acid concentration and we observe the so-called pseudo-strong acid behavior. Equation 6.89 is of course identical with Eqn. 6.84. For very strong acids, $(H^+) \approx C_{HA}$, the analytical concentration of the acid. In acetonitrile, only $HClO_4$ is thought to be completely dissociated,† so that in its solutions, $(H^+) = C_{HClO_4}$.

Hydrogen-Ion Concentration in Mixtures of Acid and Its Salt Kolthoff and Chantooni (69) derived the following relationship between (H^+) and the analytical concentrations of an undissociated acid HA, C_{HA}, and of its completely dissociated tetraalkylammonium salt, C_s. Starting with the mass balance for the salt,

$$C_s = (A^-) + (HA_2^-) \tag{6.90}$$

and for the weak acid,

$$C_{HA} = (HA) + (HA_2^-) \tag{6.91}$$

and introducing the appropriate equilibrium expressions, they obtained the following relationship for the hydrogen-ion activity:

$$a_H{}^2 f^2 C_s - a_H f K_d^{HA} \times [(C_{HA} + C_s) + K_f^{HA_2}(C_s + C_{HA})^2] + (K_d^{HA})^2 C_{HA} = 0 \tag{6.92}$$

The above equation describes all points in a titration curve of a weak acid with a tetraalkylammonium hydroxide.‡ It is significant that at the midpoint of this titration, when $C_{HA} = C_s$, Eqn. 6.92 reduces to $a_H f = K_d^{HA}$, as in the case of a simple acid dissociation. The last relationship could be derived simply by equating the expressions 6.90 and 6.91 and then introducing the

† See, however, the footnote on p. 60, Chapter 3.

‡ Strictly speaking, in a titration with a base B, C_s in the above expressions would have to be replaced by C_B, the analytical concentration of the added base.

result into Eqn. 6.82. Thus, the pa_H at the half-neutralization point in acetonitrile depends only on K_d^{HA}, but is independent of C_{HA} or of $K_f^{HA_2}$.

When the acid is in excess in the mixture with its salt and when $K_f^{HA_2}$ is very large, then at $(C_{HA} + C_s)$ of the order of 10^{-2} M or greater we observe that (H^+) will vary directly with dilution—a tenfold decrease in (H^+) results from a tenfold dilution (Eqn. 6.84).

Hydrogen-Ion Concentration in Solutions of Pure Salt In solutions of pure salt BH^+A^-, which would correspond to an equivalence point in an acid–base titration, the hydrogen-ion concentration is governed by the dissociation of the protonated base, BH^+:

$$BH^+ \rightleftharpoons H^+ + B \qquad K_d^{BH} = \frac{(H^+)(B)}{(BH^+)} \tag{6.93}$$

However, the salts BH^+A^- are incompletely dissociated in acetonitrile, their K_d magnitudes ranging from 10^{-3} to 10^{-5}. Taking into account

$$K_d^{BHA} = \frac{(BH^+)(A^-)}{(BH^+A^-)} \tag{6.94}$$

we obtain

$$(H^+) = (K_d^{BH})^{1/2} K_d^{BHA} (BH^+A^-)^{1/4} \tag{6.95}$$

Indicator Measurements and the pa_H Scale Because $HClO_4$ is completely dissociated in acetonitrile, it was possible to determine the protonation constants for indicator bases in a straightforward manner. In acid solution, a basic indicator I undergoes the protonation reaction:

$$I + H^+ \rightleftharpoons IH^+ \tag{6.96}$$

for which

$$K_I = \frac{(IH^+)f_{IH}}{(I)(H^+)f_I f_H} \tag{6.97}$$

When (I) can be determined spectrophotometrically, (H^+) is equal to the analytical concentration of $HClO_4$ and the assumption is made that $f_{IH} = f_H$ and that $f_I = 1$, the indicator constant can be evaluated. In this manner, values of K_I were obtained for *o*-nitroaniline, *o*-nitro-*p*-chloroaniline, and *o*-nitrodiphenylamine (67). Subsequently, *o*-nitroaniline ($pK_I = -4.85$) was employed in mixtures with perchloric acid to calibrate the glass electrode according to the relationship

$$a_H = \frac{(IH^+)f_{IH}}{(I)K_I} \tag{6.98}$$

in which the activity coefficient was calculated from the simple Debye–Hückel expression in the form $-\log f_{IH} = 1.53\ \mu^{1/2}$, where μ was the ionic strength.

The calibrated glass electrode was used to assign the pa_H values to a number of buffer systems. Most recommended as buffers in acetonitrile are mixtures of picric acid and tetrabutylammonium picrate, because there the homoconjugation effects can be neglected and the pa_H can be calculated using simple acid dissociation. The picrate buffers cover a pa_H range from 9.6 to 12.2. Reproducible glass electrode potentials were also obtained for mixtures of methanesulfonic acid and tetraethylammonium methanesulfonate (pa_H = 5.86–11.76) and for mixtures of 2,5-dichlorobenzenesulfonic acid with its tetraethylammonium salt (pa_H = 4.27–7.85) (69).

Effect of Conjugation on the Shape of Conductometric and Potentiometric Titration Curves When carboxylic acids and several other types of acids are titrated with amines in aprotic solvents, conductance maxima are exhibited

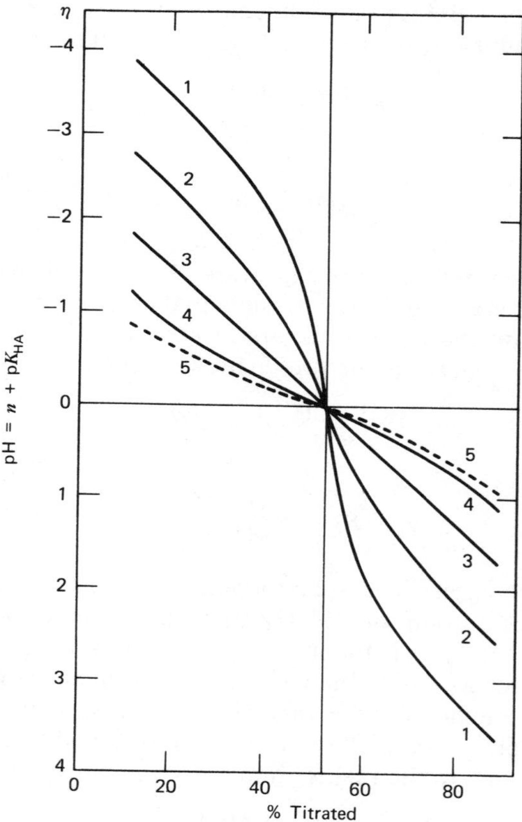

Figure 6.6 Calculated potentiometric titration curves of HA with Et_4NOH for various values of $(C_{HA} + C_s)K_f^{HA_2}$: curve 1 $(C_{HA} + C_s)K_f^{HA_2} = 10^3$; curve 2, 10^2; curve 3, 10^1; curve 4, 1; curve 5, $K_f^{HA_2} = 0$, i.e., simple acid dissociation. Reprinted with permission from *J. Amer. Chem. Soc.*, **87**, 4428 (1965). Copyright 1965. American Chemical Society.

in the neighborhood of 50% neutralization when the acid–base reaction is quantitative or nearly so. The cause for the maxima is homoconjugation. The simple product of reaction between base B and acid HA is a weakly dissociated salt BH^+A^- ($K_d = 10^{-3}$–10^{-5}). However, when the value of $K_f^{HA_2}$ is large, the salt BH^+A^- is converted to $BH^+HA_2^-$, which is a strong electrolyte. The homoconjugate anion reaches its maximum concentration at 50% neutralization and then both its concentration and the conductance of the solution decrease due to the formation of BH^+A^-. The shapes of conductometric titration curves in acetonitrile have been quantitatively described by Kolthoff and Chantooni (68).

Similarly, the potentiometric titration curves of uncharged acids with tetraalkylammonium hydroxides in protophobic aprotic solvents exhibit inflection points at 50% neutralization because of homoconjugation. Inflection points in acid–base titration curves, such as occur at the equivalence point, are known to be points of minimum buffer capacity, where the latter is defined as the change in pH (dpH) resulting from addition of an infinitesimal amount of strong base (dC_B), buffer capacity $\equiv$ (dC_B/dpH). It was shown by Kolthoff and Chantooni (69) that when the quantity $(C_{HA} + C_s)K_f^{HA_2}$ is greater than 10, the buffer capacity has a minimum at the half neutralization point, so that an inflection will occur there in a potentiometric titration curve. When the value of $K_f^{HA_2}$ is zero, as would be the case for all acids in water and for 2,6-dinitrophenol in acetonitrile, the buffer capacity reaches its maximum at the half-neutralization point and a normal titration curve is observed. These extreme cases as well as intermediate situations are illustrated in Fig. 6.6. The family of curves in Fig. 6.6 could also correspond to titration curves of the same acid at different concentrations.

Protophilic aprotic solvents, such as DMSO, compete more successfully with the anions for the acid molecules and allow little homoconjugation to occur. There the acid–base titration curves approach the same shapes as in water.

Summary Acid–base equilibria in glacial acetic acid and in acetonitrile have been interpreted quantitatively in great detail. In acetic acid, all acids and bases are weakly dissociated. The strongest acids and bases exist predominantly in the form of ion pairs H^+X^- and BH^+Ac^-, respectively. The pH of pure acid or base solutions changes by 0.5 units upon tenfold dilution. In mixtures of a base and its perchlorate, the pH decreases by 0.5 units upon tenfold dilution. In the titration of a base with an acid, the pH of a solution at the 50% neutralization point varies with the base concentration, but at the equivalence point, the pH is independent of the concentration of base and acid. In mixtures of perchloric acid with base perchlorate, the hydrogen ion concentration is inversely proportional to the square root of dilution. In the course of an acid–base titration in acetic acid, the color of a base indicator is independent of the pH and at the equivalence point it varies as the square root of the salt concentration. In acetonitrile, most acids HA undergo simple dissociation followed by homoconjugation, forming the anion HA_2^-. Homoconjugation is the cause for the maxima in the conductometric titration curves for acids and bases and the inflection points in the potentiometric titration curves of acids and bases, both appearing at the 50% neutralization points. At low acid concentrations, the equilibria in acetonitrile are characteristic of weak acids, while at high concentrations, we observe pseudo-strong acid behavior. An equation governing the hydrogen-ion concentration in mixtures of an acid and its salt is shown. At the 50% neutralization point, the a_H is equal approximately to the simple dissociation constant of the acid. Solutions of perchloric acid containing acetic acid act as completely dissociated in acetonitrile. They were used to determine the protonation constants of indicator bases, which in turn served to calibrate the glass electrode for pH measurements. pa_H values were assigned to several buffer solutions in acetonitrile.

LITERATURE CITED

1 Hantzsch, A., *Z. Elektrochem.*, **29,** 221 (1923); **30,** 194 (1924); **31,** 167 (1925); *Ber.*, **58,** 612, 941 (1925); **59,** 793, 1096 (1926); **60,** 1933 (1927); *Z. Physik. Chem.*, **125,** 251 (1927); A. Hantzsch and W. Voigt, *Ber.*, **62,** 975 (1929).

2 Brønsted, J. N., *Rec. Trav. Chim.*, **42,** 718 (1923).

3 Bell, R. P., *The Proton in Chemistry,* Cornell University Press, Ithaca, New York, 1959.

4 Lewis, G. N., *Valence and the Structure of Atoms and Molecules,* Chemical Catalog, New York, 1923.

5 Luder, W. F., and S. Zuffanti, *The Electronic Theory of Acids and Bases,* Wiley, New York, 1946.

6 Kolthoff, I. M., "Concepts of Acids and Bases," in I. M. Kolthoff and P. J. Elving, Eds., *Treatise on Analytical Chemistry,* Part I, Vol. V, Wiley, New York, 1966, Chap. 11.

7 VanDyke, R. E., and H. E. Crawford, *J. Amer. Chem. Soc.*, **71,** 2694 (1949).

8 Baaz, M., V. Gutmann, and M. Y. A. Talaat, *Monatsh. Chem.*, **91,** 548 (1960).

9 Spandau, H., and E. Brunneck, *Z. Anorg. Allgem. Chem.*, **270,** 201 (1952).

10 Kolthoff, I. M., D. Stocesova, and T. S. Lee, *J. Amer. Chem. Soc.*, **75,** 1834 (1953).

11 Zenchelsky, S. T., Y. Periale, and J. C. Cobb, *Anal. Chem.*, **28,** 67 (1956).

12 Gutmann, V., and M. Baaz, *Monatsh. Chem.*, **90,** 271, 729 (1959); M. Baaz, V. Gutmann, and L. Hübner, *Monatsh. Chem.*, **91,** 537 (1960); **92,** 272 (1961).

13 Cade, J. A., M. Kasrai, and I. R. Ashton, *J. Inorg. Nucl. Chem.*, **27,** 2375 (1965).

14 Meek, D. W., and R. S. Drago, *J. Amer. Chem. Soc.*, **83,** 4322 (1961).

15 Meek, D. W., "Lewis Acid–Base Interactions," in J. J. Lagowski, Ed., *The Chemistry of Non-Aqueous Solvents,* Vol. I, Academic Press, New York and London, 1966, Chap. 1.

16 Baaz, M., V. Gutmann, and L. Hübner, *J. Inorg. Nucl. Chem.*, **18,** 276 (1961).

17 Baaz, M., V. Gutmann, L. Hübner, F. Mairinger, and T. S. West, *Z. Anorg. Allgem. Chem.*, **311,** 302 (1961).

18 Schmulbach, C. D., *J. Inorg. Nucl. Chem.*, **26,** 745 (1964).

19 Drago, R. S., D. W. Meek, M. D. Joesten, and L. LaRoche, *Inorg. Chem.*, **2,** 124 (1963).

20 Gutmann, V., *Coordination Chemistry in Non-Aqueous Solutions,* Springer, New York, 1968.

21 Drago, R. S., and B. B. Wayland, *J. Amer. Chem. Soc.*, **87,** 3571 (1965).

22 Drago, R. S., G. C. Vogel, and T. E. Needham, *J. Amer. Chem. Soc.*, **93,** 6014 (1971).

23 Usanovich, M., *Zhur. Obshchei Khim.*, **9,** 182 (1939).

24 Bates, R. G., *Determination of pH,* Wiley, New York, 1964, Chaps. 7 and 8.

25 Bates, R. G., M. Paabo, and R. A. Robinson, *J. Phys. Chem.*, **67,** 1833 (1963).

26 Hammett, L. P., *Chem. Rev.*, **13,** 61 (1933); **16,** 67 (1935). *Physical Organic Chemistry,* McGraw Hill, New York, 1940, Chap. 9.

27 Hammett, L. P., and A. J. Deyrup, *J. Amer. Chem. Soc.*, **54,** 2721, 4239 (1932).

28 Hammett, L. P., and M. A. Paul, *J. Amer. Chem. Soc.*, **56,** 827 (1934).

29 Paul, M. A., and F. A. Long, *Chem. Rev.*, **57,** 1 (1957).

30 Boyd, R. H., "Acidity Functions," in J. F. Coetzee and C. D. Ritchie, Eds., *Solute–Solvent Interactions,* Vol. I, Marcel Dekker, New York, 1969, Chap. 3.

31 Popovych, O., and A. J. Dill, *Anal. Chem.*, **41,** 456 (1969).

32 Bowden, K., *Can. J. Chem.*, **43,** 2624 (1965).

33 Langford, C. H., Ph.D. thesis, Northwestern University, Evanston, Illinois, 1960, as cited in Ref. 30, p. 214.

34 Schwarzenbach, G., *Helv. Chim. Acta,* **26,** 418 (1943).

35 Bruckenstein, S., *J. Amer. Chem. Soc.*, **82,** 307 (1960).

36 Strehlow, H., "Electrode Potentials in Non-Aqueous Solvents," in J. J. Lagowski, Ed., *The Chemistry of Non-Aqueous Solvents,* Vol. I, Academic Press, New York, 1966, Chap. 3.

37 Folin, O., and A. H. Wentworth, *J. Biol. Chem.*, **7,** 421 (1910).

38 Conant, J. B., and N. F. Hall, *J. Amer. Chem. Soc.*, **49,** 3047, 3062 (1927).

39 Conant, J. B., and T. H. Werner, *J. Amer. Chem. Soc.*, **52,** 4436 (1930).

40 Hall, N. F., *J. Amer. Chem. Soc.*, **52,** 5115 (1930).

41 Hall, N. F., and J. B. Conant, *J. Amer. Chem. Soc.*, **49,** 3047 (1927).

42 Hall, N. F., and T. H. Werner, *J. Amer. Chem. Soc.*, **50,** 2367 (1928).

43 Moss, M. L., J. H. Elliott, and R. T. Hall, *Anal. Chem.*, **20,** 784 (1948).

44 Fritz, J. S., *Anal. Chem.*, **22,** 578, 1028 (1950); **24,** 306, 674 (1952).

45 Fritz, J. S., *Anal. Chem.*, **25,** 407 (1953).

46 Fritz, J. S., and C. A. Burgett, *Anal. Chem.*, **44,** 1673 (1972).

47 Fritz, J. S., and M. O. Fulda, *Anal. Chem.*, **25,** 1837 (1953).

48 Fritz, J. S., and R. T. Keen, *Anal. Chem.*, **24,** 308, 564 (1952).

49 Fritz, J. S., and N. M. Lisicki, *Anal. Chem.*, **23,** 589 (1951).

50 Fritz, J. S., and S. S. Yamamura, *Anal. Chem.*, **29,** 1079 (1957).

51 Fritz, J. S., A. J. Moye, and M. J. Richard, *Anal. Chem.*, **29,** 1685 (1957).

52 Marple, L. W. and J. S. Fritz, *Anal. Chem.*, **34,** 796 (1962).

53 Vespe, V., and J. S. Fritz, *J. Am. Pharm. Assoc.*, **41,** 197 (1952).

54 Pifer, C. W., and E. G. Wollish, *Anal. Chem.*, **24,** 300 (1952); *J. Am. Pharm. Assoc. (Sci. Ed.),* **40,** 609 (1951).

55 Pifer, C. W., E. G. Wollish, and M. Schmall, *Anal. Chem.*, **25,** 310 (1953).

56 Bruss, D. B., and G. E. Wyld, *Anal. Chem.*, **29,** 232 (1957).

57 Coetzee, J. F., and R. J. Bertozzi, (a) *Anal. Chem.*, **41,** 860 (1969); (b) *Anal. Chem.*, **43,** 961 (1971); **45,** 1064 (1973).

58 Coetzee, J. F., and G. R. Padmanabhan, (a) *J. Phys. Chem.*, **66,** 1708 (1962); (b) *J. Amer. Chem. Soc.*, **87,** 5005 (1965).

59 Muney, W. S., and J. F. Coetzee, *J. Phys. Chem.*, **66,** 89 (1962).

60 Bruckenstein, S., and I. M. Kolthoff, *J. Amer. Chem. Soc.*, **78,** 10, 2974 (1956); **79,** 5915 (1957).

61 Bruckenstein, S., and L. M. Mukherjee, *J. Phys. Chem.*, **66,** 2228 (1962).

62 Chantooni, M. K., Jr. and I. M. Kolthoff, (a) *J. Phys. Chem.*, **77,** 527 (1973); (b) *J. Phys. Chem.*, **78,** 839 (1974).

63 Kolthoff, I. M., *Anal. Chem.*, **46,** 1992 (1974).

64 Kolthoff, I. M., and S. Bruckenstein, *J. Amer. Chem. Soc.*, **78,** 1 (1956).

65 Kolthoff, I. M., and S. Bruckenstein, *J. Amer. Chem. Soc.*, **79,** 1 (1957).

66 Kolthoff, I. M., and S. Bruckenstein, "Acid–Base Equilibria in Nonaqueous Solutions," in I. M. Kolthoff and P. J. Elving, Eds., *Treatise on Analytical Chemistry,* Part I, Vol. V, Wiley, New York.

67 Kolthoff, I. M., S. Bruckenstein, and M. K. Chantooni, Jr., *J. Amer. Chem. Soc.*, **83,** 3927 (1961).

68 Kolthoff, I. M., and M. K. Chantooni, Jr., *J. Amer. Chem. Soc.*, **85,** 2195 (1963); **87,** 1004 (1965).

69 Kolthoff, I. M., and M. K. Chantooni, Jr., *J. Amer. Chem. Soc.*, **87,** 4428 (1965).

70 Kolthoff, I. M., and M. K. Chantooni, Jr., *J. Amer. Chem. Soc.*, **97,** 1376 (1975).

71 Kolthoff, I. M., M. K. Chantooni, Jr., and S. Bhowmik, *Anal. Chem.*, **39,** 315, 1627 (1967).

72 Fritz, J. S., *Acid–Base Titrations in Nonaqueous Solvents,* G. F. Smith Chem. Co., Columbus, Ohio, 1952.

73 Fritz, J. S., *Acid–Base Titrations in Nonaqueous Solvents,* Allyn and Bacon, Boston, Mass., 1973.

74 Gyenes, I., *Titration in Nonaqueous Media,* Van Nostrand Reinhold, New York, 1967.

75 Huber, W., *Titrations in Nonaqueous Solvents,* Academic Press, New York (1967).

76 Kucharsky, J., and L. Šafařik, *Titrations in Nonaqueous Solvents,* Elsevier, Amsterdam, 1965.

77 van der Heijde, H. B., *Anal. Chim. Acta,* **16,** 392 (1957); **17,** 512 (1957).

78 van der Heijde, H. B., and E. A. M. F. Dahmen, *Anal. Chim. Acta,* **16,** 378 (1957).

79 Hall, H. K., Jr., *J. Phys. Chem.*, **60,** 63 (1956).

80 Benoit, R. L., and P. Pichet, *J. Electroanal. Chem.*, **43,** 59 (1973).

81 Morman, D. H., and G. A. Harlow, *Anal. Chem.*, **39,** 1869 (1967).

82 Benoit, R. L., C. Buisson, and J. Choux, *Can. J. Chem.*, **48,** 2353 (1970).

83 Blumrich, K. G., and G. Bandel, *Angew. Chem.*, **54,** 374 (1941).

84 Wagner, C. D., R. H. Brown, and E. D. Peters, *J. Amer. Chem. Soc.*, **69,** 2609 (1947).

85 Streuli, C. A., *Anal. Chem.*, **30,** 997 (1958).

86 Wimer, D. C., *Anal. Chem.*, **30,** 77 (1958).

87 Courtot-Coupez, J., and M. L. Demezet, *C. R. Acad. Sci.*, **266,** 1438 (1968).

88 Price, G. C., and M. C. Whiting, *Chem. Ind.*, **1963,** p. 775.

89 Miron, R., and D. Hercules, *Anal. Chem.*, **33,** 1770 (1961).

GENERAL REFERENCES

Bates, R. G., *Determination of pH,* Wiley, New York, 1964.

Bates, R. G., "Acidity Functions for Amphiprotic Media," in J. J. Lagowski, Ed., *The Chemistry of Non-Aqueous Solvents,* Vol. I, Academic Press, New York, 1966, Chap. 3.

Bates, R. G., "Medium Effects and pH in Non-Aqueous Solvents," in J. F. Coetzee and C. D. Ritchie, Eds., *Solute–Solvent Interactions,* Vol. I, Marcel Dekker, New York, 1969, Chap. 2.

Bates, R. G., "Concept and Determination of pH," in I. M. Kolthoff and P. J. Elving, Eds., *Treatise on Analytical Chemistry,* Part I, Vol. I, 2nd ed., Wiley-Interscience, New York, 1978, Chap. 14.

Bell, R. P., *Acids and Bases: Their Quantitative Behavior,* Methuen, Science Paperbacks, London, 1969.

Davis, M. M., *Acid–Base Behavior in Aprotic Organic Solvents,* NBS Monograph 105, Washington, D.C., 1968.

Davis, M. M., "Brønsted Acid–Base Behavior in 'Inert' Organic Solvents," in J. J. Lagowski, Ed., *The Chemistry of Non-Aqueous Solvents,* Vol. III, Academic Press, New York, 1970, Chap. 1.

King, E. J., "Acid–Base Equilibria," in E. A. Guggenheim, J. E. Mayer, and F. C. Tompkins, Eds., *The International Encyclopedia of Physical Chemistry and Chemical Physics,* Vol. IV, Macmillan, New York, 1965.

King, E. J., "Acid–Base Behaviour," in A. K. Covington and T. Dickinson, Eds., *Physical Chemistry of Organic Solvent Systems,* Plenum Press, London and New York, 1973, Chap. 3.

Kolthoff, I. M., S. Bruckenstein, and R. G. Bates, *Acid–Bases in Analytical Chemistry,* An Interscience Reprint, Wiley-Interscience, New York, 1966.

Kolthoff, I. M., and P. J. Elving, Eds., *Treatise on Analytical Chemistry,* Part I, Vol. II, 2nd ed., Wiley-Interscience, New York, 1979, Chap. 19.

CHAPTER SEVEN

Transport Properties

7.1 ELECTRICAL CONDUCTANCE

Electrical conductance measurements in nonaqueous solvents provide several important parameters. The basic quantity of interest is the equivalent or molar conductance of the electrolyte at infinite dilution, Λ_o. The parameter Λ_o coupled with transference number data provides information on single-ion conductances or mobilities, λ_o^+ and λ_o^-, for the cations and anions, respectively. The mobilities of ions are basic, for example, in estimating the degree of ionic solvation and ion size, and in the understanding and development of electrolysis, and high-energy organic electrolyte batteries. A second area which has received considerable attention is the study of ion association using electrical conductance as a probe. Determination of the equivalent conductance Λ as a function of concentration of the electrolyte, followed by analysis using an appropriate equation gives rise to a value for K_A, the association constant.

The purpose of this section is to illustrate how typical nonaqueous conductance measurements can be analyzed and to discuss various ways of interpretation of the resulting data. Details of the measuring techniques for electrical conductivity can be found in several texts and monographs listed at the end of this section.

Several techniques are available for the analysis of conductance data and it is useful to review some of the more commonly used approaches at this stage.

7.1.1 The Arrhenius–Ostwald Relationship Using the assumption that the degree of dissociation of an electrolyte, γ, was equal to Λ/Λ_o, and the law of mass action, Arrhenius and Ostwald deduced the relationship

$$\frac{1}{\Lambda} = \frac{1}{\Lambda_o} + \frac{C\Lambda K_A}{\Lambda_o^2} \tag{7.1}$$

where K_A is the association constant, Λ_o is the equivalent conductance at infinite dilution, and C is the concentration of the electrolyte in equivalents per liter of solution. A plot of $1/\Lambda$ vs $C\Lambda$ is very useful in obtaining an approximate value for both Λ_o and K_A, without the need for any rigorous computer analysis. In using many of the more recent conductance equations, the Arrhenius–Ostwald approach is used as a starting point to obtain approximate values of Λ_o and K_A, both of which are required in the computer programs. The plot of $1/\Lambda$ vs $C\Lambda$ is also essential to eliminate points that are obviously inaccurate.

7.1.2 The Shedlovsky Function A modification of the Arrhenius–Ostwald function was generated by Shedlovsky. The equation is

$$\frac{1}{\Lambda S_{(z)}} = \frac{K_A C \Lambda f_{\pm}^2 S(z)}{\Lambda_o^2} + \frac{1}{\Lambda_o} \tag{7.2}$$

$S(z)$ is the Shedlovsky function, $[z/2 + (1 + (z/2)^2)^{1/2}]^2$, where

$$z = S\Lambda_o^{-3/2}(C\Lambda)^{1/2}$$

S is the Onsager coefficient, $\alpha\Lambda_o + \beta$, where

$$\alpha = \frac{0.8204 \times 10^6}{(\epsilon T)^{3/2}} \quad \text{and} \quad \beta = \frac{82.501}{\eta(\epsilon T)^{1/2}} \tag{7.3}$$

C is the molar concentration, $f_{\pm}$ is the mean ionic activity coefficient on the molar scale, ϵ and η are the dielectric constant and the viscosity of the solvent, respectively, and T is the absolute temperature.

The Shedlovsky equation has the advantage that it can be applied in almost every case and is not dependent on data of high precision as are the Fuoss–Onsager and subsequent equations discussed later.

7.1.3 The Onsager Equation This approach was based upon the Debye–Hückel theory for dilute solutions of electrolytes, where the ions were treated as point charges and the solvent as a continuum with a bulk dielectric constant and viscosity. The resulting equation for unassociated electrolytes developed by Onsager (1) is

$$\Lambda = \Lambda_o - SC^{1/2} \tag{7.4}$$

The Onsager equation is of limited application for extrapolating Λ data to obtain Λ_0 because it only yields the slope at infinite dilution and one is normally interested in analyzing data over a finite concentration range.

7.1.4 The Fuoss–Onsager Theory Several extensions of the crude model of the Debye–Hückel theory have taken into account ionic size, employing the model of the ion considered as a sphere, with its charge in the center, moving in a continuum. The most popular equation at present for unassociated electrolytes is due to Fuoss and Onsager (2, 3) [with a subsequent modification by Fuoss, Onsager, and Skinner (4)], namely,

$$\Lambda = \Lambda_0 - SC^{1/2} + EC\log C + (J - F\Lambda_0)C + J_2C^{3/2} \tag{7.5}$$

Here

$$E = E_1\Lambda_0 - 2E_2$$

$$E_1 = \frac{2.303\kappa^2a^2b^2}{24C}$$

$$E_2 = \frac{2.303\kappa ab\beta}{16C^{1/2}}$$

$$b_{,,} = \frac{e^2}{a\epsilon kT}$$

$$ab_{,,} = \frac{16.708 \times 10^{-4}}{\epsilon T}$$

$$\kappa^2 = \frac{4\pi Ne^2}{1000\epsilon kT}(\Sigma C_i z_i^2)$$

$$\kappa_{,,}/C^{1/2} = 0.50294 \times \frac{10^{10}}{(\epsilon T)^{1/2}}$$

$$\kappa = \left(\frac{8\pi e^2 N\mu}{1000\epsilon kT}\right)^{1/2}$$

where μ is the ionic strength, N is Avogadro's number, e is the charge on the electron, z_i is the charge on ion i, and k is the Boltzmann constant.

$$J = \sigma_1\Lambda_0 + \sigma_2$$

where

$$\sigma_1 = \frac{\kappa^2a^2b^2}{12C}\left(\frac{2b^3 + 2b - 1}{b^3 + 0.9074} + \ln\frac{\kappa a}{C^{1/2}}\right)$$

$$\sigma_2 = \frac{\alpha\beta + 11\beta\kappa a}{12C^{1/2}} - \frac{\kappa ab\beta}{8C^{1/2}}\left(1.0170 + \frac{\ln\kappa a}{C^{1/2}}\right)$$

$$J_2 = -(\sigma_3\Lambda_0 + \sigma_4)$$

$$\sigma_3 = \frac{\kappa a}{C^{3/2}}\frac{E_1}{2.303}\left(0.6094 + \frac{4.4748}{b} + \frac{3.8284}{b^2}\right)$$

$$\sigma_4 = \frac{\kappa a}{C^{3/2}}\frac{E_2}{2.303}\left(-1.3093 + \frac{34}{3b} - \frac{2}{b^2}\right)$$

The $F\Lambda_0C$ term corrects for viscosity changes, with F containing a value for R, the hydrodynamic radius (2). However, for small ions this term is often negligible. If required, the value of F can be estimated from the Einstein–Stokes relation

$$2.5\overline{V}_i = F \tag{7.6}$$

where $\bar{V}_i$ is the molar volume of the electrolyte in solution. The terms S, E, J, and J_2 contain contributions from the electrophoretic effect and the relaxation effect.

A few remarks should be made about these effects. The relaxation effect experienced by an ion moving through an electrolyte solution is due to the formation and decay of the ionic atmosphere that is associated with a given ion. For a stationary ion, the ionic atmosphere or charge cloud of opposite charge surrounding an ion is on the average spherically symmetrical. However, when the ion is in motion, the distribution of the ionic atmosphere becomes distorted as the ionic atmosphere will not be completely formed in front of the ion but will not be fully decayed behind the ion. The result of this effect is to place the center of the charge of the ionic atmosphere behind the moving ion. Because the two charges are of opposite sign this effect will tend to retard the motion of the ion. This process experienced by ions in motion is referred to as the relaxation effect. The electrophoretic effect arises from the fact that an ion experiences a viscous drag force as it moves through the solution. This retarding force has an effect on the drift velocity of the ion and hence its conductance.

For electrolytes that are associated in solution, the Fuoss–Onsager equation becomes

$$\Lambda = \Lambda_0 - S(C\gamma)^{1/2} + EC\gamma \log C\gamma + (J - F\Lambda_0)\, C\gamma - K_A C\gamma f_{\pm}^2 \Lambda \qquad (7.7)$$

where γ is the fraction of free ions; K_A is the association constant for the reaction $M^+ + A^- = MA$, where MA is any species that does not contribute to the conductance of the solution and $f_{\pm}$ is the mean ionic activity coefficient of the dissociated electrolyte.

The association constant can be expressed as

$$K_A = \frac{[MA]}{[M^+][A^-]} = \frac{(1 - \gamma)\, f_0}{\gamma^2 C f_{\pm}^2} \qquad (7.8)$$

where f_0 is the activity coefficient of the ion pair, which is generally taken to be unity, and the quantities in the square brackets are activities. The J and J_2 terms in Eqns. 7.5 and 7.7 are also functions of $\mathring{a}$, the so-called ion-size parameter. The validity of the equations for symmetrical univalent electrolytes extends to a concentration corresponding to $\kappa\mathring{a} = 0.2$.

7.1.5 The Pitts Equation Pitts (5) developed a conductance equation using the same model of an electrolyte solution as used by Fuoss et al., that is, taking into consideration both short- and long-range interactions. The main differences in the final results obtained by Pitts are due to the different boundary conditions used to evaluate the constants obtained by integration of several differential equations. The vigorous mathematical theory used in the development of this equation is beyond the scope of this text and only the final equation will be given.

$$\Lambda = \Lambda_0 - \frac{z^2 e\kappa F}{3\pi\eta(1 + y)} + \frac{z^4 e^3 \kappa^2 F}{3\pi\eta\epsilon_r kT(1 + y)} \frac{T_1}{3}$$
$$- \Lambda_0 \left[\frac{(ze)^2\kappa}{3\epsilon_r kT(1 + \sqrt{2})(1 + y)(\sqrt{2} + y)} + \frac{(ze)^4\kappa^2}{3(\epsilon_r kT)^2} S_1 \right] \quad (7.9)$$
$$+ \frac{(\sqrt{2} - 1)\, z^4 e^3 F \kappa^2}{9\pi\eta(1 + y)^2(\sqrt{2} + y)}$$

Here $y = \kappa \mathring{a}$ and ϵ_r is the dielectric constant of the solvent. The functions S_1 and T_1 have been tabulated by Pitts (5) for values of $\kappa\mathring{a}$ in the range 0.02–0.50. For aqueous electrolytes, the equation appears to be valid up to 0.1 M for 1:1 electrolytes. The Pitts equation includes the high-order effect of the local velocity of ions arising from the perturbation of the ionic electrostatic potential and also the contribution due to the asymmetric potentials, an effect that was not considered in the Fuoss–Onsager treatment. The Pitts equation has been modified to the form of Eqn. 7.5 by Fernandez–Prini and Prue (6).

7.1.6 The Fuoss–Hsia Equation Fuoss and Hsia (7) modified the Fuoss–Onsager equation by recalculating the relaxation field and retaining certain terms that had previously been neglected. The resulting equation is of the form

$$\Lambda = \Lambda_0 - SC^{1/2} + EC \ln C + J_1 C - J_2 C^{3/2} \quad (7.10)$$

where J_1 and J_2 are different expressions to those used in the Fuoss–Onsager equation.

7.1.7 The Justice Equation For application to associated electrolytes, Justice (8) modified the Fuoss–Onsager equation and the activity coefficient equation to include the Bjerrum critical distance, q, instead of the ion-size parameter, $\mathring{a}$, where

$$q = \frac{|z_1 z_2| e^2}{2\epsilon kT} \quad (7.11)$$

The form of the equation is

$$\Lambda(1 + BC) = \gamma(\Lambda_0 + SC^{1/2}\gamma^{1/2}) + EC\gamma \log C\gamma$$
$$+ J(R)C\gamma + J_{3/2}(R)C^{3/2}\gamma^{3/2} \quad (7.12)$$

where $K_A = 1 - \gamma/\gamma^2 C y'^2$; R is the distance parameter equated to q, the

Bjerrum critical distance, and B is the Debye–Hückel constant

$$\log y' = \frac{-\kappa q \gamma^{1/2}}{1 + \kappa R \gamma^{1/2}}$$

$$E = E_1\Lambda_0 - 2E_2$$

$$J = \sigma_1\Lambda_0 + \sigma_2$$

$$J_{3/2} = \sigma_3\Lambda_0 + \sigma_4$$

Values of Λ_0, $J_{3/2}$, and K_A are obtained from a least-squares analysis after setting $R = q$ in the expression for J and y'.

7.1.8 Revised Fuoss Analysis (1975) In a more recent development, Fuoss (25, 26) considers the Justice equation (Section 7.1.7) and the Fuoss–Hsia equations (Section 7.1.6) to be obsolete and shows that three and only three parameters, Λ_0, K_A, and R, are sufficient to generate a function $\Lambda(C)$, which reproduces observed equivalent conductivities as a function of concentration within experimental error up to concentrations that are approximately $2 \times 10^{-7}\ \epsilon^3$ equiv liter^{-1}. In this approach, R is a distance parameter, defined as the radius of the sphere inside of which a unique partner can be found for a paired ion, and outside of which continuum theory can be applied. Fuoss outlines a computer program for evaluating Λ_0, K_A, and R.

7.1.9 The Robinson–Stokes Conductance Equation Robinson and Stokes (23) developed the equation

$$\Lambda = \Lambda_0 - \frac{(\alpha + \beta\Lambda_0)C^{1/2}}{1 + B\mathring{a}C^{1/2}} \tag{7.13}$$

where B, α, and β are constants depending on the properties of the solvent.

Rearrangement of Eqn. 7.13 gives

$$\frac{(\alpha + \beta\Lambda_0)C^{1/2}}{\Lambda_0 - \Lambda} = 1 + B\mathring{a}C^{1/2} \tag{7.14}$$

This equation may be used to fit experimental data within a few tenths of a percent, but it must be pointed out that $\mathring{a}$ is an arbitrary parameter in this case, with no physical significance.

7.1.10 The Wishaw–Stokes Equation The Wishaw–Stokes equation (24) is a semiempirical equation of the form

$$\Lambda = \left[\Lambda_0 - \frac{B_2C^{1/2}}{1 + B\mathring{a}C^{1/2}}\right]\left[1 - \frac{B_1C^{1/2}F}{1 + B\mathring{a}C^{1/2}}\right]\frac{\eta_0}{\eta} \tag{7.15}$$

which introduces an explicit correction for the viscosity effect and is one of the more successful equations for concentrated solutions.

7.1.11 Other Conductance Equations The equations most commonly applied to nonaqueous electrolyte solutions are those given above. Some other approaches are discussed in the review by Barthel (9).

The problem of which equation to use for the analysis of conductance data is questionable at the present time, but all of the equations yield approximately the same value of Λ_0. Because the different equations give various values for the association constant K_A, it is advisable to use the same equation for the analysis of all the data that are being compared. The more recent equations involving ion-size parameters similar to the Bjerrum critical distance appear to be more appropriate for solvents with dielectric constants greater than about 30.

7.1.12 Limiting Ionic Conductances Earlier it was indicated that, apart from the evaluation of Λ_0 and K_A, the single-ion conductances, λ_0^+ and λ_0^- were important fundamental quantities that could be obtained from electrical conductance measurements. A comparison of single-ion conductances can provide important information on ion–solvent interactions. To evaluate single-ion conductances in a nonaqueous solvent, precise transference number data must be known for at least one electrolyte in the particular solvent, and unfortunately such data are seriously lacking at present.

To overcome this difficulty, several estimation methods have been adopted in order to generate single-ion values. In one approach, the conductance of an electrolyte composed of large ions, Λ_0, is divided equally between the cation and the anion, that is, $\lambda_0^+ = \lambda_0^- = \frac{1}{2}\Lambda_0$. A commonly used electrolyte for this approximation is $(i-Am)_4N^+(i-Am)_4B^-$. A second approach makes use of the Walden product, that is,

$$\lambda_0\eta = \text{const.} \tag{7.16}$$

The Walden product arises from Stokes's law:

$$\lambda_0\eta = \frac{|z|F^2}{6\pi r_s} \tag{7.17}$$

where r_s is the hydrodynamic radius of the entity moving through the solution and F is the Faraday; the model used considers the ion as a sphere moving through a solvent continuum.

The Stokes law approach discussed above considers the viscous frictional force opposing the motion of an ion. However, for a complete description of the mobility of an ion one must also take into account the dielectric frictional force. The effect of the orientation of the solvent dipoles around an ion as it moves through the solvent has been considered in the Zwanzig theory (21). The orientation of the solvent dipoles around the ion via an electrostatic interaction relaxes as the ion moves through the solution under an applied field, but the relaxation is not an instantaneous process because of the viscosity of the solution and the resulting time lag causes a retarding force on the ion.

The Zwanzig equation can be expressed as

$$\lambda_0\eta = \frac{|z|\,eF}{A_V\pi r + R_D\,[ze^2(\epsilon_0 - \epsilon_\infty)/r^3\epsilon_0(2\epsilon_0 + 1)]\,\frac{\tau}{\eta}} = \frac{|z|\,eF}{A_V\pi r + (B/r^3)} \qquad (7.18)$$

where τ is the dielectric relaxation time and ϵ_0 and ϵ_∞ are the dielectric constants at zero and infinite frequency, respectively. The parameters A_V and R_D can be evaluated for the two extreme cases of perfect sticking of the solvent to the ion and perfect slipping of the ion through the solvent. The Zwanzig theory is essentially a correction to Stokes's law by taking into account the dielectric frictional force and therefore the validity of the theory is governed by the validity of Stokes's law. However, the theory has been considered quite frequently in the interpretation of limiting ionic mobilities and does provide a basis for considering the two mechanisms involved in ion transport, namely, the viscous drag force and the dielectric frictional force. Evans et al. (22) have tested the theory in a variety of aprotic and hydrogen-bonding solvents.

For all solvents, the value of $\lambda_0\eta$ obtained from the Zwanzig equation passes through a maximum and then decreases. Below this maximum, for any given value of $\lambda_0\eta$, there are two values of r, the large value corresponding to the tightly solvated ion (viscous drag) and the small value corresponding to the smaller ionic entity (dielectric friction effect).

The need for refinements of the theories based on models such as Stokes' law is obvious. An alternative approach, which has been receiving attention lately, is the use of statistical mechanics.

The application of the transition-state theory to the process of conductance was advanced by Eyring who derived the following expression for λ_0, treating the quantity as a rate constant.

$$\lambda_0 = \frac{1}{6}\frac{zeF}{h}\,(\overline{l^2})\,\exp\left(\frac{\Delta S^\ddagger}{R}\right)\,\exp\left(\frac{-\Delta H^\ddagger}{RT}\right) \qquad (7.19)$$

where $\Delta S^\ddagger$ and $\Delta H^\ddagger$ are the entropy and enthalpy of activation and l is an average jump distance. The transition-state approach has found use in comparing the transition-state parameters for a variety of systems. An extension of the equation to obtain volumes of activation can be made by studying the pressure dependence of λ_0.

7.1.13 Association Constants Values of association constants (K_A) of a large number of electrolytes in a wide range of protic and dipolar aprotic solvents have been determined using one of the equations discussed above for associated electrolytes. Different values have been obtained depending on which equation has been used. For comparison purposes it is strongly recommended that as a first approach the same equation should be used for each system. Values of K_A determined from experimental electrical conductance data can be compared to values calculated from several coulombic ionic association theories.

One approach was developed by Fuoss, who, taking into account the volumes excluded by paired ions, derived the equation

$$K_F = \frac{4\pi N}{3000} a_F^3 \exp\left(\frac{z_1 z_2 e^2}{a_F \epsilon kT}\right) \tag{7.20}$$

which applies to all cases where the ions are separated by the same distance, a_F, generally taken as the sum of the crystallographic radii of cation and anion. The other approach was put forward by Bjerrum:

$$K_B = \int_{a_B}^{q} 4\pi N r^2 \, dr \exp \frac{z_1 z_2 e^2}{r \epsilon kT} \tag{7.21}$$

where a_B represents the contact distance for two spherical ions and $q = |z_1 z_2| e^2/2\epsilon kT$ is some critical distance at which ion pairing no longer is regarded to exist, because of the balance of coulombic forces and thermal energy. One limitation of the Bjerrum equation is that for solvents with high dielectric constant the sum of the ionic radii may often exceed q.

On comparing experimental values of K_A with values calculated from theory, a wide divergence in behavior is observed with experimental values being either greater or less than calculated values. The differences between the observed values and those predicted by theory can be partially accounted for in terms of a high degree of ionic solvation, dispersion forces, which can stabilize the collision complex, dielectric saturation, which would reduce the effective macroscopic dielectric constant, solvent reorganization in the vicinity of an ion pair, which would also modify the dielectric constant and the existence of two or more kinds of ion pairs, which will produce larger values for K_A.

The existence of contact and solvent-separated ion pairs can be summarized by the following equation:

$$M^+(S)_n + N^-(S)_p \underset{}{\overset{K_1}{\rightleftharpoons}} \underset{\text{solvent-separated ion pair}}{M^+SN^-(S)_{n+p-1}} \overset{K_2}{\rightleftharpoons} \underset{\text{contact ion pair}}{M^+N^-(S)_{n+p-1}} + S \tag{7.22}$$

The overall association constant is given as

$$K_{\text{total}} = K_1\left(1 + \frac{K_2}{C_S}\right) \tag{7.23}$$

where C_S is the molar concentration of the solvent.

The problem still remains that it is difficult to make an unambiguous assignment to the exact nature of the ion pair from electrical conductance measurements.

7.1.14 Conductance and Association Behavior of Electrolytes in Nonaqueous Solvents The purpose of this section is to select a few typical and more recent investigations of conductance in nonaqueous electrolyte solutions and examine how the data are used to discuss ion association and related phenomena. The intent is not to single out certain investigations as recommended data sources, but rather to indicate a cross-section of approaches used in the study of electrical conductance. Electrical conductance studies in nonaqueous solvents are numerous and the reader should refer to the research literature for further examples.

The choice of conductance equation used for analysis of data varies but the most widely used approach is that of the Fuoss–Onsager equation(s). Apparently all the equations give essentially identical values for Λ_0 and in cases where the ion pairing is considerable, a fairly close agreement exists in the values obtained for the association constant, K_A. It is important to emphasize that, when comparing the behavior of different electrolytes in the same or different solvents, it is essential to use a common equation in the initial comparison.

Alkali-Metal Perchlorates and Tetraphenylborates*† *in Anhydrous Acetonitrile Kay et al. (10) measured the electrical conductances of several alkali-metal perchlorates and tetraphenylborates in anhydrous acetonitrile at 25°C. The values obtained for Λ_0, K_A, and $\mathring{a}$ are summarized in Table 7.1. The data for the perchlorates and the second entry in Table 7.1 for cesium tetraphenylborate were analyzed using the Fuoss–Onsager equation for associated electrolytes (Eqn. 7.7). The remaining tetraphenylborates were analyzed using Eqn. 7.5, assuming complete dissociation, that is, $\gamma = 1$ and $K_A = 0$.

The presence of association can be indicated by a plot derived from a rearranged Fuoss–Onsager equation:

$$\Lambda' = \Lambda - \Lambda_0 + SC^{1/2} - EC \log C = (J - F\Lambda_0)C \tag{7.24}$$

Equation 7.24 predicts that a plot of Λ' vs C should produce a straight line with a slope of $(J - F\Lambda_0)$. In addition, the slope should increase with ionic size, because J is a function of $\mathring{a}$. For cases where the degree of association is large, curvature will appear in the Λ' vs C plot. Figure 7.1a shows a definite curvature for these plots and is evidence for the association of the perchlorates in acetonitrile. In contrast, the linear plots obtained for the tetraphenylborates shown in Fig. 7.1b indicate the complete dissociation of these salts, except for perhaps a slight amount of association for the cesium salt (cf. Table 7.1).

† Sometimes these salts are incorrectly referred to as borides.

Table 7.1 Conductance Parameters for Acetonitrile Solutions at 25°C

Electrolyte	Λ_0	$\mathring{a}$	K_A
$NaClO_4$	180.32 ± 0.03	3.7 ± 0.2	9 ± 1
	180.63 ± 0.07	3.7 ± 0.4	11 ± 2
$KClO_4$	187.41 ± 0.06	3.0 ± 0.2	13 ± 1
	187.52 ± 0.05	3.1 ± 0.1	14 ± 1
$RbClO_4$	189.55 ± 0.08	3.2 ± 0.3	19 ± 1
	189.49 ± 0.09	3.4 ± 0.3	19 ± 1
$CsClO_4$	190.98 ± 0.04	3.0 ± 0.1	20 ± 1
	191.08 ± 0.06	3.5 ± 0.2	23 ± 1
$NaPh_4B$	135.4 ± 0.1	5.1 ± 0.1	—
	135.4 ± 0.1	5.2 ± 0.1	—
KPh_4B	141.79 ± 0.06	5.5 ± 0.1	—
	141.84 ± 0.05	5.7 ± 0.1	—
$RbPh_4B$	143.86 ± 0.03	5.3 ± 0.06	—
	143.72 ± 0.05	5.2 ± 0.07	—
$CsPh_4B$	145.38 ± 0.02	2.8 ± 0.02	—
	145.31 ± 0.02	2.8 ± 0.01	—
$CsPh_4B$	145.47 ± 0.02	3.3 ± 0.1	2.7 ± 0.6
	145.36 ± 0.02	3.1 ± 0.1	1.7 ± 0.6

An alternative approach for associated electrolytes is to construct a plot based on a rearrangement of Eqn. 7.7, that is,

$$\Lambda'' = \Lambda - \Lambda_0 + S(C\gamma)^{1/2} - EC\gamma \log C\gamma + K_A f_{\pm}^2 \Lambda C\gamma = (J - F\Lambda_0)C\gamma \quad (7.25)$$

By making use of Eqn. 7.3, the term containing the association constant can be approximated as $K_A f_{\pm}^2 \gamma\,(\Lambda_0 C - SC^{3/2})$. The Λ'' vs C plot will be curved because of the $C^{3/2}$ dependence, and the curvature will increase as K_A increases.

The association constants in acetonitrile obtained by Kay indicate a slight increase in association as the crystallographic radii increase and the authors point out a similar trend in methanol, ethanol–water, and dioxane–water, in which the alkali-metal cations are known to be extensively solvated. Two factors appear to govern the extent of association of electrolytes in acetonitrile, the extent of solvation by the cations and the size of the anions.

Kay estimated limiting cation conductances (λ_0^+) using the concept of the "Fuoss–Coplan split"† which gave values of λ_0 (ClO_4^-) = 103.8 ± 0.2

† The Fuoss–Coplan split uses $(i\text{-Am})_3BuNBPh_4$ as a reference salt.

and $\lambda_0(Ph_4B^-)$ = 58.1 for the anions. The values of the limiting conductances for a number of cations in acetonitrile are shown in Table 7.2.

Limiting Walden products for the alkali-metal, the halide, the tetraalkylammonium, and the perchlorate ions in acetonitrile, methanol, and ethanol at 25°C are plotted as a function of the reciprocal crystallographic radii (measured or estimated) in Fig. 7.2. The plot indicates that decreasing the size of the R_4N^+ ions produces large differences in the Walden product. The increased mobility of the ions in methanol and ethanol, compared with acetonitrile, is attributed to the larger solvation possible with acetonitrile due to its larger dipole moment, 4.0 D, compared to 1.7 D for the alcohols. The larger size of the ethanol molecule is suggested as the reason for the greater mobility of the ions in methanol relative to ethanol.

The results for acetonitrile pose a structural problem, as the smallest alkali-metal ions have the largest Walden products. This effect is accounted for on the basis of the Lewis acid–base properties of acetonitrile, namely, that acetonitrile is less basic than the alcohols, and therefore the interaction of the lone pair of electrons of the oxygen in the alcohols with the small cation results in greater solvation and consequently lower mobilities in methanol solutions. However, as the cation size increases, the acid–base inter-

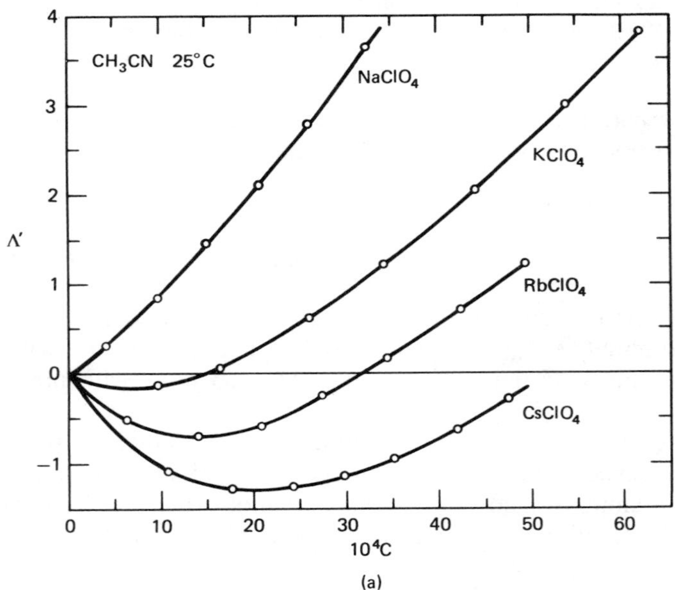

Figure 7.1 (a) Plots of Eqn. 7.22 for the alkali metal perchlorates in acetonitrile at 25°C. (b) Plots of Eqn. 7.22 for the alkali metal tetraphenylborates in acetonitrile at 25°C. Reprinted with permission from *J. Phys. Chem.*, **71**, 3925 (1967). Copyright 1967. American Chemical Society.

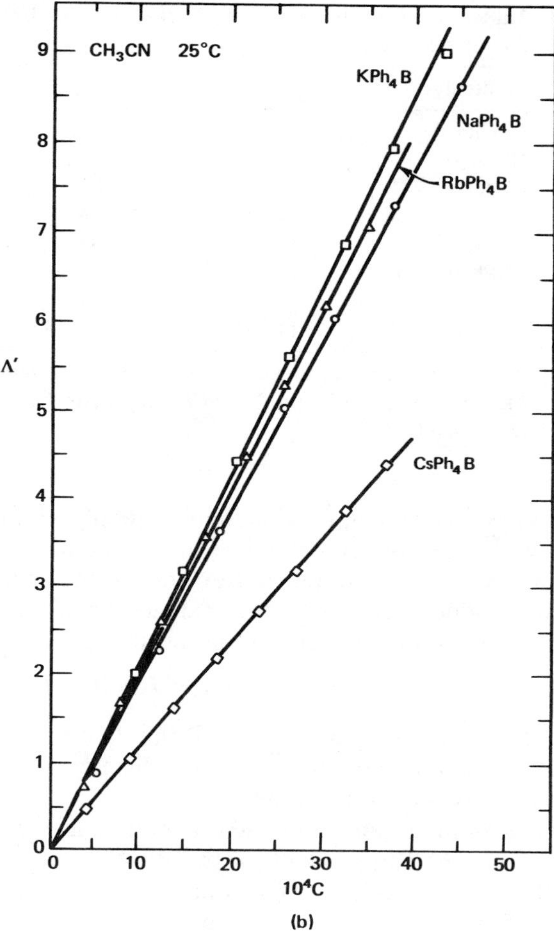

Figure 7.1 (Continued)

action is replaced by a charge–dipole interaction, and results in the crossover of the Walden product.

A similar pattern is exhibited by the anions. The smaller anions have a relatively strong interaction with the relatively strongly acidic alcohols, whereas the larger perchlorate anion is not affected by the acid properties of the solvent.

Transport Behavior in Isomeric Alcohols–Tetraalkylammonium Salts in Isobutyl and Isopentyl Alcohols at 25°C Broadwater and Douglas (12) undertook conductance measurements on Pr_4NBr, Bu_4NBr, *n*-Pe_4NBr, Pr_4NI, Bu_4NI, *n*-Pe_4NI, Bu_4NClO_4, Bu_4NBF_4, and Bu_4PClO_4 in isobutyl alcohol and on

Table 7.2 Limiting Electrolyte and Cation Conductances in Acetonitrile at 25°C

Electrolyte	Λ_0	λ_0^+
$NaClO_4$	180.4	76.6
$NaPh_4B$	135.4	77.3
$KClO_4$	187.5	83.7
KPh_4B	141.8	83.7
$RbClO_4$	189.5	85.7
$RbPh_4B$	143.8	85.7
$CsClO_4$	191.0	87.2
$CsPh_4B$	145.4	87.3

Reprinted with permission from *J. Phys. Chem.*, **71,** 3925 (1967). Copyright 1967. American Chemical Society.

Bu_4NBr and Bu_4NClO_4 in isopentyl alcohol. One of the interesting aspects of this study was to explore the effect of branching in the alkyl portion of the alcohol on the ionic mobilities and on the extent of ionic association. The equivalent conductance data were analyzed using the Fuoss–Onsager equation for associated electrolytes in the form:

$$\Lambda = \Lambda_0 - S(\gamma C)^{1/2} + E(\gamma C)\ln(\gamma C) + J(\gamma C) - K_A f_{\pm}^2 \Lambda(\gamma C) \quad (7.26)$$

A least-squares program was used to fit the data to the above equation and hence yield the parameters Λ_0, K_A, and $\mathring{a}$ (distance of closest approach). These parameters are shown in Table 7.3.

Values of the limiting ionic equivalent conductances were generated on the assumption that the $\lambda_0^+\eta$ product for the Bu_4N^+ ion is the same as that obtained in ethanol. The limiting ionic conductances are given in Table 7.4, and in Table 7.5 a comparison of Walden products is shown for several alcohols. The authors discuss the trend in the values of $\lambda_0^+\eta$ in terms of the solvation of the ions. For example, the large unsolvated Pr_4N^+ ion has an almost constant value, independent of the solvent. The increase in chain branching in going from *n*-butanol to isobutanol seems to have hardly any effect on the $\lambda_0\eta$ product for the anions, Br^-, I^-, and ClO_4^-. The lack of anion solvation is suggested as the reason for the relatively large values of the Walden products for the Br^- and I^- ions in butanone. Although the BF_4^- ion should be solvated to a greater extent than the ClO_4^-, due to the electronegative fluorine atom interacting with the solvent dipole, their Walden products were almost identical in this analysis.

A comparison was also made between the association constants of the various electrolytes in several isodielectric solvents and this is summarized in Table 7.6. The increase in branching in *i*-BuOH, compared to *n*-BuOH,

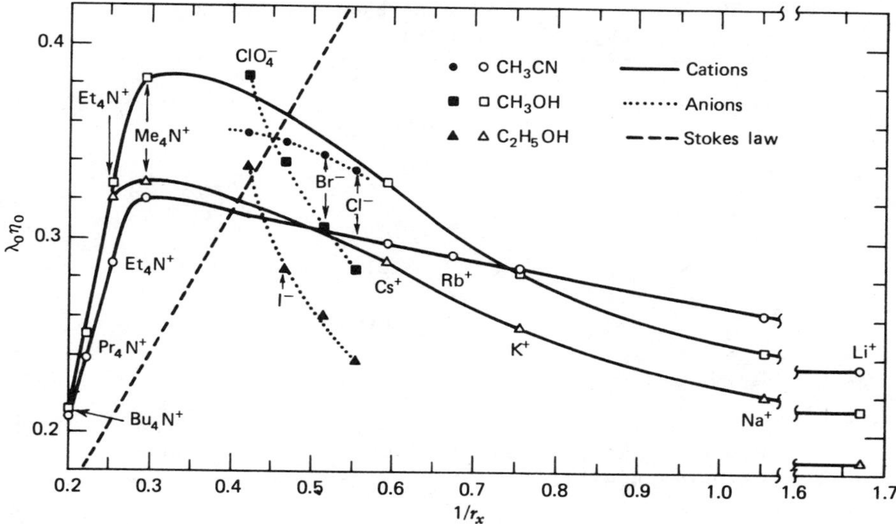

Figure 7.2 Limiting ionic conductance–viscosity ($\lambda_o\eta_o$) products as a function of crystallographic radii for cations and anions in methanol, ethanol, and acetonitrile at 25°C. Reprinted with permission from *J. Phys. Chem.*, **71**, 3925 (1967). Copyright 1967. American Chemical Society.

Table 7.3 Conductance Parameters for Electrolytes in Isobutanol and Isopentanol at 25°C

Electrolyte	Λ_0	$\mathring{a}$	K_A
	Isobutyl Alcohol		
Pr_4NBr	12.45 ± 0.08	4.11 ± 0.21	850 ± 31
Bu_4NBr	12.09 ± 0.07	5.18 ± 0.23	980 ± 28
$(n\text{-Pe})_4NBr$ (Run I)	11.77 ± 0.18	4.65 ± 0.65	1070 ± 81
$(n\text{-Pe})_4NBr$ (Run II)	11.77 ± 0.01	5.68 ± 0.12	1120 ± 9
Pr_4NI	13.74 ± 0.12	5.57 ± 0.34	1330 ± 48
Bu_4NI	12.70 ± 0.05	5.00 ± 0.13	1270 ± 20
$(n\text{-Pe})_4NI$	12.41 ± 0.03	5.26 ± 0.10	1420 ± 14
Bu_4NBF_4	13.95 ± 0.03	5.80 ± 0.06	2330 ± 13
Bu_4NClO_4	13.87 ± 0.08	4.79 ± 0.25	2350 ± 44
Bu_4PClO_4	13.69 ± 0.13	4.48 ± 0.58	2360 ± 83
	Isopentyl Alcohol		
Bu_4NBr	9.38 ± 0.04	6.00 ± 0.09	2720 ± 40
Bu_4NClO_4	10.11 ± 0.11	5.56 ± 0.47	4100 ± 144

Reprinted with permission from the *Journal of Solution Chemistry*, **4**, 485 (1975).

Table 7.4 Estimated Ionic Equivalent Conductances in Isobutanol and Isopentanol at 25°C

Ion	λ_0^+	Ion	λ_0^-
		Isobutyl Alcohol	
Pr_4N^+	6.5_7	Br^-	6.2_0
Bu_4N^+	5.8_9	I^-	6.8_1
Bu_4P^+	5.7_1	ClO_4^-	7.9_8
$(n\text{-Pe})_4N^+$	5.5_8	BF_4^-	8.0_6
		Isopentyl Alcohol	
Bu_4N^+	5.3_9	Br^-	3.9_9
		ClO_4^-	4.7_2

Reprinted with permission from the *Journal of Solution Chemistry,* **4,** 485 (1975).

gives rise to about a 10% increase in the value for K_A. This result is in contrast to the situation existing for 1-PrOH and 2-PrOH, where the difference in K_A values was a factor of 2–3, a result that was attributed to the much lower acidity in the secondary alcohol. The acidity of *i*-BuOH and *n*-BuOH is about the same.

Association behavior in hydrogen-bonded solvents has been discussed by Evans et al. (13) in terms of a two-step mechanism involving solvent-separated ion pairs, when it is assumed than only the anion is solvated:

$$R_4N^+X^-(R'OH)_n \xrightarrow{K_I} R_4N^+(R'OH)X^-(R'OH)_{n-1} \xrightarrow{K_{II}} R_4N^+X^-(R'OH)_{n-1} + ROH \quad (7.27)$$

The association constant K_I can be calculated from an equation due to Fuoss

$$K_F = \frac{4N\mathring{a}^3}{3000} \exp \frac{e^2}{\mathring{a}\epsilon kT} \quad (7.28)$$

where $\mathring{a}$ is the distance of separation of the ion pair. The measured conductance that detects both types of ion pairs is given by

$$K_A = K_I \frac{1 + K_{II}}{[R'OH]} \quad (7.29)$$

In the analysis by the authors, K_I was set equal to K_F, and the values of K_A were computed. Again it was shown that chain branching had very little effect on K_{II}, as the data in Table 7.7 indicate.

Electrical Conductance of Salts in Hexamethylphosphoramide Hanna et al. (15) measured the electrical conductance of twenty-two 1:1 electrolytes in hexamethylphosphoramide (HMPT) at 25°C.

The following equations were used to fit the data

$$\Lambda_i = \Lambda_0 - f(\kappa,d) \tag{7.30}$$

$$\alpha = \Lambda/\Lambda_i \tag{7.31}$$

$$K_A = \frac{1 - \alpha}{\alpha^2 C y_\pm^2} \tag{7.32}$$

$$-\log y_\pm = \frac{A(\alpha C)^{1/2}}{1 + Bd(\alpha C)^{1/2}} \tag{7.33}$$

The equations above consist of three adjustable parameters, namely Λ_0, K_A, and d (the association distance). The full equations of both Pitts and Fuoss–Hsia were used for the function $f(\kappa,d)$ in Eqn. 7.30. Values of Λ_0 and K_A which gave the best fit for each selected value of d were calculated. The "best value" was that which gave the minimum standard deviation in Λ_A. The best-fit parameters for the two equations are shown in Tables 7.8 and 7.9, and it should be noted that in some cases the Pitts equation produces the best fit, while in others, the Fuoss–Hsia equation is more appropriate.

It was observed that the association constants for the halides, nitrates, and perchlorates all increased in the sequence $Li^+ < Na^+ < K^+ < Rb^+ < Cs^+$. This trend was attributed to the less firm solvation of the larger cation by the negative end of the solvent dipole. The Bjerrum equation:

$$K_A = 4\pi N \int_a^d \exp \frac{e^2}{\epsilon k T r} r^2\, dr \tag{7.34}$$

was used to calculate values of a, the closest distance of approach of associated ions. It was found that in most cases the values of a were greater than a_c, the sum of the crystallographic radii of the ions, which indicated that the solvation energy of the cations was high enough for the solvent molecules to be retained in the ion pairs. In a few cases when association was strong, the value of a was less than a_c. It was observed that both $AgNO_3$ (a = 2.3 Å) and NH_4Cl (a = 1.6 Å) were quite strongly associated with extensive long-range interactions between the ions. Limiting ionic conductances, λ_0, Stokes radii, r_s, hydrodynamic radii, r_s^*, and crystallographic radii, r_c, were calculated and are tabulated in Table 7.10.

The limiting ionic conductances of the sodium ion were calculated from transport numbers using the equation

$$t_+ = \frac{\lambda_0^+ - \frac{1}{2}B_2 C^{1/2}/(1 + \kappa d)}{\Lambda_0 - B_2 C^{1/2}/(1 + \kappa d)} \tag{7.35}$$

where B_2 is the electrophoretic constant in the Onsager limiting law and t_+ is the transference number of the cation.

Table 7.5 Limiting Ionic Equivalent Conductance–Viscosity Products for Various Alcohol Solutions at 25°C

Ion	$\lambda_0^+\eta$	Ion	$\lambda_0^-\eta$
	***n*-Butanol**		
Pr_4N^+	0.228	Br^-	0.213
Bu_4N^+	0.203	I^-	0.241
		ClO_4^-	0.291
	Isobutyl Alcohol		
Pr_4N^+	0.227	Br^-	0.214
Bu_4N^+	0.203	I^-	0.235
Bu_4P^+	0.197	ClO_4^-	0.275
$(n\text{-Pe})_4N^+$	0.192	BF_4^-	0.278
	***n*-Pentanol**		
Bu_4N^+	0.212	Br^-	0.181
		I^-	0.205
	Isopentyl Alcohol		
Bu_4N^+	0.203	Br^-	0.150
		ClO_4^-	0.178
	Butanone		
Pr_4N^+	0.225	Br^-	0.328
Bu_4N^+	0.199	I^-	0.336

Reprinted with permission from the *Journal of Solution Chemistry,* **4,** 485 (1975).

Stokes-law radii (r_s) were calculated from

$$r_s = \frac{F^2}{6\pi N\eta\lambda_0^i} = \frac{0.82}{\eta\lambda_0^i} \tag{7.36}$$

where η is the viscosity in poise. Values of the hydrodynamic radii (r_s^*) were calculated from Eqn. 7.36 using the value 4 instead of 6 in the denominator. Excellent agreement between r_s^* and r_c occurs for the anions and for all the cations, except NEt_4^+, which appears to be unsolvated. The values of r_s^* are consistent with the solvation by one molecule of HMPT, which has a molecular radius of about 4 Å, assuming a spherical shape.

Table 7.6 Comparison of Association Constants for Tetraalkylammonium Salts in Isodielectric Solvents

Salt	K_A			K_F from Eqn. 7.28	K_{11} from Eqn. 7.29
	i-BuOH	*n*-BuOH	Butanone	*i*-BuOH	
Pr_4NBr	850	920	940	91	90
Bu_4NBr	980	860	787	82	120
Pe_4NBr	1100		761	76	150
Pr_4NI	1330	1160	442	86	150
Bu_4NI	1270	1180	382	78	160
Pe_4NI	1420		351	73	200
Bu_4NClO_4	2350	2200		74	330
Bu_4NBF_4	2330			77	310
Bu_4PClO_4	2360			71	340
	i-PeOH	*n*-PeOH			
Bu_4NBr	2720	2520		110	220
Bu_4NClO_4	4100			100	370

Reprinted with permission from the *Journal of Solution Chemistry*, **4**, 485 (1975).

Conductance and Ion Pairing of Hydrogen Chloride in N-Methylpropionamide at 25°C Duer, Robinson, and Bates (16) studied the association of hydrogen chloride in a solvent with a large dielectric constant ($\epsilon = 176$ at 25°C), namely, *N*-methylpropionamide. The concentration range studied was 0.0012–0.07 *M*. The data were analyzed using a technique based on the Pitts equation. A comparison of the association constants for hydrogen chloride in NMP and in water as a function of ion-size parameter is shown in Fig. 7.3. As is seen in the figure, some evidence for association was found. However, these authors emphasize the tentative nature of this finding and point out the difficulty of predicting ion pairing in solutions where the electrolyte is highly dissociated. The uncertainties arise from the choice of the

Table 7.7 Comparison of K_{II} for Tetrabutylammonium Salts in Various Alcohols

Salt	1-PrOH	2-PrOH	1-BuOH	*i*-BuOH	*n*-PeOH	*i*-PeOH
Bu_4NBr	75	240	130	120	202	220
Bu_4NI	120	350	190	160		
Bu_4NClO_4	240	540	360	330		370

Reprinted with permission from the *Journal of Solution Chemistry*, **4**, 485 (1975).

Table 7.8 Best-Fit Parameters Obtained Using the Pitts Equation

Electrolyte	Λ_0	d (Å)	K_A	σ_K	$10^3\sigma_\Lambda$
$LiClO_4$	21.233	5.69	1.6	0.6	6.4
$NaClO_4$	21.447	6.00	2.1	1.6	6.0
$KClO_4$	21.664	7.16	8.9	1.2	7.9
$RbClO_4$	21.893	7.13	10.7	1.3	9.0
$CsClO_4$	22.129	6.00	8.3	1.2	6.5
NH_4ClO_4	21.430	5.96	5.1	6.4	16.5
$LiNO_3$	25.324	6.14	5.3	1.0	10.8
$NaNO_3$	25.462	7.57	20.2	0.6	7.9
KNO_3	25.557	5.49	107	4.0	21.7
NH_4NO_3	25.412	6.93	36.6	5.3	21.5
$AgNO_3$	25.554	18.24	162	2.6	9.1
LiI	22.352	6.41	3.0	6.3	26.6
NaI	22.577	6.27	2.7	3.8	20.8
KI	22.715	5.98	6.1	2.3	18.4
NH_4I	22.500	5.67	11.2	4.2	24.2
NEt_4I	26.242	5.89	37.4	2.7	19.7
LiBr	24.164	7.39	5.7	1.6	14.6
NaBr	24.378	6.31	8.7	3.0	13.7
KBr	24.604	5.68	46.9	2.2	18.1
NH_4Br	24.337	11.10	94.7	2.7	21.8
LiCl	25.009	18.04	18.6	1.4	15.6
NH_4Cl	25.112	32.4	878	5.0	15.5

Reprinted with permission from the *Journal of Solution Chemistry*, **3**, 563 (1974).

particular conductance equation used to analyze the data and the uncertainty in selecting the value for $\mathring{a}$, the ion-size parameter.

Ion–Solvent Interactions: Conductance Behavior of $NaAlBu_4$ and Bu_4NAlBu_4 in Benzene and Tetrahydrofuran Day et al. (17) measured the electrical conductances of the tetraalkylaluminates $NaAlBu_4$ and Bu_4NAlBu_4 in two solvents with low dielectric constants, benzene and THF, from dilute solutions up to the fused-salt state. An interesting selection of systems is provided in this study, because the two electrolytes concerned have either a small or a large cation and one solvent is complexing (THF) while the other is noncomplexing (benzene). Log–log plots of Λ vs concentration of electrolyte for the two solvent systems are given in Figs. 7.4 and 7.5.

It is seen that, in benzene, Bu_4NAlBu_4 has a larger conductance than $NaAlBu_4$ at every concentration and also the minimum in the curve for $NaAlBu_4$ occurs at a higher concentration than that of Bu_4NAlBu_4. This effect is attributed to the fact that $NaAlBu_4$ is a weaker electrolyte than

Bu_4NAlBu_4, as the ion–ion interaction is dependent on cation size. The authors suggest that a high degree of aggregation occurs with $NaAlBu_4$ in benzene and that charged aggregates similar to triple ions are responsible for the increase in conductance beyond the conductance minimum for $NaAlBu_4$ in benzene. In the case of Bu_4NAlBu_4, the ion–ion interactions will be much smaller due to the larger cations.

In the case of THF, the conductance behavior is very similar for the two electrolytes, with no pronounced minima occurring in the conductance plots (Fig. 7.5). This similarity in behavior is accounted for in terms of the specific solvation of the sodium cation by the THF, to form such stable species as $4THF{\cdot}Na^+$, which has been confirmed by several studies. A species such as $4THF{\cdot}Na^+$ would be comparable in size to $NBu_4{}^+$, and therefore the resulting ionic interactions should be of the same order of magnitude as those for $Bu_4N_4{}^+$–$AlBu_4{}^-$, thus accounting for the similar concentration and temperature dependences observed for the conductance.

Table 7.9 Best-Fit Parameters Obtained Using the Fuoss–Hsia Equation

Electrolyte	Λ_0	d (Å)	K_A	σ_K	$10^3\sigma_\Lambda$
$LiClO_4$	21.243	11.22	13.5	0.7	6.8
$NaClO_4$	21.456	11.42	13.5	1.2	5.6
$KClO_4$	21.669	11.82	17.7	0.9	7.4
$RbClO_4$	21.900	11.68	19.2	1.1	8.2
$CsClO_4$	22.138	11.20	19.0	0.7	5.8
NH_4ClO_4	21.448	8.13	8.4	4.2	9.7
$LiNO_3$	25.337	11.07	15.2	0.7	9.7
$NaNO_3$	25.470	11.65	27.3	0.7	8.0
KNO_3	25.560	8.62	111	4.0	21.4
NH_4NO_3	25.419	11.19	44.0	5.0	21.1
$AgNO_3$	25.575	16.60	161	2.0	10.4
LiI	22.359	11.47	12.9	6.0	26.5
NaI	22.590	9.89	9.6	3.1	20.2
KI	22.724	11.18	16.7	2.7	18.4
NH_4I	22.530	9.08	18.6	2.1	20.6
NEt_4I	26.249	8.52	40.6	2.1	18.6
LiBr	24.172	11.74	13.6	1.9	14.9
NaBr	24.389	10.19	15.9	2.0	12.9
KBr	24.615	8.34	50.3	2.0	17.1
NH_4Br	24.342	13.32	97.6	2.6	21.2
LiCl	25.039	15.83	17.2	1.7	18.3
NH_4Cl	25.147	21.69	875	4.9	12.5

Reprinted with permission from the *Journal of Solution Chemistry*, **3**, 563 (1974).

Table 7.10 Individual Ionic Mobilities in HMPT at 25°C Based on $\lambda_0(Na^+) = 5.5$

Ion	λ_0	r_s (Å)	r_s^* (Å)	r_c (Å)
Li^+	5.3	4.8	7.1	0.6
Na^+	5.5	4.6	6.9	0.9
K^+	5.6	4.5	6.8	1.3
Rb^+	5.9	4.3	6.4	1.5
Cs^+	6.1	4.1	6.2	1.7
$NH_4{}^+$	5.4	4.7	7.0	1.5
$NEt_4{}^+$	9.1	2.8	4.2	4.0
Ag^+	5.6	4.5	6.8	1.3
CI^-	19.7	1.3	1.9	1.8
Br^-	18.9	1.3	2.0	1.9
I^-	17.1	1.5	2.2	2.2
$NO_3{}^-$	20.0	1.3	1.9	—
$CIO_4{}^-$	16.0	1.6	2.4	—

Reprinted with permission from the *Journal of Solution Chemistry,* **3,** 563 (1974).

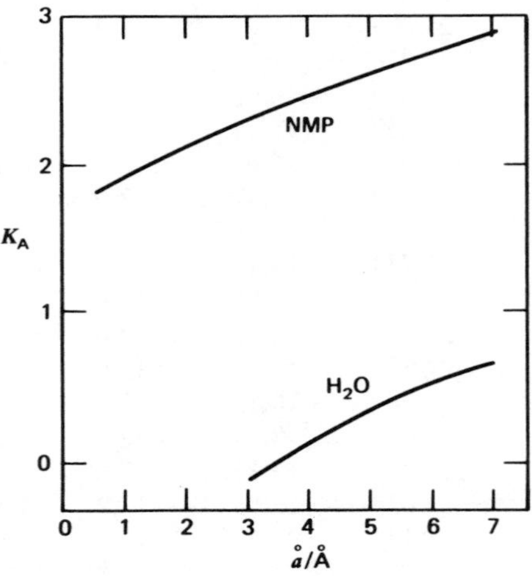

Figure 7.3 Association constant K_A for HCl in NMP and H_2O derived from conductivity measurements by the Pitts equation, as a function of the ion-size parameter $\mathring{a}$. Reprinted with permission from the *Journal of Solution Chemistry,* **5,** 765 (1976).

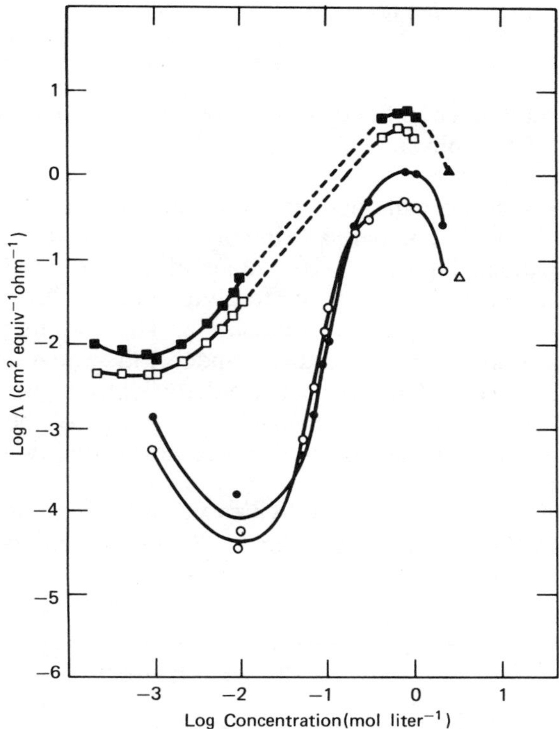

Figure 7.4 Log Λ vs log concentration of $NaAlBu_4$ at 25°C (○) and 50°C (•) in benzene and Bu_4NAlBu_4 at 25°C (□) and 50°C (■) in benzene. The broken lines in the Bu_4NAlBu_4 curves are extrapolations; two-phasing occurs in this concentration region. Equivalent conductances of the fused salts are given for $NaAlBu_4$ (△) and Bu_4NAlBu_4 (▲). Reprinted with permission from the *Journal of Solution Chemistry,* **3,** 83 (1974).

Conductance Behavior of Electrolytes in 3-Methylsulfolane at 25°C In the study by Rosenfarb et al. (18), conductance data for triisoamylbutylammonium tetraphenylborate and iodide, $NaBPh_4$, $KBPh_4$, NaSCN, KSCN, NaI, and KI were analyzed by several theoretical equations, namely, the Fuoss–Shedlovsky, Fuoss–Onsager (for associated and unassociated electrolytes), expanded Pitts, and Fuoss–Hsia equations.

The Fuoss–Shedlovsky equation yielded values of Λ_0 and K_A. The Fuoss–Onsager equations, given previously for associated and unassociated electrolytes (Eqns. 7.5 and 7.7, respectively), yielded the three parameters Λ_0, K_A, and a_J, the ion-size parameter.

Values of Λ_0, K_A, and the Bjerrum ion-size parameter, d, were also obtained from the expanded Pitts and Fuoss–Hsia equations:

$$\Lambda = \Lambda_0 - S(C\gamma)^{1/2} + EC\gamma \log (C\gamma) + J_1C\gamma - J_2(C\gamma)^{3/2} - K_AC\gamma\Lambda f^2 \quad (7.37)$$

Both S and E have been defined previously, and the coefficients J_1 and J_2 are functions of the physical properties of the solvent and the ion-size parameter.

In the Fuoss–Onsager method, the ion-size parameter a_J is considered as the contact distance, compared to the distance of closest approach of the ions in the expanded Pitts analysis. Values of d were varied (1.0–15.0 Å) in the Pitts and Fuoss–Hsia analysis to give values of Λ_0, K_A, and d corresponding to a minimum standard deviation. The various parameters obtained from the Fuoss–Onsager and expanded Pitts equations are given in Table 7.11. The Fuoss–Hsia equation yielded slightly higher values for K_A and d, but left Λ_0 essentially unchanged. The results of the analysis show that only sodium and potassium thiocyanate are distinctly associated in 3-methylsulfolane.

Limiting ionic equivalent conductivities $\lambda_0^{\pm}$ were computed by assuming that $0.5\Lambda_0(i\text{-}Am_3BuNBPh_4) = \Lambda_0(i\text{-}Am_3BuN^+)$ in 3-methylsulfolane.

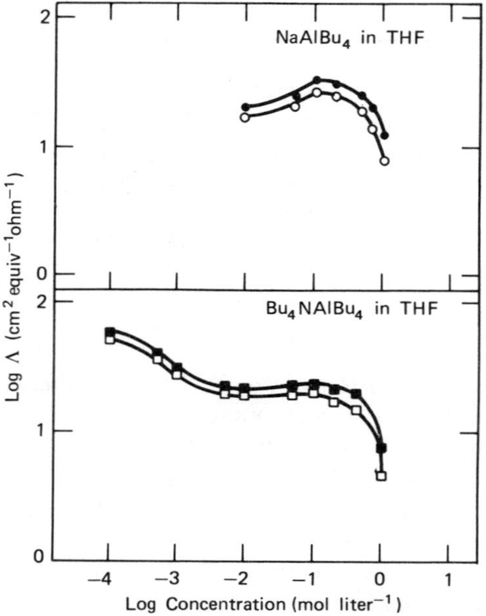

Figure 7.5 Log Λ vs log concentration of $NaAlBu_4$ at 25°C (○) and 50°C (●) Bu_4NAlBu_4 at 25°C (□) and 50°C (■) in THF. Reprinted with permission from the *Journal of Solution Chemistry*, **3**, 83 (1974).

Table 7.11 Conductance Parameters for Fuoss–Onsager and Extended Pitts Equations

Electrolyte	Parameter	Λ_0	K_A	$a_J(d)$	$\sigma\Lambda$
i-$Am_3BuNBPh_4$	FO[a]	4.00	—	4.46	0.005
	P[b]	3.94	—	4.0	0.048
i-Am_3BuNI	FO[a]	7.84	—	2.84	0.007
	P[b]	7.72	16.1	9.5	0.042
$NaBPh_4$	FO[a]	4.36		11.4	0.005
	P[b]	4.33	—	4.0	0.019
$KBPh_4$	FO[a]	4.76	—	[c]	0.011
	P[b]	4.82	10.4	3.5	0.074
NaSCN	FO[a]	10.43	252	1.48	0.022
	P[b]	10.47	300	5.5	0.018
KSCN	FO[a]	10.79	—	0.09	0.011
	P[b]	10.94	72.7	3.0	0.041
NaI	FO[a]	8.14	—	4.72	0.006
	P[b]	8.10	1.15	4.0	0.025
KI	FO[a]	8.58	—	1.03	0.008
	P[b]	8.60	2.72	2.5	0.026

[a] Fuoss–Onsager parameters.
[b] Expanded Pitts parameters corresponding to the minimum standard deviation $\sigma\Lambda$.
[c] The Fuoss–Onsager analysis gave an unreasonably high value for a_J.
Reprinted with permission from the *Journal of Solution Chemistry,* **5,** 311 (1976).

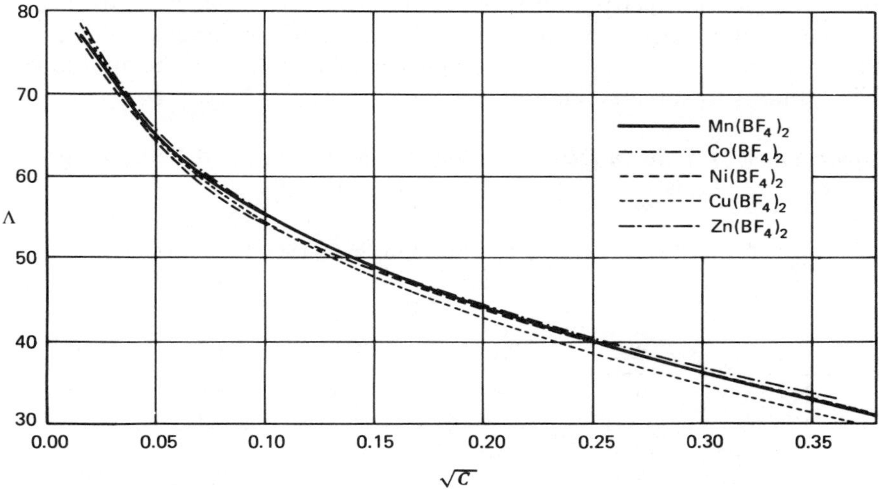

Figure 7.6 Conductometric curves of $Mn(BF_4)_2$, $Co(BF_4)_2$, $Ni(BF_4)_2$, $Cu(BF_4)_2$, and $Zn(BF_4)_2$ in *N,N*-dimethylacetamide solutions at 25°C. Λ: limiting molar conductivity of $\frac{1}{2}M(BF_4)_2$. Reprinted with permission from *Electrochimica Acta,* **22,** 181 (1977). Copyright 1977 Pergamon Press, Inc., Oxford, U.K.

Conductometric Studies of Mn(II), Co(II), Ni(II), Cu(II), and Zn(II) Tetrafluoroborates in N,N-Dimethylacetamide Association constants of various transition-metal tetrafluoroborates in DMA were obtained by evaluation of electrolytic conductance data for these salts by Kamienska and Uruska (19). One of the other features of the study was to examine the influence of the central metal ion on the transport properties of the various systems. The stoichiometry of the complexes was found to be $[M(DMA)_6]^{2+}2BF_4^-$. Shedlovsky's method was used to determine the limiting molar conductance of the ions in the solution and the results obtained were for $\frac{1}{2}Mn(BF_4)_2$, $\frac{1}{2}Co(BF_4)_2$, $\frac{1}{2}Ni(BF_4)_2$, $\frac{1}{2}Cu(BF_4)_2$, and $\frac{1}{2}Zn(BF_4)_2$: 82.3, 82.7, 81.6, 83.3, and 83.8 Ω^{-1} cm^2 mol^{-1}, respectively. Figure 7.6 shows the molar conductance curves for all the systems investigated, with the same pattern emerging for each salt, which appears to be independent of the central metal ion. There is, however, a delicate balance of effects influencing the ionic mobilities rather than a predominance of any one effect. Association constants were estimated, using the empirical Shedlovsky equation:

$$\Lambda = \Lambda_0 - \frac{\Lambda}{\Lambda_0}(\beta + \alpha\Lambda_0)I^{1/2} \qquad (7.38)$$

for the association equilibrium:

$$M^{2+} + BF_4^- \rightleftharpoons M(BF_4)^+ \qquad (7.39)$$

where α and β were defined by Eqn. 7.3. The computed association constants for $Co(BF_4)_2$ and $Cu(BF_4)_2$ in DMA solutions were 15 and 30, respectively.

The trend in the association constants is discussed in terms of a tetragonal distortion of the $[Cu(DMA)_6]^{2+}$ complex, which facilitates the approach of BF_4^- anions to solvated cations.

Conductance of Acids in Dimethylsulfoxide McCallum and Pethybridge (20)

Table 7.12 Best-Fit Parameters for Acids in DMSO; Fuoss–Hsia Equation; Acids Treated as Associated

Acid	d/Å	Λ_0	K_A	σK_A	$\sigma\Lambda$ (%)
HCl	10.18	40.226	126.5	0.03	0.004
HBr	9.11	39.987	11.5	8.3	0.27
CH_3SO_3H	10.48	35.820	42.4	0.04	0.01
CF_3SO_3H	8.32	37.584	3.92	0.01	0.005
H_2SO_4	9.82	30.454	29.2	0.03	0.02
HNO_3	9.33	42.208	27.8	0.07	0.02
$HClO_4$	6.57	40.360	3.29	0.07	0.03

Reprinted with permission from *Electrochimica Acta,* **20,** 815 (1975).

Table 7.13 Best-Fit Values of Λ_0 and K_A at $d = q$; Fuoss–Hsia Equation; Acids Treated as Associated

Acid	Λ_0	K_A
HCl	40.340	125.1
HBr	40.000	8.8
CH_3SO_3H	36.109	41.8
CF_3SO_3H	37.655	2.07
H_2SO_4	30.663	27.7
HNO_3	42.389	26.3
$HClO_4$	40.372	2.72

measured the conductance of methanesulphonic, sulfuric, nitric, perchloric, and trifluoromethanesulfonic acids in DMSO. Data were analyzed in terms of the Fuoss–Hsia equation. The values for Λ_0, K_A, and d, the association distance within which all oppositely charged ions are considered to be paired and nonconducting, are given in Table 7.12, together with data for HCl and HBr.

These authors are of the opinion that d has no physical significance and preferred to fix the value of d at q, the Bjerrum critical distance, and generate a set of association constants for comparison. The results of this approach are shown in Table 7.13. It is seen that sulfuric acid is a stronger acid than methanesulfonic acid in DMSO, and HNO_3 is comparable in strength with H_2SO_4.

7.1.15 General Discussion of Conductance and Association in Nonaqueous Electrolyte Solutions For the purposes of this discussion we will consider the conductance behavior in protic and dipolar aprotic solvents separately. It should be pointed out initially that much of the earlier data of high precision has been reanalyzed, using some of the new conductance equations. Generally the trends in association constants are more reliable in predicting ion–solvent interactions than the use of ionic mobilities. The main factors which contribute to ionic association are ion size, dielectric constant of the solvent and specific ion–solvent interactions.

Protic Solvents A large number of investigations have focused on the alcohols, particularly methanol, ethanol, propanol, and butanol, with the most widely studied electrolytes in these solvents being the alkali-metal halides, perchlorates, nitrates, thiocyanates, and a large variety of quaternary ammonium salts. Values of Λ_0 and K_A for selected electrolytes in methanol and ethanol are given in Table 7.14. It is seen that in methanol, most of the alkali-metal salts are completely dissociated, with the exception of $LiNO_3$ and CsCl. The perchlorates of potassium and cesium are associated in methanol. It should be noted that silver nitrate exhibits association in both methanol and ethanol.

Much of the discussion of association of ions in nonaqueous solvents centers around the existence of contact ion pairs vs solvent-separated ion pairs. It is generally assumed that large cations form solvent-separated ion pairs, whereas the smaller ions polarize the solvent molecules that surround them, so that contact pairs are formed.

Quaternary ammonium salts are all associated to varying degrees in ethanol, as shown in Table 7.14. The tetraalkylammonium iodides give rise to larger association constants compared to the other halides, suggesting solvation of the anions through hydrogen bonding.

The behavior of acids in alcohols indicates that the proton possesses excess mobility due to a special proton-jump mechanism of transport. Picric acid is particularly highly associated in alcohols [K_A(methanol) = 3960 mol^{-1} liter; K_A(ethanol) = 5870 mol^{-1} liter], compared to hydrogen chloride.

It might be expected that there would be little association for 1 : 1 electrolytes in amides such as *N*-methylacetamide, formamide, or *N*-methyl-

Table 7.14 Limiting Conductances and Association Constants of Selected 1 : 1 Electrolytes in Methanol and Ethanol

	Methanol		Ethanol	
Electrolyte	Λ_0	K_A	Λ_0	K_A
HCl	193.2	8	81.7	90
LiCl	92.05	0	38.94	27
$LiNO_3$	100.2	10	42.7	19
$LiClO_4$	101.0	5		
NaCl	97.4	0	42.17	44
KCl	104.9	0	45.42	95
KI	115.2	0	48.2	50
CsCl	113.4	15	48.33	158
$AgNO_3$	110.8	60	41.7	190
Me_4NCl			51.87	141
Me_4NBr			54.03	164
Me_4NPi			55.03	110
Et_4NClO_4			61.4	270
Et_4NBr			53.54	96
Et_4NI			56.34	133

Taken in part from B. Kratochvil and H. L. Yeager, *Topics in Current Chemistry,* No. 27 (1972).

propionamide, all of which have large dielectric constants, but it is found that a large number of 1:1 electrolytes show negative deviations from the Onsager limiting law in these solvents, while the alkaline-earth halides and perchlorates exhibit positive deviations in *N*-methylacetamide, which increase from Ba^{2+} to Mg^{2+}. It has proved difficult to rationalize this type of behavior.

An electrostatic model of ion interactions has been used with some success to account for association behavior, particularly with reference to the conductance of 3:3 salts, such as $LaFe(CN)_6$ and $Co(en)_3Co(CN)_6$, in formamide, with K_A values of 243 mol^{-1} liter and 605 mol^{-1} liter, respectively. (en is ethylenediamine.)

Dipolar Aprotic Solvents Recent studies of electrolytic conductance have focused on solvents such as acetonitrile, propylene carbonate, dimethylsulfoxide, sulfolane, and dimethylformamide, with particular reference to the application of these solvents in organic electrolyte batteries. However, apart from the solvents mentioned above, a large selection of other dipolar aprotic solvents have also been investigated. Some of the common electrolytes studied have been halides, perchlorates, picrates, nitrates, tetraphenylborates, and a large selection of tetraalkylammonium salts.

One of the important factors that contribute to the transport behavior in these solvents is the polarizability of the anions, which largely determines their interactions with the solvent molecules. A case in point is that the mobility of the Br^- is about double that of K^+, which is evidence that the small halide ions are poorly solvated in the dipolar aprotic solvents.

The limiting conductances and association constants of selected 1:1 electrolytes in four dipolar aprotic solvents are given in Table 7.15. It is seen that for a given R_4N^+, the association varies with the anion. Also, substituted ammonium cations can form internal hydrogen bonds with small anions, thereby increasing their association. For example, PhH_3N Pi is more associated than R_4N Pi in acetonitrile. In acetone, the association constants for tetraalkylammonium salts decrease with cationic size.

In some of the conductance studies, the values obtained for $\mathring{a}$, the ion-size parameter, in the analysis of the data by any of the theoretical approaches have been used to invoke the existence of either contact ion pairs or solvent-separated ion pairs. However, the significance of this approach is doubtful as $\mathring{a}$ is regarded by several investigators as an adjustable parameter used in the various equations in order to fit the conductance data. Normally, spectroscopic evidence for contact- or solvent-separated ion pairs is preferable.

Some interesting studies have been made on the behavior of transition-metal salts in acetonitrile. For example, it has been observed that Cu(I) salts are dissociated in acetonitrile, whereas Tl(I) salts are associated. The Λ_0 values indicate that Cu(I) is a very large solvodynamic unit, probably in the form of $Cu(AN)_4{}^+$.

Table 7.15 Limiting Conductances and Association Constants of Selected 1 : 1 Electrolytes in some Dipolar Aprotic Solvents

	Acetonitrile		Acetone		Dimethylformamide		Dimethysulfoxide	
Electrolyte	Λ_0	K_A	Λ_0	K_A	Λ_0	K_A	Λ_0	K_A
LiCl			214	3×10^5	80.2	2.9×10^2		
$LiClO_4$	173	4	187.3	5.3×10^3	77.4	0		
LiBr			194	4.5×10^3				
LiI			195	145				
NaSCN	189.8	87			89.5	8	43.0	0
NaI	179.4	0	183.6	177	82.0	0	37.6	
NaPi			163.5	680	67.3	0	31.1	
KI	186.2	0	192.6	110	82.6	0		
$AgNO_3$	192.4	70			92.5	26		
Me_4NBr	195.2	46			92.5	37	42.63	
Me_4NI	196.7	19	213	28	90.9	14	42.40	
Et_4NPi	164.6	10	176.6	45				
Pr_4NI	172.9	5	191	64	81.8	8	36.22	
Bu_4NBr	162.1	2	182.2	264			35.65	
Bu_4NI	163.7	3	180.2	6.1×10^3	77.7	8	35.39	

Taken in part from B. Kratochvil and H. L. Yeager, *Topics in Current Chemistry*, No. 27 (1972).

Generally, the anomalies in conductance behavior have been encountered most frequently in the solvents having lower cation-solvating power. According to the Parker scheme, the ability of dipolar aprotic solvents to solvate cations varies in the order dimethylsulfoxide = dimethylacetamide > dimethylformamide > acetone > acetonitrile = nitromethane > benzonitrile = nitrobenzene.

For aprotic solvents with very low dielectric constants ($\epsilon < 12$), the electrostatic interactions are very large and often ion triplets are encountered. The conductance often passes through a minimum with concentration, indicating the existence of ion triplets. Examples of this type of behavior are found with $MgCl_2$ in THF and Bu_3NHBr in *o*-dichlorobenzene.

Summary The measurement of the electrical conductance of a nonaqueous electrolyte solution provides valuable information about its transport properties. The quantity of primary interest is the equivalent or molar conductance of the electrolyte at infinite dilution, Λ_0. This quantity coupled with transference number data can be used to generate single-ion conductances, or mobilities, $\lambda_0{}^+$ and $\lambda_0{}^-$, for the cations and anions, respectively. A comparison of single-ion mobilities in a selection of nonaqueous solvents is useful in studying ion solvation. Conductance data obtained as a function of concentration are frequently used to study ion association and values of the association constant, K_A, can be obtained by using the appropriate equation. Several equations and approaches are available for the analysis of conductance data, depending on the precision required. The key equations are (1) the Arrhenius–Ostwald relation, (2) the Shedlovsky equation, (3) the Onsager equation, (4) the Fuoss–Onsager equation, (5) the Pitts equation, (6) the Fuoss–Hsia equation, (7) the Justice equation, (8) the revised Fuoss equation, (9) the Robinson–Stokes conductance equation, and (10) the Wishaw–Stokes equation. The Arrhenius–Ostwald equation is normally used as a starting point to obtain approximate values of Λ_0 and K_A. The Shedlovsky equation has the advantage that it can be applied in almost every case and is not dependent on data of high precision. The Onsager equation is of limited application for extrapolating Λ data to obtain Λ_0, because it only yields the slope at infinite dilution and one is normally interested in analyzing data over a finite concentration range. The selection of an equation for the determination of an association constant is open to debate and all the equations involving a K_A have been used. Values of association constants of a large number of electrolytes in a wide range of protic and dipolar aprotic solvents have been determined. Finally, for concentrated solutions, the Wishaw–Stokes equation is useful.

In order to generate single-ion conductances, several approaches are used. In one approach, the conductance of an electrolyte composed of large ions is divided equally between the anion and the cation. In another approach, the Walden Product is assumed to remain constant in different solvents, that is, $\lambda_0\eta$ = const. As a further refinement, the Zwanzig theory considers the effect of the orientation of the solvent dipoles around an ion as it moves through the solvent and the equation for λ_0 includes the dielectric relaxation time. The transition-state theory has also been applied to the process of conductance and an expression exists for λ_0 in terms of the entropy and enthalpy of activation.

LITERATURE CITED—ELECTROLYTIC CONDUCTANCE

1 Onsager, L., *Physik. Z.*, **27,** 388 (1926); **28,** 277 (1927).

2 Fuoss, R. M., and F. Accascina, *Electrolytic Conductance,* Wiley-Interscience, New York, 1959.

3 Fuoss, R. M., and L. Onsager, *J. Phys. Chem.*, **61,** 668 (1957).

4 Fuoss, R. M., L. Onsager, and J. F. Skinner, *J. Phys. Chem.*, **69,** 2581 (1965).

5 Pitts, E., *Proc. Roy. Soc.*, **217A,** 43 (1953).

6 Fernandez-Prini, R., and J. E. Prue, *Z. Physik. Chem.*, **228,** 373, (1965).

7 Fuoss, R. M., and L. L. Hsia, *Proc. Nat. Acad. Sci.*, **57,** 1550 (1967).

8 Justice, J. C., *J. Chim. Phys.*, **65,** 353 (1968); *Electrochim. Acta,* **16,** 701 (1971).

9 Barthel, J., *Angew. Chem. Int. Ed.*, **7,** 260 (1968).

10 Kay, R. L., B. J. Hales, and G. P. Cunningham, *J. Phys. Chem.*, **71,** 3925 (1967).

11 Coplan, M. A., and R. M. Fuoss, *J. Phys. Chem.*, **68,** 1181 (1964).

12 Broadwater, T. L., and R. T. Douglas, *J. Soln. Chem.*, **4,** 485 (1975).

13 Evans, D. F., and M. A. Matesich, *J. Soln. Chem.*, **2,** 193 (1973).

14 Fuoss, R. M., *J. Am. Chem. Soc.*, **80,** 5059 (1958).

15 Hanna, E. M., A. D. Pethybridge, J. E. Prue, and D. J. Spiers, *J. Soln. Chem.*, **3,** 563 (1974).

16 Duer, W. C., R. A. Robinson, and R. G. Bates, *J. Soln. Chem.*, **5,** 765 (1976).

17 Imhof, J., T. D. Westmoreland, and M. C. Day, *J. Soln. Chem.*, **3,** 83 (1974).

18 Rosenfarb, J., M. Martin, C. Prakash, and J. A. Caruso, *J. Soln. Chem.*, **5,** 311 (1976).

19 Kamienska, E., and I. Uruska, *Electrochim. Acta,* **22,** 181 (1977).

20 McCallum, C., and A. D. Pethybridge, *Electrochim. Acta,* **20,** 815 (1975).

21 Zwanzig, R., *J. Chem. Phys.*, **52,** 3625 (1970).

22 Evans, D. F., J. Thomas, J. A. Nadas, and M. A. Matesich, *J. Phys. Chem.*, **75,** 1714 (1971).

23 Robinson, R. A., and R. H. Stokes, *J. Amer. Chem. Soc.*, **76,** 1991 (1954).

24 Robinson, R. A., and R. H. Stokes, *Electrolyte Solutions,* Butterworths, London, 1959.

25 Fuoss, R. M., *Proc. Natl. Acad. Sci.*, **71,** No. 11, 4491 (1974).

26 Fuoss, R. M., *J. Phys. Chem.*, **79,** 525 (1975).

GENERAL REFERENCES—ELECTROLYTIC CONDUCTANCE

Coetzee, J. F., and C. D. Ritchie, Eds., *Solute–Solvent Interactions,* Vol. II, Marcel Dekker, Inc., New York, 1976.

Evans, D. F., and M. A. Matesich, "The Measurement and Interpretation of Electrolytic Conductance," in *Techniques of Electrochemistry,* Vol. II, E. Yeager and A. J. Salkind, Eds., Wiley, New York, 1973, Chap. 1.

Fernandez-Prini, R., "Conductance and Transference Numbers—Part 1: Conductance," in

Physical Chemistry of Organic Solvent Systems, A. K. Covington and T. Dickinson, Eds., Plenum Press, London and New York, 1973, Chap. 5.

Kratochvil, B., and H. L. Yeager, "Conductance of Electrolytes in Organic Solvents," *Topics in Current Chemistry,* No. 27, 1 (1972).

Shedlovsky, T., and L. Shedlovsky, "Conductometry," in *Techniques of Chemistry,* Vol. I, *Physical Methods of Chemistry, Part II. A. Electrochemical Methods,* A. Weissberger and B. W. Rossiter, Eds., Wiley-Interscience, 1971, Chap. 3.

7.2 TRANSFERENCE NUMBERS

The determination of transference numbers of electrolyte solutions is not only of basic practical importance from the standpoint of knowing the relative current-carrying capacity of an ion constituent, but also an evaluation of transference numbers is often useful in discussing ion–ion and ion–solvent interactions. A knowledge of precise transference numbers is the only unambiguous method of determining single-ion conductances. Compared to the vast amount of information available for electrical conductance in nonaqueous electrolytes, the quantity of precise data on transference numbers is fairly limited, due mainly to problems encountered with the various techniques. An excellent discussion of the experimental methods used in the determination of transference numbers has been given by Kay (1). In addition, a comprehensive review of the underlying principles of transference numbers and particularly the concentration dependence of transference numbers has been published by Spiro (2).

For special cases in which an electrolyte is completely dissociated into two ionic species only, we can refer to a transference number of a single ionic species, for example, t_{Na^+}. However, there are many cases, particularly in nonaqueous solvents, where more than two ionic species exist, such as ion pairs, ion triplets, or higher aggregates. For example, $CdCl_2$ may exist as Cd^{2+}, $CdCl^+$, $CdCl_2$, or $CdCl_3^-$. In this case, the transference number of each ionic species cannot be measured, because of the normally rapid equilibria between the species. Consequently, a more rigorous approach should refer to the transference number of an ion constituent. Therefore, the transference number of a cation constituent is defined as the net number of gram-equivalents of cation that crosses an imaginary plane in the solution, in the direction of that plane. The general expression for the transference number of a cation M^+ is

$$t_{M^+} = \frac{\sum_i (z^+/z_i) N_{+/i} C_i \Lambda_i}{\sum C_i \Lambda_i} \tag{7.40}$$

where $N_{+/i}$ is the number of moles of M^+ in 1 mole of i and C_i and Λ_i are the concentration and molar conductance of the electrolyte, respectively.

For a symmetrical electrolyte that completely dissociates into only two kinds of ions, Eqn. 7.40 reduces to

$$t^0_{M^+} = \frac{\lambda^0_{M^+}}{\lambda^0_{M^+} + \lambda^0_{A^-}} = \frac{\lambda^0_{M^+}}{\Lambda_0} \tag{7.41}$$

where $\lambda^0_{M^+}$ and $\lambda^0_{A^-}$ are the limiting ionic conductances for the cation and anion, respectively, and $t^0_{M^+}$ is the limiting transference number of the cation. Furthermore,

$$t_{M^+} + t_{A^-} = 1 \tag{7.42}$$

7.2.1 Variation of Transference Number with Concentration One approach to studying the extent of ion association or ion solvation as a function of concentration is to examine the dependence of transference numbers on the concentration of the electrolyte. Several methods have been devised in an attempt to fit the experimentally observed data to a theoretical equation. For the case of completely dissociated symmetrical electrolytes, the Onsager limiting law, that is,

$$\Lambda_i = \Lambda_i^0 - (\alpha\Lambda_i^0 + \tfrac{1}{2}\beta)C^{1/2} \tag{7.43}$$

has been applied to Eqn. 7.41. In Eqn. 7.43, α and β represent the relaxation and electrophoretic terms, respectively. The resulting expression obtained for the transference number of either cation or anion is given by

$$\begin{aligned} t_{\pm} &= \frac{\lambda^0_{\pm} - (\alpha\lambda^0_{\pm} + \tfrac{1}{2}\beta)C^{1/2}}{\Lambda_0 - (\alpha\Lambda_0 + \beta)C^{1/2}} \\ &= \frac{t^0_{\pm} + (t^0_{\pm} - 0.5)\beta C^{1/2}}{\Lambda'} \end{aligned} \tag{7.44}$$

where

$$\Lambda' = \Lambda_0 - (\alpha\Lambda_0 + \beta)C^{1/2}.$$

Subsequent treatments using the Fuoss–Onsager equation eliminated the relaxation term α, so that Λ' in Eqn. 7.44 was replaced with Λ_0, that is,

$$t_{\pm} = \frac{t^0_{\pm} + (t^0_{\pm} - 0.5)\beta C^{1/2}}{\Lambda_0} \tag{7.45}$$

However, the validity of this approximation is doubtful. The limiting Onsager slope may be calculated from Eqns. 7.45 and 7.44. As will be seen later, the experimental data for transference numbers as a function of concentration in nonaqueous solvents deviate from the limiting slopes.

At very low concentrations, the variation of transference number with concentration can be expressed as

$$\frac{dt_{\pm}}{d(C)^{1/2}_{c\to 0}} = \frac{(t^0_{\pm} - 0.5)\beta}{\Lambda_0} \tag{7.46}$$

Inspection of Eqn. 7.46 shows that t_+ should decrease with increasing concentration for uni-univalent electrolytes if $t_+ < 0.5$ and, conversely, should increase with increasing concentration if $t_+ > 0.5$. Also plots of transference numbers as a function of $C^{1/2}$ should approach infinite dilution with the limiting Onsager slope. There are a few cases for 1 : 1 electrolytes in aqueous solutions where the limiting slope is approached, but for salts of higher valence type and nonaqueous electrolytes, the equations are not obeyed.

For the concentration ranges normally encountered in transference number measurements it is advantageous to rewrite Eqn. 7.44 in the form

$$t_{\pm}^{0} = \frac{t_{\pm}\Lambda' + \frac{1}{2}\beta C^{1/2}}{\Lambda' + \beta C^{1/2}} \tag{7.47}$$

At moderate concentrations, it is observed that in fact $t_{\pm}^{0}$ is not a constant, but varies linearly with C, and to extrapolate transference number data to zero concentration, the following equation is most commonly used:

$$t_{\pm}^{0\prime} = t_{\pm}^{0} + bC \tag{7.48}$$

where the $t_{\pm}^{0\prime}$, known as the Longsworth function, is the $t_{\pm}^{0}$ calculated at each concentration in Eqn. 7.47 above.

Although several other approaches have been used that take into account finite ionic size, the fact remains that as of the present time no adequate equation exists that accounts for the deviations obtained for the limiting Debye–Hückel–Onsager slopes.

Similarly the temperature and pressure dependences of transference numbers cannot be accounted for by any theoretical treatment at present. Both the Stokes law and Zwanzig theory [see Section 7.1.12] are of very limited value in predicting the temperature dependence.

7.2.2 Examples of Transference-Number Measurements in Selected Nonaqueous Solvents The purpose of this section is to indicate some general trends in the behavior of transference numbers in a variety of nonaqueous solvents. Tabulations of other transference number data in nonaqueous solvents are given elsewhere (2, 6).

Formamide Formamide has received considerable attention as a solvent for transference-number measurements, mainly because it is an example of a solvent with a higher dielectric constant than water. The investigations to be discussed here (3–5) enable one to compare the results obtained by using the techniques of moving boundary and Hittorf.

All three investigators (3–5) undertook measurements of the cation- and anion-constituent transference numbers of potassium chloride as a function of concentration. Spiro (3) used the moving-boundary technique, while Gopal (4) and Johari (5) used some form of the Hittorf method. A comparison

Table 7.16 Comparison of Transference Numbers of KCl in Formamide at 25°C Obtained by Different Methods

Technique	Conc. range (*M*)	$t^{\circ}_{K^+}$	Ref.
Moving boundary	0.01–0.1	0.427	3
Hittorf	0.1–0.5	0.419	4
Hittorf	0.03–0.14	0.409	5

of the results obtained for the limiting transference number for K^+ is given in Table 7.16. Values of $t^0_{K^+}$ were obtained from a plot of the Longsworth function (Eqn. 7.48). An example of the Longsworth plot is given in Fig. 7.7.

The differences in the values obtained by using the two techniques can be accounted for in terms of the uncertainty involved in determining the small concentration differences encountered in the Hittorf method. The effects of diffusion, convection, and vibration on mixing of solutions in the various compartments lead to low results for the transference numbers. The moving-boundary technique is generally recognized as the most precise of all the techniques available. Spiro combined the value obtained for the limiting transference number with the data for the limiting equivalent conductance to generate a set of individual ionic conductances for a variety of ions in formamide. The results of this analysis are shown in Table 7.17. The relatively low conductances of the monatomic ions in formamide, and their greater hydrodynamic radii, are attributed to the greater size of the formamide molecules in the solvation shells, an effect that more than compensates for the decrease in solvation numbers. It should be noted that an important application of the ionic conductances in Table 7.17 is in the choice of suitable

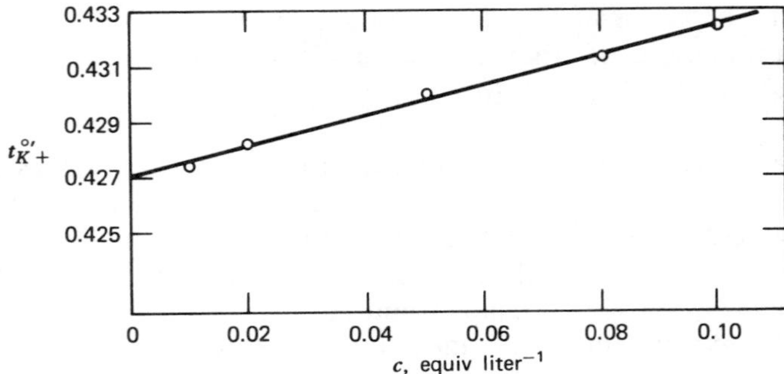

Figure 7.7 Plot of the Longsworth function $t^{\circ\prime}_{K^+}$ vs concentration of KCl in formamide. Reprinted with permission from *J. Phys. Chem.*, **70**, 1502 (1966). Copyright 1966 American Chemical Society.

Table 7.17 Limiting Ionic Equivalent Conductances in Formamide at 25°C

Cation	λ_0^+	Anion	λ_0^-
Li^+	8.5	Cl^-	17.1
Na^+	10.1	Br^-	17.2
K^+	12.7	I^-	16.6
Rb^+	12.8	SCN^-	17.2
Cs^+	13.5	NO_3^-	17.4
NH_4^+	15.6	$HCOO^-$	15.3
Me_4N^+	12.5	CH_3COO^-	11.9
Et_4N^+	10.0	$PhCOO^-$	9.8
Bu_4N^+	6.8	$PhSO_3^-$	10.4
Me_3PhN^+	10.7		
H^+	10.8		
Tl^+	15.8		

electrolytes for salt bridges in emf work, where approximate equality of anion and cation conductances is the main criterion, apart from high solubility of the electrolyte.

Gopal (4) determined the transference numbers of KCl in formamide over a range of temperatures. It was observed that t_{K^+} increased with temperature, the opposite to the behavior of t_{K^+} in aqueous solutions. Also, at any specified temperature, the transport number decreased appreciably with increasing concentration, compared to aqueous solutions, where t_+ was almost constant. These observations indicate a marked difference in ion–solvent interactions in the two solvents. As the temperature increases, there is a breakdown of the three-dimensional hydrogen-bonded water structure in the presence of large ions such as K^+, Rb^+, Cs^+, Br^-, and I^-, whereas this structure-breaking effect is of no importance for ions such as Li^+ and Na^+, which have much larger ion–dipole interactions, with a resulting high degree of solvation. However, in formamide there is no three-dimensional structure but only molecular association, and hence all the ions behave similarly in formamide irrespective of their sizes.

Johari (5) measured the transference numbers of KCl in formamide and *N*-methylacetamide.

Acetonitrile Kay (7) measured the transference number of the tetramethylammonium ion in tetramethylammonium perchlorate in anhydrous acetonitrile, using a rising-boundary technique. The value obtained for the limiting transference number was 0.4768, and it was combined with the limiting equivalent conductance of tetramethylammonium perchlorate (Λ_0 = 198.2) to yield the single-ion limiting equivalent conductances of $\lambda_0(Me_4N^+)$ = 94.5 and $\lambda_0(ClO_4^-)$ = 103.7. These values were then used to generate a set of data for limiting equivalent conductances of a variety of ions in acetonitrile using Λ_0's for electrolytes from the literature. The results of this analysis are given in Table 7.18. It is of interest to note that the single-ion conductances obtained from the transference number data differ by 0.35 conductance units from the scale based on the assumption that tetraisoamylammonium and tetraisoamylborate ions have equal mobilities. The latter assumption is based on the fact that these reference ions have practically the same size and that solvation effects should be small, since these ions are large, symmetrical and not very polarizable. The larger mobilities of the anions in acetonitrile, as seen in Table 7.18, are probably due to differences in solvation of the anion and the cation, the greater solvation of the cation causing it to move more slowly.

Dimethylsulfoxide Della Monica (8) investigated the transference number of the perchlorate ion, ClO_4^- at four concentrations of $AgClO_4$ in DMSO using the Hittorf method. The variation of transference number with concentration is shown in Table 7.19. The Longsworth extrapolation method was used to obtain a value of the limiting transference number $t^0_{ClO_4^-}$ as 0.596. Using $\Lambda_0(AgClO_4)$ = 40.47, a value for the limiting ionic conductance of ClO_4^-, λ_0 = 24.12 was obtained. A set of limiting ionic equivalent conductances of several ions in DMSO is reported in this study. The results are explained in terms of the ability of DMSO to act as an electron donor, thereby causing strong association with cations and therefore producing lower mobilities. On the other hand, the nucleophilic property of the sulfur

Table 7.18 Limiting Equivalent Conductances of Single Ions in Acetonitrile at 25°C

Cation	λ_0^+	Anion	λ_0^-
Me_4N^+	94.5	ClO_4^-	103.7
Bu_4N^+	61.4	Br^-	100.7
$(i\text{-}Am)_4N^+$	56.9	I^-	102.4
Ph_4As^+	55.8	Ph_4B^-	58.3
Ca^+	87.3	$(i\text{-}Am)_4B^-$	57.6
Rb^+	85.6	Pi^-	77.7
K^+	83.6		
Na^+	76.9		
$(EtOH)_4N^+$	64.0		
Et_4N^+	84.8		
Pr_4N^+	70.3		
Li^+	69.3		
$(i\text{-}Am)_3BuN^+$	57.8		

Table 7.19 Variation of Transference Numbers for ClO_4^- with Concentration in DMSO

Conc. (*N*)	$t_{ClO_4^-}$ (average of cathode and anode analyses)
0.02	0.611
0.05	0.620
0.06	0.623
0.1	0.632

Reprinted with permission from *Trans. Faraday Soc.*, **66**, 2872 (1970).

atom hinders any strong interaction between anions and solvent molecules, resulting in the anions being more mobile in DMSO than the cations.

Propylene Carbonate Mukherjee (9) used a concentration cell to determine the transference numbers of ClO_4^- and Li^+. The cell employed was

$$\underset{(2\text{ mol }\%)}{\text{Li(Hg)}} \mid LiClO_4(C_1) \mid LiClO_4(C_2) \mid \underset{(2\text{ mol }\%)}{\text{Li(Hg)}}$$

The emf of the cell is given by

$$E = 2t_{ClO_4^-}\frac{RT}{F}\ln\frac{a_2}{a_1} \tag{7.49}$$

where a_1 and a_2 represent the activities of the $LiClO_4$ solutions. For the unassociated case in dilute solution, Eqn. 7.49 can be rewritten as

$$E = 2t_{ClO_4^-}\frac{RT}{F}\ln\frac{C_2f_2}{C_1f_1} \tag{7.50}$$

where f_1 and f_2 are the mean ionic activity coefficients for the two solutions. The results obtained for the transference number at different concentrations of the cell components are given in Table 7.20. The limiting transference number for ClO_4^- was found to be 0.72, and using a value of Λ_0 for $LiClO_4$ in propylene carbonate as 26.08, the limiting conductances of Li^+ and ClO_4^- were found to be 7.30 and 18.78, respectively. The low value for the mobility of the lithium ion is evidence for the extensive solvation it undergoes, thereby making its effective size fairly large. It is found that the mobilities of ClO_4^-, I^-, Br^-, and Cl^- are very similar, indicating that the effective sizes of these anions are about the same and that they undergo very little solvation.

Methanol Kay and Gordon (10) used a moving-boundary technique to meas-

Table 7.20 Transference Numbers of Perchlorate Ion in Propylene Carbonate Using a Concentration Cell

C_1 (*M*)	C_2 (*M*)	*E* (V)	$t_{ClO_4^-}$
0.020	0.040	0.0245	0.757
0.025	0.050	0.0245	0.765
0.050	0.100	0.0243	0.789
0.100	0.150	0.0144	0.816
0.120	0.200	0.0181	0.820

Reprinted with permission from *J. Phys. Chem.*, **74,** 1942 (1970). Copyright 1970. American Chemical Society.

ure the transference numbers for NaCl and KCl in anhydrous methanol. The Longsworth function was used to obtain the limiting transference numbers as follows: $t^0_{Na^+} = 0.463$ and $t^0_{K^+} = 0.500$.

Dawson (11) evaluated the transference numbers of potassium bromide and potassium thiocyanate in methanol, also employing the moving-boundary method. A comparison study was also made in ethanol. Longsworth plots were found to be linear up to about 0.01 *N* and the resulting limiting transference numbers were as follows: $t^0_{K^+}$ (methanol) $= 0.479_5$ for KBr, $t^0_{K^+}$ (methanol) $= 0.455_5$ for KSCN, $t^0_{K^+}$ (ethanol) $= 0.461_2$ for KSCN.

It is seen that the potassium ion moves slower in comparison to the thiocyanate ion in both methanol and ethanol than in water, with the relative decrease in mobility being greater in methanol. It is suggested, therefore, that the greatest solvation of the cations occurs in methanol and that anion solvation is limited in both alcohols.

Acetone Brooks (12) measured the transference numbers of CNS^- in anhydrous acetone using a conductometric moving-boundary method and obtained a value for the limiting transference number $t^0_{CNS^-}$ as 0.6237. A set of limiting ionic conductances of various ions in acetone was generated using a value of Λ_0 for KCNS as 200.9 cm^2 Ω^{-1} mol^{-1}, and these are given in Table 7.21. The fact that the limiting conductances of most of the anions are larger than for the cations implies that the interaction of most cations with acetone is stronger than the interaction of acetone with the anions. The greater solvation of the cations is best understood in terms of the partial negative charge of the acetone dipole being concentrated on a single atom, the oxygen atom, compared to the positive charge being spread over the carbonyl carbon and the two methyl groups, which would hinder close approach of an anion.

Table 7.21 Limiting Ionic Conductances in Acetone at 25°C

Ion	λ_0^+	Ion	λ_0^-
H^+	100.0	F^-	90
Li^+	84.1	Cl^-	110
Na^+	87.1	Br^-	122
K^+	75.6	I^-	121
NH_4^+	116.0	Pi^-	90
Me_4N^+	93	ClO_4^-	114
Et_4N^+	86	CNS^-	125.6
n-Pr_4N^+	70	NO_3^-	125
n-Bu_4N^+	62		
n-$C_5H_{11}N^+$	53		

Reprinted with permission from *J. Chem. Soc.* **A**, 2415 (1971).

Table 7.22 Apparent Transport Numbers of Various Ions in Sulfuric Acid

Solute	m	Q (coulombs)	Equiv. gained by cathode compartment ($\times 10^4$)	Equiv. lost by anode compartment ($\times 10^4$)	t_c	t_a	t
$AgHSO_4$	0.2490	3329	9.62	8.46	0.028	0.024	0.026
	0.3056	4899	11.28	11.22	0.022	0.022	0.022
$KHSO_4$	0.6244	6166	19.18	19.42	0.030	0.030	0.030
	1.2262	6818	17.65	17.40	0.025	0.025	0.025
$NaHSO_4$	0.7918	7780	18.10	16.26	0.022	0.020	0.021
$LiHSO_4$	0.5562	5745	9.28	6.92	0.015	0.012	0.013
$Ba(HSO_4)_2$	0.1738	5175	4.94	4.80	0.009	0.009	0.009
	0.1969	5718	5.62	5.39	0.009	0.009	0.009
	0.2073	5632	5.38	3.94	0.009	0.007	0.008
	0.2292	9598	10.18	9.21	0.010	0.009	0.010
	0.2418	5644	4.41	4.39	0.008	0.007	0.008
	0.3062	7702	6.46	6.07	0.008	0.008	0.008
	0.5391	8314	4.21	3.87	0.005	0.005	0.005
	0.7981	9345	4.26	3.78	0.004	0.004	0.004
$Sr(HSO_4)_2$	0.2111	5764	4.55	4.11	0.008	0.007	0.007
	0.2788	7818	6.30	5.46	0.008	0.007	0.007
	0.5819	7620	4.06	3.34	0.005	0.004	0.005
	0.8247	9340	3.12	3.32	0.003	0.003	0.003

Reprinted with permission from *J. Chem. Soc.*, 209 (1953).

Table 7.23 Apparent Transport Numbers of Potassium Hydrogen Sulfate in Sulfuric Acid at Different Temperatures

Temp. (°C)	m	Q (coulombs)	Equiv. gained by cathode compartment ($\times 10^4$)	Equiv. lost by anode compartment ($\times 10^4$)	t_c	t_a	t
25.0	0.6244	7900	19.18	19.42	0.030	0.030	0.030
36.8	0.6536	7531	18.68	19.31	0.032	0.033	0.032
45.1	0.6388	8936	25.84	24.68	0.032	0.030	0.031
55.0	0.7098	7579	30.24	19.77	0.034	0.033	0.033
61.0	0.6556	7496	22.98	18.39	0.039	0.031	0.035

Reprinted with permission from *J. Chem. Soc.*, 209 (1953).

Sulfuric Acid Gillespie (13) measured the apparent transference numbers of lithium, sodium, potassium, silver, barium, and strontium ions in solutions of their hydrogen sulfates in sulfuric acid at 25°C. The Hittorf technique was employed in all the measurements. The results of this investigation are given in Table 7.22, where t_c and t_a represent the transference numbers determined from the cathode compartment and the anode compartment, respectively, shown as a function of concentration. The variation of the transference number of the potassium ion with temperature was also investigated and the results are shown in Table 7.23. The significant feature of these data are the very small values obtained for the transference numbers of all the cations. The conclusion is that the conductance of the solutions is due mainly to the hydrogen sulfate ion, which conducts almost entirely by a type of Grotthus chain mechanism similar to that for the proton in water. One step of such a mechanism may be represented as follows:

$$(O)_2S(OH)_2 + HO(O)_2S{-}O^- \rightarrow (O)_2S(OH)O^- + (HO)_2S(O)_2$$

Thus, by a succession of proton transfers, the charge of a hydrogen-sulfate ion is transferred through the solution without the actual movement of any ion.

From an examination of the transference numbers of the various ions at comparable concentrations it appears that the transport numbers decrease in the order

$$K^+ > Ag^+ > Na^+ > Li^+ > Ba^{2+} > Sr^{2+}$$

It should be noted that, for the univalent cations, this is the same order as that found for dilute solutions of their salts in water, the trend being due to an increase in solvation with increasing charge density on passing from K^+

to Li^+. This behavior suggests that there might be a similar increase in solvation for the same series in sulfuric acid.

There is very little effect of temperature on the variation of the transference number of K^+ in solutions of its hydrogen sulfate, as seen in Table 7.23. This behavior has been interpreted in terms of a balance between the normal increase in mobility with temperature due to decreasing viscosity of the solution and an increase in the abnormal mobility of the hydrogen sulfate ion due to the rate of molecular rotation increasing with increasing temperature and thereby favoring the Grotthus type of mechanism.

Chlorosulfuric Acid The transference numbers of some simple bases, such as alkali-metal and alkaline-earth metal chlorosulfates in chlorosulfuric acid, were measured by Robinson (14). As in the case of sulfuric acid, the anion

Table 7.24 Transport Numbers of Sodium and Barium Ions in Chlorosulfuric Acid at 25°C

Soln.[a] conc.	Meq of current	t_c [b]	t_a [c]	t_{av}
		Na^+ Cation		
0.0541	1.7000	0.108	0.122	0.115
0.0624	2.0000	0.105	0.108	0.107
0.0779	2.0860	0.114	0.130	0.122
0.0930	2.0600	0.127	0.138	0.128
0.1035	2.2720	0.119	0.114	0.117
0.1204	2.1153	0.087	0.084	0.086
0.1273	1.5401	0.119	0.088	0.104
0.1482	2.1660	0.082	0.116	0.099
		Ba^{2+} Cation		
0.0824	2.0000	—	0.061	0.061
0.0838	2.0000	0.042	0.064	0.053
0.0852	2.0000	0.043	0.021	0.037
0.1052	2.0000	0.057	0.059	0.058
0.1112	2.0000	0.059	—	0.059
0.1436	2.0000	0.058	0.038	0.048
0.1652	2.0000	0.062	0.077	0.070

[a] Solution concentration (equivalents per kilogram of solution).
[b] Cathode.
[c] Anode.
Reproduced by permission of the National Research Council of Canada from the *Canadian Journal of Chemistry*, **46,** 1719 (1968).

Table 7.25 Equivalent Conductances, Limiting Transference Numbers, and Limiting Ionic Conductances for Some Quaternary Ammonium Halides in Nitromethane

Salt	Λ_0	t^0_+	t^0_-	λ^+_0	λ^-_0
Me_4NCl	117.62	0.4674	(0.5326)	54.97	(62.65)
Me_4NBr	117.83	0.4663	(0.5337)	54.94	(62.89)
Et_4NCl	110.37	0.4320	(0.5680)	47.68	(62.69)
Et_4NBr	110.60	0.4314	(0.5686)	47.71	(62.89)
Pr_4NCl	101.88	(0.3843)	0.6157	(39.15)	62.73
Pr_4NBr	102.10	(0.3835)	0.6165	(39.16)	62.94
Bu_4NCl	96.83	(0.3526)	0.6474	(34.14)	62.69
Bu_4NBr	97.04	(0.3513)	0.6487	(34.09)	62.95

SO_3Cl^- was found to have a high mobility, compared to the mobilities of the alkali-metal cations and alkaline-earth metal cations, values of which are shown as a function of concentration in Table 7.24. The high values for the transference number of the chlorosulfate ions is accounted for by an abnormal proton transfer mechanism similar to that found for HSO_4^- ions in sulfuric acid. The mechanism is shown diagramatically below:

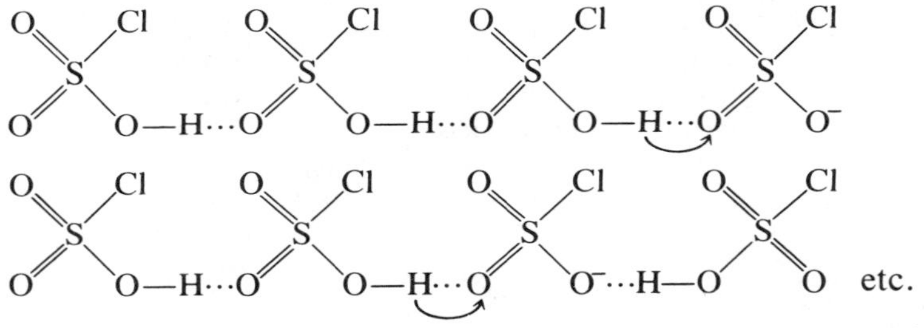

```
O   Cl      O   Cl      O   Cl      O   Cl
  S           S           S           S
O   O—H···O     O—H···O     O—H···O     O⁻

O   Cl      O   Cl      O   Cl      O   Cl
  S           S           S           S
O   O—H···O     O—H···O     O⁻···H—O    O   etc.
```

Nitromethane Blum (15) used the conductometric moving-boundary method to measure transference numbers of several quaternary ammonium chlorides and bromides in nitromethane over the concentration range 0.0002 to 0.01 *N*. The Longsworth plot was employed to obtain limiting transference numbers and, in combination with equivalent conductances, a set of single-ion mobilities was obtained. The various parameters are summarized in Table 7.25.

Amides Several investigations on transference numbers in the amides having high dielectric constants have been undertaken and these are summarized in Table 7.26. Limiting transference numbers were obtained using the Longsworth procedure and these values are summarized in Table 7.27 (see page 300). The limiting transference numbers were used to calculate ionic conductances from the appropriate conductance data. The values of the mobilities thus obtained are summarized in Table 7.28 (see page 301). The limiting conductances increase with temperature as expected. Since the tetrahedral structure present in water is missing in the amide solvents, the structure-breaking effect of the larger ions, like K^+, is missing in these solvents, and consequently all the ions behave normally.

Summary A knowledge of transference numbers for cations and anions in nonaqueous solvents can provide significant evidence for ion–solvent and ion–ion interactions. In addition, the transference numbers are needed for the evaluation of single-ion conductances and are of practical importance in, for example, nonaqueous battery systems. Transference numbers are obtained via a variety of methods including the Hittorf method, the moving-boundary technique, the emf of a concentration cell, and radioactive tracer techniques.

Transference numbers are a function of concentration. Several methods have been devised to try to fit experimentally observed data to a theoretical equation. Both the Onsager limiting law and the Fuoss–Onsager equation have been used for such treatments, but at the present time no adequate equation exists which accounts for the deviations obtained from the limiting Debye–Hückel–Onsager slopes. In addition, the temperature dependence of the transference numbers cannot be accounted for by any satisfactory theoretical treatment.

The differences in the values of transference numbers of ions in different solvents can furnish information on the relative degrees of solvation. For example, for KCl in formamide and water the transference number for K^+ increases with temperature in formamide while it decreases with temperature in water. Also, at any specified temperature the transference number in formamide decreases appreciably with increasing concentration, compared to aqueous solutions, where t_+ is almost constant. These trends indicate a marked difference in ion–solvent interactions in the two solvents. As the temperature increases, there is a breakdown of the three-dimensional hydrogen-bonded water structure in the presence of large ions such as K^+, Rb^+, Cs^+, Br^-, and I^-, whereas the structure-breaking effect is of no importance for ions such as Li^+ and Na^+, which have much larger ion–dipole interactions with a resulting high degree of solvation. However,

Table 7.26 Transference Number Studies in Amides

Solvent	Electrolyte	Conc. range	Temp. range (°C)	Technique	Ref.
N-Methylformamide	KBr	0.05–0.3 M	15–45	Hittorf	16
Dimethylformamide	LiCl	0.08–0.54 M	25	Hittorf	17
N-Methylpropionamide	KBr	0.075–0.250 M	30–50	Hittorf	18

in formamide there is no three-dimensional structure, but only molecular association, and hence all the ions behave similarly in formamide irrespective of their sizes. For $LiClO_4$ in propylene carbonate, the low mobility of the lithium ion is evidence for the extensive solvation it undergoes, thereby making its effective size fairly large. On the other hand, the mobilities of

Table 7.27 Mobilities of Ions in Amide Solvents

Solvent	Ion	Temp. °C	$\lambda_0^{\pm}$	Ref.
N-Methylformamide	Na^+	15	17.59	16
		25	21.56	
	K^+	15	18.04	16
		25	22.13	
		35	27.62	
		45	32.15	
	Cs^+	15	19.90	16
		25	24.39	
	$(C_2H_5)_4N^+$	15	21.52	16
		25	26.20	
	Pi^-	15	11.28	16
		25	13.08	
	Cl^-	15	16.88	
		25	19.70	
	Br^-	15	18.46	16
		25	21.56	
		35	25.38	
		45	28.60	
	I^-	15	19.28	16
		25	22.76	
Dimethylformamide	Li^+	25	23.62	17
N-Methylpropionamide	Na^+	30	5.06	18
		40	6.45	
		50	8.06	
		60	10.01	
	K^+	30	5.36	18
		40	6.95	
		50	8.76	
		60	10.91	
	Cl^-	30	6.24	18
		40	7.95	
		50	9.94	
		60	11.99	
	Br^-	30	7.06	18
		40	8.95	
		50	11.04	
		60	13.49	

Table 7.28 Limiting Transference Numbers in Amides

Solvent	Ion	Temp. (°C)	t^0_+	Ref.
N-Methylformamide	K^+	15	0.4945	16
		25	0.5080	
		35	0.5210	
		45	0.5290	
Dimethylformamide	Li^+	25	0.295	17
N-Methylpropionamide	K^+	30	0.4320	18
		40	0.4370	
		50	0.4425	

ClO_4^-, I^-, Br^-, and Cl^- are very similar, indicating that the effective sizes of these anions are about the same and that they undergo very little solvation.

Transference numbers of some hydrogen sulfates determined in sulfuric acid indicate that the main contributor to the current is the hydrogen sulfate ion, which conducts almost entirely by a chain mechanism involving proton transfer.

LITERATURE CITED—TRANSFERENCE NUMBERS

1 Kay, R. L., "Transference Number Measurements," in *Techniques in Electrochemistry,* Vol. II, E. Yeager and A. J. Salkind, Eds., Wiley-Interscience, New York, 1973.

2 Spiro, M., "Conductance and Transference Numbers," Part 2: "Transference Numbers," in *Physical Chemistry of Organic Solvent Systems,* A. K. Covington and T. Dickinson, Eds., Plenum Press, London, 1973, Chap. 5.

3 Notley, J. M., and M. Spiro, *J. Phys. Chem.,* **70,** 1502 (1966).

4 Gopal, R., and O. M. Bhatnagar, *J. Phys. Chem.,* **68,** 3892 (1964).

5 Johari, G. P., and P. H. Tewari, *J. Phys. Chem.,* **70,** 197 (1966).

6 Janz, G. J., and R. P. T. Tomkins, *Nonaqueous Electrolytes Handbook,* Vol. I, Academic Press, New York, 1972.

7 Springer, C. H., J. F. Coetzee, and R. L. Kay, *J. Phys. Chem.,* **73,** 471 (1969).

8 Della Monica, M., D. Masciopinto, and G. Tessari, *Trans. Faraday Soc.,* **66,** 2872 (1970).

9 Mukherjee, L. M., D. P. Boden, and R. Lindauer, *J. Phys. Chem.,* **74,** 1942 (1970).

10 Davies, J. A., R. L. Kay, and A. R. Gordon, *J. Chem. Phys.,* **19,** 749 (1951).

11 Smisko, J., and L. R. Dawson, *J. Phys. Chem.,* **59,** 84 (1955).

12 Brookes, H. C., M. C. B. Hotz, and A. H. Spong, *J. Chem. Soc., A,* 2415 (1971).

13 Gillespie, R. J., and S. Wasif, *J. Chem. Soc.,* 209 (1953).

14 Robinson, E. A., and J. A. Ciruna, *Can. J. Chem.,* **46,** 1719 (1968).

15 Blum, S., and H. I. Schiff, *J. Phys. Chem.,* **67,** 1220 (1963).

16 Gopal, R., and O. N. Bhatnagar, *J. Phys. Chem.,* **70,** 3007, (1966).

17 Paul, R. C., J. P. Singla, and S. P. Narula, *J. Phys. Chem.,* **73,** 741 (1964).

18 Gopal, R., and O. N. Bhatnagar, *J. Phys. Chem.,* **70,** 4070 (1966).

7.3 DIFFUSION

Although diffusion coefficients for electrolytes in nonaqueous solvents are of particular relevance in estimating transport limitations in various practical applications of organic electrolytes, such as lithium batteries, very few measurements of diffusion coefficients have actually been undertaken.

Various experimental methods have been employed to measure diffusion coefficients in aqueous systems; these include optical methods, electrochemical methods (chronopotentiometry, rotating disk electrode), and diaphragm (porous disk) cell techniques. Details of several methods have been carefully reviewed by Bierlein and Becsey (1). It is to be expected that some of the techniques used to gain data in aqueous systems will be extended to measurements in nonaqueous electrolyte systems.

It is sufficient in this section to present some of the fundamental principles underlying the process of diffusion to be able to understand the ideas presented in the few publications that do exist.

The spontaneous and irreversible process of diffusion occurs whenever a difference in concentration exists, that is, a concentration gradient. Diffusion is characterized by a diffusion coefficient, D, which is defined by Fick's first law as

$$J = -D\frac{\partial C}{\partial V} \tag{7.51}$$

where J represents the flux, that is, the mass of material passing across unit area in unit time in a direction that is perpendicular to the flow and $\partial C/\partial V$ is the concentration gradient, measured parallel to the flow. A point worth noting here is that D is a function of both temperature and concentration and the common appearance of the term "diffusion constant" results from the fact that in practice experimental measurements are invariably carried out at constant temperature and with a small concentration gradient.

Some of the modern experimental techniques are based on the observation of a transient process in which the concentration changes as a function of both time and distance. For this case, Eqn. 7.51 is combined with the continuity relation

$$\frac{\partial C}{\partial t} + \frac{\partial J}{\partial V} = 0 \tag{7.52}$$

to give Fick's second law:

$$\frac{\partial C}{\partial t} = \frac{\partial}{\partial V}\left(D\frac{\partial C}{\partial V}\right) = D\frac{\partial^2 C}{\partial V^2} + \frac{\partial D}{\partial V}\cdot\frac{\partial C}{\partial V} \tag{7.53}$$

To obtain a useful mathematical description of an unsteady-state diffusion process, Eqn. 7.53 must be integrated with attention to appropriate boundary conditions. This is seldom feasible unless D can be regarded as a constant, so that the second term on the right of Eqn. 7.53 vanishes. Hence the importance of avoiding large concentration variations in practice.

Essentially the determination of diffusion coefficients consists in measuring sets of simultaneous values of t, V, and C (or $\partial C/\partial V$). These measured values are then fitted to a solution of Fick's law.

For example, in the diaphragm cell method the differential diffusional coefficients, $\bar{D}$ are obtained directly from a relation such as

$$\bar{D} = \frac{1}{kt} \ln \frac{[C_1 - C_2]}{[C_3 - C_4]} \tag{7.54}$$

where C_1 and C_3 are the molar concentrations in the lower compartment of the diaphragm cell at time $t = 0$ and $t = t$, respectively; C_2 and C_4 are the corresponding concentrations in the upper compartment of the cell and k is the cell constant, obtained from calibration data.

In constant-current chronopotentiometry, a constant current density i_0 is impressed upon an electrode, and its potential is measured as a function of time. The transition time, τ, the time at which the reductant concentration is zero at the electrode surface, is given by the Sand equation:

$$\tau^{1/2} = \frac{nFC(\pi D)^{1/2}}{2i_0} \tag{7.55}$$

where C is the concentration of the electroactive species, D is its diffusion coefficient, n is the number of electrons involved in the electrode reaction, and F is the Faraday constant.

The thickness of the diffusion layer adjacent to the electrode surface affects the value of the limiting current flowing through the electrode. One way of controlling the thickness of the diffusional layer is by means of a disk rotating with a constant angular velocity, which effectively establishes a definite velocity distribution on the surface of the electrode. An equation for the rate of mass transport to unit area of the electrode surface was established by Levich:

$$K = \frac{0.554 D^{2/3} \nu^{-1/6} \omega^{1/2}}{I(\infty)} \tag{7.56}$$

where ν is the kinematic viscosity ($cm^2\ sec^{-1}$) and ω is the angular velocity of the rotating electrode ($rad\ sec^{-1}$). The limiting current can be expressed as

$$i = 0.554 \times 10^{-3} \frac{\mathrm{nFACD}^{2/3} \nu^{-1/6} \omega^{1/2}}{I(\infty)} \tag{7.57}$$

where A is the apparent area of the rotating electrode, C is the concentration of the ionic species in the bulk of the solution, and $I(\infty)$ is the value of the definite integral

$$I(\infty) = \int_0^\infty \exp\left[-x^3 + 0.885\left(\frac{D}{\nu}\right)^{1/3} x^4 + \dots\right] dx$$

The theoretical treatments governing the process of diffusion have been limited to very dilute solutions. Values for the limiting diffusion coefficient D^0 are normally calculated using the Nernst limiting equation:

$$D^0 = \frac{RT}{F^2}\frac{(z_1 + z_2)(\lambda_+^0 \lambda_-^0)}{z_1 z_2 (\lambda_+^0 + \lambda_-^0)} \tag{7.58}$$

where λ_+^0 and λ_-^0 are the limiting ionic conductances.

For single ions, a quantity known as the intrinsic ionic diffusivity, D_i^0, can be calculated as follows:

$$D_i^0 = \frac{RT}{z_1 F}\lambda_i^0$$

Another approach is to examine the driving force for diffusion in terms of the chemical potential gradient. An equation using this particular concept was developed by Nernst and Hartley:

$$D = D^0\left[1 + \frac{\partial \ln y_\pm}{\partial \ln C}\right] \tag{7.59}$$

where $y_\pm$ is the mean molar activity coefficient.

7.3.1 Diffusion Coefficients in Nonaqueous Solvents As indicated earlier, the availability of diffusion coefficient data in nonaqueous solvents is very limited. Janz et al. (2) measured the diffusion coefficients for $AgNO_3$ in acetonitrile and benzonitrile at 25°C, using the Stokes diaphragm technique. The Stokes diaphragm cell consists of a glass cell with a sintered-glass diaphragm in its center. The cell is stirred magnetically by using a series of Teflon-coated stirrers with varying densities so that a stirrer operates just above and just below the sintered disk. In this method, the diffusion process is confined to the capillary pores of the diaphragm. To measure the true length and cross-section of the diaphragm pores, the cell is calibrated by conducting diffusion experiments with an electrolyte whose diffusion coefficient has been determined by an absolute method. The diffusion measurements essentially consist of analyzing samples from each compartment of the cell as a function of time. Conductance techniques are often used for the analysis, although any applicable analytical method may be employed.

The results for the diffusion coefficients were compared to the value obtained for aqueous $AgNO_3$ and with the data for the viscosity of the three solvents. The comparison is shown in Fig. 7.8. The concentration dependence of the mobility ratio D/Λ is shown in Fig. 7.9. Marked differences are observed between the nonaqueous and the aqueous results, as seen by the

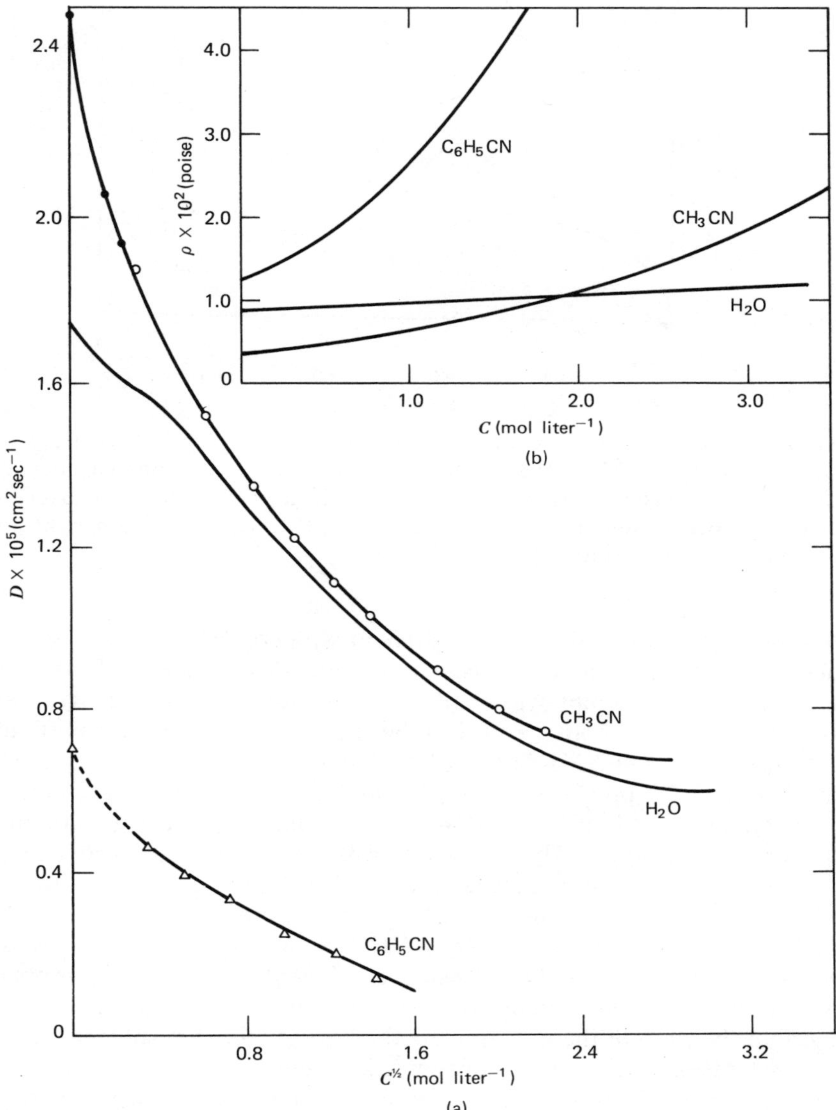

Figure 7.8 (a) Concentration dependence of diffusion coefficients for $AgNO_3$ in CH_3CN, H_2O, and C_6H_5CN at 25°C. Dark circles in the CH_3CN curve refer to the values gained by the Gouy technique. (b) Concentration dependence of viscosity for $AgNO_3$ in CH_3CN, H_2O, and C_6H_5CN at 25 °C. Reprinted with permission from *J. Phys. Chem.*, **70**, 2562 (1966). Copyright 1966 American Chemical Society.

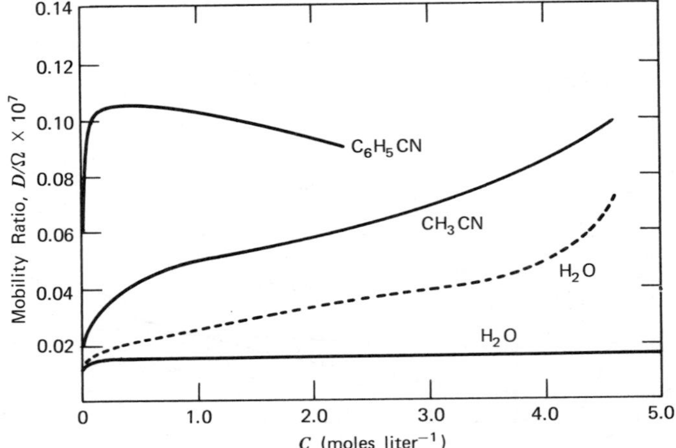

Figure 7.9 Concentration dependence of mobility ratio, D/Λ for $AgNO_3$ in CH_3CN, H_2O, and C_6H_5CN at 25°C. The dotted line refers to concentration dependence of D_{cor}/Λ (diffusion values corrected for thermodynamic nonideality) for H_2O. Reprinted with permission from *J. Phys. Chem.*, **70,** 2562 (1966). Copyright 1966 American Chemical Society.

initial rapid decrease in diffusion coefficient with concentration for the non-aqueous systems and the differences in the mobility ratio D/Λ. In this particular study, an attempt was made to correlate the diffusion data with information obtained from spectral studies of $AgNO_3$ in nitriles, in particular infrared, Raman, and NMR studies (see Section 8.3.3). The spectral results confirmed both the presence of ion pairs and marked solvation interactions, the complex species $[Ag^+(CH_3CN)_2NO_3^-]$ becoming increasingly dominant at higher concentrations. The correlation was formulated in terms of a model of a solvated kinetic unit, that is, $AgNO_3 \cdot 2RCN$, to recalculate diffusion coefficients at equivalent mole-fraction concentrations, relative to the unbound or "free" solvent for $AgNO_3$ in CH_3CN and C_6H_5CN. This approach permits a comparison of the diffusion coefficients for the most probable solvated species at equivalent concentrations in the two solvents. It was found that the ratio of the diffusion coefficients for the two solvent systems was constant for a fairly wide concentration range.

Sullivan et al. (3) used a method involving the weight change of a porous disk filled with the test solution to measure the diffusion coefficients of $LiClO_4$, LiCl, and $LiAsF_6$ in propylene carbonate (PC), dimethylformamide (DMF), acetonitrile (AN), and methyl formate (MF). The average or integral diffusion coefficient, D, was determined from the equation

$$\log [W(t) - W(\infty)] = \alpha D t + b \tag{7.60}$$

where $W(t)$ is the apparent weight of the suspended disk at time t, $W(\infty)$,

Table 7.29 Diffusion Coefficients in Some Aprotic Solvents at 25°C

Electrolyte/Solvent	Diffusion coefficient ($cm^2\ sec^{-1}$)
1 *M* $LiClO_4$/PC	2.58×10^{-6}
1 *M* $LiClO_4$/DMF	7.29×10^{-6}
1 *M* LiCl/DMF	5.87×10^{-6}
1 *M* $LiClO_4$/AN	1.71×10^{-5}
1 *M* $LiClO_4$/MF	1.68×10^{-5}
1.1 *M* $LiAsF_6$/MF	1.54×10^{-5}

Reprinted with permission of the publisher, The Electrochemical Society, Inc., from *J. Electrochem. Soc.*, **117,** 779 (1970).

the weight after equilibrium has been reached, α is an apparatus constant, and b is an integration constant. The results obtained for the diffusion coefficients are given in Table 7.29. A direct correlation was found between the diffusion coefficient and the solvent viscosity, but the validity of extending such an approach to predict diffusion coefficients for electrolyte solutions is doubtful.

The electrochemical techniques have normally focused on the study of electrochemical processes (rate constants, irreversibility, etc.) rather than ion–solvent or ion–ion interactions, and the diffusion coefficients thus obtained are used mainly in the discussion of the diffusion-controlled processes.

Giordano et al. (4–6) used chronopotentiometry and rotating-disk methods to study solutions of sodium iodide and iodine in DMSO and a rotating-disk electrode for a similar study in AN. The diffusion coefficients for triio-

Table 7.30 Diffusion Coefficients of Iodide and Triiodide Ions in DMSO and in Aqueous Solutions at 25°C Obtained from both Chronopotentiometry and the Rotating-Disk Electrode

	$D \times 10^5$ ($cm^2\ sec^{-1}$)		
Ionic species	Rotating disk	Chronopotentiometry	Ionic medium
I_3^-	0.37	0.39	0.8 *M* $KClO_4$ (DMSO)
I^-	0.69	0.71	0.8 *M* $KClO_4$ (DMSO)
I_3^-	0.42		1.0 *N* H_2SO_4(aq)
I^-	0.81		1.0 *N* H_2SO_4(aq)
I^-	2.20		1.5 *M* KI(aq)
I_3^-	1.13		0.1 *M* KI(aq)

Taken from Refs. 4–6.

dide and iodide ions in solutions of DMSO were found to be in agreement with those obtained using the rotating-disk electrode, and also the ratio of the diffusion coefficients of iodide to triiodide ion in DMSO was approximately the same as in aqueous solutions. This point is illustrated in Table 7.30. The diffusion coefficients of the reacting species in acetonitrile are larger than those in DMSO solutions as a consequence mainly of the lower viscosity of AN.

Summary Experimental data for diffusion coefficients in nonaqueous solvents are very sparse. The techniques available for measuring diffusion coefficients are optical methods, electrochemical methods (chronopotentiometry, rotating-disk electrode), and use of a diaphragm cell. The evaluation of a diffusion coefficient is based on Fick's first and second laws, and it consists essentially of measuring sets of simultaneous values of t, V, and C and fitting the values to a solution of Fick's law. Theoretical treatments governing the process of diffusion have been limited to very dilute solutions and values for the limiting diffusion coefficient, D^0, are normally calculated using the Nernst limiting equation. A comparison of diffusion coefficient data in nonaqueous and aqueous systems is useful as an insight into ion-pair formation and ion–solvent interactions. One of the major studies involved $AgNO_3$ in nitrile solvents.

LITERATURE CITED—DIFFUSION

1 Bierlein, J. A., and J. G. Becsey, "Diffusion in Electrolytes," in *Techniques of Electrochemistry,* Vol. II, E. Yeager and A. J. Salkind, Eds., Wiley-Interscience, New York, 1973, Chap. 3.

2 Janz, G. J., G. R. Lakshminarayanan, and M. P. Klotzkin, *J. Phys. Chem.,* **70,** 2562 (1966).

3 Sullivan, J. M., D. C. Hanson, and R. Keller, *J. Electrochem. Soc.,* **117,** 779 (1970).

4 Giordano, M. C., J. C. Bazán, and A. J. Arviá, *Electrochim. Acta,* **11,** 1553 (1966).

5 Giordano, M. C., J. C. Bazán, and A. J. Arviá, *Electrochim. Acta,* **12,** 723 (1967).

6 Macagno, V. A., M. C. Giordano, and A. J. Arviá, *Electrochim. Acta,* **14,** 335 (1969).

GENERAL REFERENCES—DIFFUSION

Bierlein, J. A., and J. G. Becsey, "Diffusion in Electrolytes," in *Techniques of Electrochemistry,* Vol. II, E. Yeager and A. J. Salkind, Eds., Wiley-Interscience, New York, 1973, Chap. 3.

Janz, G. J., and R. P. T. Tomkins, *Nonaqueous Electrolytes Handbook,* Vol. I, Academic Press, New York, 1972.

7.4 VISCOSITY

The investigation of the viscosity of electrolyte solutions in nonaqueous solvents is useful for evaluating ion–solvent interactions and the structure of electrolyte solutions. The cosphere† of each positive ion and the cosphere of each negative ion may contribute toward a change in the viscosity. Examination of the behavior of the viscosity of the solution as a function of concentration sometimes yields a maximum, which can be interpreted in terms of enhanced structure. The viscosity of the electrolyte solution is important in its own right, as it is a fundamental transport property that is necessary for the prediction and evaluation of other properties such as electrical conductance.

The concentration dependence of the viscosity of nonaqueous electrolyte solutions is often interpreted in terms of the semiempirical Jones–Dole equation (1):

$$\frac{\eta}{\eta_0} = 1 + \mathrm{AC}^{1/2} + \mathrm{BC} \tag{7.61}$$

where η and η_0 are the solution and solvent viscosities, respectively, C is the molar concentration, A is the Falkenhagen (2) coefficient (representing ion–ion interactions), and the B coefficient represents any specific electrostatic effect the ions may have on the viscosity of the solvent. Both A and B are characteristics of the solute.

For associated electrolytes, the Jones–Dole equation becomes

$$\frac{\eta}{\eta_0} = 1 + A(C\gamma)^{1/2} + BC\gamma + B'C(1 - \gamma) \tag{7.62}$$

where B and B' are the Jones–Dole coefficients for the free ions and the ion pair, respectively. In general, viscosity measurements must be extended to such high concentrations, owing to the small effects observed, that terms in $C^{3/2}$ and higher must be added in order to fit the data. The Jones–Dole equation has been found valid to approximately 0.1 M.

For the purpose of discussing ion–solvent interactions, the B coefficients are split into single-ion values by using the additivity relationships that exist for electrolytes with a common ion.

Ions exhibiting positive B coefficients are referred to as "structure makers," as they produce an enhancement of the structure, while ions with negative B coefficients are described as "structure breakers," as they cause a disruption in the overall structure and hence increase the fluidity of the solution.

There are several factors that contribute to the variation of B coefficients in different solvents. First, structure breaking will be much more important in water and other "three-dimensional" solvents than in less structured solvents. Second, for solvents whose solvating abilities are comparable, the higher B coefficient is normally associated with the solvent having the largest molar volume. Other factors, such as steric hindrance for small ions, dielectric constant, and dipole moment should also be taken into consideration.

† A term used to describe the region of modified solvent surrounding an ion.

In practice, the B coefficient is obtained by rearranging Eqn. 7.61 to

$$\frac{(\eta/\eta_0) - 1}{C^{1/2}} = A + BC^{1/2} \tag{7.63}$$

and plotting the left-hand side against $C^{1/2}$.

The forces that exist between ions and solvent molecules in solution are on the average stronger than in the pure solvent, with a resultant increase in the viscosity of the solution, compared to the pure solvent. The stronger forces are due to the interaction of the charge on the surface of the ion with the polar or polarized solvent molecules. The effect of increased viscosity in the solution can be approximately expressed by means of the Einstein equation:

$$\frac{\eta}{\eta_0} = 1 + 2.5\phi \tag{7.64}$$

where ϕ is the volume fraction of the solute. This equation can be used to obtain the effective molar volume of a solvated electrolyte.

It should be noted that there are a few exceptions to this general observation, where the viscosity of the solution is less than that of the pure solvent, and consequently the B values are negative. Examples include some solutions in water, ethylene glycol, and glycerol, that is, hydrogen-bonded solvents, where a competition exists between the ordering structure caused by orientation of solvent molecules around a charged ion and the ordering tendency of hydrogen bonding, which tends to keep the solvent molecules as part of a large hydrogen-bonded network.

The following discussion focuses on selected systems in which viscosity measurements have been undertaken and particularly on how the data relate to structural problems.

7.4.1 Viscosity Measurements in Nonaqueous Solvents

N-Methylformamide Some of the most precise measurements of viscosities of electrolyte solutions in nonaqueous solvents are those of Feakins and Lawrence (3), who measured the relative viscosities of sodium and potassium chlorides and bromides in *M*-methylformamide at 25, 35, and 45°C. These authors used the extended Jones–Dole equation to analyze their data, that is,

$$\frac{\eta}{\eta_0} = 1 + AC^{1/2} + BC + DC^2 \tag{7.65}$$

which can be recast as

$$\frac{\left(\dfrac{\eta}{\eta_0} - 1\right)}{C^{1/2}} = A + BC^{1/2} + DC^{3/2} \tag{7.66}$$

Table 7.31 Values of A, B, and D Obtained from Eqn. 7.66 for Various Electrolytes in N-Methylformamide

Electrolyte	Temp. (°C)	A	B	D	A (theoretical)
NaCl	25	0.0045	0.599	0.17	0.0062
	35	0.0050	0.577	0.11	0.0064
	45	0.0046	0.558	0.12	0.0067
KCl	25	0.0065	0.615	0.14	0.0061
	35	0.0065	0.589	0.10	0.0063
	45	0.0070	0.568	0.10	0.0066
NaBr	25	0.0050	0.567	0.25	0.0058
	35	0.0071	0.542	0.22	0.0062
	45	0.0092	0.511	0.22	0.0066
KBr	25	0.0039	0.584	0.21	0.0058
	35	0.0070	0.549	0.22	0.0061
	45	0.0047	0.541	0.32	0.0063

Reprinted with permission from *J. Chem. Soc.* **A,** 212 (1966).

One of the problems considered was the effect of the solvent impurities on the viscosity, and a correction for the impurity effect was given by

$$\frac{\left(\frac{\eta}{\eta_0} - 1\right)}{C^{1/2}} + A\left[1 + \left(\frac{C_1}{C}\right)^{1/2}\left(1 + \frac{C_1}{C}\right)^{1/2}\right] = A + BC^{1/2} + DC^{3/2} \qquad (7.67)$$

where C_1 is the concentration of impurity present in the solvent. Table 7.31 shows the values of A, B, and D obtained by using Eqn. 7.66, together with the theoretical value for A. The differences in B values for electrolytes with a common ion are given in Table 7.32. Within the experimental uncertainties, the B values were seen to be additive for individual ions, and the assumption was made that ion–solvent interactions are the sole factors contributing to the B values.

Table 7.32 Difference in B Values for Electrolytes with a Common Ion

Temp. (°C)	$B_{KCl}-B_{NaCl}$	$B_{KBr}-B_{NaBr}$	$B_{KCl}-B_{KBr}$	$B_{NaCl}-B_{NaBr}$
25	0.016	0.017	0.031	0.032
35	0.012	0.007	0.040	0.035
45	0.010	0.030	0.027	0.047

Reprinted with permission from *J. Chem. Soc.* **A,** 212 (1966).

Table 7.33 *B* Coefficients of KI and CsI at 25°C in Ethylene Glycol, Glycerol, Water, Methanol, and DMSO

Solvent	*B*
KI	
Methanol	0.6747
Ethylene glycol	0.0327
Water	−0.0755
Glycerol	−0.185
CsI	
DMSO	0.68
Ethylene glycol	−0.080
Water	−0.118
Glycerol	−0.408

Reprinted with permission from *J. Phys. Chem.* **73,** 2060 (1969). Copyright 1969. American Chemical Society.

The *B* values were all positive, additive for single ions and decreased with increasing temperature. The positive values indicated that any structure-breaking tendencies by these simple ions is relatively less important here than in "three-dimensional" solvents. The *B* values for all the electrolytes in NMF are much smaller than the corresponding quantities in other nonaqueous solvents.

It was noted that in *N*-methylformamide and in methanol the differences between *B* values for the halide ions appear to be of the same order as in water. For example, at 25°C, $B_{Cl} - B_{Br} = 0.031, 0.035$, and 0.035 for *N*-methylformamide, methanol, and water, respectively, and $B_{Cl} - B_{I} = 0.098$ and 0.062 for the last two solvents, respectively. This trend was accounted for in terms of the equality of the structure-breaking contribution to the *B* coefficients for Cl^-, Br^-, and I^- in water, compared to the falloff in the structure-making contribution from Cl^- to I^-. To account for the observation that the *B* values in NMF are lower than in, say, ethanol it was suggested that fewer NMF molecules are bound to a particular ion than ethanol molecules. The higher dielectric constant of NMF causes a lower ionic field, thus the contribution to the value of *B* from long-range ordering of solvent molecules will be less in NMF than in ethanol.

Ethylene Glycol and Glycerol A study of the viscosities of ethylene glycol and glycerol solutions of potassium iodide and cesium iodide at 25°C was undertaken by Crickard and Skinner (4). The resulting *B* coefficients obtained from the Jones–Dole equation are given in Table 7.33, with a comparison of the values in methanol, water, and dimethylsulfoxide. The lower *B* coefficients for KI in ethylene glycol and glycerol compared to methanol suggest a considerable weakening of the hydrogen bonding in both the glycolic solvents, particularly in glycerol. Cesium iodide is evidently a structure breaker in both glycolic solvents.

Dimethylsulfoxide As part of a general investigation of the behavior of cesium iodide in DMSO, Archer and Gasser (5) measured the viscosities of CsI–DMSO solutions over a fairly large concentration range. It was observed that the viscosity varied markedly with the concentration of cesium iodide, particularly above 1 *M*. The extended Jones–Dole equation was used to fit the data. The value obtained for the *B* coefficient was 0.68, a much larger positive value compared to that obtained in aqueous solutions.

Sulfuric Acid Gillespie (6) has provided the only viscosity data in an acid such as sulfuric acid. Viscosities of solutions containing water, sulfur trioxide, alkali- and alkaline-earth metal hydrogen sulfates, nitric acid, sulfuryl chloride, and a number of organic compounds were measured. The results are shown graphically in Figs. 7.10–7.12. Because all the electrolytes except disulfuric acid and sulfuryl chloride ionize to produce the hydrogen sulfate ion, it is the effect of the cation that is reflected in the trends in viscosity. Examination of Fig. 7.10 shows that inorganic cations generally increase the viscosity of sulfuric acid, particularly the bivalent cations, while the organic cations, with the exception of *o*-phenylenediammonium, exhibit very small changes with increasing concentration.

The viscosity changes were attributed to structural changes in the solvent caused by the cations. An increase in viscosity is caused by the small inorganic cations, which tend to orient solvent molecules around them, thereby causing the solvent molecules to be closer together than in the bulk solvent, thus "tightening" the solvent. On the other hand, the large organic cations tend to disrupt the solvent structure, giving rise to a decrease in viscosity.

Other Studies There are many examples in which viscosities of electrolyte solutions in nonaqueous solvents have been measured, with no discussion of the data being attempted. Often the viscosity results are coupled with conductance data and examined in terms of Walden products ($\Lambda\eta$). Selected examples of this type of approach can be found in Refs. 7–11.

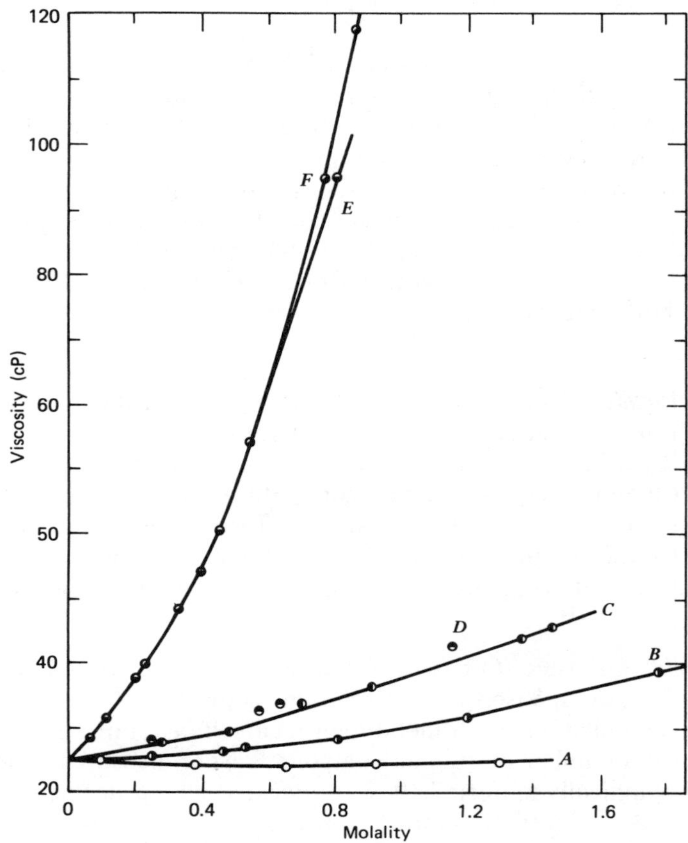

Figure 7.10 Viscosities of solutions in sulfuric acid at 25° C. *A*, NH_4HSO_4; *B*, $KHSO_4$; *C*, $NaHSO_4$; *D*, $LiHSO_4$; *E*, $Ba(HSO_4)_2$; *F*, $Sr(HSO_4)_2$. Reprinted with permission from *J. Chem. Soc.*, 215 (1953).

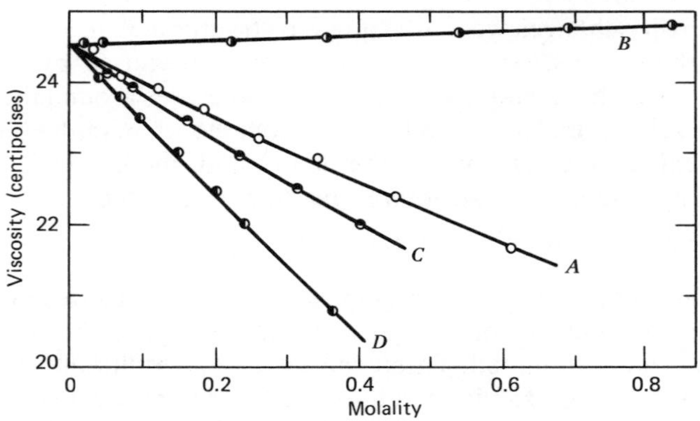

Figure 7.11 Viscosities of solutions in sulfuric acid. *A*, H_2O; *B*, $H_2S_2O_7$; *C*, SO_2Cl_2; *D*, HNO_3. Reprinted with permission from *J. Chem. Soc.*, 215 (1953).

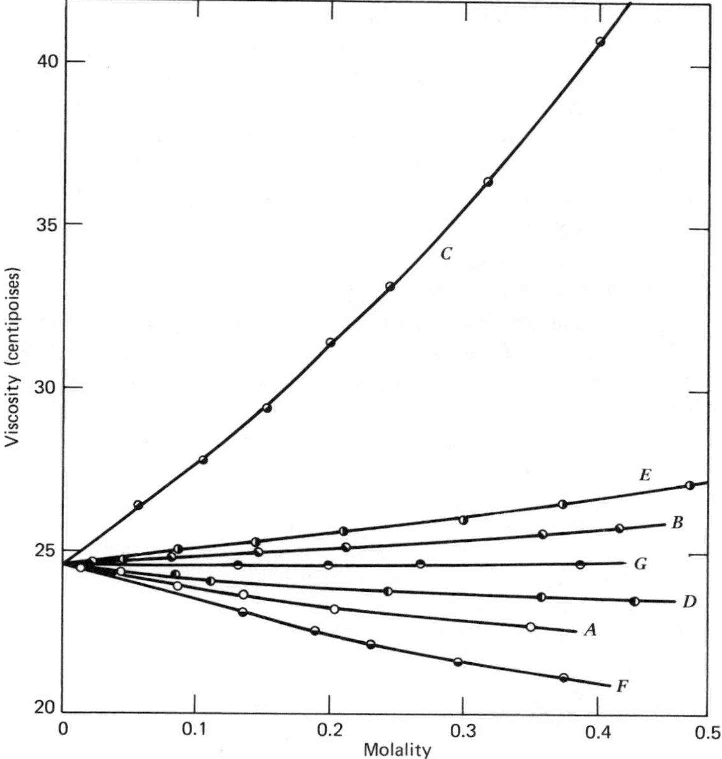

Figure 7.12 Viscosities of solutions in sulfuric acid. *A*, NH_2Pr^n; *B*, NH_2Ph; *C*, o-$C_6H_4(NH_2)_2$; *D*, AcOH; *E*, Ph·CO_2H; *F*, $COMe_2$; *G*, p-C_6H_4Me-NO_2. Reprinted with permission from *J. Chem. Soc.*, 215 (1953).

Summary Precise viscosity measurements of electrolytes in nonaqueous solvents are useful for evaluating ion–solvent interactions. In particular, a study of the viscosity as a function of concentration for a mixed solvent system often yields a maximum due to an enhanced structural region. The concentration dependence of the viscosity is frequently treated in terms of the semiempirical Jones–Dole equation and, in particular, evaluation of the *B* coefficient in this equation provides information on the structure-making or structure-breaking abilities of the individual ions. Values for the individual ions are obtained by using additivity relationships. Several factors contribute to the variation of *B* coefficients in different solvents. Structure breaking will be much more important in "three-dimensional" solvents such as water compared to less structured solvents. The higher *B* coefficient is normally associated with the solvent with the largest molar volume. Other factors such as steric hindrance for small ions and dipole moments also play a role.

LITERATURE CITED—VISCOSITY

1 Jones, G., and M. Dole, *J. Amer. Chem. Soc.*, **51,** 2950 (1929).

2 Falkenhagen, H., and M. Dole, *Physik. Z.*, **30,** 611 (1929).

3 Feakins, D., and K. G. Lawrence, *J. Chem. Soc.* (*A*), 212 (1966).

4 Crickard, K., and J. F. Skinner, *J. Phys. Chem.*, **73,** 2060 (1969).

5 Archer, M. D., and R. P. H. Gasser, *Trans. Faraday Soc.*, **62,** 3451 (1966).

6 Gillespie, R. J., and S. Wasif, *J. Chem. Soc.*, 215 (1953).

7 Sears, P. G., and L. R. Dawson, *J. Chem. Eng. Data*, **13,** 124 (1968).

8 Hoover, T. B., *J. Phys. Chem.*, **68,** 876 (1964).

9 Janz, G. J., A. E. Marcinkowsky, and I. Ahmad, *J. Electrochem. Soc.*, **112,** 104 (1965).

10 Mercier, P. L., and C. A. Kraus, *Proc. Nat. Acad. Sci. U.S.A.*, **42,** 487 (1956).

11 Longo, F. R., J. D. Kerstetter, T. F. Kumosinski, and E. C. Evers, *J. Phys. Chem.*, **70,** 431 (1966).

CHAPTER EIGHT

Spectroscopy

8.1 INTRODUCTION

Thermodynamic and transport data as well as measurements of colligative properties generally provide information on the bulk characteristics of solutions, from which the nature of physicochemical interactions at the molecular level can only be inferred. This is so because quantities such as heats of solution, viscosities, or freezing points are average properties of macroscopic quantities of a solution and are not characteristic of individual molecules or ions and their interactions. Spectroscopic measurements, on the other hand, can reflect more directly the energetics and structure of individual molecules and ions, their aggregates and solvation shells, thus complementing the bulk data in the elucidation of solvent–solute and solvent–solvent interactions. Spectroscopy provides information about the energy levels within molecules and ions and the manner in which they vary with the solvent medium, intermolecular forces between solvent molecules, as well as the energies of solvent–solute interactions. Spectroscopic methods are used to identify, in solution, molecules, ion pairs and complexes, the existence of hydrogen bonding, and of solvation shells around ions, including their solvation numbers. New species in solution can be detected and identified directly by spectroscopic methods.

In some areas of application, thermodynamic and spectroscopic measurements may provide overlapping information, as in the qualitative and quantitative analysis of dissolved species and the determination of certain equilibrium constants. However, an almost unique advantage of spectroscopy is its ability in many cases to provide quantitative measures of properties of single ionic species, independent of the counterion present. Unlike thermodynamic measurements, which yield average quantities characteristic of electrically neutral combinations of ions, spectroscopic data can be specific to a given ionic species, since spectral energy levels are often entirely independent of the counterions present. Particularly striking in this respect are the electron-spin resonance (ESR) spectra of radical ions, although information about the solvation of single ionic species is obtainable also from nuclear magnetic resonance (NMR), infrared (IR), Raman, and charge-transfer-to-solvent (CTTS) spectra. It should be noted that another technique capable of providing data on single ions in solution is the measurement of ionic transference numbers.

In this chapter we are primarily interested in the spectroscopic evidence for the interaction between nonaqueous solvents and solutes, mainly electrolytes and ions, and the physicochemical information obtainable from it. Spectroscopy in the ultraviolet (UV) and visible region has been a valuable tool in determining the strength of coordination between transition-metal ions and the molecules of nonaqueous solvents. Similarly, the nature of ion pairing is sometimes reflected in electronic spectra. In the case of many anions, CTTS spectra are observed, which are a function of rather intimate anion–solvent interactions. Spectral absorption due to donor–acceptor charge-transfer (CT) complexes has formed the basis for defining empirical scales of solvent "polarity." Nonspecific solvent effects have been instrumental in making spectroscopic assignments for UV and visible spectral transitions. Deviations from expected solvent effects have been used as indices of hydrogen bonding and specific solvation.

Even more information on the intermolecular forces, the structure of solvents and the nature of solvation can be derived from IR and Raman spectroscopy. Solute–solvent interactions not only affect vibrational spectra of the solvents, but also lead to the appearance of new vibrational bands due to solute–solvent complexes. The identification of complexes and ion pairs, the detection of hydrogen bonding, and the measurement of acid–base strength are some of the accomplishments of nonaqueous IR spectroscopy. Raman spectroscopy has also been applied to studies of acid–base strength, of equilibria between free and associated ions, and to specific solvation in mixed solvents.

Another handle on solvation and association phenomena is afforded by NMR spectroscopy. In addition to techniques based on the resonance of the ubiquitous protons, other nuclei, such as ^{17}O, ^{19}F, ^{23}Na, ^{31}P, and those of the paramagnetic ions, find application as probes in the studies of solvation. Solvent effects on the free radicals and radical ions in solution are, of course, uniquely suited for investigation by ESR spectroscopy. Frequently the information obtained from any given spectroscopic technique is inconclusive, in which case the chemical system must be described by a combination of spectroscopic and other techniques.

8.2 ELECTRONIC ABSORPTION SPECTROSCOPY

When electromagnetic radiation in the UV and visible region of the spectrum is absorbed, a transition takes place in which an electron is promoted from the ground state to an excited state. An electronic transition can be *intra*molecular, when the electron remains bound to the excited molecule, or *inter*molecular, as in the case of CT spectra. The energy difference between the ground and the excited state, ΔE, is related to the frequency of radiation ν via the Planck constant, h, by the expression

$$\Delta E = h\nu \tag{8.1}$$

The magnitude of ΔE determines the location of the absorption maximum. In Eqn. 8.1, the frequency is in Hertz and the energy, in ergs. More commonly, spectral maxima and shifts are expressed in wave numbers $\bar{\nu}$ having units of cm^{-1}, which are directly proportional to energy, or in the units of wavelength λ, such as Ångstroms (1 Å $= 10^{-8}$ cm) or nanometers (millimicrons in earlier literature), where 1 nm $= 10^{-7}$ cm. The near-UV region, in which measurements can be made in air, extends roughly from $\lambda = 200$ to 400 nm (50,000–25,000 cm^{-1}) and the visible region, from 400 to 800 nm (25,000–12,500 cm^{-1}).

The intensity of absorption is governed by the Beer–Lambert law:

$$A = abc \tag{8.2}$$

where A is the absorbance, b is the path length, c is the concentration in mol $liter^{-1}$, and a is the concentration-independent measure of spectral intensity, known as the molar absorptivity.† The most intense absorption bands in the UV–visible region are characterized by a_{max} values of the order of 10^4–10^5 cm^{-1} M^{-1}. These represent the most probable spectral transitions.

Electronic spectra of polyatomic molecules in solution comprise broad bands resulting from superimposition of vibrational and rotational energy levels. Solvents generally affect the position, intensity, and shape of absorption spectra of solutes because of solvent–solute interactions. When these interactions result in the formation of coordination- or H-bonded complexes or lead to ion association, we might observe spectral changes due to the appearance of a new absorbing species. However, solvent effects can be appreciable even when no new absorbing species are formed. Below we discuss briefly several types of solvent effects on electronic absorption spectra.

8.2.1 Nonspecific Solvent Effects A UV spectrum of benzene vapor exhibits a great deal of fine structure, which is blurred when the same spectrum is observed in a polar solvent. It is a general phenomenon that many vibrational bands that a solute might exhibit in the vapor state or in a nonpolar solvent become obliterated in hydrogen-bonding and other polar media, where strong solvent–solute interactions disrupt the quantization of vibrational energy levels. Such interactions result in the formation of several new absorbing species and the rapid exchange among them produces smoother average bands instead of distinct sharp peaks. However, from a quantitative viewpoint, more informative is the effect of solvents on the position of absorption bands.

Both polar and nonpolar solvents cause a shift in the position of maximum absorption relative to the vapor state. From the quantitative treatments of solvent shifts by Bayliss and McRae (1–3), Lippert (4) and later by Suppan (5), it is possible to derive certain qualitative guidelines for predicting solvent effects on the position of spectral maxima. Here the determining spectroscopic parameters are the momentary transition dipole, which develops in the process of absorption, and the difference between the permanent dipole moment of the solute in the ground and the excited state. In all solvents, there is a contribution from a small shift toward longer wavelengths, known

† The modern symbol for *molar* absorptivity is ϵ. Here we use a in order to avoid confusion with the dielectric constant ϵ.

Table 8.1 Examples of Solvent Effects on the Position of UV Maxima

	$\bar{\nu}_{max}$, cm^{-1}		
Solvent	NO_3^- [a]	*p*-Nitrophenol[a]	8-Quinolinol[b]
Water	33,110	31,550	32,790
Methanol	33,000	32,050	31,950 (also $CHCl_3$)
Ethanol	33,000–32,890	31,850	31,850 (also NMF[c])
1-Propanol	—	—	31,750
Acetonitrile	32,050–31,950	32,470	—
Dimethylformamide	31,950–31,850	—	31,650 (also isopropyl alcohol)
Isooctane	—	—	31,400
Cyclohexane	—	34,840	31,350 (also *n*-heptane)
Carbon tetrachloride	—	—	31,250

[a] Selected data from a compilation by Lantzke (7), with wavelengths converted to wave numbers. A range indicates differences among literature values.
[b] From Ref. 6.
[c] *N*-Methylformamide.

as the polarization red shift, which for a given solute is a function of the solvent refractive index n. When the solute is nonpolar, this red shift is the only effect on the wavelength of an absorption maximum. Generally, the red shift varies linearly with the function $(n^2 - 1)(2n^2 + 1)^{-1}$ (1). When the solute is polar, the above function may be obeyed in nonpolar solvents, as was demonstrated on the example of 8-quinolinol (6) (see Table 8.1). More generally, however, the overall shifts for polar solutes are subject to an additional contribution, the direction of which depends in part on whether the solute dipole moment increases or decreases during the transition. If the solute dipole increases in the transition, the excited state† is stabilized by solvation relative to the ground state, and a red shift results. If the solute dipole is smaller in the excited state than in the ground state, the relative stabilization of the latter will contribute a blue shift (i.e., toward shorter wavelengths). However, the net solvent shift in such cases depends on the relative magnitude of the above blue shift and the polarization red shift. In polar solvents, the blue shift generally prevails.

The above criteria apply to intramolecular transitions and to intermolecular transitions other than CTTS. In the case of the CTTS spectra of anions, strongly solvating and H-bonding solvents cause blue shifts, while aprotic solvents cause red shifts. Obviously, in order to make even qualitative predictions about solvent shifts, it is necessary to know the nature of the electronic transition involved and how the dipole moment of the solute changes upon excitation. On the other hand, information on the nature of

† This is the so-called Franck–Condon excited state of the solute, in which the nuclear geometry of the solute and the associated solvation shell, if any, is the same as in the ground state.

solvation can be derived from a comparison of observed and theoretically predicted solvent shifts. Thus, in the absence of specific solvation, most polar solutes follow the linear relationship between frequency shifts and the functions $f(\epsilon) = [2(\epsilon - 1)(2\epsilon + 1)^{-1}]$ or $\phi(\epsilon) = [(\epsilon - 1)(\epsilon + 2)^{-1}]$ (7). In addition, McRae (2) proposed a combination of refractive-index and dielectric-constant functions in a single equation. Substantial deviations of observed solvent shifts from those predicted by the above functions are interpreted as evidence of more profound solvent–solute interactions, such as H-bonding and various types of specific solvation.

8.2.2 Hydrogen Bonding and Preferential Solvation In the section immediately preceding, it was mentioned that a plot of spectral shifts as a function of $f(\epsilon)$, $\phi(\epsilon)$ or a related expression could serve as a source of information about the nature of interactions between the absorbing species and the solvent. When the plot is linear, it is taken as evidence that nonspecific ion–dipole, dipole–dipole, and dispersion interactions prevail. Deviations from linearity are interpreted as arising from specific solute–solvent interactions, of which H-bonding is most common.

The effect of H-bonding on the 300–315 nm UV band of the NO_3^- ion has been studied in many solvents. In the absence of ion association, this nitrate band experiences a red shift in solvents of decreasing H-bonding ability (Table 8.1), which is interpreted in terms of decreasing stabilization of the ground state. Also, the values of a_{max} or the integrated intensity of the band decrease with decreasing dielectric constant and H-bonding ability of the solvent.

It was noted by Suppan (5) that deviations due to H-bonding are different for each band of a given solute species, being largest for the bands corresponding to transitions of an electron that is specifically H-bonded in the ground state. A striking example of such band specificity is observed in the effect of H-bonding on the absorption bands of 3-amino-4-methoxyacetophenone(I), whose spectrum is a composite of the bands of 3-aminoacetophenone(II) and 4-methoxyacetophenone(III). When the solvent shifts of the CT bands of the monosubstituted acetophenones are plotted vs $f(\epsilon)$, the band of (II) experiences a blue shift from the straight line in H-bonding solvents (water, ethanol), while the band of (III) experiences a red shift. Significantly, the corresponding bands in (I) show the same shifts (i.e., in opposite direction from each other) in H-bonding solvents as they do in the monosubstituted acetophenones (5). These effects are depicted in Fig. 8.1.

The enhancement of solvent shifts due to H-bonding has formed the basis for defining a quantitative scale of solvent hydrogen-bond acceptor basicities known as the β scale (8). It is derived from the solvent shift of the bands of 4-nitroanilinium relative to those of *N,N*-diethyl-4-nitroaniline and of 4-nitrophenol relative to those of 4-nitroanisole in the same solvents.

In mixed solvents, the nonlinearity of a plot of $\Delta\bar{\nu}_{max}$ vs the solvent mole

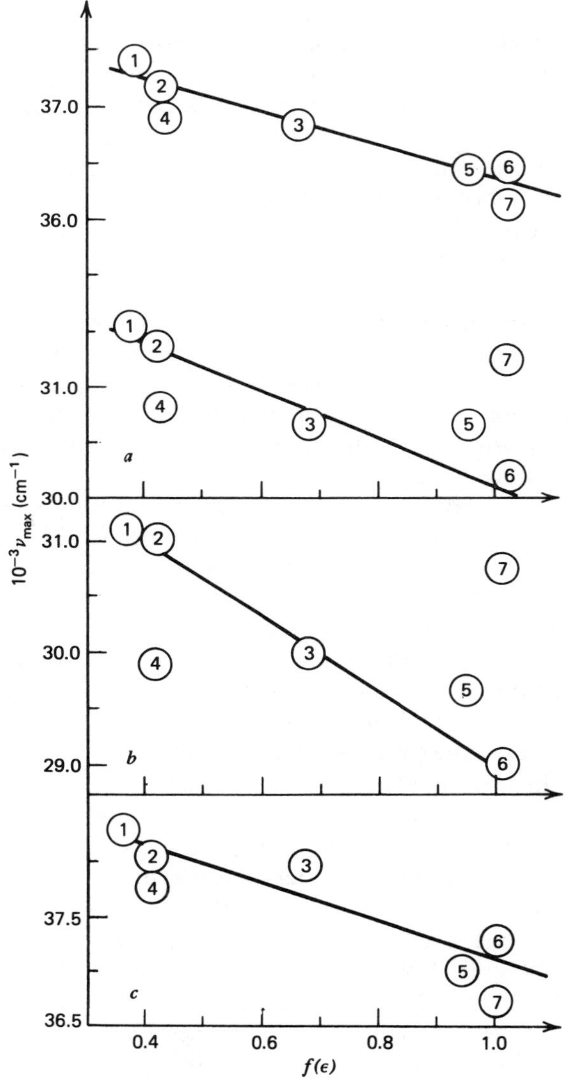

Figure 8.1 (a) f(ϵ) Plot of solvent shift of 3-amino-4-methoxyacetophenone; (b) the charge-transfer band of 3-aminoacetophenone; (c) the charge-transfer band of 4-methoxyacetophenone. Solvents: *1*, isopentane; *2*, cyclohexane; *3*, ether; *4*, dioxane; *5*, ethanol; *6*, dimethylformamide; *7*, water. Reprinted with permission from *J. Chem. Soc.* (**A**), 3125 (1968). Copyright 1968. The Chemical Society.

fraction is interpreted as evidence of preferential solvation by one of the components. It was found that in mixtures of DMF with ether and of ethanol with cyclohexane, the functions $f(\epsilon)$ varied almost linearly with the solvent mole fraction, but the values of $\Delta\bar{\nu}_{max}$ for five polar organic compounds did not (9). This is believed to be the result of enrichment of the polar solvent in the solvation shell of the polar solute, otherwise known as preferential solvation. This effect is particularly pronounced when the dipole moment of the solute is large, and there is an appreciable difference between the dielectric constants of the two solvents in the mixture. In the above study, such conditions existed for solutions of 4-nitroaniline (μ = 6.2 D) in ethanol–cyclohexane mixtures.

8.2.3 Charge-Transfer Spectra When an electron is promoted from a high-energy filled orbital of a donor to a low-energy vacant orbital of an acceptor, the process is called a charge-transfer (CT) transition and is usually characterized by the appearance of a new spectral band. Intramolecular CT transitions are known to occur in transition-metal complexes, but most are intermolecular. Solvents can act either as electron donors or acceptors. Some dipolar aprotic solvents are good donors, forming so-called contact CT complexes with dissolved oxygen which absorb in the near-UV region for most solvents, although in anilines the oxygen bands extend well into the visible region (450–500 nm). Obviously these oxygen bands represent a general source of interference in spectral work in donor solvents.

More common is the formation of charge-transfer-to-solvent (CTTS) complexes by anions which act as electron donors. This subject has been reviewed by Blandamer and Fox (10). For example, CT absorption by the iodide ion is interpreted as producing a state where the iodine atom and an electron reside in what is known as the same solvent "cage," that is, the process is $I^-(\text{solv.}) \xrightarrow{h\nu} (I + e)_{\text{solv.}}$. The CTTS bands are usually of high intensity, for example, a_{max} for the I^- in water is 1.35×10^4 at 20°C.

The CTTS spectra of anions, which at low concentrations are generally independent of the cation, have been studied primarily in aqueous and mixed solutions. However, the CTTS spectra of the iodide ion have been determined in many solvents. For example, λ_{max} for the I^- ion occurs at 226 nm in water and at 245 nm in acetonitrile (the corresponding maxima for the Br^- ion occur at 198 nm and 219 nm, respectively). The energy of the maximum of the CTTS transition, E_{max}, of the I^- ion has been proposed as a basis for a scale of solvent "polarity" (11). Presumably the rationale for it lies in the fact that the excited state of the I^- ion is determined in this case primarily by the solvent medium so that its energy can serve as a measure of the ion–solvent interaction.

In kcal mol^{-1}, the value of the E_{max} corresponding to the lower-frequency band of the I^- is lowest in liquid ammonia (108.1) and highest in glycerol (134.12). The values for other solvents are ethanol (130.50), methanol (130.00), water (126.40), and THF (116.4). Evidently there is no correlation between the E_{max} and the dielectric constant or the acid–base properties of the solvent. In Chapter 3, we mentioned Kosower's (12) Z scale of solvent "polarity," based on the E_{max} of the CT absorption of 1-ethyl-4-carbomethoxypyridinium iodide. It should be noted, however, that there is no apparent correlation between the Z scale and the I^- scale.

When *t*-butyl alcohol, *n*-propyl alcohol, ethanol, and methanol are added to aqueous solutions of I^-, the E_{max} of the iodide ion is shifted rapidly to higher energies. The magnitude of this shift decreases in the above order of the alcohols. At 25°C, the E_{max} is a linear function of the alcohol mole fraction, X_2, over the range $0 < X_2 < 0.03$, but at higher mole fractions of *t*-butyl alcohol (up to $X_2 = 0.5$), E_{max} is relatively insensitive to changes in solvent composition. These spectral changes have been interpreted as arising from the effect of the alcohols on the structure of water. Small amounts of alcohol are believed to reinforce water–water interactions (cf. discussion of solvation in Chapters 2 and 5) and this effect can be felt even at the lowest alcohol concentrations because of its long-range influence on the water structure (10). The interpretation of E_{max} changes, at alcohol mole fractions higher than about 0.04, is unclear. It has been pointed out, however, that many physicochemical properties of alcohol–water mixtures show inflection points or extrema in the vicinity of that critical composition (10).

The effects of ion association may or may not show up in CTTS spectra, depending on the type of ion pairing involved. When the anion and the cation form a solvent-separated ion pair, there may be no observable effect on the spectra, even when a (generally small) degree of association is detected from electrolytic conductance of such systems. This seems to be the case for the I^- ion in acetone, acetonitrile, methanol, and *n*-butanol. A different situation arises in the case of contact ion pairs, in which there are no intervening solvent molecules between the anion and the cation. For contact ion pairs, greater electrostatic stabilization leads to higher values of E_{max}. Furthermore, in these systems, the CTTS spectra of the anion are dependent on the nature of the cation. Indeed, in solvents like dioxane, THF, methylene chloride, *t*-butyl alcohol, and *t*-pentyl alcohol, the CTTS spectra of the I^- ion are dependent both on the solvent and on the cation, and E_{max} generally decreases with an increase in cation size.

8.2.4 Ion Association Electronic spectra are sensitive to ionic association only in systems where there is some orbital overlap of the associating ions, which introduces new energy levels into the system. This occurs, for example, when cationic transition-metal complexes associate with anions, in which case new bands characteristic of the ion pairs develop at shorter wavelengths relative to the bands of the corresponding unassociated complexes (7). Spectrophotometric and conductometric studies of association between bisethylenediaminecobalt(III) complexes and anions in dipolar aprotic solvents were reviewed by Watts (13).

In general, electronic spectra cannot distinguish between contact- and solvent-separated ion pairs. A notable exception, however, was reported in a series of studies of the spectra of alkali-metal carbanions and radical anions, where an equilibrium between the two types of ion pairs was described quantitatively. It was first discovered by Hogen-Esch and Smid (14) that the spectrum of fluorenylsodium (Na^+Fl^-)† in tetrahydrofuran (THF) shows a sharp absorption peak at 355 nm at room temperature, which gradually diminishes upon cooling, while a new peak develops at 373 nm at its expense. Below −50°C only the 373-nm peak exists. The above spectral changes are reversible with the temperature, but the peak-height ratios at any given temperature are not affected by dilution or common-ion effects. These phenomena were explained by the existence of a rapid equilibrium between *contact-* or *intimate ion pairs* (Fl^-,Na^+) and *solvent-separated ion pairs*, ($Fl^- \parallel Na^+$),‡ of which the latter forms at lower temperatures. From the temperature dependence of the equilibrium constant for the process

$$(Fl^-,Na^+) \rightleftharpoons (Fl^- \parallel Na^+),$$

the gain in solvation enthalpy ΔH was calculated to be ~ −6 kcal mol^{-1}.

For a given anion the relative concentration of contact- and solvent-separated ion pairs is a sensitive function of the temperature, solvent, and counterion. We mentioned already that lowering the temperature favors the formation of the solvent-separated ion pairs. At constant temperature in a given solvent, the fraction of solvent-separated ion pairs increases as the tendency of the cation to be solvated increases. In the alkali-metal series the order of increasing solvation is $Cs^+ < K^+ < Na^+ < Li^+$, so that Li^+ tends to form predominantly solvent-separated ion pairs. Indeed, in a spectral study of the 4,5-methylenephenanthrene anion (15), it was shown that in THF at 0°C, the fraction of the solvent-separated ion pairs is 0.80 for Li^+, 0.15 for Na^+, and approximately zero for K^+ and Cs^+. Even at −70°C, cesium seems to form no solvent-separated ion pairs. For a given cation and temperature, the fraction of solvent-separated ion pairs increases as the ability of the solvent to solvate the cation increases. Thus, it was found that the spectrum of 9-fluorenyllithium in toluene and dioxane solutions at room temperature exhibited only the peak characteristic of contact ion pairs (λ_{max} = 346 nm); in THF at 25°C, both the peak corresponding to contact ion pairs (λ_{max} = 349 nm) and the solvent-separated ion pairs (λ_{max} = 373 nm) were observed and the fraction of the latter was 0.80. In more powerful solvents, such as pyridine, 1,2-dimethoxyethane, and DMSO, only solvent-separated ion pairs (λ_{max} = 373 nm) were detected (16). Similarly, in the case of 4,5-methylenephenanthrenelithium at 0°C, the fraction of solvent-separated ion pairs was 1.0 in 1,2-dimethoxyethane, 0.80 in THF, 0.25 in 2-methyl-THF, and 0.15 in tetrahydropyran (15). Thus, the extent of for-

† Fluorenyl is the anion of fluorene, or diphenylmethane.

‡ This notation (with anion first) is retained from the original references.

mation of solvent-separated ion pairs has been recognized as another method of studying the relative solvation of cations.

While the existence of an equilibrium between two thermodynamically distinct ion-pair species was most clearly demonstrated by spectroscopy, it was also corroborated by electrolytic conductance (e.g., Refs. 17–19) (cf. Eqn. 7.22 and discussion), as well as invoked in the interpretation of solvolytic processes (e.g., Ref. 20) and ESR spectra (21). A theoretical formulation for the energetics of a system with two types of ion pairs was given by Grunwald (22).

8.2.5 Transition-Metal Complexes Spectra arising from transitions between the electronic *d* and *f* levels of transition-metal ions and their complexes may show very pronounced solvent dependence. For example, the absorption maxima of the two *d–d* bands of hexaquochromium(III) shift from 407 and 575 nm in aqueous solutions to 444 and 634 nm in DMSO, respectively. This solvent shift, which is proportional to the mole fraction of DMSO in aqueous mixtures, is believed to be the result of coordination of DMSO molecules to the Cr^{3+} ion, which gradually displaces water molecules from the complex. When the concentration of DMSO in water reaches 14 *M*, each chromium atom is already coordinated to six DMSO ligands (23).

Changes in the energies of *f–f* transitions of trivalent lanthanide ions were used to establish the relative strength of interaction between these ions and the nitrate ion as well as several solvents as ligands. The interaction decreased in the order DMF > tributylphosphate > NO_3^- ~ H_2O > C_2H_5OH > dioxane. In DMF solutions, the lanthanides are coordinated to six solvent molecules, while in dioxane, the rubidium lanthanum nitrate employed in that study was insoluble (24).

In Chapter 3, we had occasion to mention the fact that the solvent shifts of the *d–d* transitions of the Ni^{2+} ion are so large that a scale of solvent strength was based on their numerical values.

In summary, what we have attempted to highlight in this section is the type of information on solute–solvent interactions that is obtainable from electronic spectra. Whereas nonspecific interactions lead to shifts in spectral bands that can be rationalized on the basis of the dielectric constant and the refractive index of the medium, more pronounced shifts are interpreted as indices of specific solvent–solute interactions, including H-bonding, coordination, and charge transfer. Solvent effects on spectra are particularly pronounced in the case of transition-metal ions and charge-transfer complexes. In both systems, these effects have been used as measures of solvent "strength." Occasionally, electronic spectra can detect ion pairing and, in the exceptional case of alkali-metal carbanions and radical anions, have been used to distinguish between contact- and solvent-separated ion pairs.

8.3 VIBRATIONAL SPECTROSCOPY

When absorption of radiation causes transitions to occur between vibrational energy levels of a molecule, it may be possible to observe an infrared or a Raman spectrum of the absorbing species. Direct absorption of infrared radiation occurs when the dipole moment of the absorbing group changes during the vibration. In such cases we observe an IR spectrum. A Raman spectrum is observed only when the polarizability of the absorbing species changes during the vibration. Thus the two forms of vibrational spectroscopy provide complementary information. In Raman spectroscopy, the sample is excited by visible or UV radiation and the frequency of the scattered radiation is analyzed. Most of the exciting radiation is scattered at the incident frequency ν_0 (Rayleigh scattering), but in addition, lines of much lower intensity occur both at lower frequencies ν_S (Stokes lines) and at higher frequencies ν_{aS} (anti-Stokes lines). The frequency differences $(\nu_0 - \nu_s)$ and $(\nu_{as} - \nu_0)$, which are known as the Stokes shifts and anti-Stokes shifts, respectively, represent the amount of energy transferred from the incident beam to the molecule and vice versa. The shifts are independent of the frequency of the incident radiation and are characteristic of the absorbing system, corresponding to the spacings between the vibrational levels of the molecule.

Solute–solvent interactions are generally studied in the frequency range of 4000–100 cm^{-1}. In modelistic terms, the frequency of a vibrational band is approximated by that of a harmonic oscillator,

$$\nu = \frac{1}{2\pi}\left(\frac{k}{\mu}\right)^{1/2} \tag{8.3}$$

where μ is the reduced mass of the absorbing species and k is the force constant, which is a measure of the bonding strength of the vibrating group. Equation 8.3 is sometimes applied quantitatively, such as in predicting the effect of isotopic substitution, but more frequently it serves as a qualitative guide.

In contrast to visible and near-UV spectroscopy, where most solvents are transparent above a certain wavelength, vibrational spectroscopy suffers from the complication that many solvents themselves absorb throughout the vibrational region. A compilation of literature references on the vibrational spectra of nonaqueous liquids is available (25). Overlap between the bands due to solvent vibrations and those of the solvent–solute complexes often present a problem. Isotopic substitution and computer techniques for the resolution of overlapping bands are some of the remedies (see, e.g., Ref. 26). In this connection, one should be aware of the fact that IR spectrometers with improved resolution have been available only since the early 1960s. Where resolution is critical, earlier data should be carefully scrutinized. In Raman spectroscopy, the application of laser sources of high intensity has increased substantially the sensitivity as well as the resolution of the technique.

Solvent–solute interactions manifest themselves in vibrational spectra in two general ways. (1) A solute may cause the solvent bands to shift, change in intensity, broaden, develop shoulders, or split into two bands. (2) A complex between the solvent and the solute species may form, leading to the appearance of new vibrational bands. Furthermore, ion association is generally detectable from vibrational spectra, so that spectral changes as a function of the solvent may be the result of changes in the equilibrium between free and associated ions. Below, each type of these solvent effects on vibrational spectra is illustrated with specific examples.

8.3.1 Far-Infrared Solvation Bands of Cations In many solvents, the far-infrared spectra of alkali-metal and ammonium salts exhibit bands that are not present in either the pure solvent or in the pure salt. Except in solvents of low dielectric constant and low solvation ability (donor capacity), these bands are generally characteristic of the cation and the solvent, but are independent of the anion. Thus, when first observed on solutions of alkali-metal salts of the $Co(CO)_4^-$ anion in tetrahydrofuran (THF), they were assigned to a vibration of the cation in a solvent cage (27). This assignment has been corroborated since through studies of the dependence of the corresponding frequencies on isotopic substitution of the solvent and the metal ion (e.g., Refs. 28–33). Some representative examples of the cation–solvent bands are collected in Table 8.2. For a complete listing of frequencies for individual salts published up to 1970, see a review by Irish (25).

In DMSO, which is known for its high solvating ability for cations, the cation–solvent bands are completely independent of the anion. Furthermore, the bands of deuterium-substituted DMSO-d_6 and NH_4-d_4 exhibit spectral shifts as predicted from the changes in the mass of the vibrating species (Eqn. 8.3). Also, the observed shift of the S—O stretching frequency to a lower value points to a coordination between the oxygen atom and the cation. A similar anion independence was reported for sulfolane (34). Some anion dependence, however, is observed in solvents of lower solvating ability, such as acetone and 1-methyl-2-pyrrolidone. Here, the bands of lithium salts of the smaller halide anions (Cl^-, Br^-) appear at lower frequencies (shown in parentheses in Table 8.2) than the remaining lithium salts. Evidently the small anions can penetrate the inner solvation shell, replacing a solvent molecule and forming a solvated contact ion pair with the cation. In such systems, the cation is believed to be vibrating in a solvent cage that also contains the anion. This phenomenon was first reported by Evans and Lo (35), who observed anion-dependent bands of tetraalkylammonium salts in benzene and assigned them to cation–anion vibrations.

Pronounced anion dependence of the cation–solvent bands is evident in solvents of low dielectric constant or of low solvating ability for cations (or

Table 8.2 Examples of Far-Infrared Cation-Solvation Bands[a]

	DMSO[b]	Acetone[c]	1-Methyl-2-pyrrolidone[d]	Propylene carbonate[e]	Sulfolane[f]
Li^+	429	425 (409)	398 (377)	397 (383)	—[g]
NH_4^+	214	—	207	184	196
Na^+	200	195	204	186	186
K^+	153	148	140	144	155
Rb^+	125	—	106	115	—[g]
Cs^+	110	—	—	112	—[g]

[a] Average values for several salts. The values in parentheses are for Cl^- and Br^- salts. The frequencies are in cm^{-1} and their experimental precision is reported to be in the range of $\pm(2–6)$ cm^{-1}. For the frequency values for individual salts, see original references and the compilation by Irish (25).
[b] References 28 and 29.
[c] Reference 32.
[d] Reference 30.
[e] Reference 38.
[f] Reference 34.
[g] These bands are probably overlapped by solvent bands.

both). As can be seen from the data in Table 8.3, the frequencies of the cation–solvent bands vary with each anion both in THF ($\epsilon = 7.6$) and in nitromethane (36). The latter has an appreciable dielectric constant (35.9 at 30°C), but has a low donor ability. The anion dependence is interpreted as evidence for the existence of contact ion pairs. Conversely, the independence of a cation–solvent band of the nature of the anion in a solvent of low dielectric constant suggests strongly that the ion pairs, which must exist in that medium, are solvent separated or solvent shared. This appears to be the case in glacial acetic acid (37).

Table 8.3 Examples of Far-Infrared Cation-Solvation Bands Showing Evidence of Contact Ion Pairs

Tetrahydrofuran (27b)	$\bar{\nu}$ (cm^{-1})	Nitromethane (36)	$\bar{\nu}$ (cm^{-1})
$LiCo(CO)_4$	413	$LiClO_4$	368
$LiBPh_4$	412	LiI_3	340
$LiNO_3$	407		
LiCl	387		
LiBr	378		
LiI	373		
$NaBPh_4$	198		
$NaCo(CO)_4$	192		
NaI	184		

A general conclusion drawn from these studies was that the far-infrared cation–solvent spectra can differentiate between contact ion pairs on the one hand, and cations containing only solvent molecules in the inner solvation shell, on the other hand. Thus, they cannot distinguish among solvent-shared ion pairs, solvent-separated ion pairs, and free solvated cations (38). Solvent-separated ion pairs have one or more solvent molecules situated between the paired ions. However, in the case of separation by one solvent molecule, the term "solvent-shared ion pair" is sometimes used instead. In general, the far-infrared cation–solvent bands are Raman inactive, indicating that the corresponding cation–solvent or cation–anion interactions are predominantly electrostatic in nature. An exception, however, was reported by Tsatsas and Risen (39), who observed a Raman-active band in cyclohexane solution of sodium tetrabutylaluminate at 200 cm^{-1}. This band, which is not observed in THF, was interpreted as an indication of a partially covalent character of the ion–solvent interaction.

The frequency of a cation–solvent band increases with the degree of solvation of the cation. Thus, the best solvator in Table 8.2 (DMSO) and the most strongly solvated cation (Li^+) are characterized by the highest frequencies. Coordination (solvation) numbers of alkali-metal cations can be determined by plotting the intensity of the spectral absorption as a function of the solvent:cation mole ratio in some "inert" medium. Dioxane and nitromethane have been the traditional "inert" media adopted for determining the stoichiometry of solvation. For example, mole-ratio studies in dioxane yielded for the Na^+ ion a solvation number of 6 with DMSO and 4 with 1-methyl-2-pyrrolidone (31). In the same study, the Li^+ ion was found to be solvated by 2 molecules of DMSO and 4 molecules of 1-methyl-2-pyrrolidone. Similar studies of the acetone–Li^+ stoichiometry in nitromethane led to a solvation number of 4 for Li^+. More recently, in the case of Na^+, it has been possible to employ sodium tetrabutylaluminate, a salt that dissolves in saturated hydrocarbons, which, unlike THF or nitromethane, do not compete with coordinating solvents at all.

8.3.2 Solvation and Mid-Infrared Spectra Information on ion–solvent and ion–ion interactions in solution can be obtained from the vibrational spectra of higher frequencies as well. For example, unperturbed THF exhibits an asymmetric C–O–C stretching band at 1071 cm^{-1} and a symmetric C–O–C vibration at 913 cm^{-1}. In experiments in which THF was added in increasing concentrations to $NaAl(Bu)_4$ in cyclohexane, the 1071 cm^{-1} band was absent at a THF : salt ratio of less than 1 : 1, while new bands appeared at 1048 and 900 cm^{-1}. An increase in the ratio shifted the 1048 cm^{-1} band to 1053 cm^{-1} and intensified it, while the 1071 cm^{-1} band of the free THF reappeared too. These results were interpreted (40) by the existence of the equilibrium

$$\mathrm{Na{\cdot}THF^+X^-} + 3\mathrm{THF} \rightleftharpoons \mathrm{Na{\cdot}4THF^+X^-}$$

Similarly, the vibrational bands of acetone at 390, 528, and 1224 cm^{-1} are split by addition of Li^+ salts, with new bands appearing at 369, 539, and 1239 cm^{-1}, respectively. The new bands are assigned to cation–acetone vibrations (33). Mole-ratio studies of the intensity of these bands in nitromethane medium confirmed the solvation number of Li^+ as being 4. It is interesting that in solutions of $LiClO_4$ in nitromethane–acetone mixtures, the characteristic frequency of the perchlorate ion at 934 cm^{-1} appears as a narrow symmetrical band at acetone–Li^+ mole ratios of 4 or greater. Below this ratio, the band broadens and shifts, suggesting that ClO_4^- enters the inner solvation shell of the Li^+ ion, evidently forming a contact ion pair.

Interaction between metal ions and propylene carbonate (PC) causes the solvent bands at 1182 and 1798 cm^{-1} to split, with the appearance of new bands at 1205 and 1773 cm^{-1}, respectively. From mole-ratio studies in nitromethane, it was found that Li^+, Na^+, K^+, and Ag^+ ions were solvated by six PC molecules. Significantly, no solvation was detected for the $(Bu)_4N^+$ ion (41).

8.3.3 Ion Pairs and Complexes The vibrational spectrum of a free unperturbed nitrate ion, such as exists in dilute solutions of dissociating solvents, consists of four lines, of which two are both Raman (R) and IR active (~720 and ~1380 cm^{-1}), one is IR active only (~830 cm^{-1}) and one, Raman active only (~1050 cm^{-1}). When nitrate ion forms a contact ion pair, additional lines, characteristic of the bound nitrate, appear in its vibrational spectrum. If the intensity ratio of a pair of lines characteristic of the free and the bound nitrate, respectively, is concentration dependent, it is possible to calculate an association constant for the ion pair from such spectral data. These conditions are fulfilled by the vibrational spectra of zinc nitrate in anhydrous methanol, which consist both of the lines characteristic of the free nitrate ion [at 715 (R), 828 (IR), and 1044 (R) cm^{-1}] as well as those of the nitrate ion bound in a contact ion pair [at 754 (R, IR), 817 (IR), 1310 (R,IR) and 1500 (IR) cm^{-1}]. Furthermore, the intensity ratio of the 817 and 828-cm^{-1} lines showed a concentration dependence, from which the association constants for the pairing between zinc and nitrate ions were calculated to be $K_1 = 2.6$ and $K_2 = 0.05$ at 25°C (42).

Janz and his associates (43, 44) investigated the Raman spectra of $AgNO_3$ in acetonitrile in the concentration range of 0.01–9 *M*. Particularly useful from a diagnostic viewpoint proved to be the pair of very intense nitrate lines at 1041 and 1036 cm^{-1}. The 1041-cm^{-1} line was assigned to a symmetric stretching vibration of the free NO_3^- ion, while the 1036-cm^{-1} line, to the same vibration in a contact ion pair, perturbed by the cation. At concentrations of the order of 10^{-2} *M*, only the 1041-cm^{-1} line was evident; at ~0.2 *M* concentration, the relative intensities of the two lines were about equal, while at higher concentrations, the 1036-cm^{-1} line predominated (Fig. 8.2). The high intensity of these lines made it possible to study their intensity

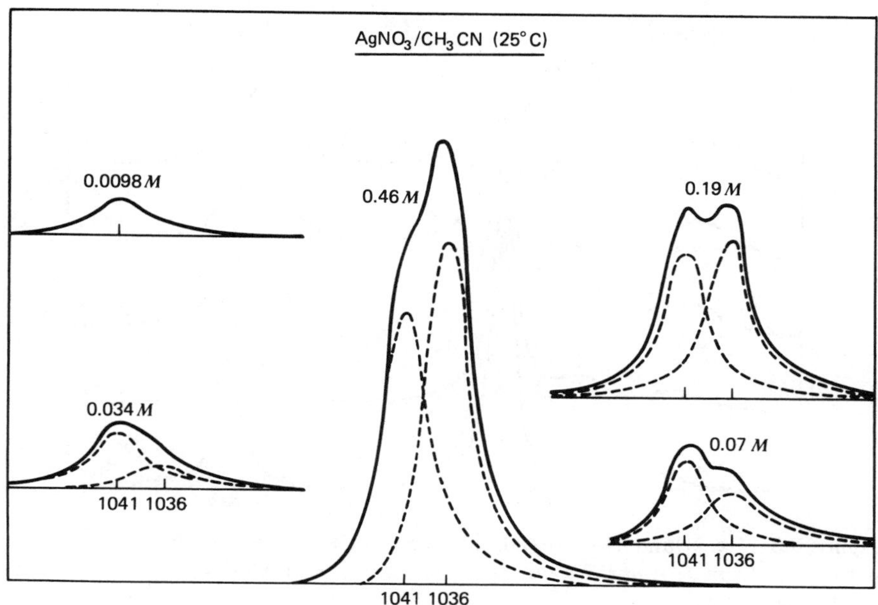

Figure 8.2 NO_3^- bands in the 1050-cm^{-1} region in $AgNO_3$–CH_3CN solutions at low concentrations. Reprinted with permission from *J. Amer. Chem. Soc.*, **92,** 4189 (1970). Copyright 1970. American Chemical Society.

ratios down to the concentration region where the degree of ion pairing derived from Raman data could be compared with that obtained from electrolytic conductance. Good agreement was reported between the results based on the two techniques in the concentration range of 0.03–0.25 *M*.

In the same study, it was also shown that the silver ion is coordinated to four acetonitrile molecules in its primary solvation shell in dilute solutions, but the coordination number drops to 2 in the ~0.5–5 *M* range and to one at higher concentrations. Coordination between Ag^+ and acetonitrile was noted even earlier (43) from IR data. The C≡N and C—C stretching bands, which in pure acetonitrile appear at 2253 and 919 cm^{-1}, respectively, are diminished in the presence of $AgNO_3$, while new bands grow at their expense at 2272 and 929 cm^{-1}, respectively. These effects are illustrated in Fig. 8.3. From the concentration dependence of the intensity ratio of these bands, the coordination numbers of the Ag^+ ion were determined to be the same as those obtained from Raman spectroscopy. Similarly, the coordination between acetonitrile (solvent) and zinc (45, 46) as well as cadmium (46) ions was deduced from Raman spectra on the basis of frequency differences between the bands of free and complexed acetonitrile.

Another area where vibrational spectroscopy has proved valuable is in the identification of metal–halide complexes that exist in nonaqueous media,

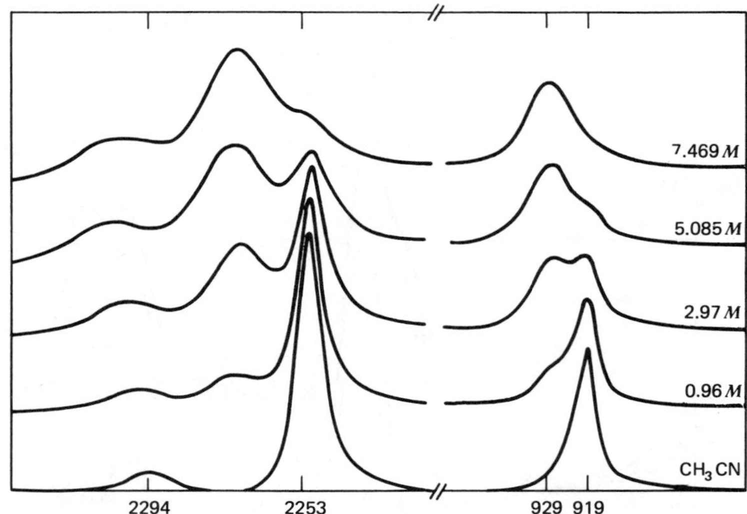

Figure 8.3 C≡N and C—C bands in $AgNO_3$–CH_3CN solutions. Reprinted with permission from *J. Amer. Chem. Soc.*, **92,** 4189 (1970). Copyright 1970. American Chemical Society.

primarily as a result of solvent extraction from aqueous solutions. For example, when the corresponding halides are extracted from aqueous acid solutions into the ethyl ether phase, the species in the nonaqueous phase were identified by Raman spectroscopy to be $InBr_4^-$, $InCl_4^-$, $GaCl_4^-$, $FeCl_4^-$, $SnCl_3^-$, $SnBr_3^-$, $CuCl_2^-$, and $CuBr_2^-$. The literature on these and related studies was reviewed by Irish (25).

8.3.4 Hydrogen Bonding and Acid–Base Reactions Vibrational spectroscopy has been the classical method for detecting the existence of a hydrogen bond and determining its strength as well as the degree of association. The vibrational spectrum of a proton donor X—H (where X is generally O, N, S) is affected by hydrogen bonding as follows (25): (1) The stretching frequency (usually in the vicinity of 3000–3500 cm^{-1}) shifts to a lower value, while the corresponding half-intensity band width and the integrated absorption of the band increase. Both the frequency and the intensity are sensitive functions of the concentration, temperature, and solvent. (2) The bending frequency exhibits a slight shift to higher values, but there are no pronounced changes in band width or intensity. (3) New vibrations due to the hydrogen-bonded complexes appear.

There has been a wealth of spectroscopic studies of hydrogen bonding, primarily devoted to association within pure substances and to H-bonding with covalent acceptors. A noteworthy example of this type is the determination of the relative basicities of aromatic hydrocarbons from the magnitude of the shifts in the infrared H—X stretching frequencies of proton donors dissolved in them. By analyzing their own as well as literature data, Olah et al. (47) determined the basicity order to be $C_6H_6 > C_6H_5I > C_6H_5Br > C_6H_5Cl > C_6H_5F$. As an example, the frequency shift for HCl (gas to solution) was -130 cm^{-1} in benzene and -76 cm^{-1} in fluorobenzene. Similarly, shifts in the O—H frequencies of phenol and *p*-fluorophenol were used as a measure of the H-bonding interaction between these proton donors and unsaturated aliphatic hydrocarbons acting as acceptors (48). In the case of 2-methyl-1-butene, which was one of the strongest bases studied, the shift was 108 cm^{-1} for phenol and 116 cm^{-1} for *p*-fluorophenol.

Chloroform and other chlorinated aliphatic hydrocarbons containing C—H groups have been the favorite hydrogen-bond donors to be employed in infrared studies. Equilibrium constants for the association between chloroform and a variety of electron donors have been calculated from spectral data. For example, both acetone and cyclohexanone were found to form 1:1 H-bonded complexes with chloroform in cyclohexane solution, both having the same association constant, within experimental error (K = 1.2 liter mol^{-1} at 30°C) (49). The association constants of chloroform with aniline and cyclohexylamine were reported to be 0.51 and 1.10 liter mol^{-1}, respectively, at 25°C (50). These are only a few examples from a vast body of available data on H-bonding interactions investigated by vibrational spectroscopy. Several reviews (49–56) as well as a monograph (57) on the subject are available.

However, of even greater general interest in solution chemistry is spectroscopic evidence for the hydrogen bonding between proton donors and anions. It was first observed by Lund (58) that the vibrational stretching frequencies of several N—H and O—H proton donors (at ~3300 cm^{-1}) shifted to lower values and their bands intensified upon addition of small amounts of tetrabutylammonium bromide. These spectral changes were attributed to the formation of $XH \cdots Br^-$ bonds. Bufalini and Stern (59) studied the effect of a number of quaternary ammonium salts on the infrared spectra of such H-bond donors as methanol, *n*-butanol, *t*-butanol, and *N*-methylacetamide in dilute benzene solutions. For example, in the absence of added electrolyte, methanol in benzene solution exhibits bands attributed to monomer and dimer. Addition of electrolyte causes the dimer bands to disappear, while new bands due to the $CH_3OH \cdots A^-$ complexes form at lower frequencies. The frequency shift is a function of the anion and increases with increasing charge density of the latter. Thus, solvation via H-bonding increases in the order: Pi^- (no shift) $< NO_3^- < Br^- < Cl^-$. The same order of relative H-bonding ($I^- < Cl^- < F^-$) was confirmed in a study of frequency shifts for alkali-metal halides reported by Waldron (60). On the other hand, Allerhand and Schleyer (52), in their investigations of infrared spectral shifts for methanol, propargyl bromide, and deuterated haloforms in inert solvents, observed that the strength of the H-bond increased in the order $I^- < Br^- < F^- < Cl^-$. It is also interesting that cations had a pronounced effect on the shift, but only in the case of the chloride. Significantly, the conclusion that the strongest hydrogen bond (as indicated by the largest frequency shift) forms with the smallest anion, derived from an element of highest electronegativity, is valid only in the case of strong H-bonds to anions, but not for weaker H-bonding, such as exists to covalently bonded electronegative atoms. Thus, H-bonding of methanol in carbon tetrachloride solutions seems to be stronger to alkyl iodides than to alkyl fluorides (61).

A different perspective on anion–methanol interactions was provided by the Raman studies of saturated solutions of electrolytes in methanol (62). In liquid methanol, extensive hydrogen bonding causes the O—H stretching band to appear at a frequency of ~3380 cm^{-1}, as opposed to 3600–3700 cm^{-1} observed for the methanol monomer. Different electrolytes break the H-bonded structure of methanol to varying degrees, thus shifting the 3380-cm^{-1} band to higher (increasingly "monomeric") frequencies. It is not surprising that the ability to raise the frequency (structure breaking) decreases in the order $ClO_4^- > NO_3^- > SO_4^{2-} > Cl^- > F^- > H_2O$, which is the opposite from the order of hydrogen-bonding strength between the anions and methanol. The stronger the H-bonding, the lower the frequency. More recently, the ~3300-cm^{-1} frequency has been assigned to alcohol-anion complexes, based on NMR as well as infrared evidence (63).

It was pointed out in Chapter 3 (Eqn. 3.5) that proton transfer from an acid to a base can be visualized as proceeding through a number of intermediate stages involving varying degrees of H-bonding and ionic bond character. In the case of very weak acid–base interactions, there may be no proton transfer, only the formation of a hydrogen-bonded complex. We have seen examples of this in the complexation between chloroform (acid) and weak bases such as acetone. Stronger acid–base interactions in non-dissociating solvents (solvents of low dielectric constant and poor solvation ability for ions) usually result in the formation of hydrogen-bonded ion pairs. Since H-bonding is such an integral part of Brønsted acid–base reactions, vibrational spectroscopy has been a valuable technique for determining the extent of such reactions as well as the nature and the structure of the resulting products.

This has been illustrated, for example, in the extensive infrared studies by Barrow and his associates (64–67), in which amines were reacted with acetic acid in chloroform and carbon tetrachloride media. Since the relative concentrations of the bases and the acid were varied, the composition of the adducts at various relative and total concentrations of acid and base as well as the equilibrium constants for the association could be determined. For example, when 0.1 *M* acetic acid is half-neutralized with triethylamine (Et_3N), a stable acid salt is formed believed to have the structure (I). Complete neutralization of the acid produces a 1 : 1 acid–base adduct, believed to have the structure (II) in carbon tetrachloride, but the same adduct is solvated in chloroform (III).

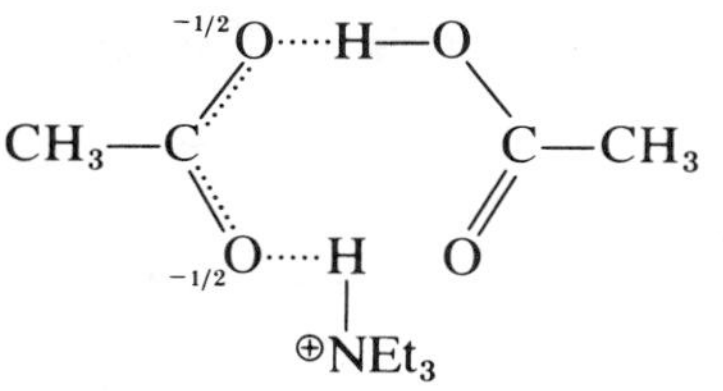

(I) 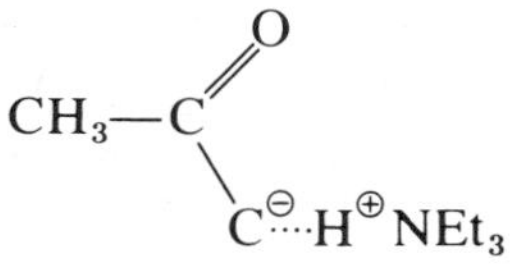 (II) 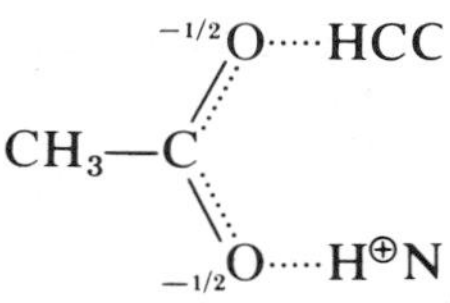 (III)

Note that all the products are hydrogen-bonded ion pairs. In the case of reaction between acetic acid and primary or secondary aliphatic amines, where the ammonium ion has more than one proton available for bonding, multiple hydrogen bonding to a single oxygen atom is believed to occur. This is the case for diethylammonium acetate in carbon tetrachloride, as shown in structure (IV).

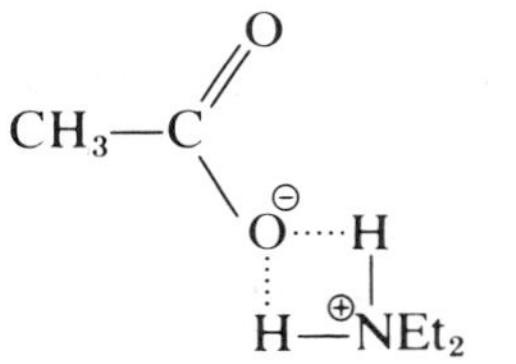

(IV)

The results of the IR investigations of Barrow et al. are summarized in Table 8.4. Many other studies of reactions between acid–base pairs have been carried out in inert solvents by vibrational spectroscopy, and the reader will find them summarized in the excellent reviews by Davis (53).

The strengths of very weak bases have been estimated also from Raman spectroscopic measurements. Deno and Wisotsky (68) noted that the C—O stretching band of methanol at 1021 cm^{-1} gradually disappeared as methanol became protonated and took advantage of this phenomenon to determine the basicity of methanol. Finding that methanol was 50% protonated in 39% H_2SO_4, they calculated a $pK_{BH^+} = -2.2$ for the methoxonium ion. In a similar manner, they determined the basicities of acetamide, acetone, dioxane, and methylamine. In the case of methanol, however, the interpretation of the Raman data was apparently not unambiguous, as Weston et al. (69) have subsequently reevaluated the pK_{BH^+} in HCl solution to be $-4.86 \pm 0.3_7$. For 2-propanol, the value of pK_{BH^+} was calculated to be $-4.7_2 \pm 0.9_5$ for HCl solutions and $-5.1_6 \pm 0.4_5$, for $HClO_4$ solutions (69).

In summary, a limited number of representative studies have been cited that illustrate the scope and power of vibrational spectroscopy in providing information on intermolecular interactions in nonaqueous media. The extent of cation–solvent and anion–solvent interaction (solvation), the nature of the resulting solvation shell as well as the type and the degree of ion pairing and complexation are some of the data derivable from vibrational spectra. The quantitative elucidation of intermolecular association via H-bonding

Table 8.4 Values of Log K (25–30°C, Molar Units) for Some Acid–Base Reactions, from Infrared Absorbance Data[a,b]

Stoichiometric concns.		Reacting species				
HOAc	Base	(1)	(2)	Product	Solvent	Log K
			Base = Et_3N (64)			
0.001	0.001–0.025	HA	B	BHA	CCl_4	2.90
0.10, 0.30	0.05–0.50	$B(HA)_2$	B	2BHA	CCl_4	0.20
0.001	0.0005–0.005	HA	B	BHA	$CHCl_3$	3.48
0.3	0.15–0.50	$B(HA)_2$	B	2BHA	$CHCl_3$	1.34
			Base = Et_2NH (65)			
0.0008	0.0002–0.004	HA	B	BHA	CCl_4	3.45
0.0008–0.01	0.0016–0.020		2BHA	$(BHA)_2$	CCl_4	2.38
0.10, 0.30	0.075–0.40	$B(HA)_2$	B	$(BHA)_2$	CCl_4	1.20
0.0008	0.0004–0.004	HA	B	BHA	$CHCl_3$	3.48
0.10, 0.30	0.10–0.60	$B(HA)_2$	B	2BHA	$CHCl_3$	2.00
			Base = *n*-$BuNH_2$ (66)			
0.0008	0.0008–0.008	HA	B	BHA	CCl_4	2.78
0.0008	0.0008–0.0064	HA	$B.HCCl_3$	$BHA{\cdot}HCCl_3$	$CHCl_3$	3.45

[a] From average values of K given in tables of Barrow and Yerger (64–66).
[b] HA signifies acetic acid; B signifies Et_3N, Et_2NH, or *n*-$BuNH_2$.
Reprinted with permission from M. M. Davis, *Acid-Base Behavior in Aprotic Organic Solvents,* NBS Monograph 105, Washington, D. C., 1968, p. 65. Table courtesy of the National Bureau of Standards.

within pure substances and between different acid–base pairs ranging from loose association, through a hydrogen-bonded ion pair, all the way to complete proton transfer, represents another area where application of vibrational spectroscopy has been particularly fruitful.

LITERATURE CITED—ELECTRONIC AND VIBRATIONAL SPECTROSCOPY

1 Bayliss, N. S., *J. Chem. Phys.,* **18,** 292 (1950).

2 Bayliss, N. S., and E. G. McRae, *J. Phys. Chem.,* **58,** 1002 (1954).

3 McRae, E. G., *J. Phys. Chem.,* **61,** 562 (1957).

4 Lippert, E., *Z. Elektrochem.,* **61,** 962 (1957).

5 Suppan, P., *J. Chem. Soc.* (*A*), 3125 (1968).

6 Popovych, O., and L. B. Rogers, *J. Amer. Chem. Soc.,* **81,** 4469 (1959).

7 Lantzke, I. R., "Spectroscopic Measurements," Part 1: "Electronic Absorption Spec-

troscopy," in *Physical Chemistry of Organic Solvent Systems,* A. K. Covington and T. Dickinson, Eds., Plenum Press, London and New York, 1973, Chap. 4.

8 Kamlet, M. J., and R. W. Taft, *J. Amer. Chem. Soc.,* **98,** 377 (1976).

9 Midwinter, J., and P. Suppan, *Spectrochim. Acta,* **25A,** 953 (1969).

10 Blandamer, M. J., and M. F. Fox, *Chem. Rev.,* **70,** 59 (1970).

11 Griffiths, T. R., and M. C. R. Symons, *Trans. Faraday Soc.,* **56,** 1125 (1960).

12 Kosower, E. M., *J. Amer. Chem. Soc.,* **80,** 3253 (1958); *J. Amer. Chem. Soc.,* **82,** 2188 (1960).

13 Watts, D. W., *Record of Chemical Progress,* **29,** 131 (1968).

14 Hogen-Esch, T. E., and J. Smid, *J. Amer. Chem. Soc.,* **87,** 669 (1965).

15 Casson, B., and B. J. Tabner, *J. Chem. Soc.* (*B*), 572 (1969).

16 Hogen-Esch, T. E., and J. Smid, *J. Amer. Chem. Soc.,* **88,** 307 (1966).

17 Carvajal, C., K. J. Tölle, J. Smid, and M. Szwarc, *J. Amer. Chem. Soc.,* **87,** 5548 (1965).

18 Savedoff, L. G., *J. Amer. Chem. Soc.,* **88,** 664 (1966).

19 Nicholls, D., C. Sutphen, and M. Szwarc, *J. Phys. Chem.,* **72,** 1021 (1968).

20 Winstein, S., E. Clippinger, A. H. Fainberg, and G. C. Robinson, *J. Amer. Chem. Soc.,* **76,** 2597 (1954).

21 (a) Hirota, N., *J. Phys. Chem.,* **71,** 127 (1967); (b) M. C. R. Symons, *J. Phys. Chem.,* **71,** 172 (1967).

22 Grunwald, E., *Anal. Chem.,* **26,** 1696 (1954).

23 Ashley, K. R., R. E. Hamm, and R. H. Magnuson, *Inorg. Chem.,* **6,** 413 (1967).

24 Abrahamer, I., and Y. Marcus, *J. Inorg. Nucl. Chem.,* **30,** 1563 (1968).

25 Irish, D. E., "Spectroscopic Measurements," Part 2: "Infrared and Raman Spectroscopy," in *Physical Chemistry of Organic Solvent Systems,* A. K. Covington and T. Dickinson, Eds., Plenum Press, London and New York, 1973, Chap. 4.

26 Janz, G. J., and J. R. Downey, Jr., "Digital Methods in Raman Spectroscopy," in *Advances in Infrared and Raman Spectroscopy,* Vol. I, R. J. H. Clarke and R. C. Hester, Eds., Heyden, London, 1975, Chap. 1.

27 (a) Edgell, W. F., A. T. Watts, J. Lyford, and W. M. Risen, *J. Amer. Chem. Soc.,* **88,** 1815 (1966); (b) Edgell, W. F., J. Lyford, R. Wright, W. M. Risen, and A. Watts, *J. Amer. Chem. Soc.,* **92,** 2240 (1970).

28 Maxey, B. W., and A. I. Popov, *J. Amer. Chem. Soc.,* **89,** 2230 (1967).

29 Maxey, B. W., and A. I. Popov, *J. Amer. Chem. Soc.,* **91,** 20 (1969).

30 Wuepper, J. L., and A. I. Popov, *J. Amer. Chem. Soc.,* **91,** 4352 (1969).

31 Wuepper, J. L., and A. I. Popov, *J. Amer. Chem. Soc.,* **92,** 1493 (1969).

32 McKinney, W. J., and A. I. Popov, *J. Phys. Chem.,* **74,** 535 (1970).

33 Wong, M. K., W. J. McKinney, and A. I. Popov, *J. Phys. Chem.,* **75,** 56 (1971).

34 Buxton, T. L., and J. A. Caruso, *J. Phys. Chem.,* **77,** 1882 (1973).

35 Evans, J. C., and G. Y-S. Lo, *J. Phys. Chem.,* **69,** 3223 (1965).

36 Baum, R. G., and A. I. Popov, *J. Solution Chem.,* **4,** 441 (1975).

37 Wong, M. K., and A. I. Popov, *J. Inorg. Nucl. Chem.,* **33,** 1203 (1971).

38 Popov, A. I., *Pure Appl. Chem.,* **41,** 275 (1975).

39 Tsatsas, A. T., and W. M. Risen, *J. Amer. Chem. Soc.,* **92,** 1789 (1970).

40 Höhn, E. G., J. A. Olander, and M. C. Day, *J. Phys. Chem.,* **73,** 3880 (1969).

41 Yeager, H. L., J. D. Fedyk, and R. J. Parker, *J. Phys. Chem.,* **77,** 2407 (1973).

42 Al-Baldawi, S. A., M. H. Brooker, T. E. Gough, and D. E. Irish, *Can. J. Chem.,* **48,** 1202 (1970).

43 Janz, G. J., K. Balasubrahmanyan, and B. G. Oliver, *J. Chem. Phys.*, **51**, 5723 (1969).
44 Balasubrahmanyan, K., and G. J. Janz, *J. Amer. Chem. Soc.*, **92**, 4189 (1970).
45 Evans, J. C., and G. Y.-S. Lo, *Spectrochim. Acta*, **21**, 1033 (1965).
46 Addison, C. C., D. W. Amos, and D. Sutton, *J. Chem. Soc. (A)*, 2285 (1968).
47 Olah, G. A., S. J. Kuhn, and S. H. Flood, *J. Amer. Chem. Soc.*, **83**, 4581 (1961).
48 West, R., *J. Amer. Chem. Soc.*, **81**, 1614 (1959).
49 Whetsel, K. B., and R. E. Kagarise, *Spectrochim. Acta*, **18**, 329 (1962).
50 Whetsel, K. B., and J. H. Lady, *J. Phys. Chem.*, **68**, 1010 (1964).
51 Josien, M.-L., J.-P. Leicknam, and N. Fuson, *Bull. Soc. Chim. France*, 188 (1958).
52 Allerhand, A., and P. von R. Schleyer, *J. Amer. Chem. Soc.*, **85**, 1233, 1715 (1963).
53 (a) Davis, M. M., *Acid–Base Behavior in Aprotic Organic Solvents*, National Bureau of Standards Monograph 105, 1968. (b) Davis, M. M., "Brønsted Acid–Base Behavior in "Inert" Organic Solvents," in *The Chemistry of Non-Aqueous Solvents*, Vol. III, J. J. Lagowski, Ed., Academic Press, New York, 1970, Chap. 1.
54 Hallam, H. E., in *Infrared Spectroscopy and Molecular Structure*, M. Davies, Ed., Elsevier, Amsterdam, 1963, Chap. 12.
55 Jakobsen, R. J., J. W. Brasch, and Y. Mikawa, *Appl. Spectr.*, **22**, 641 (1968).
56 Murthy, A. S. N., and C. N. R. Rao, *Appl. Spectr. Rev.*, **2**, 69 (1968).
57 Pimentel, G. C., and A. L. McClellan, *The Hydrogen Bond*, Freeman, San Francisco, Calif. 1960.
58 Lund, H., *Acta Chem. Scand.*, **12**, 298 (1958).
59 Bufalini, J., and K. H. Stern, *J. Amer. Chem. Soc.*, **83**, 4362 (1961).
60 Waldron, R. D., *J. Chem. Phys.*, **26**, 809 (1957).
61 Schleyer, P. R., and R. West, *J. Amer. Chem. Soc.*, **81**, 3164 (1959).
62 Hester, R. E., and R. A. Plane, *Spectrochim. Acta*, **23A**, 2289 (1967).
63 Green, R. D., J. S. Martin, W. B. McG. Cassie, and J. B. Hyne, *Can. J. Chem.*, **47**, 1639 (1969).
64 Barrow, G. M., and E. A. Yerger, *J. Amer. Chem. Soc.*, **76**, 5211 (1954).
65 Yerger, E. A., and G. M. Barrow, *J. Amer. Chem. Soc.*, **77**, 4474 (1955).
66 Yerger, E. A., and G. M. Barrow, *J. Amer. Chem. Soc.*, **77**, 6206 (1955).
67 Barrow, G. M., *J. Amer. Chem. Soc.*, **78**, 5802 (1956).
68 Deno, N. C., and M. J. Wisotsky, *J. Amer. Chem. Soc.*, **85**, 1735 (1963).
69 Weston, R. E., Jr., S. Ehrenson, and K. Heinzinger, *J. Amer. Chem. Soc.*, **89**, 481 (1967).

8.4 NUCLEAR MAGNETIC RESONANCE

8.4.1 An Overview of Applications Nuclear magnetic resonance, usually referred to as NMR, has been used for a variety of applications, including the study of chemical bonds and the kinetics of certain reactions, and it plays a vital role in the identification of organic compounds. The use of NMR in studying the kinetics and activation processes associated with ligand-exchange reactions and certain organic reactions in nonaqueous solvents is described in some detail in Chapter 10.

In addition, NMR can be used with advantage in studying solution properties of electrolytes in nonaqueous solvents as, for example, in ionic solvation, ionic association, selective solvation, solvation (coordination) numbers, solvent exchange rates, ionization equilibria and outer-sphere solvation. This section illustrates how NMR is used for studies such as these. A very thorough review of the subject of NMR studies of ions in pure and mixed solvents has been given by Hinton and Amis (1), and for those with special interest in this area it is strongly recommended that this review be used as a springboard for further reading. One of the more recent discussions of the application of NMR to studies in nonaqueous solvents is that by Lantzke (2). Details of the theory of NMR spectroscopy are provided in several texts (3–7), and it will be sufficient here to review only the salient points.

One of the most important sources of information in NMR that is used as a probe for studying ion–solvent, ion–ion interactions, and relative solvation numbers is the chemical shift value δ, which is an indication of the displacement of the resonance peak of the proton (or other nuclei) from the resonance peak of a reference substance and is given by

$$\delta = \frac{H_{\text{reference}} - H_{\text{sample}}}{H_{\text{reference}}} \times 10^6 \tag{8.4}$$

where H is the magnetic field strength.

The values of δ are expressed in parts per million (ppm). The chemical shift results from a change of electron density which causes the effective magnetic field to be altered. Chemical shifts are frequently modified by kinetic processes, the most common situation being one where the nucleus under observation (X) exchanges rapidly between two environments A and B. In this case:

$$\delta_{\text{obs}} = \rho_A\delta_A + \rho_B\delta_B \tag{8.5}$$

where ρ_A, ρ_B are the fractions of X in environments A and B, respectively.

Some individual substances exhibit a simple NMR spectrum, which is easy to interpret, but for most cases the spectra are considerably more complex due to a phenomenon known as spin–spin splitting. This complication results from the splitting of the resonance peaks into multiplets of two or more closely spaced lines by the interaction of the spins of nonequivalent neighboring protons. Important information concerning the number and types of protons which are coupled is provided by the spin–spin splitting or coupling effect. The coupling of two sets of protons is characterized by the coupling constant J, which is equal numerically to the separation of adjacent lines in each multiplet, expressed in units of Hertz. First-order spin–spin splitting is said to occur if the value of J is less than approximately one sixth of the difference between the chemical shift values of two sets of coupled protons. The relative peak intensities within the multiplet deviate increasingly from the values predicted using the first-order rules. The protons in the two sets are then said to be strongly coupled. The magnitudes of coupling constants can provide valuable structural information.

NMR may be used to determine the lifetime of solvent molecules in the solvation sphere as well as the lifetime for solute translation. The derived parameters are, τ_d, the average time between two translational jumps and τ_c, the correlation time. Separate NMR signals can be obtained for coordinated and free solvent molecules when the exchange of the molecules of the primary solvation sphere of the cation with the bulk solvent molecule is sufficiently slow. The following discussion illustrates some typical applications of NMR spectroscopy in the study of the properties of nonaqueous electrolyte solutions and is intended to show how parameters such as chemical shifts and coupling constants are used to interpret NMR data.

8.4.2 Solvent Coordination Numbers Fratiello et al. (8) undertook a study of the coordination number of $AlCl_3$ in aqueous mixtures of acetone, *N,N*-dimethylformamide, dimethylsulfoxide, dioxane, tetrahydrofuran, and tetramethylurea using proton magnetic resonance over a range of solvent compositions. The solvents investigated in this study were selected to provide a wide range of dipole moment and basicity and, thereby, offer some insight into those properties of a molecule which enhance its solvating ability. Figure 8.4 shows the proton magnetic resonance spectrum for a 2 *M* $AlCl_3$ solution in a 10:1 mole ratio mixture of water to DMSO, which illustrates the four resonance peaks due to bulk solvent (B_{H_2O} and B_{DMSO}) and solvent in the Al(III) coordination shell (C_{H_2O} and C_{DMSO}). Also shown in this figure is a recording of the electronic integration. It was observed that, at the low temperatures employed in this study (−20°C), proton exchange is slowed

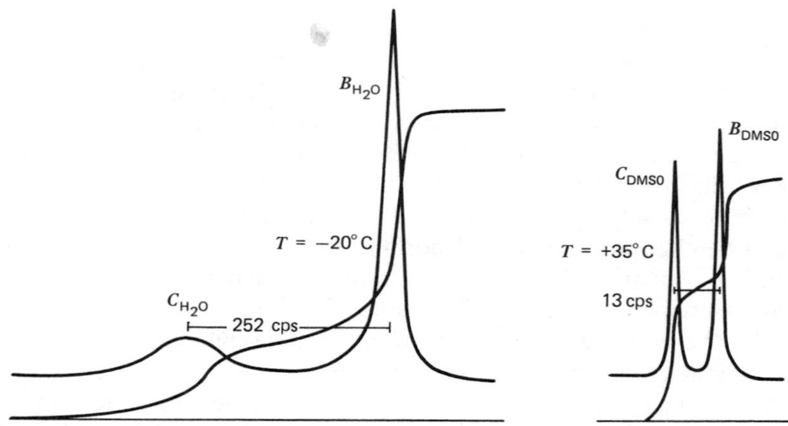

Figure 8.4 The proton magnetic resonance spectrum of a 2 *M* $AlCl_3$ solution in a 10:1 mole-ratio mixture of water to dimethylsulfoxide (DMSO), illustrating the signals due to bulk solvent (B_{H_2O} and B_{DMSO}) and solvent in the Al(III) coordination shell (C_{H_2O} and C_{DMSO}). Reprinted with permission from *J. Chem. Phys.*, **47**, 4951 (1967). Copyright 1967 American Institute of Physics.

Table 8.5 Al(III) Coordination Numbers in Some Aqueous–Nonaqueous Solvent Mixtures

Solvent mole ratios	$AlCl_3$ concentration (mol liter^{-1})	Al(III) coordination numbers	
H_2O/DMSO		**H_2O**	**DMSO**
10:1	2.00	5.01	0.92
10:1	1.00	5.63	0.53
5:1	1.80	4.12	1.69
5:1	1.00	4.97	1.28
3.5:1	1.50	3.56	2.02
3.5:1	1.00	4.06	1.92
2:1	1.00	3.08	2.61
H_2O/DMF		**H_2O**	**DMF**
10:1	2.00	4.79	1.21
10:1	1.50	4.71	1.29
10:1	1.00	4.87	1.13
7.5:1	1.50	4.47	1.53
7.5:1	1.00	4.43	1.57
H_2O/Acetone		**H_2O**	**Acetone**
10:1	2.00	6.3	None ↓
10:1	1.75	5.7	
10:1	1.50	5.8	
10:1	1.00	6.1	
7.5:1	1.50	5.8	
7.5:1	1.00	6.0	
5:1	0.70	5.8	
5:1	0.40	5.6	
3.5:1	0.50	5.7	
2.5:1	0.30	6.0	
H_2O/Tetramethylurea		**H_2O**	**Tetramethylurea**
10:1	1.00	5.8	None ↓
10:1	0.40	6.1	
7.5:1	1.00	5.7	
7.5:1	0.40	6.3	
5:1	1.00	5.6	
5:1	0.40	6.0	
2.5:1	0.80	5.4	
2.5:1	0.50	6.1	
1:1	0.40	5.6	
1:1	0.20	6.1	

Table 8.5 (*Continued*)

Solvent mole ratios	$AlCl_3$ concentration (mol liter^{-1})	Al(III) coordination numbers	
H_2O/Dioxane		**H_2O**	**Dioxane**
10:1	2.00	6.0	None
10:1	1.75	6.1	↓
10:1	1.50	6.1	
10:1	1.00	6.1	
7.5:1	1.50	6.1	
7.5:1	1.00	6.2	
H_2O/THF		**H_2O**	**THF**
10:1	2.00	6.3	None

Reprinted with permission from *J. Chem. Phys.* **47,** 4951 (1967). Copyright 1967. American Institute of Physics.

to such an extent that separate resonance signals are observed for bulk water and water molecules in the Al(III) solvation shell. Coordination numbers were measured by direct integration of the two water signals. The calculated coordination numbers for the various systems are given in Table 8.5 and for the aqueous mixtures of DMSO and DMF they represent average numbers of water and nonaqueous solvent in the first solvation shell of Al(III). The strong solvating ability of DMSO and DMF is seen by examination of the coordination-number data where it is observed that, even in the presence of large amounts of water, the nonaqueous solvent does, in fact, contribute to the first coordination shell of Al(III), that is, mixed solvation occurs in the first solvation shell. It appears that DMSO and DMF are the only solvents considered in this study that are capable of solvating Al(III) in the presence of water, at least in the concentration ranges that were investigated. In the aqueous mixtures of acetone and tetramethylurea and, over a small concentration range, of dioxane and THF, only water was seen to solvate Al(III), as indicated by the value of 6 for the coordination number of water and the fact that no nonaqueous resonance signal appeared in the NMR spectra. One factor which may account for this behavior is the difference in basicities of the solvents, as DMSO and DMF are more basic by several orders of magnitude than are acetone, dioxane, and THF.

8.4.3 Preferential Solvation The use of NMR in the study of preferential solvation and the accompanying use of thermodynamic models to account for the data has been the subject of several recent investigations. Frankel et al. (9) investigated the solvation of tris-acetylacetone complexes $(acac)_3$ of Co(III) and Cr(III) in several solvent mixtures. Preferential solvation was detected using two techniques, both of which are involved with short-range interactions. One method utilized the solvent dependence of the chemical shift of a nucleus in a diamagnetic neutral solute, while the other method examined the effect of a paramagnetic solute on the transverse relaxation time (T_2) of the solvent nuclei.

The chemical-shift method is only applicable when the solute contains a nucleus that is particularly solvent-sensitive. In this respect, the ^{59}Co nucleus is a good probe, as the low-lying excited electronic states cause the chemical shift to be very sensitive to the surroundings. The results for the solvation of $Co(acac)_3$ in chloroform–carbon tetrachloride mixtures, using the ^{59}Co chemical-shift method, are presented in Fig. 8.5. Strong evidence for the preferential solvation by chloroform is indicated by the marked deviation from a straight line, as when no preferential solvation occurs, a plot of ^{59}Co chemical shift vs bulk solvent composition will yield a straight line. The equisolvation point shown in Fig. 8.5 gives the composition at which the chemical shift lies midway between the values for the pure solvents when both solvents participate equally in the contact solvation shell.

In the relaxation-time method, the relaxation time (T_2) is given by

$$\frac{1}{T_2} = \frac{\rho_A}{T_{2A}} + \frac{\rho_B}{T_{2B}} \tag{8.6}$$

where ρ_A is the probability of finding a solvent molecule in the diamagnetic environment (bulk solvent), T_{2A} is the relaxation time in that environment, and ρ_B and T_{2B} are the corresponding quantities for the paramagnetic environment (solvation shell). The values of T_2 are determined from the line

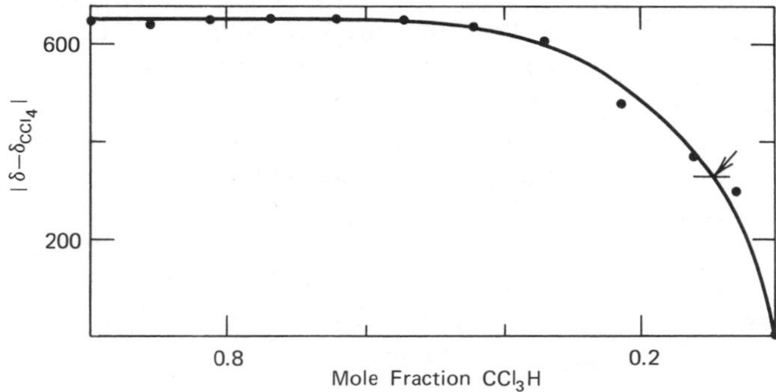

Figure 8.5 Solvation of $Co(acac)_3$ in chloroform–carbon tetrachloride mixtures; ^{59}Co chemical-shift method. The equisolvation point is indicated by the arrow. Reprinted with permission from *J. Phys. Chem.* **74,** 1376 (1970). Copyright 1970 American Chemical Society.

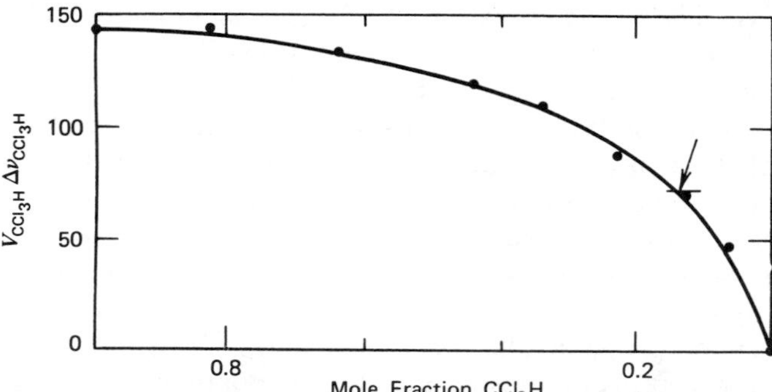

Figure 8.6 Solvation of $Cr(acac)_3$ in chloroform–carbon tetrachloride mixtures; solvent relaxation-time method. The equisolvation point is indicated by the arrow. Reprinted with permission from *J. Phys. Chem.* **74,** 1376 (1970). Copyright 1970 American Chemical Society.

width at half-height. Preferential solvation causes a deviation from linearity in a plot of the solvation-shell composition vs bulk-solvent composition. Figure 8.6 shows the preferential solvation occurring with $Cr(acac)_3$ in chloroform–carbon tetrachloride mixtures. The product of the line width ($\Delta\nu$) of CCl_3H and the volume fraction (V) of CCl_3H in the solvent mixture is plotted vs the mole fraction of CCl_3H in the bulk solvent. It should be pointed out that the analysis of these data was based on one important assumption, namely, that the magnetic coupling between the central metal atom and a given molecule in the solvation sphere is independent of the overall composition of the solvation sphere. This particular assumption appeared to be quite valid for the systems studied in this investigation, but will not necessarily be true for every system. It is readily seen by comparing Figs. 8.5 and 8.6 that the equisolvation points determined by the chemical-shift method and the relaxation-time method were in close agreement for the chloroform–carbon tetrachloride system and for solvent mixtures such as chloroform–acetone, chloroform–benzene, chloroform–methanol, and chloroform–dimethylformamide. Frankel et al. (9) discussed the NMR data obtained in their investigation using a thermodynamic model in which the solvent was considered to be distributed between two phases, the bulk solvent and the solvation shell of the solute, and the solvation number was taken to be the same for the two solvents. The detailed thermodynamic treatment is given in Ref. 9, but the essential result for the standard free energy change for the process is given by

$$\Delta G^\circ = -RT \ln K \tag{8.7}$$

where

$$K = \frac{y_A Y_B}{y_B Y_A} \tag{8.7a}$$

Here, y_A and y_B refer to the mole fractions of A and B in the solvation shell, and Y_A and Y_B refer to the bulk solvent. This particular treatment predicts a simple relationship between the composition of the bulk solvent and the solvation shell along an isotherm, that is, $y_a/y_B = K(Y_A/Y_B)$. A plot of the correct ratios should give a straight line of slope K. A plot of this type for the system $Co(acac)_3$ in chloroform–benzene mixtures at three different temperatures is given in Fig. 8.7. It is observed that the solvation becomes less preferential at higher temperatures. For some systems, the solvation isotherms exhibit curvature due to either nonregular solution behavior or a breakdown of the assumption of equal solvation number.

Langford and Stengle (10) have examined the chlorine-35 NMR of the chloride ion in DMSO–water and CH_3CN–water mixtures. The variation of chemical shifts for this system are shown in Fig. 8.8, which indicates that the ^{35}Cl chemical shifts are a function of solvent composition. The results for the CH_3CN–H_2O mixtures appear to support the idea of hydrogen bonding of the anion in that there is a strong preference of Cl^- for small amounts of water. However, in the case of the DMSO–H_2O mixtures, this tendency is not observed, with the competition for solvation sites being approximately equal for the DMSO and H_2O. A more detailed examination of the results postulated that in fact even competition for the Cl^- solvation sites may occur in both systems and that the Cl^- solvent preference should be related to the variation of bulk solvent activity as the solvent components vary. The straight line in Fig. 8.8 shows the correlation of ^{35}Cl chemical shift with bulk

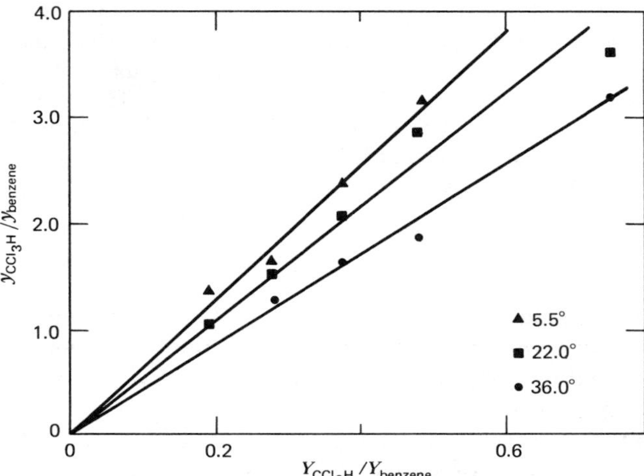

Figure 8.7 Temperature dependence of the solvation isotherm of $Co(acac)_3$ in chloroform–benzene mixtures; ^{59}Co chemical-shift method. Reprinted with permission from *J. Phys. Chem.* **74,** 1376 (1970). Copyright 1970 American Chemical Society.

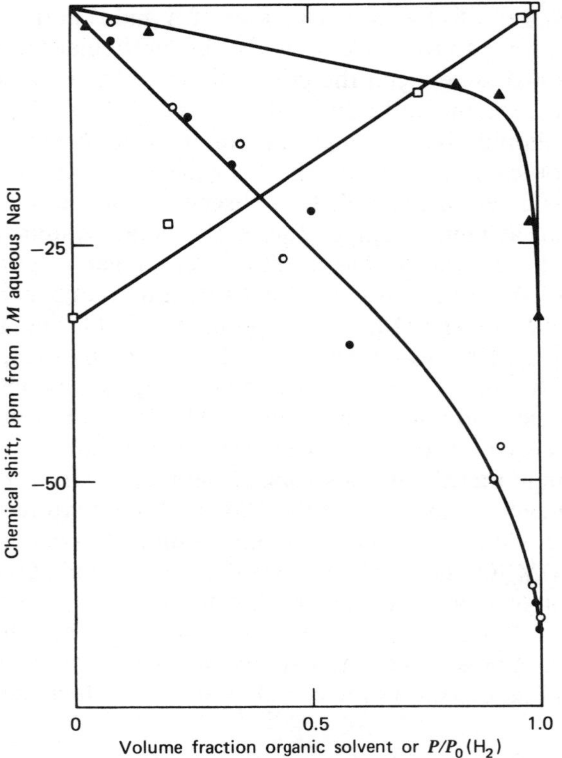

Figure 8.8 The variation of ^{35}Cl chemical shifts of chloride salts in mixtures of CH_3CN and DMSO with water. Open circles represent 0.25 *M* DMSO–water. Closed triangles represent 0.25 *M* $(C_2H_5)_4NCl$ in CH_3CN–water. All of these curves are chemical shift as a function of volume fraction of the organic component. The last curve (open squares) represents 0.25 *M* $(C_2H_5)_4NCl$ in CH_3CN–water, this time showing chemical shift as a function of the relative partial pressure of water (approximately relative activity) compared to pure water (P/P_O). Reprinted with permission from *J. Amer. Chem. Soc.*, **91,** 4014 (1969). Copyright 1969. American Chemical Society.

water activity. Using the hydrogen-bonding idea, water should remain the principal solvating species of the chloride ion until the bulk water activity has been reduced by a factor comparable to the activity change for transfer of the ion from a protic to an aprotic solvent. However, the results of this study show that the immediate environment of Cl^- is related to the long-range structural aspects of solvent mixtures. One reason suggested for the fact that water is not strongly preferred over the aprotic solvent in the layer in contact with Cl^- might well be that a protic solvent molecule hydrogen-bonded to Cl^- is strongly polarized so that it presents a "lyate-ion-like" aspect to the middle solvent region and this "lyate-ion-like" species is poorly solvated by the aprotic solvent.

Schneider and Strehlow (11) studied the proton NMR spectra of acetonitrile in $AgNO_3$ solutions ranging in molality from 0.1 to saturation over the entire range of acetonitrile–water solvents. The variation of the chemical shift of acetonitrile protons as a function of the mole fraction of acetonitrile in the mixture confirmed the existence of the $[Ag(CH_3CN)_2]^+$ complex. Furthermore, as the proton shift was a linear function of the $AgNO_3$ molality at low concentrations, it was possible to determine the solvation numbers of acetonitrile for the Ag^+ and the NO_3^- ions from the known formation constants of the silver–acetonitrile complexes. It turned out that at mole fractions of acetonitrile of 0.9 or greater, the solvation numbers were $n_{Ag^+} = 4.0$ and $n_{NO_3^-} = 2.0$. At very low acetonitrile contents, they approached 2 and zero, respectively.

8.4.4 Ion Association Concentration effects on the NMR spectra of electrolytes can be used to study ion association. Buckson and Smith (12) studied the effect of concentration on the NMR spectra of tetra-*n*-butylammonium chloride, bromide, iodide, perchlorate, and picrate in nitrobenzene over a salt concentration range of 0.005–0.25 *M*. A plot of Δ_a, the position of resonance of the α-methylene protons relative to tetramethylsilane, is illustrated in Fig. 8.9 for the various salts. It is readily observed that both the concentration of electrolyte and the nature of the anion affect the value of Δ_a. The most marked effect is observed with chloride, while the picrate ion is relatively insensitive to concentration. Also, a limiting value is obtained for Δ_a as the concentration of salt decreases. The data were used to gain values for the ion-pair dissociation constants for the halide salts using the following approach.

The observed values of Δ_a were taken as the weighted average of the proton resonance from dissociated ions, Δ_i and ion pairs, Δ_ρ, where

$$\Delta_a = \alpha\Delta_i + (1 - \alpha)\,\Delta_\rho \tag{8.8}$$

(α = the degree of dissociation).

Using the relation for the ion-pair dissociation constant

$$K = \frac{(Bu_4N^+)(X^-)}{(Bu_4N^+X^-)} \tag{8.9}$$

one obtains

$$C = (\Delta_\rho - \Delta_i)K\frac{\Delta_a - \Delta_i}{(\Delta_\rho - \Delta_a)^2} \tag{8.10}$$

where C is the stoichiometric concentration. A plot of $(\Delta_a - \Delta_i) \times (\Delta_\rho - \Delta_a)^{-2}$ vs C for Bu_4NCl at various values of Δ_ρ is shown in Fig. 8.10. Using the value of 208 Hz for Δ_i and of 232 Hz for Δ_ρ produces a linear plot

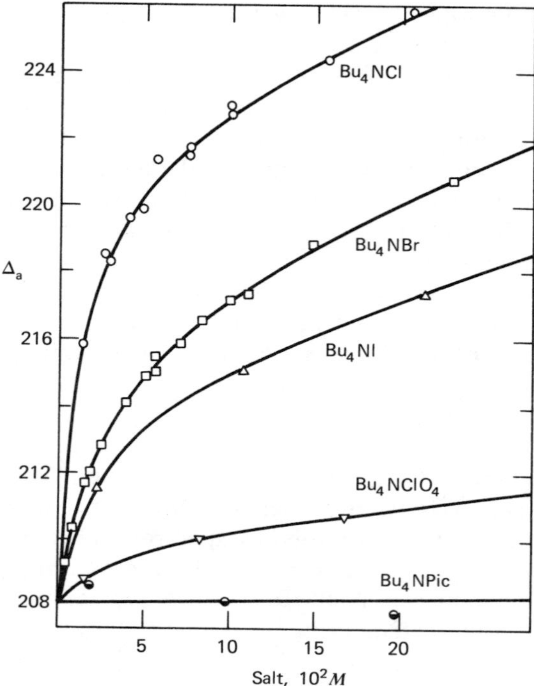

Figure 8.9 Plot of Δ_a vs concentration for various salts in nitrobenzene at 35.9°C. Reprinted with permission from *J. Phys. Chem.*, **68,** 1875 (1964). Copyright 1964. American Chemical Society.

for the chloride salt. In Table 8.6, the ion-pair dissociation constants obtained using this method are compared with their counterparts determined from electrolytic conductance. It is seen that the values obtained for the ion-pair dissociation constants are fairly similar regardless of whether the activity coefficient factor is included or omitted. The values obtained using the conductance method are lower, but this may have been due to the higher concentrations employed in the NMR studies.

8.4.5 Ligand- and Solvent-Exchange Rates An example of a study of ligand exchange rates and determination of solvation numbers using NMR techniques is that by Meiboom et al. (13–15). The exchange rates of methanol molecules in the solvation shells of magnesium, cobalt, and nickel ions were determined in acidic anhydrous methanol, and in the case of cobalt and nickel, methanol–water mixtures were used. The exchange rates were obtained from the broadening of the OH proton magnetic resonance signals of the free and bound methanol, these signals being well separated from each other. Methanol possesses a distinct advantage as a solvent for this type of investigation, since the solutions may be cooled to a low enough temperature to slow down the exchange rate into a range that is accessible to the NMR technique.

Focusing on the solvation of the magnesium ion, the NMR spectrum of a 1 *M* solution of $Mg(ClO_4)_2$ in anhydrous methanol at $-75°C$ is shown in Fig. 8.11. The detail of the OH quadruplet of the methanol in the Mg^{2+} solvation shell is given in Fig. 8.12 on an expanded frequency scale and under conditions of smaller linewidth. This well-defined quadruplet structure indicates that the exchange of the OH protons of the solvation methanol is slower than about 1 sec^{-1}, this being true both for exchange with the bulk solvent and for exchange within a solvation complex. This type of behavior is to be contrasted with the proton exchange between bulk methanol molecules, which is very rapid; the bulk OH and CH_3 signals do not exhibit

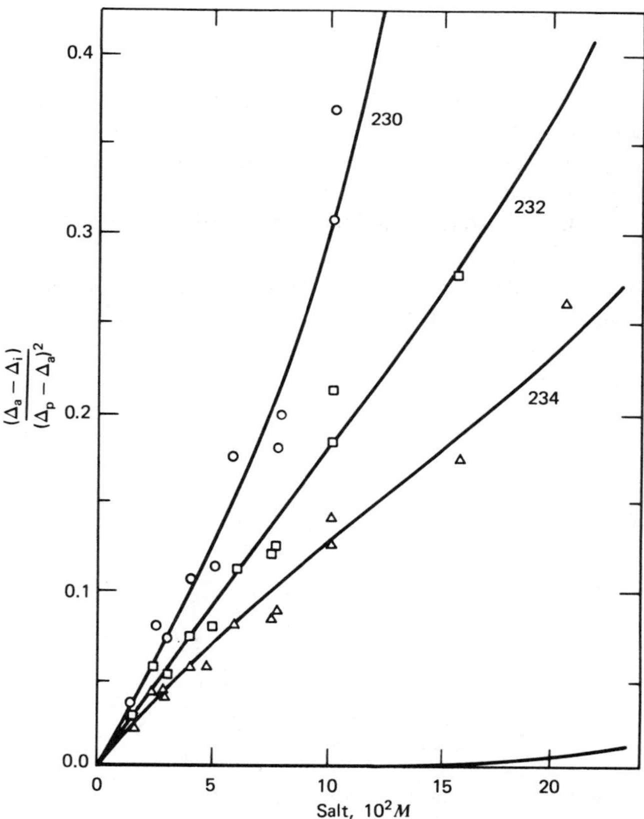

Figure 8.10 Plot of $\Delta_a - \Delta_i/(\Delta_p - \Delta_a)^2$ vs C for Bu_4NCl and various values of Δ_p. Reprinted with permission from *J. Phys. Chem.*, **68,** 1875 (1964). Copyright 1964. American Chemical Society.

Table 8.6 Summary of Ion-Pair Dissociation Constants in Nitrobenzene Obtained by Various Methods

	10^2K liter mol^{-1}		
Salt	NMR[a]	NMR[b]	Conductance[c]
$(n\text{-Bu})_4NCl$	2.2	2.7	
$(n\text{-Bu})_4NBr$	4.4	3.9	1.75
$(n\text{-Bu})_4NI$	6.8	5.3	3.7

[a] Temp. = 39.5°C.
[b] Debye–Hückel limiting law activity coefficients used in calculations.
[c] Temp. = 25°C.
Reprinted with permission from *J. Phys. Chem.* **68,** 1875 (1964). Copyright 1964. American Chemical Society.

multiplet structure. It is found that only below 0°C and in an acidified solution is the separate solvation shell signal observed. The solvation and the bulk OH resonances merge above about 10°C, which indicates the increasingly rapid exchange between solvation and bulk OH. The CH_3 signals in both the coordinated methanol and the bulk methanol are almost identical so that no separate signals are observed.

A technique using selective broadening of the bulk methanol by the addition of paramagnetic Cu^{2+} ions makes it possible to observe two peaks

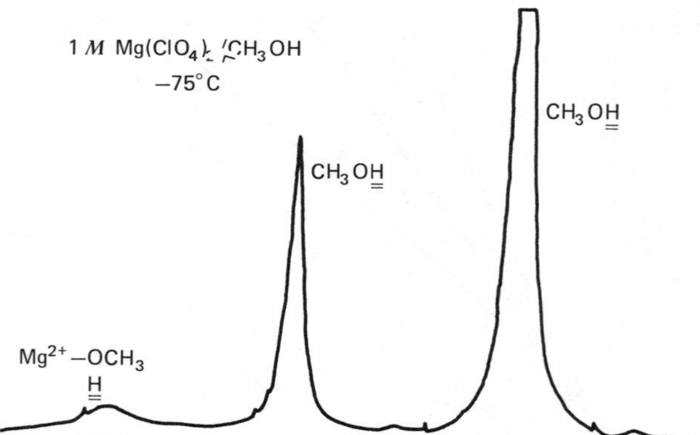

Figure 8.11 NMR spectrum of a 1 *M* solution of $Mg(ClO_4)_2$ in anhydrous methanol at −75°C. The field increases from left to right. The interpretation of the different peaks is indicated by underlining the protons involved. The top of the methyl peak is off scale. Reprinted with permission from *J. Amer. Chem. Soc.*, **89,** 1765 (1967). Copyright 1967. American Chemical Society.

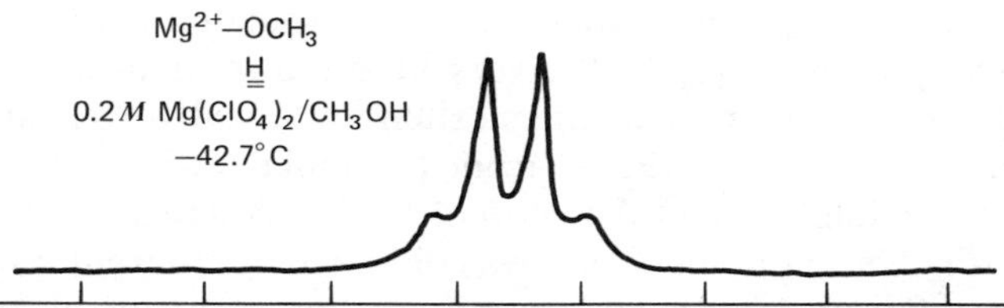

Figure 8.12 Detail of the OH quadruplet of the methanol in the Mg^{2+} solvation shell. The spectrum is of a 0.2 *M* $Mg(ClO_4)_2$ solution in anhydrous methanol at about −42.7°C. Reprinted with permission from *J. Amer. Chem. Soc.*, **89,** 1765 (1967). Copyright 1967. American Chemical Society.

separately and thus establish that the exchange in acidified solutions is predominantly ligand- rather than proton exchange. The results obtained for the addition of cupric perchlorate to this system are shown in Fig. 8.13. The CH_3 signal of the coordinated methanol, which was hidden behind the bulk CH_3 line in the previous spectrum, can now be observed. At the lower temperatures, the separate OH and CH_3 peaks of the methanol coordinated with Mg^{2+} are evident. From the temperature behavior of this signal, it can be concluded that ligand exchange, rather than simple OH proton exchange, is the dominant process. The methanol molecules coordinated with the paramagnetic Cu^{2+} exchange rapidly with the bulk methanol, but not directly with those in the magnesium solvation shell. Thus the bulk OH and CH_3 signals are strongly broadened, but the Mg^{2+} solvation shell signals, only

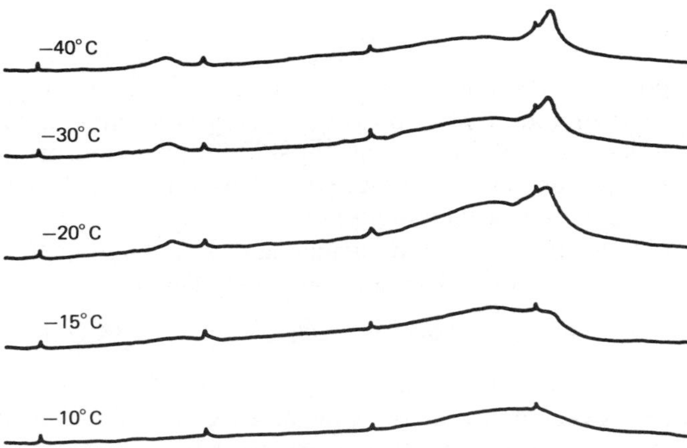

Figure 8.13 Spectra of a solution containing 0.4 *M* $Mg(ClO_4)_2$ and 0.0006 *M* $Cu(ClO_4)_2$ in methanol. The very broad line extending over the right half of the spectra is due to methanol exchanging rapidly between bulk solvent and the Cu^{2+} solvation shell. Reprinted with permission from *J. Amer. Chem. Soc.*, **89,** 1765 (1967). Copyright 1967. American Chemical Society.

slightly so. As a result, the signals of the coordinated methanol appear as relatively sharp lines on top of the very broad bulk signals.

The solvation number of the magnesium ion in anhydrous methanol was determined at -80 and -19.2°C from the molar ratio of $Mg(ClO_4)_2$ and methanol in the solution and the ratio of the areas under each signal in the spectrum. A solvation number of 6 was obtained at different concentrations and the two different temperatures, an indication of an octahedral geometry of the coordination compound as evidenced by the simple quadruplet signal. In addition to solvation-number studies, the first-order rate constant for solvent exchange and the activation parameters, $\Delta H^\ddagger$ and $\Delta S^\ddagger$ were obtained. The details of this approach are discussed in Chapter 10.

Proton magnetic resonance can be used successfully to study the anion effect on solvent exchange rates and chemical shifts of bound solvent molecules. O'Brien and Alei (16) studied solutions of $AlCl_3$ in acetonitrile from -40 to 80°C. A solvation number of 1.5 was obtained by integration of free and bound proton resonances. This result is to be compared to solvation numbers for Al(III) of 6 in water, DMSO, DMF, and DMSO–H_2O mixtures recorded by other investigators. In the solvents where a solvation number of 6 is attained, the anionic species Cl^- or ClO_4^- is completely displaced by solvent from the metal-ion coordination sphere. Obviously in acetonitrile, the anion competes with the solvent for coordination sites. The solvation number obtained for aluminum perchlorate is 2.9. This solvation number can be explained by the presence of all aluminum in the form $Al(CH_3CN)(ClO_4)_3$ or by a mixture of species containing various ratios of acetonitrile and perchlorate in the coordination sphere of the Al(III). The various species existing in this system give rise to an average solvation number of 2.9.

8.4.6 Solvation of "Onium" Salts Taylor (17) studied ionic interactions in solution by examining NMR spectra of quaternary ammonium salts in a variety of solvents. Magnitudes of solvent shift for given ions in an aromatic solvent were discussed in terms of factors such as ion size and shape, solvent polarity, and the nature of the counterion.

Table 8.7 shows the data for the magnetic resonance frequency of methyltributylammonium picrate in various solvents. The quantity ν is given by

$$\Delta = \nu - \nu_{CDCl_3} \tag{8.11}$$

where Δ, the aromatic-solvent-induced shift (ASIS) for a given nucleus, is the difference between its magnetic resonance frequency in that solvent and in the reference solvent $CDCl_3$. It is observed that the α-methyl resonance frequency of this electrolyte is strongly solvent dependent. Cation solvation accounts for these effects, as the large differences in solvent shifts are not reduced at lower electrolyte concentrations, which indicates that ion association cannot be responsible for them. The fact that no correlation exists

Table 8.7 Magnetic Resonance Frequency of $CH_3N^+(C_4H_9)_3Pi^-$ [a] in Various Solvents

Solvent	$-\nu$	Dielectric constant	Solvent	$-\nu$	Dielectric constant
$C_6H_5NO_2$	201	32.0	CH_3NO_2	184	34.2
C_5H_5N	196	12.0	o-$C_6H_4Cl_2$	173	9.5
$C_6H_5COCH_3$	196	17.0	CH_3CN	171	34.3
CD_3COCD_3	195	19.3	C_6H_5Cl	162	5.4
C_6H_5CHO	194	15.0	C_6H_5Br	162	5.2
$CDCl_3$	191	4.5	C_6H_6	145	2.2
CH_2Cl_2	187	8.0	$C_6H_5CH_2OH$	128	12.0
C_6H_5CN	187	24.0	1-$C_{10}H_7Br$	109	4.7

[a] Pi = picrate.
Reprinted with permission from *J. Amer. Chem. Soc.* **92,** 4813 (1970). Copyright 1970. American Chemical Society.

between the chemical shift and the dielectric constant of the solvent is indicative that reaction field effects are unimportant. Large upfield shifts are produced by benzyl alcohol, chlorobenzene, bromobenzene, benzene, and 1-bromonaphthalene. This effect is due to the proximity of the positive charge on the central atom to the electron-rich face of the aromatic ring, with a result that the α-methyl group is in a favorable position to experience a large upfield shift due to the diamagnetic anisotropy of the ring (18). In contrast, aromatic solvents such as nitrobenzene and benzonitrile do not give rise to substantial upfield effects, which is probably due to the fact that the polar end of these molecules is in close proximity to the positive charge, and the aromatic ring is held either further from or at a different angle to the α-methyl group of the cation.

Ion-size effects were examined by measuring Δ's for the series $CH_3N^+R_3X^-$ where R = alkyl and X^- = picrate or tetrabutylborate. Inspection of the data in Table 8.8 shows that, for either anion, Δ decreases as the alkyl chain length increases. This trend was interpreted in terms of the large chains acting to increase the mean separation between the positive charge and the solvent.

In addition, a series of tolyl phosphonium salts were studied to demonstrate the importance of the positive charge on phosphorus. This effect is seen in Table 8.9, where results for the phosphonium salts are compared with those for the phosphines. Hydrogens which are further from the positively charged heteroatom give rise to smaller Δ's. The *ortho* methyl group, being closest to the positive phosphorus atom, experiences a larger shift than the *meta* or *para* ring methyls. The *meta* and *para* methyl groups of the phosphines experience a measurable ASIS, while only slight effects are seen for the *ortho* group.

Table 8.8 Magnetic Resonance Frequency and Δ's for $CH_3N^+R_3X^-$ in Various Solvents[a]

	Solvent					
R	$CDCl_3$ $-\nu$	CH_3CN Δ	C_6H_5Br Δ	C_6H_6 Δ	$C_6H_5CH_2OH$ Δ	1-$C_{10}H_7Br$ Δ
(1) CH_3	202	19	54	79	81	136
(2) C_2H_5	185	15	44	70	75	126
(3) Quin	185	15	40	60	68	116
(4) C_3H_7	191	18	33	53	66	93
(5) C_4H_9	191	18	29	46	63	82
(6) C_3H_{11}	191	19	26	40	56	71
(7) C_6H_{13}	191	18	24	36	53	67

[a] R = alkyl; X^- = picrate for 1, 2, and 3 or tetrabutylborate for 4–7; Quin = quinuclidinium, C_7H_{13}.

Ion association effects in nonaqueous solvents were examined by measuring the NMR spectra of a series of ammonium, phosphonium, and arsonium salts and noting the anion dependence of the α-methyl resonance frequencies. The results are summarized in Table 8.10. It is seen that, in general, the smaller the anion, the more downfield is the cation, suggesting

Table 8.9 Ring Methyl Magnetic Resonance Frequencies of Tolyl Phosphonium Salts, $CH_3P^+(PhCH_3)_3X^-$

	Solvent		
	$CDCl_3$ $-\nu$	1-$C_{10}H_7Br$ Δ	$C_6H_5CH_2OH$ Δ
X^- = iodide			
ortho	145	55	38
meta	150	36	26
para	149	45	22
X^- = picrate			
ortho	142	67	36
meta	145	36	22
para	149	42	21
Ring methyl resonances of triaryl phosphines $[P(PhCH_3)_3]$[a]			
ortho	142	−4	0
meta	136	16	15
para	138	16	13

[a] 0.05 *M*.

Table 8.10 Chemical Shifts of Methyltriphenylphosphonium Salts and Other Onium Salts in a Variety of Solvents

A. Chemical Shifts of Methyltriphenylphosphonium Salts ($CH_3P^+Ph_3X^-$)

	Solvent						
	$C_6H_5NO_2$ $-\nu$	C_6H_5CN $-\nu$	$CDCl_3$ $-\nu$	CH_2Cl_2 $-\nu$	CH_3NO_2 $-\nu$	CH_3OH $-\nu$	CH_3CN $-\nu$
Salt Concentration, 0.05 *M*							
$X^- = Cl^-$	209	204	203	197	179	179	175
$X^- = Pi^-$	187	177	182	174	176	179	169
Difference	22	27	21	23	3	0	6
Salt Concentration, 0.004 *M*							
$X^- = Cl^-$	189	195	206	196	177	179	169
$X^- = Pi^-$	185	175	185	173	176	179	169
Difference	4	20	21	23	1	0	0

B. Chemical Shifts of Onium Salts in $CDCl_3(CH_3M^+X^-)^a$ [M = N, P, or As]

Cation	Cl^-	Br^-	I^-	Pi^-	CH_3B^-(cyclohexyl)$_3$
$CH_3N^+(C_4H_9)_3$			198	191	186
$CH_3P^+(C_4H_9)_3$			127	120	115
$CH_3P^+Ph_3$	203	200	194	182	(172)[b]
$CH_3As^+Ph_3$	199	197	193	180	(170)[b]

[a] 0.05 *M*.

[b] Compound decomposed in $CDCl_3$; resonance frequency estimated from trends observed for other salts; estimated uncertainty ±3 Hz.

Reprinted with permission from *J. Amer. Chem. Soc.* **92,** 4813 (1970). Copyright 1970. American Chemical Society.

ion association. The smaller anions are more efficient at displacing solvent molecules or the anions "shield" the cations from the solvent. The anion effect is greatly magnified in the solvent 1-bromonaphthalene, as seen in Table 8.11. The magnitude of Δ is dependent upon the anion size, a more upfield shift being associated with a larger anion.

A further study examined the possibility of the coordination of anions by using a strong complexing agent that can remove anions from close proximity to cations and thereby provide a means of decreasing the effects of ion association. Alcohols have been observed to hydrogen-bond with halide ions. The enhanced cation–solvent interaction arising from the addition of methanol to 1-bromonaphthalene solutions of various quaternary ammonium salts is seen in Table 8.12. Cations that are well "shielded" by

Table 8.11 Chemical Shifts of Onium Salts in 1-Bromonaphthalene

Cation	($CH_3M^+R_3X^-$)		
	X^-	$-\nu$	Δ
$CH_3N^+(C_4H_9)_3$	I^-	162	36
	Pi^-	111	80
	CH_3B^-(cyclohexyl)$_3$	83	103
$CH_3P^+(C_4H_9)_3$	I^-	95	32
	Pi^-	46	74
	CH_3B^-(cyclohexyl)$_3$	19	96
$CH_3P^+Ph_3$	Cl^-	223	−20
	Br^-	204	−4
	I^-	170	24
	Pi^-	125	57
	CH_3B^-(cyclohexyl)$_3$	66	106
$CH_3As^+Ph_3$	Pi^-	129	51
	CH_3B^-(cyclohexyl)$_3$	74	96

anions, and hence have small ASIS's, experience the largest changes when methanol is added. The most striking observation of this particular study was that the small ions were more efficient "shielders" than larger ones. The effect was discussed by considering the two types of associated species that exist in these aromatic solvents of low dielectric constant ($\epsilon < 15$), namely, the contact- and the solvent-separated ion pairs. The chemical shifts

Table 8.12 Effect of Methanol on Cation ASIS[a,b]

Salt	$\Delta_{1\text{-}C_{10}H_7Br}$	Methanol effect (Hz)
$CH_3N^+(C_4H_9)_3Pi^-$	80	−7
$CH_3P^+(C_4H_9)_3Pi^-$	76	−5
$CH_3P^+Ph_3Pi^-$	60	11
$CH_3P^+(C_4H_9)_3I^-$	34	22
$CH_3N^+(C_4H_9)_3I^-$	36	24
$CH_3P^+Ph_3I^-$	21	38

[a] Salts are 0.04 *M*.
[b] 3% methanol (by volume).

of the contact species are very sensitive to the counterion, whereas the chemical shifts of the solvent-separated species are relatively independent of the counterion and are dominated by solvent effects. The "shielding" capabilities of ions can be interpreted in terms of the relative number of contact- and solvent-separated species that they form.

8.4.7 Molecular Complexes Finally, mention should be made of the use of NMR in evaluating association constants for nonaqueous solvents. An example of this approach is that by Foster and Fyfe (19). For the case of a 1:1 molecular complex in solution, the association constant, K, for the interaction between an electron-donor molecule D and an electron-acceptor molecule A to form a charge-transfer complex AD is given by

$$K = \frac{[AD]}{[A][D]} \tag{8.12}$$

For the condition where one of the component species is in large excess

$$\delta^{A}_{obs} - \delta^{A}_{o} = \frac{[D]}{1 + [D]K}(\delta^{A}_{AD} - \delta^{A}_{o}) \tag{8.13}$$

where δ^{A}_{obs} is the observed shift of the acceptor protons in the complexing medium, δ^{A}_{o} is the shift of the acceptor protons in the uncomplexed state and δ^{A}_{AD} is the shift of the acceptor protons in the pure complex. If $\Delta = \delta^{A}_{obs} - \delta^{A}_{o}$ and $\Delta_o = \delta^{A}_{AD} - \delta^{A}_{o}$, then

$$\Delta = \frac{[D]K\Delta_o}{1 + [D]K} \tag{8.14}$$

This equation assumes that δ^{A}_{obs} is the result of molecular complexing between A and D, and that there are no significant solvent effects on chemical shifts of the various species.

Values of K are obtained from the slope of the relation

$$\frac{\Delta}{[D]} = -\Delta K + \Delta_o K \tag{8.15}$$

by plotting $\Delta/[D]$ vs Δ. In this study, the association constants for a series of aromatic hydrocarbon and *N*-alkylaniline complexes with 1,4-dinitrobenzene and 1,3,5-trinitrobenzene were determined. Typical plots of $\Delta/[D]$ vs Δ for the complexes of 1,4-dinitrobenzene are shown in Fig. 8.14. The values of K obtained using the NMR technique were comparable with those determined by optical methods.

8.4.8 Sodium Magnetic Resonance Sodium magnetic resonance studies have been the subject of several investigations (20–22). The relative sensitivity of 0.1 with respect to the proton indicates that measurements can be carried out in fairly dilute solutions. Herlem and Popov (21) examined

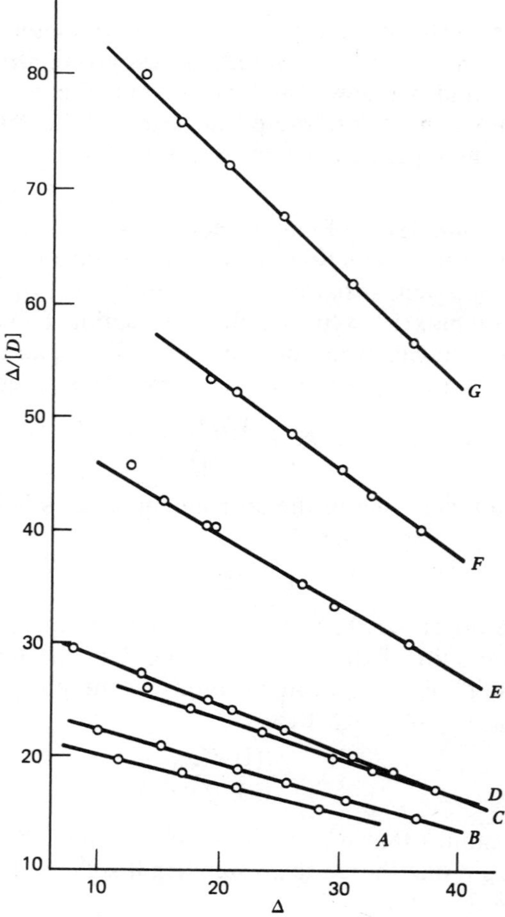

Figure 8.14 Plots of Δ/[D] *vs* Δ for the complexes of 1,4-dinitrobenzene with (*A*), benzene; (*B*), toluene; (*C*) *p*-xylene; (*D*), mesitylene; (*E*), durene; (*F*) pentamethylbenzene; (*G*) hexamethylbenzene in carbon tetrachloride at 33.5°C. Reprinted with permission from *Trans. Faraday Soc.*, **61,** 1626 (1965).

the chemical shifts of the sodium-23 nucleus in sodium iodide and tetraphenylborate solutions as a function of concentration in the strongly basic solvents, liquid ammonia, ethylenediamine, ethylamine, isopropylamine, *tert*-butylamine, hydrazine, and 1,1,3,3-tetramethylguanidine. The aim of this study was to examine the nature of the ionic species in liquid ammonia. The plots of the chemical shifts of ^{23}Na with respect to aqueous saturated sodium chloride solutions as a function of electrolyte concentration are given in Fig. 8.15. The strong solvating ability of these solvents is reflected by all the chemical shifts occurring downfield from the reference. The chemical shifts for the two salts in hydrazine, ammonia, and ethylenediamine solutions are approximately the same at the lower concentrations. This result appears to indicate that, in dilute sodium iodide solutions (≤1 M), the cation is completely solvated by ammonia, ethylenediamine, and hydrazine, while in the other solvents, contact ion pairs are formed to some extent. In the case of solutions of sodium tetraphenylborate, no evidence for contact ion pairs was found. In an earlier study, Erlich and Popov (20) examined the influence of different anions on the ^{23}Na chemical shifts. The salts sodium tetraphenylborate, perchlorate, iodide, and thiocyanate were used as probes and it was readily observed that the nature of the anion had a marked influence on the chemical shift. For sodium iodide and sodium thiocyanate solutions, the chemical shifts were concentration dependent with the degree of dependence being roughly an inverse function of the dielectric constant of the medium. In contrast, the ^{23}Na resonances of the tetraphenylborate and perchlorate solutions did not show a concentration dependence in the concentration range of 0.5–0.1 *M*. The results were interpreted in terms of the relative ease with which the iodide and thiocyanate ions form contact ion

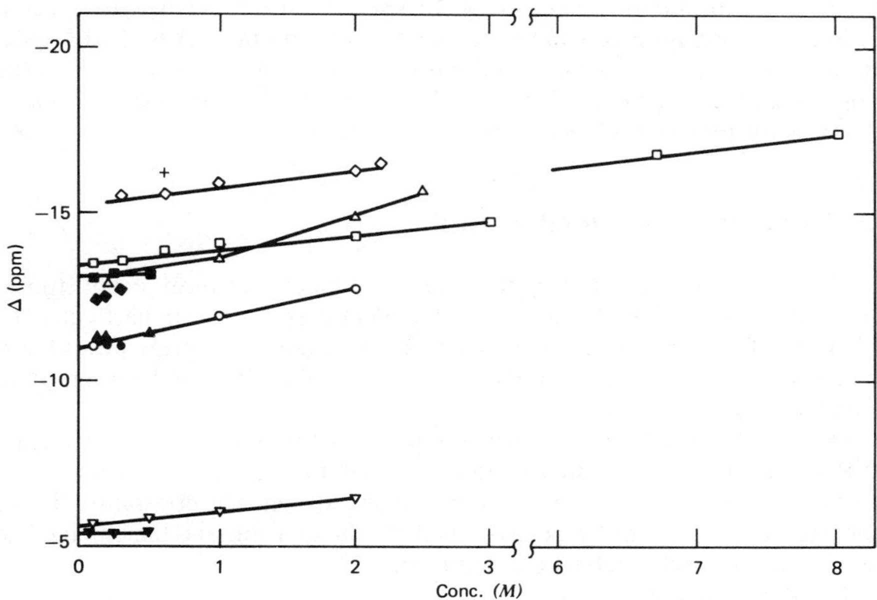

Figure 8.15 Variation in the ^{23}Na chemical shifts with concentration in basic solvents: (+) NaI in *tert*-butylamine; (◇) NaI in isopropylamine; (◆) $NaPh_4B$ in isopropylamine; (□) NaI in ammonia; (■) $NaBPh_4$ in ammonia; (△) NaI in ethylamine; (▲) $NaBPh_4$ in ethylamine; (○) NaI in ethylenediamine; (●) $NaBPh_4$ in ethylenediamine; (▽) NaI in hydrazine; (▼) $NaBPh_4$ in hydrazine. Reprinted with permission from *J. Amer. Chem. Soc.*, **94,** 1431 (1972). Copyright 1972. American Chemical Society.

pairs by replacing a solvent molecule in the inner solvation shell of a metal ion, whereas such a tendency is largely absent in the large symmetrical anions, such as perchlorate or tetraphenylborate. In addition, the magnitude of the chemical shift for the ^{23}Na nucleus in different solvents appears to be directly related to the electron-donor abilities of these solvents, provided that the shifts are not complicated by the formation of contact ion pairs.

In summary, we have discussed some representative studies of ion–solvent and ion–ion interactions in which nuclear magnetic resonance was employed as the experimental tool. By far the most widely applied has been the proton NMR, which has been successful in determining such solution properties as the following: (1) Solvent coordination numbers for ions, when separate signals could be observed for bulk solvent molecules and those in the first solvation shell. In related studies, the rates of exchange of solvent molecules in the solvation shells of cations and the effect of anions on these processes have been determined. (2) Preferential solvation of solutes by one component of a solvent mixture, which has been demonstrated either by the chemical-shift method or by the effect of a paramagnetic solute on the transverse relaxation time of the solvent nuclei. (3) Ion-pair association constants, from the concentration dependence of NMR spectra on the electrolyte concentration. (4) Formation constants for charge-transfer complexes. NMR spectra of solutes have been correlated with ionic size and shape as well as the properties of the solvents. Other nuclei, such as ^{35}Cl and ^{23}Na, have also served as probes in studies of ionic solvation.

8.5 ELECTRON SPIN RESONANCE

Electron spin resonance (ESR) has been the subject of several recent studies which have focused on the behavior of solvent systems and in particular the solvation of paramagnetic solute species. The theoretical principles underlying ESR have been thoroughly treated in several reviews and authoritative texts (23–27).

One of the most important aspects in the resolution of an experimental ESR spectrum of an organic free radical in solution is the recognition of the isotropic hyperfine splitting and the associated splitting constants. These splitting constants can be related to the spin densities at the nucleus of interest and at each nucleus bonded to it.

For the fragment

$$\begin{matrix} B_2 \diagdown & \\ & D - B_1 \\ B_3 \diagup & \end{matrix}$$

$$a^{\mathrm{d}} = [S^{\mathrm{d}} + \sum Q^{\mathrm{d}}_{\mathrm{db}_i}]\, \rho^{\pi}_{\mathrm{d}} + \sum Q^{\mathrm{d}}_{\mathrm{b}_i\mathrm{d}}\, \rho^{\pi}_{\mathrm{b}_i} \tag{8.16}$$

where a is the hyperfine splitting constant, S and Q are constants, and ρ^{π}_{i} is the spin density at the ith atom.

Another important concept in ESR is that of the g factor, as this is an indication of the magnetic field at which the unpaired electron will resonate with the fixed frequency microwaves of an ESR spectrometer. The factor g is given by

$$g = \frac{3}{2} + \frac{S(S+1) - L(L+1)}{2J(J+1)} \tag{8.17}$$

where S is the spin angular momentum of the electron, L is the orbital angular momentum, and J is the total angular momentum. For many aromatic radicals, $g = 2$ (free spin) because the unpaired electron has no orbital angular momentum. However, in practice the situation is complicated by spin–orbit coupling and various approaches for calculations of g values for organic radicals have been postulated and are described in most texts on this subject. In this chapter, some selected examples of the use of ESR for studying nonaqueous solvents are discussed to convey the approach used in this method and the type of results that can be expected. A fairly comprehensive review of ESR studies in nonaqueous solvents is that by Gough (28).

8.5.1 Solvent Effects on Hyperfine Splitting Constants Symons and Gough with their associates (29–32) have studied the ESR behavior of several quinones in various solvents. For example, one study examined the spectrum of monoprotonated durosemiquinone. It was found that two different methyl

Table 8.13 Hyperfine Splitting Constants of the Electron Spin Resonance Spectrum of Monoprotonated Durosemiquinone in Various Solvents[a]

Solvent	Z	$\lvert a \rvert_{-\mathrm{OH}}$	$a_{\mathrm{Me(I)}}$	$a_{\mathrm{Me(II)}}$	$\frac{1}{2}[a_{\mathrm{Me(I)}} + a_{\mathrm{Me(II)}}]$	$\frac{a_{\mathrm{Me(I)}}}{a_{\mathrm{Me(II)}}}$
Benzene	60	0.38	+5.49	−1.15	+2.17	−4.77
Tetrahydrofuran	63	0.27	+5.24	−1.04	+2.10	−5.03
1,4-Dioxane	64	0.38	+5.38	−1.08	+2.15	−4.99
Acetone	65.7	0.42	+5.20	−0.99	+2.10	−5.26
t-Pentanol	76	0.10	+4.91	−0.79	+2.06	−6.22
Isopropanol	76.3	0.22	+4.83	−0.68	+2.07	−7.09
Ethanol	79.6	0.38	+4.80	−0.68	+2.06	−7.06
Methanol	83.6	0.62	+4.80	−0.62	+2.09	−7.75
tri-*n*-Butyl phosphate		0.10	+5.04	−0.96	+2.04	−5.25
Sulfolane		0.63	+5.20	−1.02	+2.09	−5.10
Acetone–methanol mixtures						
(i)		0.48	+5.12	−0.93	+2.08	−5.45
(ii)		0.53	+4.96	−0.80	+2.08	−6.21
(iii)		0.53	+4.90	−0.71	+2.08	−6.91

[a] All splittings are quoted in gauss and are accurate to ±0.05 G.
Reprinted with permission from *Trans. Faraday Soc.*, **62**, 279 (1966).

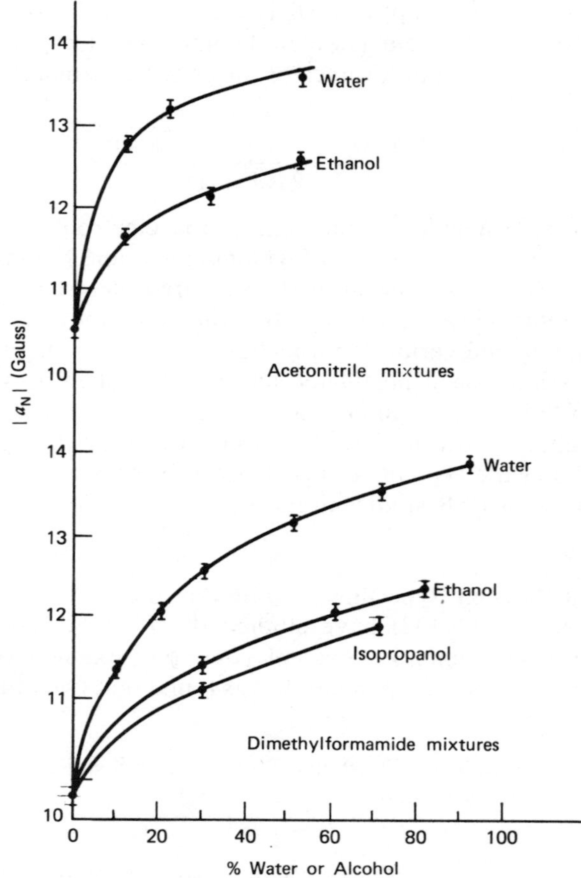

Figure 8.16 Solvent dependence of a_N in nitrobenzene anion. Reprinted with permission from *J. Amer. Chem. Soc.*, **86,** 4568 (1964). Copyright 1964. American Chemical Society.

splittings were observed. The general aim of the study was to examine the variation of the splitting constants with solvent and compare the results with predictions made from valence-bond calculations. The hyperfine splitting constants of the ESR of the quinone in various solvents are given in Table 8.13, which indicates the solvent dependence of the splittings. It is seen that as one set of methyl group hyperfine splittings decreases, the other increases.

Ludwig et al. (33) undertook a study to investigate the effect of various protic solvents on the nitrogen coupling constant (a_N) of some aromatic and aliphatic nitro radical anions that were generated electrolytically *in situ*. The results are summarized in Fig. 8.16, which shows the solvent dependence of a_N in the nitrobenzene anion. It is seen that the value of a_N increases markedly with increasing water or alcohol content. The solvent effect on a_H is less pronounced. It is observed that, in the region where the concentrations of water or alcohol are small, the greatest change in a_N with solvent depends very much on the particular solvent, as shown in Fig. 8.17. For *o*-nitrobenzoic acid in ethanol, there is very little change with solvent, while in the case of nitromesitylene, there is a pronounced change. Hydrogen bonding plays a significant role in the interaction between a radical ion and the solvent, and it is observed experimentally that, for those cases when the solvent can act as a proton donor, a large variation of a_N occurs, as the H-bonding is much stronger than a normal solvation interaction. In addition,

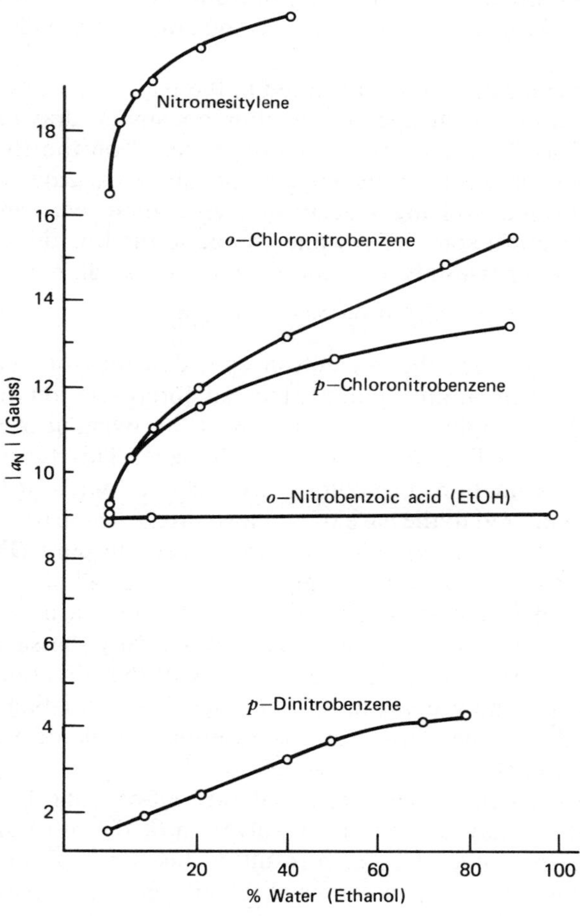

Figure 8.17 Solvent dependence of a_N for substituted nitrobenzene anions. Reprinted with permission from *J. Amer. Chem. Soc.*, **86,** 4568 (1964). Copyright 1964. American Chemical Society.

the radical ion must act as an acceptor to produce large changes in coupling constants. An illustration of the hydrogen-bonding effect can be seen in Fig. 8.17 for solvent effects on *ortho*-substituted nitrobenzene anions where intra-H-bonding is possible. The increased localized charge on the nitro group of the anion radical makes it a good candidate as an acceptor in a H-bond pair. For example, no shift in the value of a_N for *o*-nitrobenzoic acid is seen in going from pure DMF to 100% ethanol. When *o*-nitrobenzoic acid is electrolytically reduced in ethanol, the ESR spectrum corresponds to the monoanion with the intra-H-bonded carboxylate proton retained, as the spectrum yields a proton doublet with a rather large a_H. On the other hand, when the electrolysis is carried out in water or D_2O the doublet splitting of the carboxylic acid proton disappears. For *p*-nitrobenzoic acid, where intra-H-bonding is not possible, the values of a_N shift markedly with additions of alcohol or water.

An interesting discussion is presented in the paper by Ludwig et al. (33) on the interpretation of changes in coupling constants, drawing also upon some earlier ideas (34). The interpretation is based on the formation of a localized complex between the solvent and the functional group of the radical anion. Because rapid exchanges occur in such complexes, a marked redistribution of π-electron spin density and charge in the functional group may occur. A useful expression formulated by Fraenkel et al. (35, 36) is

$$a_N = 99.0\rho_N^\pi - 71.6\rho_O^\pi \tag{8.18}$$

where the ρ^π values are the π-electron spin densities on the respective nitrogen and oxygen atoms. One suggested interpretation for the large change in the ^{14}N coupling constant involved the twisting of the solvated nitro group around a line in the plane of the ring. This twisting effect is illustrated by examination of *o*-chloronitrobenzene, where a large change in a_N occurs, compared to the case of *p*-chloronitrobenzene (6.5 G compared to 4.2 G, where G = gauss). For *p*-dinitrobenzene in pure DMF, the two equivalent ^{14}N nuclei show a relatively small a_N = 1.47 G, whereas when large quantities of water are added, the a_N value increases to about 4 G. However, the ring proton couplings remain practically the same. The conclusion to this observation is that the formation of the solvation complex in this case does not show evidence of any significant twisting of the nitro group, provided that the nitro group is essentially isolated spatially from other ring substituents.

A further model proposed concerned the out-of-plane bending of the nitro group, which may be caused by solvation of the nitro anion radical. The increase in the value of a_N as a result of this bending mode is thought to be due to an increasing *S* character of the nitrogen. This particular effect could be pronounced in the case of the nitrobenzene anion and other sterically unhindered nitro anions (e.g., *para*-substituted nitrobenzenes).

In contrast to the aromatic nitro compounds, the ^{14}N couplings of aliphatic nitro compounds were found to be practically solvent independent.

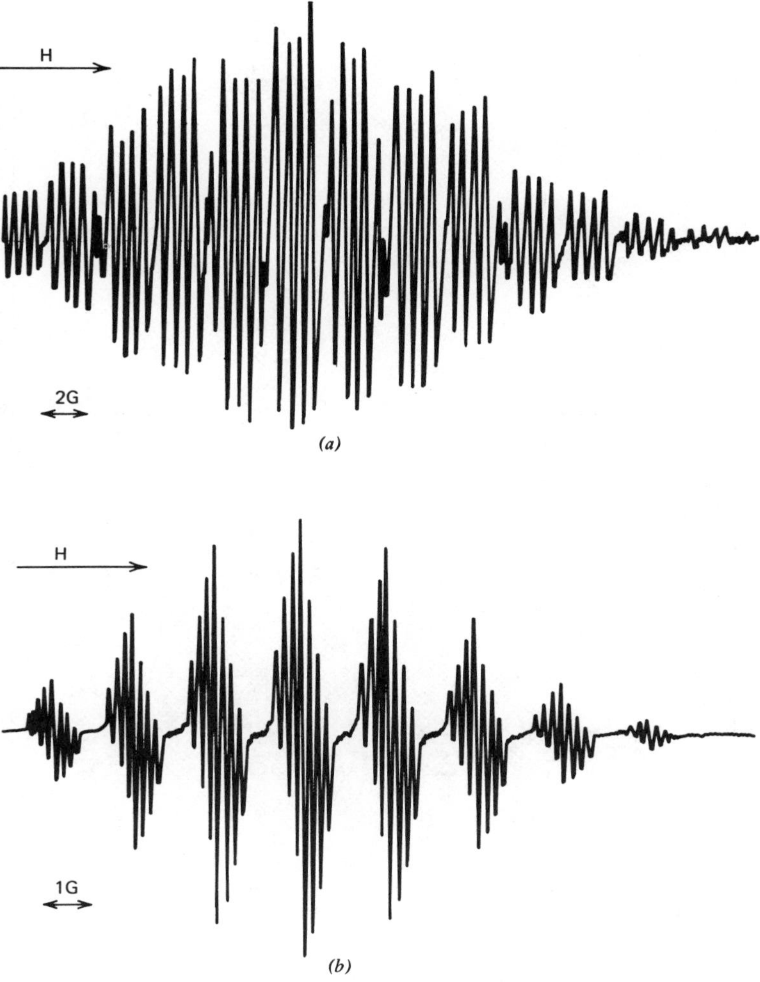

Figure 8.18 (a) ESR spectrum of the solution obtained after the reduction of duroquinone by sodium in tetrahydrofuran; sample temperature $-20°C$. (b) ESR spectrum of the solution obtained after a similar reduction in the presence of sodium tetraphenylborate; sample temperature $-20°C$; the low intensity lines between the main groups are caused by a ^{13}C splitting of 1.42 G from the methyl groups. Reproduced by permission of the National Research Council of Canada from the *Canadian Journal of Chemistry,* **47,** 1698 (1969).

The reasons for this behavior are not well understood, but probably the unrestricted rotation of the nitro groups about the CN bond for the aliphatic compounds plays a significant role, which means that an aliphatic nitro compound should be relatively insensitive to distortions of molecular geometry which might accompany solvation.

Table 8.14 Coupling Constants of Hydrocarbon Anion Radicals

Compound	Metal	Solvent	Temp. (°C)	a_1	a_2	a_{other}
Naphthalene (positions 1, 2)	K	NH_3	−50	5.02	1.88	
	K	THF	Room temp.	4.95	1.87	
	Na	THF	+22.5	4.90	1.83	1.05 (Na)
Anthracene (positions 9, 1, 2)	K	NH_3	−50	2.78	1.51	5.56
	K	DME	—	2.726	1.513	5.346
	K	THF	Room temp.	2.74	1.51	5.56
Benzophenone (positions 1, 2, 3)	K	NH_3	−50	2.63	0.84	3.48
	K	DME	+20	2.53	0.84	3.45, 0.39 (K)
	Na	DME	+20	2.58	0.86	3.44, 1.125 (Na)
Cyclooctatetraene	K	NH_3	+22	3.28		
	Li	THF	+22	3.209		

8.5.2 Detection of Triple Ions ESR finds a useful application in the detection of the presence of triple ions in solution. As an example, Gough and Hindle (37) discovered the triple ion (durosemiquinone–Na_2)$^+$ resulting from the reduction of duroquinone by sodium in tetrahydrofuran in the presence of sodium tetraphenylborate. The ESR spectrum obtained with and without the use of sodium tetraphenylborate is shown in Fig. 8.18. The spectrum in Fig. 8.18a may be analyzed in terms of six equivalent protons with a hyperfine splitting constant of 2.66 G, six equivalent protons with a splitting of 1.20 G, and one sodium with a splitting of 0.25 G. For the 1:1 molar mixture of duroquinone and sodium tetraphenylborate (Fig. 8.18b), the spectrum is analyzed in terms of 12 equivalent protons with a hyperfine splitting constant of 1.94 G, and two equivalent sodium nuclei with a splitting of 0.17 G. This spectrum is attributed to a triple ion rather than an ionic quadrupole, because the addition of sodium ions in excess of those introduced by the reduction process is necessary for the formation of this species. In terms of structure, the triple ion was considered as a durosemiquinone anion, which was ionically associated with a sodium cation at each oxygen, so that all four methyl groups were equivalent. Additional evidence for the existence of the triple ion is given by the smaller sodium splitting observed when two cations interact, as each cation of the triple ion will be attracted to a lesser extent than the cation of the ion pair.

8.5.3 Anion Radicals The study of anion radicals in liquid ammonia was the subject of an investigation by Smentowski and Stevenson (38). Anion radicals were generated by the reduction of organic substances with alkali-metal solutions in liquid ammonia and the radicals were observed by ESR. The coupling constants of several representative anion radicals are given in Table 8.14. For anion radicals in liquid ammonia, the coupling constants of the hydrocarbon anion radicals vary by only about 5% from the values obtained in aprotic solvents such as DMF and THF. In contrast, the coupling constants of anion radicals with functional groups are very solvent depend-

Table 8.15 Coupling Constants of Nitrobenzene Anion Radicals

Metal or method of preparation	Solvent	Temp. (°C)	a_N	a_o	a_m	a_p
Li	NH_3	−50	10.40	3.25	1.03	3.83
Na	NH_3	−50	10.3	3.24	1.03	3.82
K	NH_3	−50	10.62	3.23	1.03	3.84
Alkaline dithionite	H_2O	Room temp.	13.30	3.40	0.90	3.40
Electrolytic	CH_3CN	—	10.32	3.39	1.09	3.97
Electrolytic	DMF	—	10.33	3.46	1.13	3.86
Electrolytic	DMSO	—	9.87	3.37	1.07	4.02

Reproduced by permission of the National Research Council of Canada from the *Canadian Journal of Chemistry*, **47**, 1698 (1969).

ent. In Table 8.15 are the coupling constants of the nitrobenzene anion radical formed both chemically and electrolytically in various solvents. The effects of ion pairing on the ESR spectral parameters of the anion radical were very minor.

To summarize, electron spin resonance spectroscopy has found application as an experimental tool for the study of paramagnetic solute species, particularly organic free radicals and radical anions. Solvent effects on the hyperfine-splitting, or -coupling, constants of radical anions have been interpreted in terms of complexes formed with the solvent in general and with H-bonding in some systems. Anion radicals generated in liquid ammonia and triple ions with unpaired electrons have been some of the systems studied by ESR.

LITERATURE CITED

Nuclear Magnetic Resonance

1 Hinton, J. F., and E. S. Amis, *Chem. Rev.*, **67,** 367 (1967).

2 Lantzke, I. R., "Spectroscopic Measurements," Part 1: "Electronic Absorption Spectroscopy," in *Physical Chemistry of Organic Solvent Systems,* A. K. Covington and T. Dickinson, Eds., Plenum Press, London and New York, 1973, Chap. 4.

3 Jackman, L. M., and S. Sternhell, *Nuclear Magnetic Resonance Spectroscopy in Organic Chemistry,* 2nd ed., Pergamon Press, Oxford, 1969.

4 Roberts, J. D., *Nuclear Magnetic Resonance. Application to Organic Chemistry,* McGraw-Hill, New York, 1959.

5 Roberts, J. D., *An Introduction to the Analysis of Spin-Spin Splitting in High Resolution Nuclear Magnetic Resonance Spectra,* Benjamin, New York, 1961.

6 Mathieson, D. W., Ed., *Nuclear Magnetic Resonance for Organic Chemists,* Academic Press, London, 1967.

7 Abragam, A., *The Principles of Nuclear Magnetism,* Oxford University Press, London, 1961.

8 Fratiello, A., R. E. Lee, V. M. Nishida, and R. E. Schuster, *J. Chem. Phys.*, **47,** 4951 (1967).

9 Frankel, L. S., C. H. Langford, and T. R. Stengle, *J. Phys. Chem.*, **74,** 1376 (1970).

10 Langford, C. H., and T. R. Stengle, *J. Amer. Chem. Soc.*, **91,** 4014 (1969).

11 Schneider, H., and H. Strehlow, *Z. Physik. Chem. (Frankfurt)*, **49,** 44 (1966).

12 Buckson, R. L., and S. G. Smith, *J. Phys. Chem.*, **68,** 1875 (1964).

13 Nakamura, S., and S. Meiboom, *J. Amer. Chem. Soc.*, **89,** 1765 (1967).

14 Luz, Z., and S. Meiboom, *J. Chem. Phys.*, **40,** 1058 (1964).

15 Luz, Z., and S. Meiboom, *J. Chem. Phys.*, **40,** 1066 (1964).

16 O'Brien, J. F., and M. Alei, Jr., *J. Phys. Chem.*, **74,** 743 (1970).

17 Taylor, R. P., and I. D. Kuntz, Jr., *J. Amer. Chem. Soc.*, **92,** 4813 (1970).

18 Laszlo, P., "Solvent Effects and Nuclear Magnetic Resonance," in *Progress in Nuclear Magnetic Resonance Spectroscopy,* Vol. III, J. W. Emsley, J. Feeney, and L. N. Sutcliffe, Eds., Pergamon Press, London, 1967, Chap. 6.

19 Foster, R., and C. A. Fyfe, *Trans. Faraday Soc.,* **61,** 1626 (1965).

20 Erlich, R. H., and A. I. Popov, *J. Amer. Chem. Soc.,* **93,** 5620 (1971).

21 Herlem, M., and A. I. Popov, *J. Amer. Chem. Soc.,* **94,** 1431 (1972).

22 Bloor, E. G., and R. G. Kidd, *Can. J. Chem.,* **46,** 3425 (1968).

Electron Spin Resonance

23 Carrington, A., *Quart. Rev.,* **17,** 67 (1963).

24 Atherton, N., *Sci. Prog.,* **56,** 179 (1968).

25 Bersohn, M., and J. C. Baird, *An Introduction to Electron Paramagnetic Resonance,* Benjamin, New York, 1966.

26 Ayscough, P. B., *Electron Spin Resonance in Chemistry,* Methuen, London, 1967.

27 Carrington, A., and A. D. McLachlan, *Introduction to Magnetic Resonance,* Harper and Row, New York, 1967.

28 Gough, T. E., "Spectroscopic Measurements" Part 3: "E. S. R. Spectroscopy," in *Physical Chemistry of Organic Solvent Systems,* A. K. Covington and T. Dickinson, Eds., Plenum Press, New York, 1973, Chap. 4.

29 Claxton, T. A., T. E. Gough, and M. C. R. Symons, *Trans. Faraday Soc.,* **62,** 279 (1966).

30 Gough, T. E., *Trans. Faraday Soc.,* **62,** 2321 (1966).

31 Claxton, T. A., J. Oakes, and M. C. R. Symons, *Trans. Faraday Soc.,* **63,** 2125 (1967).

32 Oakes, J., and M. C. R. Symons, *Trans. Faraday Soc.,* **64,** 2579 (1968).

33 Ludwig, P., T. Layloff, and R. N. Adams, *J. Amer. Chem. Soc.,* **86,** 4568 (1964).

34 Gendell, J., J. H. Freed, and G. K. Fraenkel, *J. Chem. Phys.,* **37,** 2832 (1962).

35 Freed, J. H., P. H. Rieger, and G. K. Fraenkel, *J. Chem. Phys.,* **37,** 1881 (1962).

36 Rieger, P. H., and G. K. Fraenkel, *J. Chem. Phys.,* **39,** 609 (1963).

37 Gough, T. E., and P. R. Hindle, *Can. J. Chem.,* **47,** 1698 (1969).

38 Smentowski, F. J., and G. R. Stevenson, *J. Amer. Chem. Soc.,* **90,** 4661 (1968).

CHAPTER NINE

Electrode Processes

9.1 VOLTAMMETRY

9.1.1 Introduction Voltammetry encompasses a variety of related techniques in which current is measured as a function of the applied potential in electrolysis cells under conditions where only the surface layer at the indicator electrode is depleted of electroactive material. These conditions are achieved when the electrolysis is carried out under controlled conditions of diffusion or convection, the electrode at which oxidation or reduction takes place (indicator, or working electrode) has a very small area, the reference electrode is relatively large, and the electroactive species (usually at millimolar concentrations) is electrolyzed in the presence of about a hundredfold excess of an inert supporting electrolyte. The last requirement ensures that the electroactive species does not contribute measurably to the migration of current through the solution, so that it is transported to the electrode surface primarily by diffusion or controlled convection. In voltammetry we measure the current as a function of a systematically varied potential applied to the indicator electrode.

Indicator electrodes in common use in voltammetry are the dropping mercury electrode (DME) and the rotating electrodes (e.g., platinum or carbon). The reference electrodes are usually the saturated calomel electrode (SCE) or the mercury pool. When the DME is employed, the technique is known as *polarography*. It includes classical, or direct current (DC) polarography, amperometric titrations, derivative polarography, single-sweep, or oscillographic polarography, alternating-current (AC) polarography, cyclic voltammetry, and a variety of other modern techniques. It is beyond the scope of this book to present a general discussion of the theory, instrumentation, methodology, and the wide range of applications for these techniques. The reader is alerted to some excellent monographs and chapters on the subject cited under "General References" at the end of this chapter. Modern developments can be followed in the multivolume serial publications *Progress in Polarography* (1), *Advances in Electrochemistry and Electrochemical Engineering* (2), and *Electroanalytical Chemistry* (3). General literature reviews on both theory and applications can be found in the biennial Fundamental Reviews in the April issues of *Analytical Chemistry*.

In this chapter we intend to deal only with those voltammetric phenomena where the change from aqueous to nonaqueous medium causes a significant change in behavior, where physicochemical events occur in the nonaqueous medium that could not occur in water and where a comparison of electrochemical properties in different solvents is involved.

What are the voltammetric parameters that might show a solvent dependence? In classical polarography the current–voltage curve yields two parameters: the diffusion current, i_d, and the half-wave potential, $E_{1/2}$. The diffusion current is the limiting current on the plateau of the current–voltage curve (polarogram). The half-wave potential is the value of the potential of the indicator electrode at the point where the current is equal to one half the diffusion current. From standard texts on polarography we recall that the average diffusion current, i_d, at the DME is governed by the Ilkovič equation:

$$i_d = 607nD^{1/2}Cm^{2/3}t^{1/6} \tag{9.1}$$

In the above equation the current is in microamperes, n is the number of electrons transferred to or from the diffusing species at the electrode, D is the diffusion coefficient (in cm^2 sec^{-1}), C is the concentration of the electroactive species (in mmol liter^{-1}), m is the rate of flow of mercury (in g sec^{-1}), and t is the lifetime (in sec) of a mercury drop. Thus, under specified experimental conditions, the diffusion current is directly proportional to the concentration of the electroactive species, so that the Ilkovič equation forms the basis for quantitative polarographic analysis. However, it is not the current–concentration relationship that would normally change with the solvent. Instead, we would expect some solvent effect on the diffusion coefficient, D, as it is inversely proportional to the viscosity of the medium. Furthermore, the nature of the medium might alter the reaction mechanism in such a way as to change the value of n for the electrode process.

The voltammetric parameter which is expected to experience greater and fundamentally more important solvent effects is the half-wave potential. For a reversible reaction in which both the oxidized and the reduced forms are soluble in the solvent, but not in mercury, the (reduction) half-wave potential is related to the corresponding standard potential, $E°$, by the expression

$$E_{1/2} = E° + \frac{\mathrm{RT}}{\mathrm{nF}} \ln \frac{D_{\mathrm{red}}^{1/2} f_{\mathrm{ox}}}{D_{\mathrm{ox}}^{1/2} f_{\mathrm{red}}} \tag{9.2}$$

where f_{ox} and f_{red} are the activity coefficients of the oxidized and the reduced forms of the electroactive species and the D's are the corresponding diffusion coefficients. It is a common assumption that the logarithmic term in Eqn. 9.2 can be neglected because of cancellations, so that

$$E_{1/2} \cong E° \tag{9.3}$$

We recall from Chapter 5 that for a given oxidation–reduction process the standard free energy change, $\Delta G°$, is related to the corresponding $E°$ via

$$\Delta G° = -nFE° \tag{9.4}$$

where F is the Faraday. It follows (as will be shown explicitly later, see Eqns. 9.36–9.39) that the difference between the standard potentials in two solvents is a measure of the difference between the solvation energies of the electroactive species in the two solvents, or its transfer free energy. For example, in the case of a metal ion M^{n+} being reduced to the metal, the transfer free energy, $\Delta G_t^\circ(M^{n+})$, is given by

$$\Delta G_t^\circ(\mathrm{M}^{n+}) \equiv {}_{\mathrm{s}}G_{\mathrm{M}}^\circ - {}_{\mathrm{w}}G_{\mathrm{M}}^\circ = \frac{nF}{2.3RT}({}_{\mathrm{s}}E_{\mathrm{M}^{n+}/\mathrm{M}}^\circ - {}_{\mathrm{w}}E_{\mathrm{M}^{n+}/\mathrm{M}}^\circ) \tag{9.5}$$

where subscripts s and w denote nonaqueous solvent and water, respectively (see Eqns. 5.15 and 5.24). Frequently, however, values of standard potentials are unavailable or unobtainable for certain redox systems, for example, when one of the forms is unstable in solution. In such cases the voltammetric half-wave potentials are used as approximations of the corresponding $E°$'s (Eqn. 9.3). Thus, the $E_{1/2}$'s are another experimental measure of the solvation energies and transfer free energies of electroactive ions and molecules.

A different type of information can be obtained from AC polarography. The latter is a technique in which an AC voltage of small amplitude (a few millivolts) is superimposed on the DC potential applied to the indicator electrode. When the AC current is measured as a function of the applied DC voltage, a polarogram is obtained which has a symmetrical peak rather than an inflection point at the $E_{1/2}$ and resembles a derivative of the DC polarogram (Fig. 9.1). When the rate of electron transfer is very high compared to the rate of diffusion, rate constants for the transfer of electrons between an electrode and an electroactive species can be derived from AC polarograms. The subject is much too specialized to be developed here, but the rate constants are evaluated from the frequency dependence of the impedance of the DME at the half-wave potential. The rate constants have a direct bearing on the reversibility of electrode processes and represent an important datum in the kinetics of electrode reactions that do vary with the solvent medium.

It is not difficult to see why nonaqueous solvents have expanded the scope of voltammetry as they did the scope of other areas of solution chemistry because the reasons are usually the same. By changing the solvent it is possible to vary the dielectric constant, the proton-transfer and the coordination ability of the medium as well as a variety of other properties, which we collectively call the solvation ability. Moreover, by resorting to solvent mixtures, it is possible to vary these properties in a continuous man-

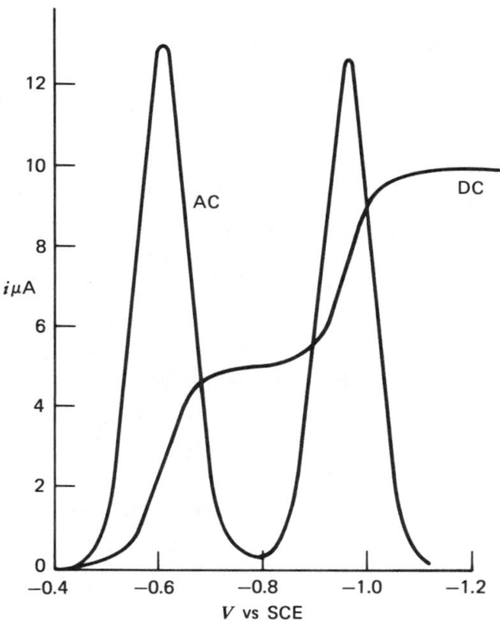

Figure 9.1 DC and AC polarograms of *p*-dinitrobenzene in dimethylformamide, 0.1 *M* NEt_4ClO_4, illustrating fast electron transfer for successive one-electron steps; AC frequency 35 cps Univector. Reprinted with permission from A. J. Bard, Ed., *Electroanalytical Chemistry, A Series of Advances*, Vol. 2, Marcel Dekker, New York, 1967, p. 9, by courtesy of Marcel Dekker, Inc.

ner, thus observing various degrees of intermediate behavior. In organic polarography nonaqueous liquids were first introduced in mixtures with water in order to increase the solubility of organic and metal–organic compounds. The choice of solvents for voltammetry included first, amphiprotic liquids, such as the simple alcohols, glycols, acetic acid, and the Cellosolves (monoalkyl ethers of ethylene glycol), in aqueous mixtures or in the anhydrous state. Later, mixtures of amphiprotic with aprotic solvents, such as dioxane–water and benzene–methanol, and eventually the pure aprotic liquids—acetonitrile, pyridine, sulfolane, dimethylsulfoxide (DMSO), and the formamides—found increasing application.

It is the change from water to aprotic solvents in particular that expanded significantly the working range of voltammetry and the type of information that can be obtained from it. The advantages go far beyond solubility considerations. One is the compatibility of the solvent with the solute species of interest. Some substances that are unstable in aqueous solution (alkali metals, free radicals) can be studied in aprotic solvents. The other advantage is the extended potential range that is offered by aprotic solvents. In any solvent the range of measurable electrode potentials is limited by the oxidation and reduction of the solvent or of the supporting electrolyte. Quaternary ammonium salts are popular supporting electrolytes because of their electrochemical stability and solubility in both aqueous and nonaqueous media. At the DME in aqueous solutions the cathodic potential range is effectively limited by the reduction of quaternary ammonium ions, which in basic solutions occurs at about -2.6 V vs the SCE. On the anodic side, the polarographic range in aqueous solution is limited by the oxidation of the mercury electrode, which occurs at about $+0.4$ V in noncomplexing media. At a platinum electrode, however, the usable voltage range in aqueous solutions extends only from about $+1$ to -1 V (vs the SCE). The considerably longer cathodic range at the DME is the result of the high overvoltage for the evolution of hydrogen on mercury. Overvoltage is the difference between the actual potential at which the evolution occurs and the thermodynamic or equilibrium value of the potential that would be predicted from the Nernst equation. Aprotic solvents may be less easily oxidized or reduced than water and their potential ranges tend to be longer. Examples of voltammetric ranges for some aprotic solvents are shown in Table 9.1. Properties of nonaqueous solvents relevant to their use in electrochemistry were reviewed by Mann (4).

A further advantage of aprotic media is that the absence of any appreciable hydrogen-ion concentration avoids not only the limitation imposed on cathodic potentials by the evolution of hydrogen, but also eliminates the complications caused by the hydrogen ion in the electrode processes. For example, in aqueous solutions, all organic reductions occur with the participation of the H^+ ion, which is believed to be a major cause of the irreversibility of such processes, for example,

$$RNO_2 + 4H^+ + 4e \longrightarrow RNHOH + H_2O$$

$$CH_2{=}CHCN + 2H^+ + 2e \longrightarrow CH_3CH_2CN$$

Also illustrative of these types of processes are the general reactions represented by Eqns. 9.9–9.12 as well as the reduction of benzophenone to benzhydrol shown by Eqns. 9.17–9.20.

In aprotic media, the rate of charge transfer is usually very high, so that the same reduction processes are characterized by reversible one-electron waves. Accordingly, the thermodynamic significance of the half-wave potentials measured in aprotic solvents is generally greater than in water.

The products of reduction of organic substances are often radical anions, which could not exist in amphiprotic media, but have sufficiently long half-lives in aprotic solvents to be studied by electron spin resonance (ESR) spectroscopy. One important class of organic compounds that are reduced to relatively stable radical anions are aromatic hydrocarbons and their derivatives, such as quinones. For example, the half-life of the acetophenone radical anion in dimethylformamide (DMF) is about 20 sec. Another example of a species that could not exist in an amphiprotic medium is the superoxide anion, which can be prepared in aprotic media by the electrolytic reduction of oxygen. Oxidation of hydrocarbons to radical cations is also known.

The possiblity of generating in aprotic solvents relatively long-lived radical ions by electrolytic methods has led to a number of interdisciplinary studies involving voltammetry, ESR spectroscopy, and theoretical chemistry. Theoretical interest derives from the fact that, as is shown in Section 9.1.3, molecular energy levels are related to oxidation and reduction potentials. ESR spectroscopy is a highly sensitive technique for the study of spe-

Table 9.1 Voltammetric Potential Ranges of Some Aprotic Solvents

Solvent	DME[a] (V vs SCE)	Rotating Platinum (V vs SCE in 0.1 *M* $LiClO_4$)
Acetone	+0.7 to −2.5	—
Acetonitrile	+0.6 to −2.8	+1.8 to −1.5
Benzonitrile	+0.5 to −2.1	—
N,N-Dimethylformamide	+0.4 to −2.8	+1.7 to −1.9
Dimethylsulfoxide	+0.3 to −2.8	+0.7 to −1.9
Propylene carbonate	+0.6 to −2.9	+2.2 to −3.2
Sulfolane	+0.6 to −2.9	+2.7 to −1.9

Condensed from the compilation in Ref. 5.

[a] The supporting electrolyte, Et_4NClO_4, was 0.1 *M* in all solvents except for dimethylsulfoxide, where it was 1 *M*.

cies containing unpaired electrons (see Chapter 8), and it was natural to apply it to electrolytically generated radical ions. Such interdisciplinary studies have done much to elucidate the mechanism of electrode processes and to identify the nature of the intermediates and the products in the oxidation–reduction reactions.

However, the practice of voltammetry in nonaqueous solvents is not without its disadvantages and pitfalls. All the general considerations involved in choosing a solvent for any use apply to voltammetry as well. One has to consider its cost and toxicity, the need for purification, and susceptibility to contamination by atmospheric moisture and carbon dioxide. Also the viscosity and the liquid range of the solvent may be factors in the selection. Special precautions may be required in handling highly volatile solvents.

There are also problems specific to voltammetry. Nonaqueous solutions of conventional supporting electrolytes generally have higher resistances than the corresponding aqueous solutions, and in traditional polarography the resulting iR drop in solution caused the polarograms to be drawn out, that is, the voltage at the DME was partly a function of the current flowing through the solution. The higher resistance of nonaqueous solutions is partly a function of the lower solubilities of many electrolytes and partly due to ion pairing in solvents of lower dielectric constant. Of course, when the electrolyte is adequately soluble and the solvent has a dielectric constant of about 40 or higher, there is no reason why a nonaqueous solution cannot be highly conducting.

Another source of resistance in a polarographic cell is the salt bridge of the reference electrode. One way to reduce this source of error in the potential is to use the mercury pool as the reference electrode, instead of the SCE, but the potential of the mercury pool varies with the composition of the solution and must be referred in each case to the SCE. Nowadays, the iR drop is being minimized by introducing a third, current-carrying, electrode, and measuring the DME potential against the SCE placed very close to it in solution. Another complication common to all electrochemistry in nonaqueous solvents is the extent of ion pairing and the magnitude of the salt-effect activity coefficients, which are not always known in solvents of lower dielectric constant (see Chapter 4). Investigators frequently ignore these effects, or assume that they remain constant when solutions of different supporting electrolytes are compared in the same solvent or even in different solvents.

Most half-wave potentials are measured against the aqueous SCE regardless of the solvent in which the electrolysis is being carried out and thus include an unknown liquid-junction potential at the aqueous–nonaqueous boundary. It is generally assumed that the liquid-junction potential is reproducible and constant regardless of the nature of the supporting electrolyte, which is not always the case. In their efforts to compare the half-wave potentials measured in different solvents against the aqueous SCE, electrochemists began to wonder about the magnitude of the variable liquid-junction potential and, more generally, about the possibility of correcting or measuring the half-wave potentials in such a manner as to express them on a single solvent-independent scale. This problem in voltammetry is of course identical in principle with that of correlating the series of standard electrode potentials in different solvents, to which we address ourselves in Chapter 5. However, the methods by which electrochemists have attacked the problem are in some ways unique and deserve a separate review. In one approach the transfer free energies, ΔG_t°, for electroneutral pairs of ions have been determined from the changes in the $E_{1/2}$'s observed to accompany the transfer between two solvents (Eqn. 9.41). These were then split into ΔG_t° values for individual ions by means of an extrathermodynamic calculation, exemplified by Eqn. 9.48. Values of the transfer free energies for single ions thus estimated were then used to correlate aqueous and nonaqueous electrode potentials by expressing them on a single scale (Eqn. 9.5). Alternatively, certain redox couples based on metal–organic complexes (e.g., ferrocene–ferricinium, cobaltocene–cobalticinium) were proposed as electrodes whose potential would be independent of the solvent. These efforts have contributed greatly to our knowledge of relative solvation in different media and helped advance the field of transfer free energies.

9.1.2 The Rates and Mechanisms of Some Organic Reactions

Electrolytic Reduction of Aromatic Compounds In aprotic solvents many aromatic hydrocarbons and their derivatives are reduced electrolytically in two successive one-electron steps, of which the first, reversible, process leads to the formation of a radical anion, and the second, often irreversible, step produces the dianion

$$M + e \rightleftharpoons M^{\overset{\cdot}{-}} \tag{9.6}$$

$$M^{\overset{\cdot}{-}} + e \rightleftharpoons M^{2-} \tag{9.7}$$

As a result a polarogram of such a hydrocarbon solution consists of two one-electron waves, of which the second wave generally occurs at a potential about 0.5 V more negative than the first (Fig. 9.2). A more negative cathodic potential indicates a lesser tendency toward reduction, and it is not surprising that a radical anion should be more difficult to reduce than the neutral parent molecule.

These types of processes were investigated systematically by Hoijtink et al. (6–12) on solutions of polycyclic aromatic hydrocarbons in dioxane–water mixtures and confirmed by subsequent studies in DMF (13–16) and acetonitrile (16, 17). The lifetime of the radical anions was found to be long in comparison with the lifetime of the mercury drop, so that the potential remained well defined during the measurements. The stability of the anion

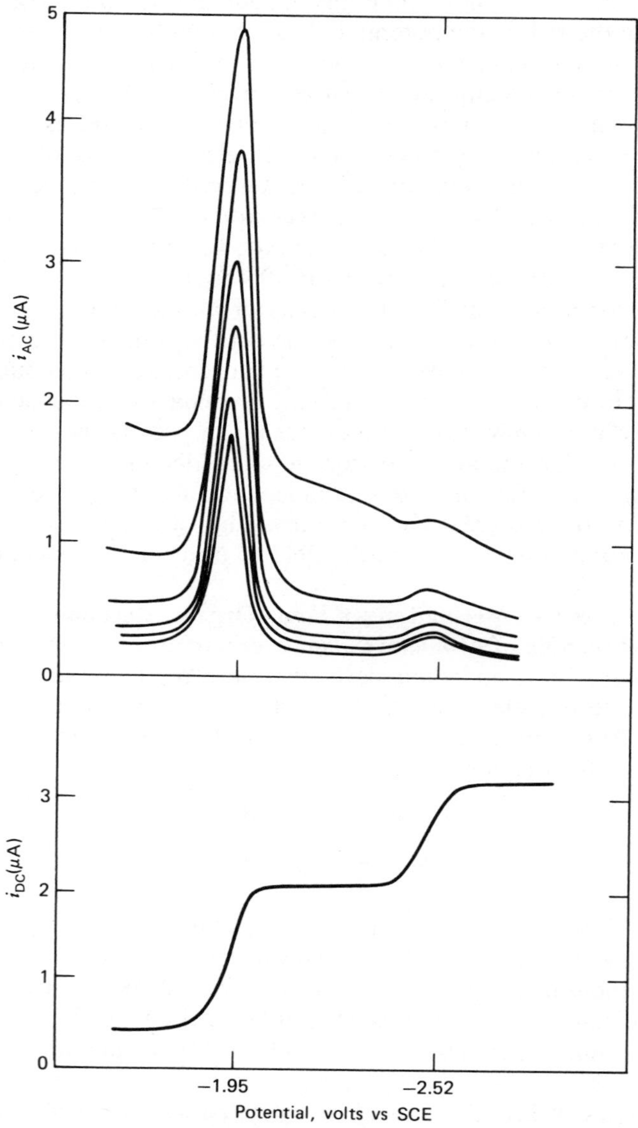

Figure 9.2 Direct-current and alternating-current polarograms of anthracene in dimethylformamide. Potential in volts vs SCE. Reprinted with permission from *Trans. Faraday Soc.*, **55**, 324 (1959).

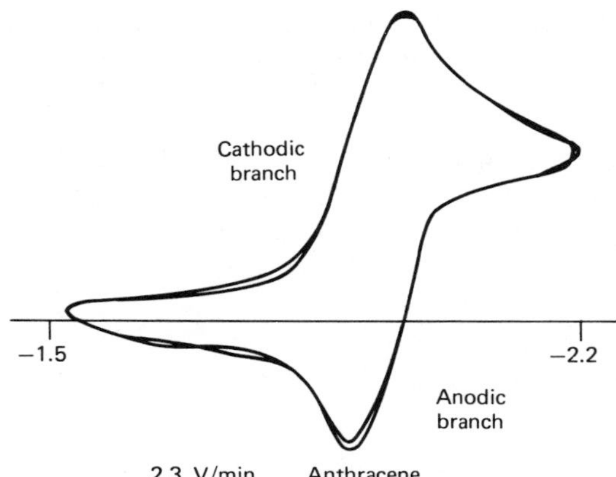

Figure 9.3 Cyclic voltammogram for the first reduction step of anthracene in acetonitrile. Hanging mercury-drop electrode, 0.1 *M* NEt_4ClO_4; frequency 0.1 cps. Potential in volts vs SCE. Reprinted with permission from *Polarography—1964*. G. J. Hills, Ed., Interscience, New York, 1966, p. 1017, by courtesy of John Wiley & Sons, Inc.

radicals and the reversibility of the reaction by which they were produced was demonstrated by a variety of techniques, among them most graphically by cyclic voltammetry (18). In this technique a hanging mercury drop or a rotating platinum electrode are used. A fast triangular-wave potential (changing linearly as a function of time) is applied to the electrode; in the case of the DME, the application lasts only over a lifetime of a single drop. The electroactive species is first reduced in a rapid forward sweep of potential and then reoxidized in the backward sweep, the potential returning to the starting point. If the process is reversible the anodic and cathodic current–voltage curves are virtually symmetrical, and the two peaks are separated by $80/n$ mV, each peak occurring $40/n$ mV beyond the $E_{1/2}$ value in the direction of the sweep. For irreversible processes, the locations of the cathodic and the anodic peaks are farther apart and their shapes are very different. In Fig. 9.3 we reproduce a cyclic voltammogram for the first reduction step of anthracene in acetonitrile:

$$\text{(anthracene)} + e \rightleftharpoons [\text{(anthracene)}]^{\overline{\cdot}}$$

We can see that the generation of the anthracene radical anion is indeed a reversible process in this medium.

The polarographic half-wave potentials for the reduction of aromatic hydrocarbons (Table 9.2) correlate well with the reduction potentials de-

Table 9.2 AC-Polarographic Peak Potentials and Rate Constants for Electron Transfer in Aromatic Hydrocarbons

Compound		$-E_s$ [a]	log k [b]
Dibiphenylene ethylene (bifluorenylidene)	I	1.00	0.8
	II	1.50	0.23
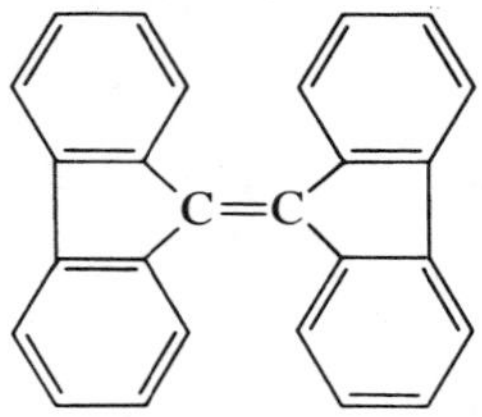 Dibiphenylene butadiene	I	1.10	0.8
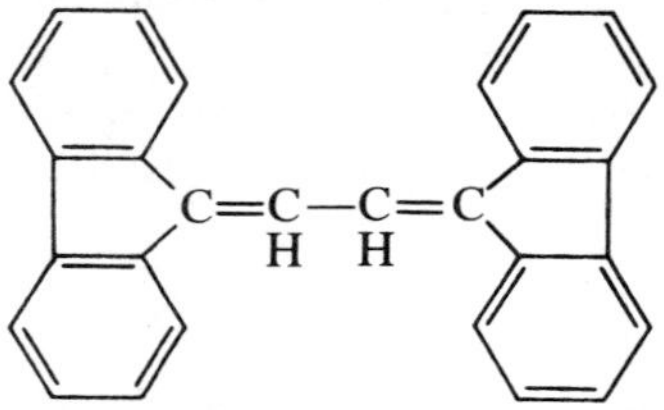	II	1.43	0.35
1-Phenyl-4-biphenylene butadiene	I	1.49	0.6
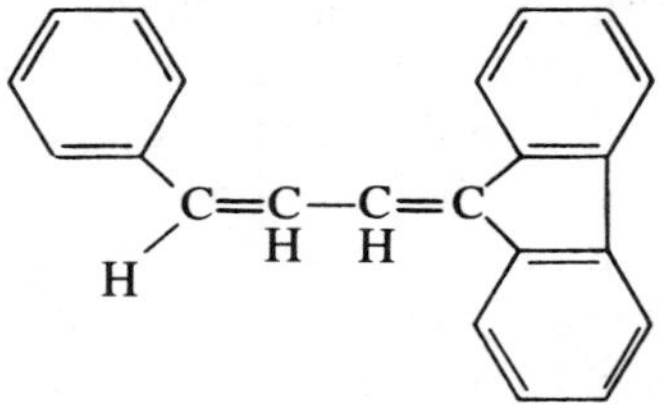	II	1.89	−2.19
Perylene 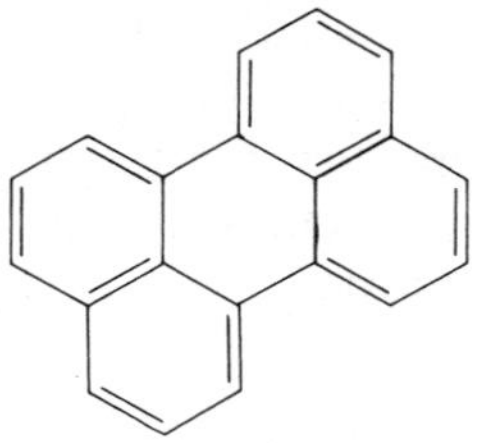	I	1.67	0.5
	II	2.26	−2.20
Anthracene 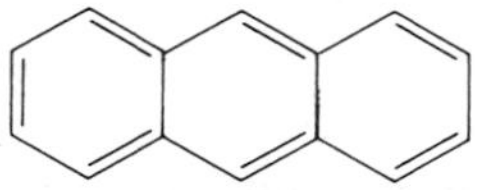	I	1.95	0.6
	II	2.55	−2.22
Tetracene 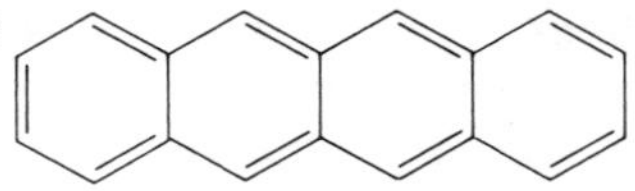	I	1.58	0.5

Table 9.2 *(Continued)*

Compound		$-E_s$ [a]	log k [b]
Stilbene $\phi{-}\overset{H}{C}{=}\overset{H}{C}{-}\phi$	I	2.22	0.4

Data from Ref. 7. ϕ = phenyl radical, C_6H_5. Dimethylformamide medium containing 0.1 M Bu_4NI at 25°C.
[a] Potential at the peak of an AC polarogram, in volts vs SCE. It corresponds to the $E_{1/2}$ in a DC polarogram. I and II correspond to addition of the first and second electron, respectively.
[b] The rate constants, k, are in cm sec^{-1}.

termined from the potentiometric titration of the hydrocarbons with sodium biphenyl in dimethoxyethane and tetrahydrofuran. This implies that the $E_{1/2}$ values represent an electrochemical series arranged in such an order that electron transfer will occur spontaneously from a lower to a higher member of the series, a fact that was also demonstrated by chemical displacement (19).

Fast, reversible electrode reactions are also characterized by symmetrical AC polarograms where the magnitude of the current indicates that the electron transfer is fast enough to follow the AC field. This is the case with the process of electrolytic generation of radical anions (Eqn. 9.6). However, the reversibility of the second reduction step (Eqn. 9.7) depends on the potential at which the reduction of the radical anion to the dianion occurs. When the second reduction step occurs at relatively positive potentials, the corresponding wave is reversible, but if it occurs at potentials more negative than -1.5 V, the electron transfer to the radical anion becomes slow and the process, irreversible. These phenomena are illustrated by the polarograms of anthracene shown in Fig. 9.3: the first peak is reversible and the second, irreversible. The irreversibility of the second reduction step is believed to be due to the protonation of the dianion by the DMF itself or by some more acidic impurities in that solvent. Protonation confers irreversibility because the oxidation of the protonated dianion would not occur at the same $E_{1/2}$ as the reduction process by which the dianion was generated in the first place.

From the measurement of the impedance of the DME at the $E_{1/2}$ as a function of the frequency of the AC current superimposed on the DC polarizing potential, rate constants for electron transfer to aromatic hydrocarbons were evaluated in DMF (7). The values of the rate constants, shown in Table 9.2, are believed to be at least two orders of magnitude greater than would be required for a diffusion-controlled process. In water, on the other hand, the electron transfer to organic compounds is generally a slow process, for example, the rates reported for quinones in acid buffers are of the order of 10^{-3} cm sec^{-1} (19).

Anion radicals of aromatic compounds other than the hydrocarbons have also been prepared electrolytically and the mechanism of their reduction elucidated. Examples of such mechanisms are represented by Eqns. 9.13–9.21. The studies included quinones, nitrobenzene, benzonitriles, ketones, and acid anhydrides (19). Many investigations involved the application of ESR spectroscopy, in conjunction with voltammetry, particularly in the determination of the rates of electron transfer between the neutral molecules and the corresponding anions. The solvent media employed were mostly acetonitrile and DMF, although pyridine (20) was also used. For extensive tabulations of the voltammetric oxidation and reduction potentials of organic compounds see the monograph by Mann and Barnes (21).

Interesting is the effect of the cation of the supporting electrolyte on the $E_{1/2}$ of the hydrocarbons and their derivatives. In acetonitrile and DMF solutions the $E_{1/2}$ values become increasingly positive in the presence of Cs^+, K^+, Na^+, and Li^+ ions, in that order, apparently governed by the crystal radius of the cation. It is believed that the shift is due to the formation of ion pairs between the radical anions and the cations, the extent of ion pairing being greatest for the Li^+ ion. Presumably the positive shift occurs because ion pairing ties up the anions, which are the products of the reduction, thus facilitating the reduction process. The shift in $E_{1/2}$ resulting from addition of Li^+ ions formed the basis for calculating the ion-pair formation constant for the Li^+ ion and the anthrasemiquinone anion (22).

The cation dependence of the $E_{1/2}$ is much greater for the second reduction step, which produces a dianion, than for the first reduction step, which yields a mononegative ion. Again, the largest shift is observed in the presence of lithium, while the tetraalkylammonium ions cause the smallest shift. Also very large $E_{1/2}$ shifts occur for the quinones in the presence of multivalent cations, such as Ba^{2+}, Mg^{2+} and Nd^{3+} ions. Apparently, both anions and cations of higher charge interact more strongly with their counterions than do univalent ions. It is significant that ion pairing between the cations of the supporting electrolyte and the aromatic anions (also called "complexation") can cause a reversible process to become irreversible (by removing the cation from electrochemical equilibrium) or vice versa. The latter situation occurs when the irreversibility is caused by a very negative reduction potential in the uncomplexed state. When the complexation shifts the potential in the positive direction, the reduction may become reversible. In some instances, a perturbation in the charge-density distribution of the anion as revealed by ESR spectroscopy accompanies the ion pairing with the cation.

Another form of intermolecular interaction that causes shifts in the $E_{1/2}$ values is the formation of donor–acceptor complexes of the charge-transfer type (see Chapter 2). Aromatic compounds can be either donors or acceptors. In chloroform and methylene chloride solvents the half-wave potentials of certain strong electron acceptors are shifted negatively when such electron

Table 9.3 Charge-Transfer Interaction Between Aromatic Compounds and Hexamethylbenzene or Pyrene (Donors) in Methylene Chloride at 25°C [a]

Acceptor	$E_{1/2}$ (V vs SCE)	Donor[b]	$E_{1/2}$ (mV)[c]	K (liter mol^{-1})
2,3-Dichloro-5,6-dicyano-*p*-quinone 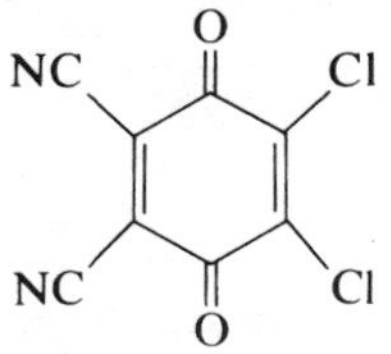	0.49	HMB	72	K_1 = 10 K_2 = 42[d]
Tetracyanoethylene 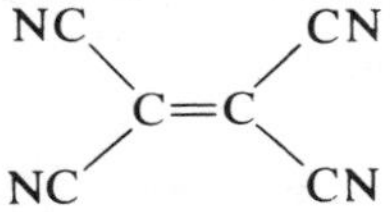	0.25	HMB	39	7.2
		HEB	6	~0.5
7,7,8,8-Tetracyanoquinodimethane 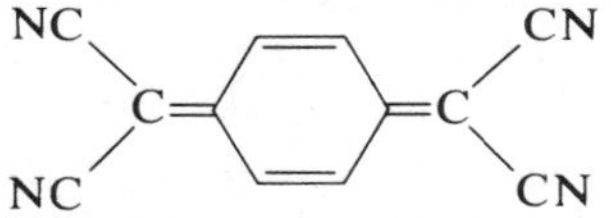	0.18	HMB	5	~0.4
		Pyrene	40	K_1 = 4.6 K_2 = 6[d]
Chloranil 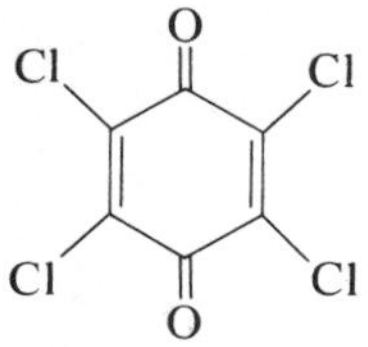	0.06	HMB	26	3.6
		Pyrene	23	3.0
1,2,4,5-Tetracyanobenzene 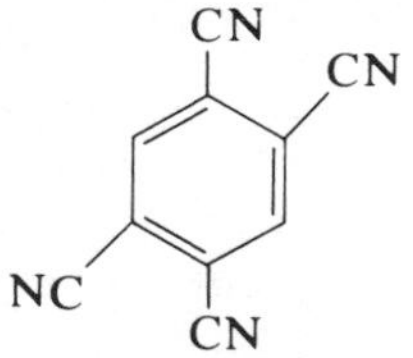	−0.64	HMB	13	1.3
		Pyrene	33	5.3

[a] Solutions contained 0.5 M Bu_4NClO_4.
[b] HMB = hexamethylbenzene; HEB = hexaethylbenzene.
[c] Shift in $E_{1/2}$ for a donor concentration of 0.5 M, calculated on the assumption that $I_s = I_c$ (Eqn. 9.8).
[d] In liters^2 mol^{-2}. Data from Ref. 23.

donors as hexamethylbenzene or pyrene are added (Table 9.3). The formation of charge-transfer complexes is often accompanied by color changes and the corresponding formation constants can be determined spectrophotometrically. The relationship between the shift in the $E_{1/2}$, the concentration of the donor (D), and the formation constants of the complexes, K_1, K_2, etc. is given by (23)

$$F_0 \equiv \text{antilog}_{10}\left[\frac{0.4343}{RT}\,\Delta E_{1/2} - \log_{10}\frac{I_s}{I_c} + \log_{10}\frac{i_d}{i_l}\right] \tag{9.8}$$
$$= 1 + K_1(\mathrm{D}) + K_2(\mathrm{D})^2 + \cdots$$

In Eqn. 9.8, I_s and I_c are the experimental diffusion–current constants for the uncomplexed and the complexed systems, respectively. A diffusion current constant can be calculated from Eqn. 9.1 in the form

$$I = i_d\,(Cm^{2/3}t^{1/6})^{-1}$$

and is equal to $607nD^{1/2}$. It is used to correlate voltammetric data when the values of n and D are unknown. The ratio of the diffusion current to limiting current, $i_d i_l^{-1}$, disappears when the process is diffusion controlled. When $(F_0 - 1)$ is plotted vs (D), straight lines are obtained which are parallel to the abscissa for 1:1 complexes and have finite slopes for higher complexes. The complexation constants, shown in Table 9.3, indicate that the complexes involved here are rather weak. However, the products of electrolytic reduction themselves, the radical anions and dianions, are powerful electron donors and as such could affect the behavior of electron acceptors in solution.

Coupled Chemical Reactions Although radical anions and dianions of aromatic compounds generated by electrolytic reduction are stable relative to the lifetime of a mercury drop (several seconds), they do undergo chemical reactions on a longer time scale. These so-called coupled chemical reactions are commonly abstractions of a proton from the solvent or another proton donor and dimerization of the radical. It is particularly informative in this connection to study the changes in the polarographic reduction process, which occur when an amphiprotic solvent or a solute proton donor are added to an aprotic medium. Through gradual enrichment of the medium in the protic component, it is possible to bring about a gradual transition in the polarographic behavior ranging from that characteristic of an aprotic medium to one characteristic of a protic medium. An example of such gradual transition is illustrated by the effect of proton donors on the polarograms of anthracene, shown in Fig. 9.4. We can see that in going from the nearly aprotic 96% dioxane in water to the appreciably aqueous 75% dioxane, the second wave gradually disappears and is eliminated completely when an even stronger proton donor, HI, is added.

The effect of adding protons in the form of water or acids to the reduction

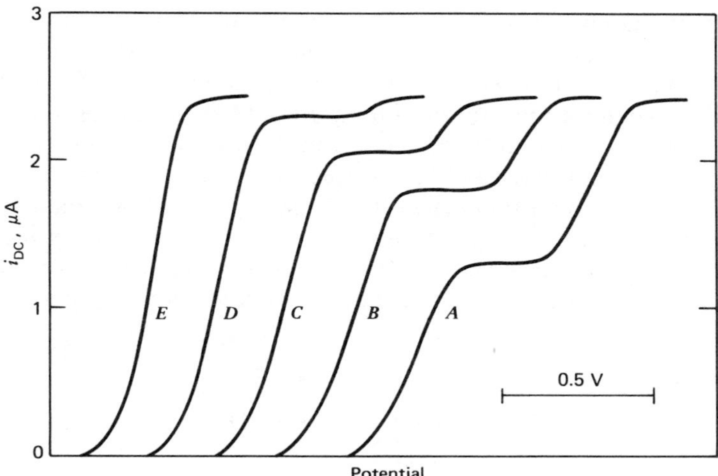

Figure 9.4 The effect of proton donors on the polarograms of anthracene in aqueous dioxane. Dioxane content: *A*, 96%; *B*, 92%, *C*, 89%, *D*, 75%, *E*, 96% + 0.02 *M* HI. The waves have been shifted on the potential axis for the sake of clarity. Reprinted with permission from G. J. Hoijtink, J. van Schooten, E. deBoer and W. Y. Aalbersberg, *Recl. Trav. Chim. Pays-Bas,* **73,** 355 (1954).

products of aromatic hydrocarbons and their derivatives depends primarily on the basicity of the proton acceptor. The overall reaction mechanism for the reduction and the coupled chemical reactions can be divided into four distinct categories (19). In the first category protonation of the dianion occurs and it manifests itself in a shift of the second wave to positive potentials with a rapid reduction in the height of the AC wave. The AC current is diminished because the electron-transfer step is followed here by a fast irreversible reaction, so that the electrode reaction proceeds essentially in one direction only. The second wave is irreversible due either to a slow step in the electron-transfer process or to the coupled chemical reaction (protonation). The first wave remains unchanged. This type of behavior was just exemplified by anthracene in dioxane–water mixtures without added HI (curves *A*–*D* in Fig. 9.4). Here the second wave is irreversible because the dianion, M^{2-}, is protonated rapidly and removed from equilibrium 9.7. If the protonation of the dianion leads to formation of a new polycyclic system that is further reducible, the result could be an *increase* in the height of the second wave and the appearance of new waves. This has been observed, for example, with perylene, anthraquinone, and phenanthraquinone in DMF with added phenol.

In cases where water is added to solutions of very basic anions (e.g., dibiphenylene ethene, stilbene) or when stronger proton donors, such as HI, are added to less basic anions (e.g., anthracene, naphthalene), the addition of the proton donors to an aprotic medium causes the first wave to increase at the expense of the second, until the former reaches a two-electron height. Hoijtink offered the following mechanism to account for the above phenomenon (12). When strong proton donors are added to radical ions $M^{\overline{\cdot}}$, the latter become protonated before they can diffuse away from the electrode surface. Thus, a new radical, $MH\cdot$, is formed at the electrode surface which has a higher electron affinity than the parent hydrocarbon M and therefore can be reduced at the same potential. Additional waves or an increase in the wave height are possible if the product MH_2 is reducible at the available potentials. The sequence of reactions involved can be summarized as follows:

$$M + e \rightleftharpoons M^{\overline{\cdot}} \tag{9.9}$$

$$M^{\overline{\cdot}} + HX\ (\text{or } H^+) \longrightarrow MH\cdot + X^- \tag{9.10}$$

$$MH\cdot + e \rightleftharpoons MH^- \tag{9.11}$$

$$MH^- + HX\ (\text{or } H^+) \longrightarrow MH_2 + X^- \tag{9.12}$$

A gradual variation in the solvent medium from aprotic to protic confirms the applicability of the above mechanism. Thus, anthracene exhibits two one-electron waves in aprotic media, one two-electron wave in Cellosolve and an intermediate behavior in dioxane–water mixtures (the last illustrated in Fig. 9.4).

A third type of behavior is observed when an acid (e.g., benzoic, acetic) is added to a solution of benzophenone, xanthone, or benzaldehyde in acetonitrile or DMF. At first a new wave grows at the expense of the original two one-electron waves at a more positive potential, until a single two-electron wave is obtained. The new wave is believed to be due to an acid–ketone complex, which has a higher electron affinity than the parent molecule. At higher concentrations of acetic acid, the new wave can be shifted far enough in the positive direction so as to produce two separate waves again. The following mechanism was proposed to account for these phenomena (24):

$$M + HX \longrightarrow MHX \quad \text{(acid–ketone complex)} \tag{9.13}$$

$$MHX + e \longrightarrow MHX^- \tag{9.14}$$

$$MHX^- \longrightarrow MH\cdot + X^- \tag{9.15}$$

$$MH\cdot + e \longrightarrow MH^- \tag{9.16}$$

Prior to addition of acid, the ketone, for example, benzophenone, is reduced by the same mechanism as hydrocarbons, yielding the two one-electron waves (Eqns. 9.9–9.12). Addition of acid at first produces the one two-electron wave presumably because of a superimposition of reactions 9.14 and 9.16. At very high acid concentrations, the potential for reduction 9.14 is shifted to a value even more positive than that for reduction 9.16, and the wave again splits into two components.

In the absence of acids, electrolysis of benzophenone at the potential of the first wave, forms benzhydrol by the mechanism shown in Eqns. 9.9–9.12, but in 4 *M* acetic acid in DMF, benzopinacol is the product. Presumably, the radical MH· (Eqn. 9.16) is now dimerizing in preference to further reduction (25, 26). The production of benzhydrol and benzopinacol from benzophenone can be explained by the following sequence of reactions:

$$\underset{\text{benzophenone}}{(C_6H_5)_2C{=}O} + e \rightleftharpoons [(C_6H_5)_2\dot{C} - O]^- \quad (9.17)$$

$$[(C_6H_5)_2\dot{C} - O]^- + H^+ \text{ (or HX)} \longrightarrow (C_6H_5)_2\dot{C}OH \quad (9.18)$$

$$(C_6H_5)_2\dot{C}OH + e \rightleftharpoons [(C_6H_5)_2COH]^- \quad (9.19)$$

$$[(C_6H_5)_2COH]^- + H^+ \text{ (or HX)} \rightleftharpoons \underset{\text{benzhydrol}}{(C_6H_5)_2CHOH} \quad (9.20)$$

$$2(C_6H_5)_2\dot{C}OH \rightleftharpoons \underset{\text{benzopinacol}}{(C_6H_5)_2COHCOH(C_6H_5)_2} \quad (9.21)$$

Finally, in the case of benzaldehyde, xanthone and purpurogallin, addition of proton donors results in the formation of a new wave between the first and the second waves at the expense of the second. This is believed to be the result of competition between the dimerization and the protonation of the radical anion. At potentials where protonation prevails, the intermediate wave due to reduction to MH_2 is observed (24).

Electrolytic Oxidation of Aromatic Compounds Oxidation of aromatic hydrocarbons and their derivatives to form radical cations has been accomplished at platinum electrodes in acetonitrile. The oxidation potentials are generally too positive for the use of a mercury electrode. A two-step oxidation mechanism analogous to that of reduction is exhibited by 9,10-diphenylanthracene in acetonitrile. The first one-electron wave corresponds to a fast reversible electron transfer, without any coupled reactions. The second wave, at a potential more positive by about 0.5 V, is irreversible. Among hydrocarbons, however, the above mechanism is rather the exception than the rule as the cation radicals are generally more reactive than anion radicals. Thus, for other unsubstituted aromatics (e.g., anthracene) the oxidation is irreversible under normal conditions. Wave heights corresponding to more than one electron are the rule, because of the coupled reactions. The first one-electron oxidation step can be isolated by fast sweep rates (10–100 V sec^{-1}), and a reversible reaction is observed under those conditions (27).

The stability of aromatic cation radicals can be improved by substitution, particularly when an aromatic group is substituted into a reactive position. An example where this occurs is the enhanced stability of the cation radical of 9,10-diphenylanthracene as compared to that of 9-methylanthracene. Also aromatic amines form cation radicals that are more stable than those of the parent hydrocarbons. Thus, *p*-phenylenediamine and tetramethyl-*p*-phenylenediamine yield two one-electron waves, of which the first is reversible according to criteria of AC polarography and cyclic voltammetry (19). In general, molecules capable of greater charge delocalization produce the more stable cation radicals. Structural effects on the stability of aromatic radical cations were reviewed by Adams (28).

Rather original is the behavior of rubrene, 5,6,11,12-tetraphenyltetracene, which forms a radical cation at anodic potentials and a radical anion at cathodic potentials. When the potential is cycled rapidly from an anodic to a cathodic value (or if the radical anion is simultaneously produced at a nearby electrode), the two radical ions combine to form an excited molecule which emits light. The process is known as *electrochemiluminescence* (10, 29–31).

9.1.3 Correlation Between the Half-Wave Potentials and Other Properties of Aromatic Hydrocarbons A fruitful partnership was developed between voltammetry and theoretical chemistry when it was recognized that the half-wave potentials of aromatic hydrocarbons were suitable parameters with which the calculated or spectroscopically observed molecular energy levels could be correlated. Such correlations are not particularly surprising when one considers that according to molecular-orbital theory, the energy of the lowest unoccupied orbital is related to the electron affinity of the molecule and the energy of the highest occupied orbital is related to the ionization potential of the molecule, while the reduction and the oxidation potentials are also related to electron affinity and ionization potential, respectively. We mentioned in discussing Eqns. 9.2 and 9.3 that it is common practice in voltammetry to equate $E_{1/2}$ and $E°$. Even when such an approximation is valid, the $E_{1/2}$ (or $E°$) is obviously not an absolute potential reflecting the electron-transfer tendency of the electroactive species, because its measured value includes a reference-electrode potential and in the case of $E_{1/2}$, a liquid-junction potential as well. In the following equations these two potentials are accounted for by a constant factor C. The relationship between the $E°$ (or $E_{1/2}$) and the electron affinity or the ionization potential is given by

$$\text{Reduction:} \qquad E° = \mathrm{A} + \alpha_{\mathrm{M}} - \alpha_{\mathrm{M}^-} + \mathrm{C} \tag{9.22}$$

$$\text{Oxidation:} \qquad E° = \mathrm{I} + \alpha_{\mathrm{M}^+} - \alpha_{\mathrm{M}} + \mathrm{C} \tag{9.23}$$

Here A and I represent the electron affinity and the ionization potential, respectively, and α stands for the *real* solvation energy of the hydrocarbon M and its oxidized (M^+) and reduced (M^-) forms. In the case of individual ions real solvation energy differs from the *chemical* solvation energy (such as given by the Born equation, Eqn. 2.16) in that it comprises not only ion–solvent interactions in the bulk of the solution, but also the electrical work required to transport the ion across the vacuum–solution interface. For electrically neutral combinations of ions the electrical work cancels out, and the real and chemical solvation energies are equal.

Indeed, it was found that the reduction and the oxidation half-wave potentials of polynuclear aromatic hydrocarbons correlate fairly well with the electron affinities and the ionization potentials, respectively (32). Furthermore, a good correlation was observed between the polarographic half-wave potentials for the reduction of aromatic hydrocarbons and the energy of the lowest unoccupied molecular orbital as calculated from molecular-orbital theories (9). (Fig. 9.5). A variety of other, often interrelated, correlations

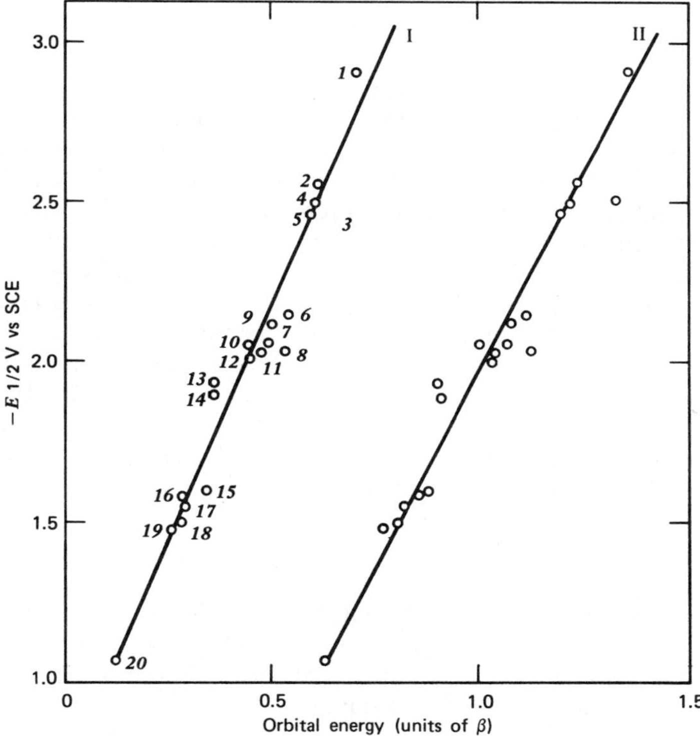

Figure 9.5 Correlation between the half-wave potential for the reduction of aromatic hydrocarbons in 96% dioxane and the root a of the molecular-orbital secular equation of aromatic hydrocarbons. Root a is a measure of the energy of the lowest unoccupied π-orbital. I, Hückel theory; II, Wheland theory, shifted 0.5 units; *1*, diphenyl; *2*, naphthalene; *3*, triphenylene; *4*, phenanthrene; *5*, terphenyl; *6*, quaterphenyl; *7*, 1,2-diphenylethene; *8*, coronene; *9*, 1-2-benzpyrene; *10*, pyrene; *11*, 1,2,5,6-dibenzanthracene; *12*, 1,2-benzanthracene; *13*, 3,4-benzpyrene; *14*, fluoranthene; *15*, perylene; *16*, acenaphthylene; *17*, tetracene; *18*, decacyclene; *19*, 1-phenyl-4-diphenylenebutadiene; *20*, 1,4-dibiphenylenebutadiene. Reprinted with permission from A. J. Bard, Ed., *Electroanalytical Chemistry, A Series of Advances*, Vol. 2, Marcel Dekker, New York, 1967, p. 44, by courtesy of Marcel Dekker, Inc.

have been reported between the $E_{1/2}$'s and the energies of spectral transitions for aromatic hydrocarbons and their derivatives (19).

9.1.4 The Reduction of Oxygen and Hydrogen In aqueous solutions the polarographic reduction of oxygen yields two irreversible two-electron waves, the first corresponding to the formation of peroxide and the second to the reduction of the peroxide to hydroxyl ion. In aprotic media oxygen is reduced in two one-electron steps, the first leading to the formation of the superoxide radical anion, and the second to the peroxide ion:

$$O_2 + e \rightleftharpoons \dot{O}_2^- \tag{9.24}$$

$$\dot{O}_2^- + e \rightleftharpoons O_2^{2-} \tag{9.25}$$

Maricle and Hodgson (33) studied the above processes in DMF and DMSO both at the DME and at a platinum electrode. With Bu_4NOH as the supporting electrolyte, the first $E_{1/2}$ appeared at -0.73 V and the second at -2.40 V (vs aqueous SCE). By electrolytic reduction, a solution of tetrabutylammonium superoxide was prepared from which insoluble metal superoxides could be precipitated. The superoxide anion cannot be prepared in protic media, where it is instantly protonated, followed by decomposition to oxygen and H_2O_2. The process represented by Eqn. 9.24 was proved to be a one-electron reduction by controlled-potential coulometry (34). It is interesting that the first stage in the reduction of oxygen in aprotic media is reversible and occurs in the same potential range in different solvents (-0.7 to -0.9 V), whereas the second reduction step is irreversible and the $E_{1/2}$ corresponding to it depends on the nature of the metal ion present. This is believed to be due to the formation of insoluble metal peroxides in the second step:

$$\dot{O}_2^- + 2M^+ + e \longrightarrow M_2O_2 \downarrow \tag{9.26}$$

In the case when a tetraalkylammonium salt is the supporting electrolyte, the superoxide ion may engage in more complex chemical reactions. For example, solutions of Et_4NClO_4 in DMSO are believed to be decomposed by the superoxide ion as follows (35):

$$\dot{O}_2^- + (C_2H_5)_4N^+ \longrightarrow HO_2^- + CH_2{=}CH_2 + (C_2H_5)_3N \tag{9.27a}$$

$$HO_2^- + (CH_3)_2SO \longrightarrow (CH_3)_2SO_2 + OH^- \tag{9.27b}$$

It is well known that solutions of strong acids in water give identical polarograms corresponding to the reduction of the hydronium ion to hydrogen gas. This is so because water is a leveling medium for acids and bases (Chapter 6). Acetonitrile, on the other hand, is a differentiating medium for acids and bases. Coetzee and Kolthoff (36) confirmed the differentiating nature of acetonitrile also by observing that a much greater spread exists for the $E_{1/2}$ values of acids in acetonitrile than in water. Their results are

Table 9.4 Differentiating Effect of Acetonitrile on the Polarographic Half-Wave Potentials of Acids

Acid[a]	$-E_{1/2}$ (V vs aq. SCE)
Perchloric	0.70
Hydrobromic	0.90
Hydrochloric	1.06
p-Toluenesulfonic	1.20
2,5-Diethylanilinium ion	1.43
Oxalic	1.55
Phosphoric	1.75
Benzoic	2.1
Acetic	2.3

[a] 10^{-3} *M* solutions.

Reprinted with permission from *J. Amer. Chem. Soc.*, **79,** 6112 (1957). Copyright 1957. American Chemical Society.

reproduced in Table 9.4. We can see that the $E_{1/2}$ values become increasingly negative as the acid strength decreases (reduction of the H^+(solv.) becomes more difficult), and that acids which are identical in strength in water are well differentiated in acetonitrile (e.g., perchloric vs *p*-toluenesulfonic acids). The differentiating feature of acetonitrile on the $E_{1/2}$ of acids stems not only from the low basicity of the solvent, but perhaps also from the low overvoltage of hydrogen as compared to its value in water. However, the order of $E_{1/2}$ values as shown in Table 9.4 does not necessarily reflect the correct order of acid strengths because the reduction of acids in acetonitrile is not completely reversible.

In contrast to acetonitrile, DMSO levels the strong acids $HClO_4$, HCl, and H_2SO_4, all of them being reduced at an $E_{1/2}$ of -1.1 V, but the weaker acids are differentiated: H_2S (-1.66 V), HAc (-2.3 V), and NH_4^+ (-2.45 V). The polarographic reduction of hydrogen in nonaqueous media was reviewed by Elving and Spritzer (37).

9.1.5 Voltammetric Studies of Ionic Solvation

The Relationship Between $E_{1/2}$, $E°$ and the Solvation Energy of a Metal Ion Forming an Amalgam with the DME In the absence of reliable values for standard potentials, $E°$, voltammetric half-wave potentials, $E_{1/2}$, can be used as indices of the relative solvation energies of ions (Eqns. 9.3–9.5). However, before we apply $E_{1/2}$ to the studies of solvation, it is useful to know precisely how it differs from the $E°$. We mentioned earlier that one reason why the $E_{1/2}$ of a polarographic half-cell differs from the $E°$ is because it is determined in the presence of supporting electrolyte, so that activity coefficients and possibly ion pairing must be accounted for, and because the diffusion coefficients of the oxidized and reduced forms of the electroactive species may not be identical (Eqns. 9.2 and 9.3). These are the sources of discrepancy in the case of redox couples where both forms of the electroactive species are soluble in the solvent, but not in the mercury electrode. When the reducible species is a metal ion which forms an amalgam, the relationship between $E°$ and $E_{1/2}$ becomes even more complicated. Here the electrode reaction for the reduction of the metal ion M^{n+} is

$$M^{n+} + ne + Hg \rightleftharpoons M(Hg) \tag{9.28}$$

and if the reaction is reversible, the half-wave potential of the DME will be

$$E_{1/2} = E_a^\circ + \frac{RT}{nF} \ln a_{Hg} + \frac{RT}{nF} \ln \frac{D_{red}^{1/2} f_{ox}}{D_{ox}^{1/2} f_{red}} \tag{9.29}$$

Equation 9.29 differs from Eqn. 9.2 in that the reduced form refers now to the metal in the amalgam phase, the activity of mercury, a_{Hg}, has been introduced and the E_a° is of course the standard potential of the amalgam electrode, corresponding to process (9.28). We are interested, however, not in E_a°, but in $E^\circ_{M^{n+}/M}$, which reflects more directly the solvation energy of the metal ion M^{n+} (cf. Eqn. 9.5). The relationship between E_a° and $E^\circ_{M^{n+}/M}$ can be derived from the measured potential of a metal–amalgam cell in a solvent that is compatible with the pure metal; in the case of alkali metals, ethylamine has been used:

$$M(s) \mid \underset{\text{in } C_2H_5NH_2}{M^{n+} \text{ solution}} \mid M(Hg)M \tag{9-I}$$

The potential of cell 9-I, which is independent of the activity of the metal ion in solution, is given by

$$E = E_a^\circ - E^\circ_{M^{n+}/M} - \frac{RT}{nF} \ln \frac{a_{M(Hg)}}{a_{Hg}} \tag{9.30}$$

For a given amalgam all terms in Eq. 9.30 other than $E^\circ_{M^{n+}/M}$ are independent of the solvent or ion activity in solution and can be collected into a constant, sometimes referred to as the energy of amalgamation. Thus in conjunction with Eq. 9.29:

$$E_{1/2} = E^\circ_{M^{n+}/M} + \frac{RT}{nF} \ln f_{ox} + \frac{RT}{nF} \ln \frac{D_{red}^{1/2}}{D_{ox}^{1/2}} + \text{const.} \tag{9.31}$$

If the same assumption is made as for Eqn. 9.2, namely, that the logarithmic terms in Eqn. 9.31 can be neglected, we obtain

$$E_{1/2} = E^\circ_{M^{n+}/M} + \text{const.} \tag{9.32}$$

Obviously when the same metal ion is compared in two solvents, the solvent-independent constant will drop out, so that

$$\Delta E_{1/2} \cong \Delta E^\circ \tag{9.33}$$

It should be noted, however, that a series of $E_{1/2}$ values for different metal ions in the same solvent might be distorted relative to the E° series due to differences in the amalgamation energies among the metals. Differences between E° and $E_{1/2}$ are known to be greatest for alkali-metal ions.

So far in our comparison of $E_{1/2}$ and E° we have focused only on the indicator half-cell. However, voltammetric $E_{1/2}$ values, as all other electrode potentials, must be measured against a reference electrode, which in the case of $E_{1/2}$ values is almost invariably complicated by a liquid-junction potential as well. The most common reference electrode in voltammetry is the aqueous SCE connected to the polarographic half-cell by a bridge of aqueous KCl. In dipolar aprotic solvents Ag^+/Ag electrodes in the given solvent are also popular reference electrodes. Thus the measured $E_{1/2}$ usually contains a reference-electrode potential and a liquid-junction potential, E_j. When the half-wave potentials for a series of ions are measured in one solvent against the same reference electrode, the E_j is likely to be fairly constant, provided the same supporting electrolyte is present in a large enough excess in all cases. Under such conditions, the relative $E_{1/2}$ values will be devoid of significant error due to the E_j, though the error due to differences in amalgamation energies would persist. On the other hand, when the $E_{1/2}$'s are measured in different nonaqueous solvents against the aqueous SCE, the E_j at the interface of the aqueous KCl bridge and the nonaqueous solutions is likely to vary significantly with the solvent. Therefore, in order to make a meaningful comparison of $E_{1/2}$ values measured in different solvents, it is necessary either to eliminate the liquid-junction potential in the measuring process or to evaluate its magnitude.

Liquid-junction potentials can be virtually eliminated if one confines comparisons only to the *differences* between the $E_{1/2}$ of each cation and some reference cation in a given solvent. It has been customary to adopt the $E_{1/2}$ of the rubidium ion as the reference potential and to express other half-wave potentials relative to it, that is, $\Delta E_{1/2}(\mathrm{M}) \equiv E_{1/2}(\mathrm{M}) - E_{1/2}(\mathrm{Rb})$.† When such relative values, $\Delta E_{1/2}(\mathrm{M})$, are compared in water, $\Delta E_{1/2}(\mathrm{M,W})$, and in a nonaqueous solvent, $\Delta E_{1/2}(\mathrm{M,S})$, the differences between the amalgamation energies of the two metals and the differences between the liquid-junction potentials will cancel,‡ so that in a *double comparison* of this type,

† The choice of rubidium originated with Pleskov's (38) proposal to adopt the rubidium electrode as the solvent-independent zero point for electrode potentials. While the idea that rubidium has the same potential in all solvents cannot be correct, inspection of $E_{1/2}$ data in many solvents reveals that there is not much difference between the $E_{1/2}$ values of Rb^+, Cs^+, and K^+, so that either one of them can be (and has been) used as a reference ion relative to which the $E_{1/2}$'s of more reactive ions can be expressed.

‡ Actually, a fraction of the E_j might remain uncanceled, because the transfer activity coefficient of each cation contributes (differently) to the overall E_j.

an approximate equality between the *changes* in $E_{1/2}$ and $E°$ can be achieved:

$$\Delta\Delta E_{1/2}(\mathrm{M}) = \Delta E_{1/2}(\mathrm{M,S}) - \Delta E_{1/2}(\mathrm{M,W}) \cong \Delta E°(\mathrm{M,S}) - \Delta E°(\mathrm{M,W}) \quad (9.34)$$

How is the $\Delta\Delta E°$ quantity on the right-hand side of Eqn. 9.34 related to changes in the solvation energy of the metal ions? We know that for the same process, $\Delta G° = -nF\,\Delta E°$ (Eqn. 9.4). However, the process which defines solvation energy for a metal ion is

$$\mathrm{M}^{n+}(\mathrm{g}) \longrightarrow \mathrm{M}^{n+}(\mathrm{solv.}) \quad (9.35)$$

whereas the process which defines the standard (reduction) potential of a metal-ion/metal electrode is

$$\mathrm{M}^{n+}(\mathrm{solv.}) + ne \rightarrow \mathrm{M(metal)} \quad (9.36)$$

The standard free energy change for the process in (9.36) can be expressed as

$$nFE°_{\mathrm{M}^{n+}/\mathrm{M}} = \bar{G}°_{\mathrm{solv.}}(\mathrm{M}^{n+}) + \bar{G}°_{\mathrm{e}} - \bar{G}°_{\mathrm{M}} \quad (9.37)$$

where $\bar{G}°_{\mathrm{e}}$ and $\bar{G}°_{\mathrm{M}}$ are the standard chemical potentials of the electron and the metal, respectively, both in the metal phase. Obviously, the last two terms on the right-hand side of Eqn. 9.37 are properties of the metal phase and therefore independent of the solvent. Thus, in a comparison of the $\Delta E°$'s in two solvents, they would cancel out. As an example, let us apply Eqn. 9.36 to the difference between the potentials of lithium and rubidium in methanol and in water and then combine them with Eqn. 9.37 to obtain an expression for the difference between the transfer free energy of lithium and rubidium in methanol. Remembering that $\Delta E°(\mathrm{Li}) \equiv (E°_{\mathrm{Li}} - E°_{\mathrm{Rb}})$ in each solvent, and substituting in Eqn. 9.34 for each $E°$ an expression from Eqn. 9.37, we obtain

$$_{\mathrm{m}}G°(\mathrm{Li}^+) - {}_{\mathrm{w}}G°(\mathrm{Li}^+) - {}_{\mathrm{m}}G°(\mathrm{Rb}^+) + {}_{\mathrm{w}}G°(\mathrm{Rb}^+) = \frac{nF}{2.3RT}[\Delta E°(\mathrm{Li,m}) - \Delta E°(\mathrm{Li,W})] \quad (9.38)$$

where m and w (or W) refer to methanol and water, respectively, and the subscript "solv." has been dropped from the $G°$'s.† Of course, the right-hand side of Eqn. 9.38 represents a measure of the difference between the transfer free energies of lithium and rubidium ions in methanol. We had occasion to define transfer free energy in Eqn. 9.5, for example, for the transfer from water to methanol:

$$\Delta G°_{\mathrm{t}}(\mathrm{Li}^+) \equiv {}_{\mathrm{m}}G°(\mathrm{Li}^+) - {}_{\mathrm{w}}G°(\mathrm{Li}^+) \quad (9.39)$$

† Normally ionic charges are omitted from $G°$ symbols. Here they were retained in order to avoid confusion with the properties of uncharged metals just discussed.

In terms of transfer free energies Eqn. 9.38 would be rewritten as

$$\Delta G_t^\circ(\mathrm{Li}^+) - \Delta G_t^\circ(\mathrm{Rb}^+) = \frac{nF}{2.3RT}[\Delta E^\circ(\mathrm{Li,m}) - \Delta E^\circ(\mathrm{Li,W})] \quad (9.40)$$

Substituting into the above equation the literature values (16) for $\Delta E^\circ(\mathrm{Li,m}) = -0.183$ V and $\Delta E^\circ(\mathrm{Li,W}) = -0.120$ V, we obtain for the difference between the transfer free energies of Li^+ and Rb^+ in methanol the value of -1.4 kcal mol^{-1}. The corresponding $\Delta E_{1/2}$ values are -0.27 V and -0.20 V in methanol and water, respectively (39) on the basis of which the same transfer free energy difference would be $-1._6$ kcal mol^{-1}.

Half-Wave Potentials and Relative Solvation of Metal Ions In the preceding section, equations were derived which related half-wave potentials to standard electrode potentials and the latter, in turn, to solvation energies of ions. In the present section we will apply these relationships to a comparison of experimental values of $E_{1/2}$ in water and nonaqueous solvents. Particularly interesting in this respect is a comparison of ionic solvation in water and in dipolar aprotic media.

The polarographic behavior of metal ions in acetonitrile was subjected to extensive studies by Kolthoff and Coetzee (36, 40), followed by several investigations by Coetzee et al. (39, 41–43) covering a number of dipolar aprotic solvents. In Table 9.5 some of the $E_{1/2}$ values reported in the above studies for water, acetonitrile, and sulfolane are intercompared. A more negative $E_{1/2}$ value indicates stronger solvation and corresponds to a lower, or a larger *negative*, value of the solvation energy.† On this basis one can generalize that, in the absence of specific solvation, the cations in Table 9.5 are solvated more strongly by water than either by acetonitrile or sulfolane (the quantities $\Delta\Delta E_{1/2}$ are generally positive). Note that the $E_{1/2}$ values in sulfolane were measured against a different reference electrode than the rest of the data, so that only their differences are directly comparable with the corresponding data in other solvents.

Within the alkali–metal series, the Li^+ ion, which is known to be most strongly hydrated, suffers the greatest loss of (negative) solvation energy when transferred to dipolar aprotic media (it has the largest positive $\Delta\Delta E_{1/2}$ values). For example, the fact that $\Delta\Delta E_{1/2}(\mathrm{Li}) = 0.23$ V in acetonitrile means that the solvation energy of the Li^+ ion relative to that of the Rb^+ ion is lower (more negative) in water than in acetonitrile by 5.3 kcal mol^{-1}. Acetonitrile and sulfolane are not unique in accentuating the solvation energy of Li^+ and its change upon transfer to an aprotic medium. Thus, in dimethylformamide, the $E_{1/2}$ of Li^+ is -1.81 V, whereas the $E_{1/2}$'s of the other

† It is frequently stated that solvation energy is "low" when ion–solvent interactions are weak, and "high" when they are strong. But when both the magnitude and the *sign* of the energy are taken into account, it is clear that precisely the opposite is true.

Table 9.5 Comparison of Half-Wave Potentials in Water, Acetonitrile, and Sulfolane[a]

Ion	$(-E_{1/2})_W$ [b]	$(\Delta E_{1/2})_W$ [c]	$(-E_{1/2})_{AN}$ [b]	$(\Delta E_{1/2})_{AN}$ [c]	$(\Delta\Delta E_{1/2})_{AN-W}$ [d]	$(-E_{1/2})_{SL}$ [e]	$(\Delta E_{1/2})_{SL}$ [c]	$(\Delta\Delta E_{1/2})_{SL-W}$ [d]
Li^+	2.33	−0.20	1.95	0.03	0.23	2.67	0.00	0.20
Na^+	2.12	0.01	1.85	0.13	0.12	2.56	0.11	0.10
K^+	2.14	−0.01	1.96	0.02	0.03	2.66	0.01	0.02
Rb^+	2.13	0.00	1.98	0.00	0.00	2.67	0.00	0.00
Cs^+	2.09	0.04	1.97	0.01	−0.03	2.66	0.01	−0.03
Ba^{2+}	1.90	0.23	1.63	0.35	0.12	2.42	0.25	0.02
Zn^{2+}	1.00	1.13	0.70	1.28	0.15	1.13	1.54	0.41
Cd^{2+}	0.59	1.54	0.27	1.71	0.17	0.81	1.86	0.32
Ag^+	−0.57[f]	2.70	−0.32	2.30	−0.40	0.09	2.58	−0.12
Cu^+	−0.28[f]	2.41	0.36	1.62	−0.79	—	—	—

[a] Unless otherwise stated, the data are from Refs. 39, 41–43. AN = acetonitrile; SL = sulfolane.
[b] Volts vs aqueous SCE; $E_{1/2} = E_{1/2}(M) - E_{SCE}$.
[c] Volts vs $E_{1/2}$ of Rb^+ in the same solvent; $\Delta E_{1/2}(M) = E_{1/2}(M) - E_{1/2}(Rb)$.
[d] $\Delta\Delta E_{1/2}(M) = \Delta E_{1/2}(M,S) - \Delta E_{1/2}(M,W)$ (Eqn. 9.34).
[e] Volts vs Ag/0.1 M $AgClO_4$ in sulfolane. Data in sulfolane at 30°C, at 25°C in other solvents.
[f] Values of E°, not $E_{1/2}$. Data from Ref. 36.

alkali–metal ions are bunched in the narrow range of -1.52 to -1.55 V (vs the mercury pool) (44).

Stronger solvation by acetonitrile than by water is evident in the case of Ag^+ and Cu^+ ions, for which $(\Delta\Delta E_{1/2})_{AN\text{-}W}$ values are appreciably negative. Silver and cuprous ions are known to coordinate with molecules containing electron-donor nitrogen atoms and what we observe here is polarographic evidence of this type of specific solvation by acetonitrile. The stabilization of Cu^+ ions relative to Cu^{2+} ions in acetonitrile is so great that the polarographic reduction wave corresponding to the process $Cu^{2+} + e = Cu^+$ begins in the area of anodic dissolution of mercury, which in noncomplexing media occurs at about $+0.6$ V. On the basis of other measurements, the standard potential for the Cu^{2+}/Cu^+ couple in acetonitrile was reported to be $+1.0$ V (45). The peculiarities of silver–copper chemistry in acetonitrile can be verified by chemical displacement experiments. Solutions of cupric salts are rapidly decolorized by metallic silver, yielding quantitative conversion to Cu^+ and Ag^+ ions. Similarly, metallic copper added to Ag^+ solutions in acetonitrile will deposit metallic silver and liberate an equivalent amount of Cu^+ ions. These phenomena can be accounted for quantitatively in terms of the large (negative) transfer free energies of Ag^+ and Cu^+ ions in acetonitrile, which was done in Chapter 5. We recall that specific interaction between Ag^+ and acetonitrile has been confirmed by a variety of physicochemical data.

Also interesting are the changes in the $\Delta E_{1/2}$ values between acetonitrile and sulfolane. It turns out that the quantity

$$(\Delta\Delta E_{1/2})_{SL\text{-}AN} \equiv (\Delta E_{1/2})_{SL} - (\Delta E_{1/2})_{AN}$$

is negative for most cations, suggesting that in general, cations prefer oxygen to nitrogen donors. Significantly, Ag^+, Zn^{2+}, and Cd^{2+}, which are known to coordinate preferentially with nitrogen ligands, have positive values of $(\Delta\Delta E_{1/2})_{SL\text{-}AN}$.

Solvation of Individual Ions: Proposals for Solvent-Independent Electrodes We have seen so far how voltammetric data can provide information on the effect of solvent on the difference between two electrode potentials and on the difference between the transfer free energies of pairs of cations. Chemists, however, are much more interested in determining how the potential of a *single* electrode changes upon transfer from water to another solvent, which involves the determination of the transfer free energies for *single* ions.

In Chapter 5 we devote considerable attention to the classical dilemma confronting a chemist who tries to correlate electrode potentials in different solvents or to estimate the transfer properties of individual ions. For example, if we were to measure directly the potential difference between standard silver electrodes in water and in acetonitrile, the measured voltage would be determined not only by the transfer free energy of the Ag^+ ion, but also by an unknown and probably appreciable liquid-junction potential at the interface of the two solutions. A similar result would be obtained if we were to measure the $E_{1/2}$ of, say, the lithium ion in water and in acetonitrile, in each case against the aqueous SCE: the difference between the two $E_{1/2}$ values would be distorted by the liquid-junction potential, E_j, at the interface of the acetonitrile solution and the aqueous bridge solution. If, on the other hand, the E_j is eliminated by comparing the solvent effects on the potential of the electrode of interest measured against a reference electrode in each solvent, we introduce in the result also the solvent effect on the reference-electrode potential, or the transfer free energy of the ion to which the reference electrode responds.

The best that can be done in the latter situation is to choose a reference electrode the potential of which is least likely to vary with the solvent. In the preceding section the effect of solvent on the $E_{1/2}$ and the transfer free energies of ions which experience strong interactions with solvents (e.g., Li^+, Ag^+, Cu^+) was expressed quantitatively on the (tacit) assumption that the Rb^+ ion itself was immune to solvent effects. This was tantamount to assuming that the $E_{1/2}$ and the solvation energy of the Rb^+ ion are independent of the solvent. While in terms of insensitivity to solvent effects the rubidium electrode does constitute a considerable improvement over the hydrogen electrode, electrochemists never accepted the "rubidium scale" of potentials as the final answer and continued their search for electrodes that might be even less sensitive to solvent effects that the rubidium electrode.

Koepp, Wendt, and Strehlow (46) examined 18 metal–organic redox systems in which the metal can exist in two adjacent oxidation states with the objective of finding one that could provide the basis for a nearly solvent-independent electrode potential. A redox couple suitable for this purpose should consist of an uncharged molecule as the reduced form and a unipositive cation as the oxidized form. The two components should be large, spherical, and undergo no structural changes in the redox process. Reversibility of the redox reaction at an inert electrode, compatibility of both forms with common solvents and a reasonable solubility are the other desired criteria. Only two redox couples were found to fulfill the above criteria. These were ferricinium–ferrocene (ferrocene = dicyclopentadienyl iron(II), $Fe(C_2H_5)_2$, a sandwich compound formed between ferrous ion and cyclopentadienyl anions) and its cobalt analog, the cobalticinium–cobaltocene couple. The ferrocene electrode can be used both in voltammetry and in potentiometry. Application of the cobaltocene electrode is restricted to polarographic $E_{1/2}$ measurements because cobaltocene is unstable in solution.

Once we assume that the $E^\circ_{Fc^+/Fc}$ (Fc = ferrocene) is independent of the solvent, the potential of any electrode can be compared in two solvents by measuring the voltage of a cell composed of the electrode of interest and the ferrocene electrode first in one solvent and then in the other. For example, the standard potential of the hydrogen electrode in a nonaqueous solvent, $E°(H,S)$, relative to the SHE in water (for which $E°(H,W) = 0$) is given by the difference between the emf of cell 9-II in water, $E(W)$, and in the nonaqueous solvent, $E(S)$,

$$\mathrm{Pt(s); H_2(g), 1\ atm, H^+}\ (a = 1),\ \mathrm{Fc, Fc^+\ (1:1); Pt(s)} \tag{9-II}$$

When $a_H = 1$ both in water and in the nonaqueous medium,

$$E(W) - E(S) = E°(H,S) - E°(H,W) \tag{9.41}$$

$$= 2.3\frac{RT}{F}\log {}_{m}\gamma_H \tag{9.42}$$

$$= \frac{1}{F}\Delta G_t^\circ(H^+) \tag{9.43}$$

The last equation shows that the transfer free energy of the proton follows automatically from the comparison of standard hydrogen electrode potentials in water and the nonaqueous solvent (cf. Eqn. 5.31). Of course, the potential changes and transfer free energies of other ions can be evaluated in an analogous manner.

The ferrocene and cobaltocene electrodes have been the favorites of Strehlow (47–49) and also of many French electrochemists. In water, the $E°$ is 0.400 V for the ferrocene electrode and −0.918 V for the cobaltocene electrode. Encouraging was the finding that the difference between the $E°$'s of the two electrodes was fairly constant in different solvents, as was the difference between them and the $E°$ of the rubidium electrode. When the ferrocene electrode is assumed to be independent of the solvent, the $E°(H)$ (relative to $E°(H,W) = 0$) is calculated to be −0.01 V, −0.15 V, and +0.15 V in methanol, formamide, and acetonitrile, respectively (47). The corresponding transfer free energies for the proton, $\Delta G_t^\circ(H^+)$, are −0.2, −2.5, and +2.5 kcal mol^{-1}, in that order.

Unfortunately, these and other values of the transfer free energies for single ions obtained by the ferrocene assumption do not agree with those estimated by the presently popular tetraphenylborate assumption (Table 5.5). In Table 9.6, the values of $\Delta G_t^\circ(Ag^+)$ estimated by the two assumptions are compared for several solvents. The ferrocene results are negative with respect to the tetraphenylborate results by anywhere from 1 to 5.5 kcal mol^{-1} and it has been suggested that the ferrocene assumption may give results of the wrong order of magnitude (50). Specific interactions between water and both the ferrocene molecule and the ferricinium ion have been suggested (41, 54) as the possible drawbacks of that redox couple and the criticism was corroborated by a solvation-energy calculation (55). However, even in the absence of specific solvent effects, the *type* of assumption underlying the application of the ferrocene and similar electrodes will always incorporate a formal error. The error inherent in the calculation of the transfer

Table 9.6 Comparison of Transfer Free Energies for the Silver Ion, $\Delta G_t^\circ(Ag^+)$, Estimated by the Ferrocene Assumption and by the Tetraphenylborate Assumption (25°C, Molar Scale)

	$\Delta G_t^\circ(Ag^+)$, kcal mol^{-1}	
Solvent	Ferrocene assumption[a]	Tetraphenylborate assumption[b]
Propylene carbonate	2.2	3.8
Methanol	−3.1	1.5[c]
Ethanol	−4.8	0.95[c]
Formamide	−4.7	−3.7
Dimethylformamide	−6.8	−4.1
Acetonitrile	−8.6	−5.2
N-Methyl-2-pyrrolidone	−8.2	−7.2
Dimethylsulfoxide	−10.6	−8.0
Hexamethylphosphoramide	−12.8	−9.3[a]

[a] Calculated from the data in Ref. 51.
[b] Calculated from the data in Ref. 52, unless otherwise stated.
[c] Calculated from the data in Ref. 53.

free energy of an ion by the ferrocene assumption can be shown by the example of deriving the value of $\Delta G_t^\circ(H^+)$ from cell 9-II. Here, the transfer free energy for the cell reaction is given by

$$F[E(W) - E(S)] = \Delta G_t^\circ(H^+) + \Delta G_t^\circ(Fc) - \Delta G_t^\circ(Fc^+) \qquad (9.44)$$

Obviously, the ferrocene assumption requires that

$$\Delta G_t^\circ(Fc) = \Delta G_t^\circ(Fc^+) \qquad (9.45)$$

We recall from Eqn. 5.34 that the ΔG_t° of any ion can be expressed as the sum of electrostatic and nonelectrostatic contributions. Because the ferrocene molecule is fairly large, the nonelectrostatic component of the $\Delta G_t^\circ(Fc^+)$ can be approximated by $\Delta G_t^\circ(Fc)$ (cf. Eqn. 5.35b):

$$\Delta G_t^\circ(Fc^+) \cong \Delta G_t^\circ(Fc) + \Delta G_t^\circ(Fc^+,el) \qquad (9.46)$$

When Eqn. 9.46 is introduced into Eqn. 9.44, it follows that the term $\Delta G_t^\circ(Fc^+,el)$ does not vanish, so that the ferrocene assumption evaluates the quantity $[\Delta G_t^\circ(H^+) - \Delta G_t^\circ(Fc^+,el)]$, instead of $\Delta G_t^\circ(H^+)$. This may account for the consistently negative deviations of the ferrocene results as compared to the tetraphenylborate results for cations. An advantage of the ferrocene electrode lies in the fact that the electrostatic contribution to the transfer free energy of the ferricinium ion could be calculated from a model, such as the Born equation. However, if specific interactions (not accounted for by the Born model) do take place, they would be included in the formal term $\Delta G_t^\circ(\text{Fc}^+, \text{el})$ contributing a component that would be difficult to evaluate.

Ferrocene and cobaltocene are not the only metal–organic redox couples proposed for the correlation of emf series in different solvents. Their chromium analog, bisdiphenylchromium(I) (chromocene), was employed in polarographic studies by Gutmann and Schmidt (56). Substituted 9,10-phenanthrolines (ferroins) have been used occasionally (45, 57), but these are couples of the $+2/+3$ charge type and are therefore believed to be inherently inferior to the $+1/0$ charge types both in terms of the magnitude of their electrostatic interactions and specific interactions with the solvents.

Another approach to the correlation of electrode potentials which has been popular in the field of voltammetry is based on the estimation of the transfer free energies of single ions by a method originally proposed by Latimer, Pitzer, and Slansky (58). In this method, the Born equation (Eqn. 2.16) for the electrostatic solvation energy for an electroneutral pair of ions is modified by adding empirical corrections R_+ and R_- to the crystallographic radius of the cation, r_+, and of the anion, r_-, respectively, until agreement is obtained between the experimental and the calculated values of the solvation energy. For a 1:1 electrolyte, the Born equation modified in this manner assumes the form

$$-G_\pm^\circ = \left[\frac{Ne^2z^2}{2}\right]\left[1 - \frac{1}{\epsilon}\right]\left[\frac{1}{r_+ + R_+} + \frac{1}{r_- + R_-}\right] \qquad (9.47a)$$

The assumption is then made that the solvation energy of the cation can be adequately represented by Eqn. 9.47b and that of the anion, by Eqn. 9.47c:

$$-G_+^\circ = \left[\frac{Ne^2z^2}{2}\right]\left[1 - \frac{1}{\epsilon}\right]\left[\frac{1}{r_+ + R_+}\right] \qquad (9.47b)$$

$$-G_-^\circ = \left[\frac{Ne^2z^2}{2}\right]\left[1 - \frac{1}{\epsilon}\right]\left[\frac{1}{r_- + R_-}\right] \qquad (9.47c)$$

An analogous, but separate, radius-modification procedure is carried out for the data in water and in each nonaqueous solvent of interest. Finally, the transfer free energy of, for example, the cation is calculated from the difference between the expressions 9.47b representing nonaqueous solvation and hydration:

$$-\Delta G_t^\circ\,(\text{cation}) = \frac{Ne^2z^2}{2}\left[\frac{1 - \dfrac{1}{\epsilon_s}}{r_+ + R_+(s)} - \frac{1 - \dfrac{1}{\epsilon_w}}{r_+ + R_+(w)}\right] \qquad (9.48)$$

In Eqn. 9.48 the empirical radius corrections $R_+(s)$ and $R_+(w)$ are specific to the nonaqueous solvent and to water, respectively, and ϵ_s and ϵ_w are the dielectric constants of the nonaqueous solvent and of water. The above procedure was used to estimate the transfer free energies of the rubidium ion in several nonaqueous solvents, from which the E°'s of both the rubidium and the hydrogen electrodes were calculated relative to the aqueous SHE (47).

Subsequently, the precision of this method was improved by excluding anions from the calculation and applying the modified Born equation directly to *differences* between the transfer free energies of two cations (i.e., differences of the right-hand side of Eqn. 9.48). This approach is particularly suitable to polarographic data in the form of $\Delta\Delta E_{1/2}$ (Eqn. 9.34). In this manner Coetzee et al. (41, 43) evaluated the differences between the standard electrode potentials of the alkali-metal ions in water and both acetonitrile and sulfolane. Some of their results are reproduced in Table 9.7. In Table 9.8 we compare the transfer free energies of single ions obtained by the radius-modified Born equation and by the tetraphenylborate assumption discussed in Chapter 5. The agreement between these two methods is very poor. It should be noted, however, that in comparing results obtained by two extrathermodynamic methods one should not ascribe all of the discrepancy to differences in the methods. Unless the individual ionic contributions are evaluated independently, the differences between ΔG_t° values for ions of like charge and their sums for anions and cations should be the same in any given solvent, independent of the assumption. The fact that it is not the case reflects the poor accuracy of some of the thermodynamic data which the extrathermodynamic calculations use as the starting point. This situation serves to emphasize how difficult it is to obtain reliable physicochemical data in nonaqueous solutions.

Nevertheless, the discrepancies in Table 9.8 go beyond even gross experimental errors. In sulfolane the estimates differ even in sign. The data in methanol were obtained by the original calculation method, which in-

Table 9.7 Standard Potentials of Alkali–Metal Electrodes in Acetonitrile and Sulfolane Relative to Their Values in Water

	$\Delta E^{\circ} = E^{\circ}(S) - E^{\circ}(W)^{a}$	
Electrode	Acetonitrile	Sulfolane[b]
Li^+	0.40	0.31
Na^+	0.29	0.21
K^+	0.20	0.13
Rb^+	0.17	0.11
Cs^+	0.15	0.09

Data from Ref. 43.
[a] In volts; both referred to $E^{\circ}(H,W) = 0$.
[b] At 30°C.

Table 9.8 Comparison of Transfer Free Energies of Single Ions Estimated by the Modified Born Equation and by the Tetraphenylborate Assumption

	Acetonitrile		Sulfolane		Methanol	
Ion	Modified Born[a]	Ph_4B[b]	Modified Born[a]	Ph_4B[b]	Modified Born[c]	Ph_4B[b]
Li^+	9.22	—	7.14	—	—	0.9
Na^+	6.68	3.3	4.92	−0.7	−0.41	2.0
K^+	4.61	1.9	3.00	−1.0	−0.03	2.4
Rb^+	3.91	1.6	2.54	−2.1	+0.06	2.4
Cs^+	3.46	—	2.08	−2.4	0.15	2.3
Cl^-	7.6[d]	10.1	—	12.6	5.60	3.0
Br^-	5.0[d]	7.6	—	9.5	4.95	2.7
I^-	1.9[d]	4.5	—	4.9	4.17	1.6

[a] Unless otherwise stated, the data are from Ref. 43. All values in kcal mol^{-1}.

[b] Based on the assumption that $\Delta G_t^\circ(Ph_4B) = \Delta G_t^\circ(Ph_4As)$. Data from Table 5.5.

[c] Data from Ref. 47.

[d] Data from Ref. 41.

cluded anions. For acetonitrile, that method yielded a negative value for the transfer free energy of the rubidium ion (not shown in Table 9.8), which is contrary to everything we know about the relative solvation of alkali–metal ions in water and acetonitrile.

Any extensive critique of the radius-modified Born equation would be beyond the scope of this text, but the reader can find it in the specialized reviews (59–62).

Summary The most significant contributions of nonaqueous voltammetry lie in two areas: (1) In aprotic media the elucidation of the chemistry of certain radicals that could not exist in protic media. (2) Solvation- and transfer-free energies of cations. Voltammetric studies in aprotic media of the reduction of aromatic compounds to form radical anions and dianions and their oxidation to form radical cations have been highlighted. The rates and mechanisms of these processes and of the coupled chemical reactions, which usually involve abstraction of protons from donors, have also been studied by voltammetry. In aprotic media the reduction of oxygen produces the superoxide anion. In acetonitrile strong acids can be differentiated according to their $E_{1/2}$'s for the reduction of the hydrogen ion. Voltammetric half-wave potentials and their differences between different solvents have been used as measures of the relative solvation of cations. From such studies emanated some proposals for the establishment of solvent-independent emf series, such as based on the ferrocene electrode or on the radius modification of the Born equation.

9.2 ELECTRICAL DOUBLE LAYER

9.2.1 Introduction When we say that certain processes occur at the surface of an electrode, what we really mean by the "surface" is a very narrow (~10–100 Å)† but three-dimensional interphase region sandwiched in between the electrode and the solution phases. In this region the arrangement of solvent and solute molecules and ions differs greatly from that in the bulk of the solution. Adsorption of ions and uncharged solute species as well as of solvent molecules occurs here and the solvent dipoles will tend to be completely or partially oriented in the field of a charged electrode. Significantly, each side of the electrode–solution interface acquires a net electrical charge, for example, if the surface of the metal electrode becomes negatively charged, an equal net positive charge will develop on the solution side of the interface. The term *electrical double layer*, or simply *double layer* stems from the older, oversimplified, model of the interphase region, which was depicted as an arrangement of only two layers of equal and opposite charge. In the modern view, however, the structure of the interphase region is usually more complex (Fig. 9.6). Electric fields of the order of 10^7 V cm^{-1} exist in the double-layer (DL) region as a result of which the surface of the electrode is generally covered with a layer of oriented adsorbed solvent molecules. Furthermore, those ions which have no primary solvation shells or are able to lose them at least on the side facing the metal electrode can displace some of the adsorbed solvent molecules and envelop the electrode surface themselves. This process, which in aqueous solutions is characteristic of most anions and of large cations, is variously called *contact adsorption, specific adsorption*, or *superequivalent adsorption*. The term "specific" refers to the fact that contact adsorption is largely governed by specific chemical properties of the given ion rather than by the sign of the charge on the electrode. Similarly, the term "superequivalent" implies that as a result of specific chemical interactions, more ions might adsorb on the electrode surface than would correspond to the excess charge density on the electrode.

The array of contact-adsorbed molecules and ions is known as the *inner layer* and the locus of the centers of the contact-adsorbed ions, as the *inner Helmholtz plane* (IHP). Solvated ions cannot contact adsorb, and their distance of closest approach to the electrode is the so-called *outer Helmholtz plane* (OHP), located approximately 2 diameters of a water molecule plus the ionic radius from the electrode surface (Fig. 9.6). Because the long-range

† The estimated thickness of the diffuse double layer (see text) varies significantly with electrolyte concentration and charge type and can be greater in very dilute solutions.

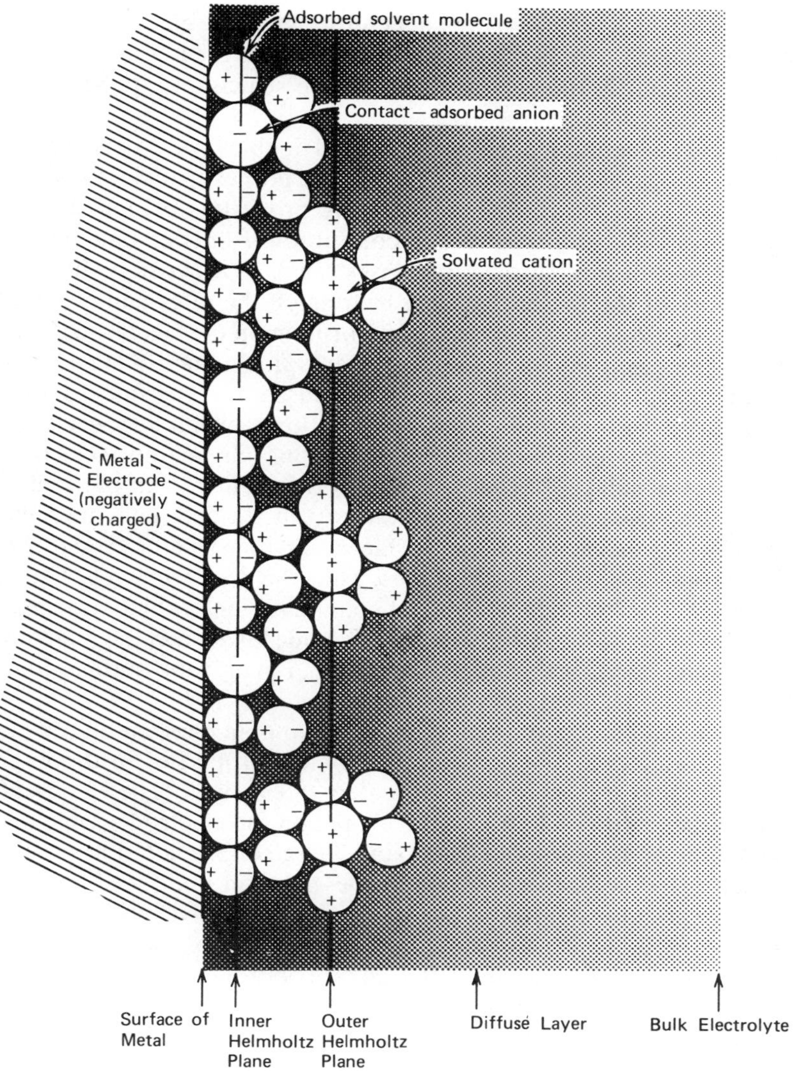

Figure 9.6 A modern model of the electrical double layer.

electrostatic interactions between the electrode and the solvated ions in the OHP are essentially independent of the chemical properties of the ions, the latter are referred to as being *nonspecifically adsorbed*. However, the nonspecifically adsorbed ions are located in or near the OHP only in concentrated solutions; as the solution grows dilute, the competition between the electrical and the thermal forces smears out the solution charge into a *diffuse layer*, extending from the OHP into the bulk of the solution. Adsorption in the DL region means that solute concentrations at the electrode surface may be higher than in the bulk of the solution. When the solutes are ionic, ion–ion interactions both of the electrostatic and the structural type are the rule in the DL region.

It should be mentioned that an electrified double layer exists at the boundary of any two phases and is not an exclusive property of electrode–solution interfaces. Chemists are interested in the properties of double layers for many reasons. DL properties are important toward understanding of colloid chemistry and of any surface processes, such as corrosion. In particular, the structure of the DL has a direct bearing on the rates and mechanism of electrode reactions, on the extent of adsorption on the electrode surface and its effects on electrochemical measurements, including trace analysis.

In this chapter we restrict the DL discussion to the interface between the mercury electrode and electrolyte solutions, which has received by far the most attention in the literature. The reasons why mercury electrodes are preferred over solid electrodes in DL work are many. In particular, each mercury drop presents to the solution a fresh surface, uncontaminated by impurities. Furthermore, liquid surfaces are nonstructured and therefore more reproducible than solid surfaces. Thus the disadvantages of solid electrodes lie in the irreproducibility and contamination of their surfaces due to adsorption of impurities, the existence of oxide layers and the nonuniformity in the surface structure of the metal itself. Finally, the surface tension of liquid metals is relatively easy to measure.

The general objective of DL research is to obtain a quantitative description of the electrical and structural–chemical properties in the interphase region. The two are, of course, intimately intertwined. The electrical properties characteristic of a DL are the charge on the electrode (which is equal in magnitude and opposite in sign to the charge on the solution side) and the potential difference across the electrode–solution interface. Furthermore, the electrical equivalent of a DL of charge separated by a dielectric is a parallel-plate condenser and for a triple layer of charge, two such condensers connected in series. Therefore, the DL capacity and its variation with the applied potential is a frequently measured electrical property. The chemical information most frequently sought in DL studies is the nature of the solute species, ions as well as uncharged molecules, that contact adsorb on the electrode and the relative strength of their adsorption vs solvation. A quantitative measure of adsorption in the DL is the *surface excess* of the solutes in the interphase. The surface excess is the amount of a given species adsorbed per unit area in excess of what the adsorbed amount would be in the absence of a DL.

There are several experimental handles on the properties of the DL just enumerated. Although it is impossible to measure the potential difference across a single electrode–solution interface, it is possible to measure changes in that potential difference across a polarizable interface if the interface of the second electrode in the cell is nonpolarizable. At this point it might be advantageous to review briefly the difference between polarizable and nonpolarizable electrodes. The familiar electrodes in galvanic cells, at which oxidation–reduction processes occur, are sometimes called *charge-transfer electrodes*, or said to have *nonpolarizable* interfaces. Such electrodes can exist in thermodynamic equilibrium with solutions only at a single potential specified by the Nernst equation. *Ideally polarized*, or simply *polarizable* electrodes, are those for which no charge transfer occurs across the metal–solution interface, so that they can exist in thermodynamic equilibrium with a solution at any potential applied from an external source. Consequently, a polarized electrode responds exactly to changes in applied potential. Of course, an ideally polarized electrode does not exist, but over a certain range of potentials, the DME in contact with a solution devoid of oxidizable or reducible species acts as a polarizable electrode.

To summarize, one reason why the DME has been so popular in DL research is its polarizability. Another is the fact that each mercury drop offers a fresh electrode surface, uncontaminated by previous events in the solution. Finally, since mercury is a liquid, it is easy to measure its interfacial tension in contact with solutions.

The interfacial tension of mercury, which varies both with the applied potential and with the composition of the solution, is the oldest, but extremely versatile, source of information on the DL properties. Qualitatively, the lowering of interfacial tension in a solution (relative to pure solvent) is proportional to the extent of contact adsorption. When the interfacial tension, γ_i, is plotted as a function of applied voltage, V, the resulting curve, which is nearly a parabola, is known as the *electrocapillary curve*, which passes through the *electrocapillary maximum* (ECM). At any applied potential the charge on the electrode, q_M, is equal to the negative slope of the electrocapillary curve, $-(d\gamma_i/dV)$.† Thus the ECM corresponds to a potential at which the electrode has zero charge, denoted by *pzc*, or E_z. Moreover, double differentiation of the electrocapillary curve yields the differential capacity of the double layer,

$$C = (dq_M/dV) = -(d^2\gamma/dV^2).$$

Finally, when the interfacial tension is plotted as a function of log $a_\pm$, where $a_\pm$ is the mean activity of the electrolyte, the slope of the resulting curve at constant applied potential is a measure of the surface excess of a single ionic species, Γ. It turns out that the surface excess of the cation, Γ_+, is determined when the nonpolarizable electrode in the cell is reversible to anion, and vice versa, for example,

$$(d\gamma/2RTd \ln a_\pm)_{\text{const. } V-} = -\Gamma_+ .$$

The minus sign on the symbol $V-$ indicates that the nonpolarizable electrode is reversible to an anion. By application of DL models, it is further possible

† All partial differentials are at constant solution composition.

to calculate individually the fraction of the surface excess that corresponds to contact adsorption and to the OHP, respectively. While double differentiation of the electrocapillary curve was the traditional method of determining DL capacitance, this technique has been largely supplanted by the more accurate direct measurements of differential capacitance with an AC impedance bridge. The magnitude of excess charge density on the electrode can then be obtained from the capacitance by integration.

What has been presented here is a most rudimentary qualitative introduction to a highly specialized branch of electrochemistry dealing with phenomena not yet completely understood. In view of the very limited application of DL theory to nonaqueous solvents to date, no attempt is made here to discuss the various quantitative treatments, on which there exists a formidable literature. A reader new to this subject would find most instructive the detailed step-by-step development of the DL theory and applications in the very readable teaching text by Bockris and Reddy (63). Several in-depth reviews of the DL models with the relevant mathematical treatment and data interpretation are available (e.g., 64–69). Understandably, most of the DL work has focused on aqueous solutions. With a few well-known exceptions, the use of nonaqueous solvents in DL research is comparatively new and the existing findings are tentative. The limited results have been reviewed by Payne (70, 71) and Schiffrin (72).

9.2.2 Studies in Nonaqueous Solvents

A General Overview The change from water to a nonaqueous solvent can affect several properties which determine the structure of the DL. One such solvent-dependent property is the surface tension at the mercury–solvent interface, or the interfacial tension, γ_i. It is related to the work of adhesion of the solvent on mercury, W_a, by the expression

$$\gamma_i = \gamma_{Hg} + \gamma_s - W_a \tag{9.49}$$

where γ_{Hg} and γ_s are the surface tensions at the mercury–air and the solvent–air interfaces, respectively. Interfacial tension is particularly relevant to elucidating the effect of solvent on the process of contact adsorption and the structure of the resulting inner layer. An important energy step in the process of contact adsorption of solute species is the desorption of the solvent molecules from the electrode surface. Before a solute molecule can occupy a site on the electrode surface, it must first displace from it a solvent molecule. This requires an input of energy equal to W_a. The displaced solvent molecule must then recombine with the bulk solvent, a process determined by the internal cohesion of the liquid, the energy of which is measured by the surface tension, γ_s. Greater internal cohesion favors desorption of the solvent. Thus the overall work required to desorb a solvent molecule from the electrode surface is $(\gamma_{Hg} - \gamma_i)$. The lower the interfacial tension, the harder it is to desorb the solvent. However, as can be seen from Eqn. 9.49,

Table 9.9 Surface Tension and Adhesion of Solvents on Mercury[a]

	Surface tension	Interfacial tension	Work of adhesion	μ (Debye)
Water	72 (25°C)	425 (25°C)	131	1.84
Methanol	22 (25°C)	393	114	1.70
Ethanol	23[b]	387	120	1.69
n-Propanol	24	384	124	1.68
n-Butanol	25[b]	383	126	1.66
Formic Acid	28	399	123	1.52
Formamide	58	389 (25°C)	154	3.75
Dimethylformamide	36	376 (20°C)	144	3.82
Dimethylsulfoxide	45 (25°C)	371 (25°C)	158	4.3
Sulfolane	53 (30°C)	374 (30°C)	164	4.81
Aniline	43	357	170	1.53
Pyridine	38	363	159	0

[a] Unless otherwise referenced, the data are from the compilation by Payne (71). Values are at 18°C, unless otherwise specified. The energy values are in erg cm^{-2}.
[b] From Ref. 70.

a lower γ_i does not necessarily indicate stronger adsorption of the solvent on mercury because of the differences in the internal cohesion among different liquids.

The relationship between interfacial tension and the work of adhesion for some liquids is illustrated in Table 9.9. Significantly, water is more strongly adsorbed on mercury than the alcohols (higher W_a), but because of the strong internal cohesion of liquid water (greater γ_s), its interfacial tension is also higher. As a result water is generally easier to displace from the mercury–electrode surface than are the alcohols and other organic solvents so that adsorption of solute species is strongest from aqueous solutions. However, as the data on the work of adhesion indicate, many organic solvents interact with mercury more strongly than water. The large work of adhesion observed for pyridine and aniline has been attributed to specific interactions of the π-electrons with mercury. In the case of the formamides, dimethylsulfoxide, and sulfolane we are dealing with solvent molecules possessing large dipole moments and polarizabilities. Even when a metal electrode is uncharged, dipolar substances can induce in the metal so-called image dipoles, with resulting dipole-dipole interactions. Also dispersion forces contribute to electrode–solvent interactions. It has been reported, however, that the DL properties of amides seem to be independent of the dipole moment.

Solvent–electrode and solvent–solvent interactions are not the only determining forces in contact adsorption. Obviously the solvation energy of the ion enters into consideration, because the ion must shed its solvation shell on the side facing the electrode before it can come in contact with it. Consequently, the poorly solvated ions have the greatest tendency toward contact adsorption. For example, although dimethylsulfoxide and sulfolane adsorb strongly on mercury (Table 9.9), anion adsorption from their solutions is also very strong, presumably because of weak solvation.

Another frequently measured property that depends on the solvent is the DL capacity and its variation with the applied potential. Qualitatively, the solvent effect on the DL capacity might be expected to follow the expression for the capacity of a parallel-plate condenser:

$$C = \frac{\epsilon}{4\pi d} \tag{9.50}$$

where d is the distance between the plates. However, the macroscopic dielectric constant of the solvent cannot apply here. The high electric field in the vicinity of a charged electrode leads to an orientation of the solvent dipoles, resulting in so-called dielectric saturation. Furthermore, the solvent molecules adsorbed at the electrode surface are depolymerized. As a result of both these effects, the effective value of ϵ in the inner layer approaches a low limiting value (e.g., $\epsilon_{H_2O} \cong 6$) which does not differ much with the solvent. The solvent molecules in layers more distant from the electrode are only partially oriented and for aqueous solutions the average dielectric constant in this region was calculated to be about 40 (roughly the mean value between 78 and 6). If the entire double layer is then represented by two capacitors connected in series (one having $\epsilon = 6$ and the other, $\epsilon = 40$), it is the low-dielectric region which largely determines the overall DL capacity (63).

It seems that DL capacity is not a function of the bulk dielectric constant of the solvent. Instead, it has been suggested by Payne (70, 71, 73) that the nature of the solvent determines the average thickness of the inner layer, which in model terms would correspond to the distance parameter d in Eqn. 9.50. Indeed, in amides the capacity is inversely proportional to the molecular weight of the solvent (see Table 9.11 and discussion).

Much information about the structural changes in the DL is derived from the study of capacity as a function of applied potential. The capacity–potential curves (Fig. 9.7) in water and in most nonaqueous solvents are similar in shape in that most of them consist of a constant–capacity region, a hump, and sometimes a steep rise at extreme anodic or cathodic potentials, or both (Fig. 9.8). The interpretation of the capacity hump is the subject of continuing controversy. One school of thought ascribes it to a minimum in the orientation polarization of adsorbed solvent dipoles (73–75). According to it the dipoles of adsorbed solvent molecules are free to rotate in the potential region of the hump, but are permanently oriented at other potentials. It is further argued that the dipoles possess a preferred orientation, which is positive (toward the metal) when the hump occurs on the anodic side of the

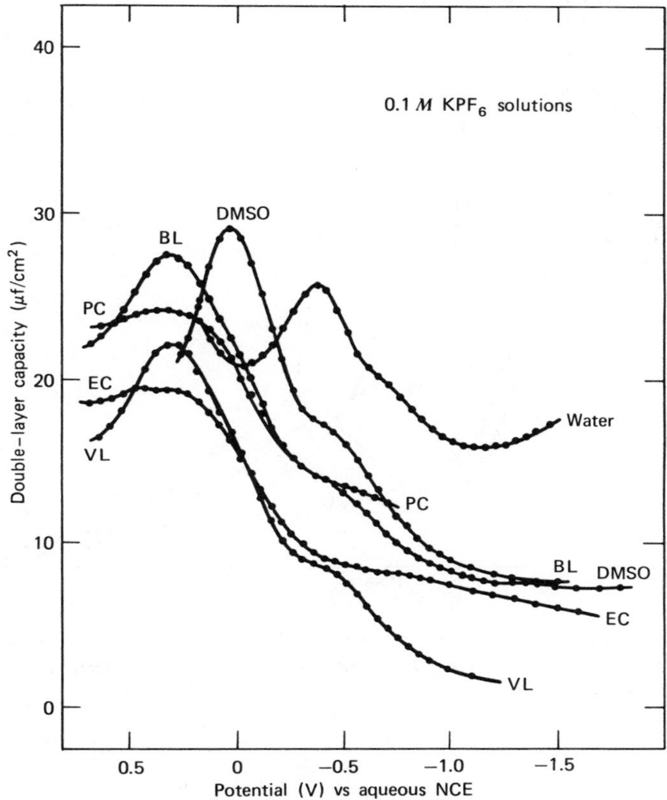

Figure 9.7 Capacity-potential curves for 0.1 *M* KPF_6 solutions in water, dimethylsulfoxide (DMSO), propylene carbonate (PC), ethylene carbonate (EC), 4-butyrolactone (BL), and 4-valerolactone (VL) at 25°C. The curves for the EC and VL solutions are displaced downward by 5 $\mu f/cm^2$ for clarity. Reprinted with permission from *J. Phys. Chem.*, **71,** 1548 (1967). Copyright 1967. American Chemical Society.

pzc and negative for cathodic humps. Another view interprets the hump as the result of contact adsorption of ions, more specifically as the point where there is a maximum rate of growth in the amount of the contact-adsorbed charge as a function of the potential or the charge of the electrode (63, 76). The problem is further complicated by findings that no capacity hump seems to occur in some solvents (acetonitrile, dimethylformamide), while two humps occur in others (*N*-monoalkylacetamides, *N*-methylpropionamide). These and other DL effects will be considered next for specific classes of solvents.

Alcohols Electrocapillary research in the lower alcohols dates back at least

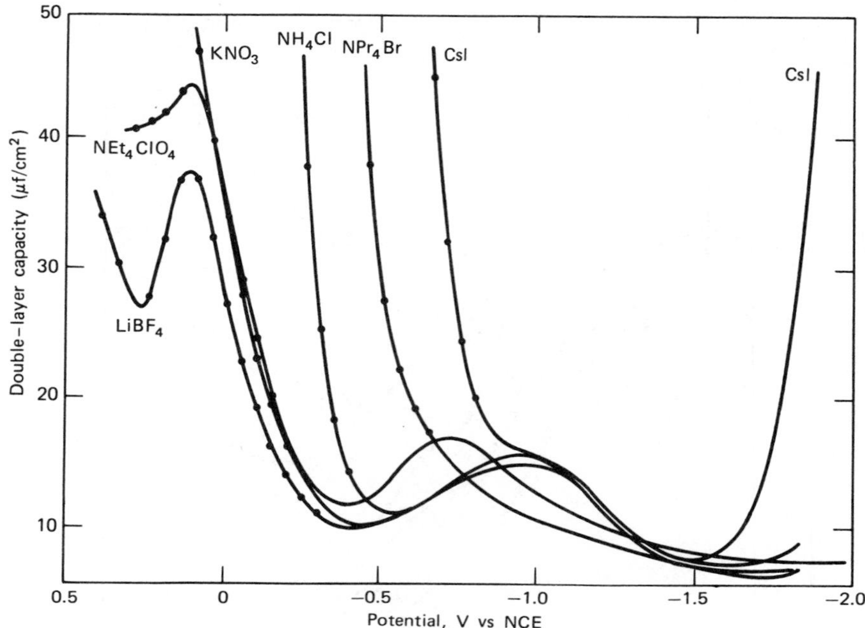

Figure 9.8 Double-layer capacity for 0.1 *M* solutions in *N*-methylacetamide at 30°C. Reprinted with permission from *J. Phys. Chem.*, **73**, 3598 (1969). Copyright 1969. American Chemical Society.

Table 9.10 Some Double-Layer Properties of the Alcohols[a]

Alcohol	W_a (erg cm^{-2})	γ_i (erg cm^{-2})	C_{min} [b] (μF cm^{-2})	Length of molecule (Å)	ϵ (25°C)
Methanol	114	393	8.7[c]	4.4	32.6
Ethanol	120	387	6.8[c]	5.9	24.3
Butanol	126	383	6.5[d]	8.4	17.1
Octanol	137	375	3.1	13.3	10.3 (20°C)

[a] Condensed from a compilation by Payne (70). Data are at 25°C.
[b] Minimum capacity on the cathodic side.
[c] In 0.1 *N* LiCl.
[d] In 0.53 *N* LiCl.

to the work of Gouy in 1906 (77) and of Frumkin in 1923 (78). More recent are the direct differential-capacity measurements by Grahame (79), Garnish and Parsons (80), and many others, whose results have been reviewed by Payne (70). The relationship between several DL parameters and some physical properties of alcohols are evident from Table 9.10. As the length of the alkyl chain of the alcohol increases, there is a drop in the interfacial tension and in the DL capacitance as well as an increase in the work of adhesion on mercury. The DL capacity seems to correlate with the macroscopic dielectric constant of the solvent and with the size of the alcohol molecule, which controls the thickness of the inner layer.

Adsorption of ions from methanolic solutions is fairly strong, but weaker than from aqueous solutions, as would be predicted from the lower interfacial tension of methanol. Adsorption grows progressively weaker as the chain length of the alcohol increases. Halide ions other than the F^- ion as well as Cs^+ and NH_4^+ ions are contact adsorbed from methanolic solutions at potentials other than the most negative (beyond -1.0 V vs the aqueous SCE). This fact was demonstrated by a virtual overlap of the capacity–voltage curves for 1 M solutions of KF and KI in methanol (80). In methanol, as in water, the capacity within the alkali series increases in the order $Li^+ < Na^+ < K^+ < Cs^+$. The lower capacity for the smaller ions could be a function of an increase in their effective radii due to primary solvation (70). No capacity humps such as are found in most other solvents (Figs. 9.7–9.9) have been detected in alcohols. The question of the preferred orientation of the alcohol dipoles at the electrode surface is still controversial, though current opinion is leaning toward a negative orientation, that is, with the oxygen atom pointing toward the electrode.

Amides Several studies of the DL capacity as a function of potential in formamide, and a series of *N*-alkylformamides, acetamides, and propionamides have been reported and analyzed (70, 71, 73). It was generalized that a single capacity hump in the cathodic branch of the capacity–potential curve is characteristic of amides having at least one unsubstituted *N*-hydrogen atom (formamide, *N*-methylformamide). The cathodic hump is absent in fully substituted amides, such as dimethylformamide and dimethylacetamide, but a different hump appears in their solutions on the anodic side of the ECM. In *N*-monoalkylacetamides and *N*-methylpropionamide, both an anodic and a cathodic hump appear. The cathodic humps have been interpreted as a property of the solvent. Specifically, they are considered as an indication of a negative orientation of the solvent dipoles at a mercury electrode. No unambiguous assignment could be made for the anodic humps, but the possibility of specific adsorption of anions has been advanced as the origin. Figure 9.9 shows an example of a system with two humps.

Capacity–potential curves also serve as an index of specific (contact) adsorption of ions, which is manifested by the steep rise in the anodic branch

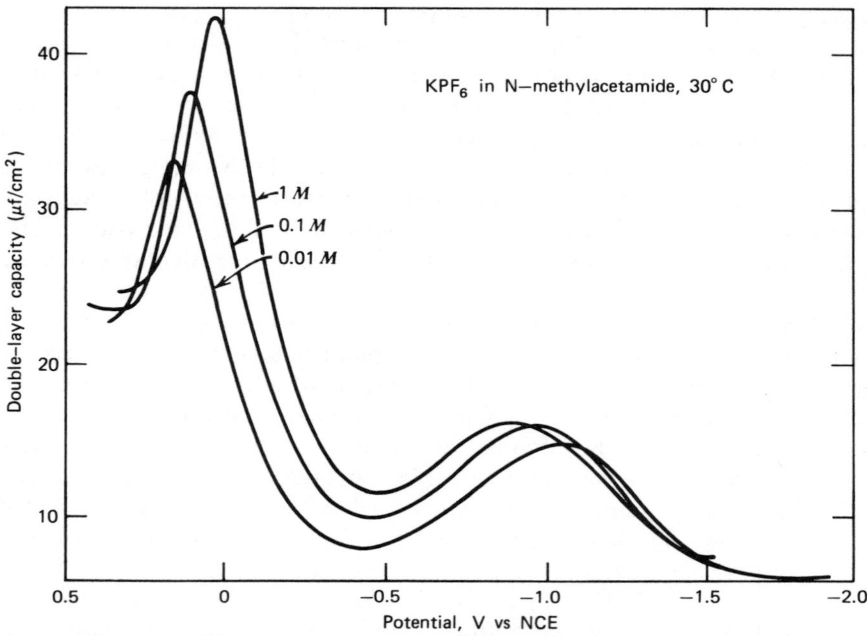

Figure 9.9 Double-layer capacity for KPF_6 solutions of indicated concentration in *N*-methylacetamide ar 30°C. Vertical lines indicate potential of zero charge. Reprinted with permission from *J. Phys. Chem.*, **73,** 3598 (1969). Copyright 1969. American Chemical Society.

for the adsorption of anions, and in the cathodic branch, for the adsorption of cations. The relative degree of adsorption is proportional to the relative negative shift of the *pzc*. For example, from the data in Fig. 9.8, the order of contact adsorption from *N*-methylacetamide for anions is $I^- > Br^- > Cl^- > PF_6^- > ClO_4^- > NO_3^- > BF_4^-$. Among the cations only Cs^+ and NH_4^+ (not shown in Fig. 9.8) are contact adsorbed from NMA. In the case of the Cs^+ ion the adsorption is also confirmed by a sharp drop in the interfacial tension. Contact adsorption of Cs^+ is a general phenomenon in amide solvents.

Differences in the specific adsorption of anions show a marked dependence on their solvation energies. Simple anions with localized charge are preferentially solvated by H-bonding solvents, as compared to aprotic solvents. As a result, the specific adsorption of the halide ions from solvents like formamide and water is much weaker than from DMF or DMSO. Measurements of the lowering of the interfacial tension at the ECM (relative to pure solvent) indicate that the strength of halide adsorption increases in the order formamide < NMF < DMF. The reverse is true for the adsorption of polyatomic ions, such as PF_6^-, NO_3^-, SCN^-, or ClO_4^-, which are more polarizable and tend to be preferentially solvated by the more polarizable DMF; these adsorb most strongly from aqueous solutions. Significantly, the strongly solvated F^- ion exhibits no specific adsorption from formamide or *N*-methylformamide. For similar reasons the specific adsorption of tetraalkylammonium cations from formamides is considerably less than from aqueous solutions.

The dependence of DL capacity on the properties of amide solvents in the absence of specific adsorption of ions was examined by Payne (73). As shown by his data reproduced in Table 9.11, the capacity is not a function of the macroscopic dielectric constant, but appears to be related to the molecular weight of the solvent and to the length of the solvent molecule, suggesting a proportionality with the thickness of the solvent layer in the inner region (71). The relationships are particularly striking in the case of isomers, such as dimethylformamide and *N*-methylacetamide, which have dielectric constants of 36.7 and 178.9, respectively, but have virtually equal DL capacities. It is interesting, however, that the DL capacity of CsCl solutions in NMF (81) could be accounted for by adopting for the solvent an effective dielectric constant value of 83.6 (as opposed to the bulk value of 182.4) in

Table 9.11 Double-Layer Capacity and Physical Properties of Amides at 25°C (73)[a]

Solvent	C_{min} [b] (μF cm^{-2})	Mol Wt	Length of molecule (Å)	ϵ
Formamide	12.5[c]	45	5.6	109.5
N-Methylformamide	8.5[c]	59	6.4	182.4
Dimethylformamide	6.8	73	6.7	36.7
N-Methylacetamide	6.7 (30°C)	73	6.8	178.9
N-Ethylformamide	9.5	73	7.7	—
Dimethylacetamide	6.0	87	6.9	37.8
N-Methylpropionamide	5.6	87	7.8	172.2
N-Ethylacetamide	5.6	87	7.7	129.9
N-*t*-Butylformamide	8.4	101	7.5	—
N-*n*-Butylacetamide	5.2	115	9.6 or 7.0[d]	101.7

[a] The length of a molecule is measured along the axis of the dipole assuming a planar molecule.
[b] Minimum capacity on the cathodic side measured in a solution of 0.1 *M* KPF_6, unless otherwise stated.
[c] 0.1 *M* KCl.
[d] *n*-Butyl group and oxygen atom in *cis* (7.0 Å) or *trans* (9.6 Å) configuration.

the diffuse DL, a value which would correspond to partial dielectric saturation.

Other Solvents DL work has been reported in dimethylsulfoxide, sulfolane, formic acid, acetone, pyridine, propylene carbonate and other cyclic esters, and lactones (71). Anions are strongly contact adsorbed from DMSO in the order $I^- > Br^- > Cl^- > ClO_4^- > PF_6^-$, but there is virtually no adsorption of cations. In both DMSO and sulfolane capacity humps occur on the anodic side of the *pzc*, from which it was inferred that the preferred orientation of the solvent dipoles is with the positive end facing the electrode. It must be stressed, however, that potential shifts recorded for different solvents always contain an unknown and probably appreciable liquid-junction potential between the aqueous calomel electrode and the nonaqueous solution. It seems that estimates of liquid-junction potentials at the boundaries of solutions in different solvent media, such as is discussed in Chapter 5, could improve interpretation of some DL data.

An example of a detailed analysis of the DL properties in a nonaqueous medium is the article by Kim et al. (82) dealing with LiCl solutions in DMSO. It is much too specialized to be included here, but is mentioned as possible advanced reading material.

Undoubtedly we are at the present time in the initial stages of DL research in nonaqueous media, and there is every reason to believe that this frontier of electrochemistry will see significant developments in the future.

Summary Studies of the electrical double layer (DL), or the interphase region between an electrode (usually mercury) and a solution, aim to measure and interpret primarily such properties as the charge on the electrode, the DL capacity and its variation with applied potential, and the electrocapillary curve, which is a plot of the interfacial tension of mercury as a function of applied voltage. From such measurements, the extent of adsorption on the electrode, the nature of the adsorbed species, and other information on the structure of the DL is obtained. Contact adsorption of ions is governed by an interplay of solvent–electrode, solvent–solvent, and ion–solvent interactions. Adsorption is generally strongest from aqueous solutions and it varies inversely with the strength of ionic solvation. Capacity–potential curves have been analyzed for electrolyte solutions in water, alcohols, amides, and other solvents. They have been interpreted in terms of the relative adsorption and solvation of ions as well as the preferred orientation of solvent dipoles at the electrode surface.

LITERATURE CITED

1 Zuman, P., L. Meites, and I. M. Kolthoff, Eds., *Progress in Polarography,* Vol. III, Wiley-Interscience, New York, 1972.

2 Delahay, P., and C. W. Tobias, Eds., *Advances in Electrochemistry and Electrochemical Engineering,* Wiley-Interscience, New York, 1970.

3 Bard, A. J., Ed., *Electroanalytical Chemistry, A Series of Advances,* Marcel Dekker, New York, 1967.

4 Mann, C. K., "Nonaqueous Solvents for Electrochemical Use," in A. J. Bard, Ed., *Electroanalytical Chemistry, A Series of Advances,* Vol. III, Marcel Dekker, New York, 1969.

5 Headridge, J. B., *Electrochemical Techniques for Inorganic Chemists,* Academic Press, New York, 1969.

6 Aten, A. C., and G. J. Hoijtink, *Z. Physik. Chem.* (*Frankfurt*) **21,** 192 (1959); *Advances in Polarography,* Pergamon Press, Oxford, 1961.

7 Aten, A. C., C. Büthker, and G. J. Hoijtink, *Trans. Faraday Soc.,* **55,** 324 (1959).

8 Aten, A. C., J. Dieleman, and G. J. Hoijtink, *Discussions Faraday Soc.,* **29,** 182 (1960).

9 Hoijtink, G. J., *Rec. Trav. Chim.,* **74,** 1525 (1955).

10 Hoijtink, G. J., *Discussions Faraday Soc.,* **45,** 4 (1968).

11 Hoijtink, G. J., E. de Boer, P. H. Van der Mey, and W. P. Weyland, *Rec. Trav. Chim.,* **75,** 487 (1956).

12 Hoijtink, G. J., J. van Schooten, E. de Boer, and W. Y. Aalbersberg, *Rec. Trav. Chim.,* **73,** 355 (1954).

13 Given, P. H., *J. Chem. Soc.,* **1958,** 2684.

14 Pointeau, R., *Ann. Chim.,* **7,** 669 (1962).

15 Streitwieser, A., and I. Schwager, *J. Phys. Chem.,* **66,** 2316 (1962).

16 Wawzonek, S., R. Berkey, E. W. Blaha, and M. E. Runner, *J. Electrochem. Soc.,* **102,** 235 (1955).

17 Allison, A. C., T. A. Gough, and M. E. Peover, *Nature,* **197,** 764 (1963).

18 Gough, T. A., and M. E. Peover, *Polarography—1964,* G. J. Hills, Ed., Interscience, New York, 1966. p. 1017.

19 Peover, M. E., "Electrochemistry of Aromatic Hydrocarbons and Related Substances," in A. J. Bard, Ed., *Electroanalytical Chemistry, A Series of Advances,* Vol. II, Marcel Dekker, New York, 1967.

20 Cisak, A., and P. J. Elving, "Electrochemistry in Pyridine II, Polarographic Behavior of Benzophenone," in T. Kambara, Ed., *Modern Aspects of Polarography,* Plenum Press, New York, 1966.

21 Mann, C. K., and K. K. Barnes, *Electrochemical Reactions in Nonaqueous Systems,* Marcel Dekker, New York, 1970.

22 Peover, M. E., and J. D. Davies, *J. Electroanal. Chem.,* **6,** 46 (1963).

23 Peover, M. E., *Trans. Faraday Soc.,* **60,** 417 (1964).

24 Given, P. H., and M. E. Peover, *J. Chem. Soc.,* **1960,** 385.

25 Austen, D. E. G., D. E. J. Ingram, P. H. Given, and M. E. Peover, *Nature,* **182,** 1784 (1958).

26 Given, P. H., and M. E. Peover, *Collection Czech. Chem. Commun.,* **25,** 3195 (1960).

27 Peover, M. E., and B. S. White, *J. Electroanal. Chem.,* **12,** 93 (1967).

28 Adams, R. N., *Acct. of Chem. Res.,* **2,** 175 (1969).

29 Bader, J., and T. Kuwana, *J. Electroanal. Chem.,* **10,** 104 (1965).

30 Feldberg, S. W., *J. Amer. Chem. Soc.,* **88,** 390 (1966).

31 Hercules, D. M., *Science,* **145,** 808 (1964).

32 Case, B., N. S. Hush, R. Parsons, and M. E. Peover, as cited in Ref. (19), p. 42.

33 Maricle, D. L., and W. G. Hodgson, *Anal. Chem.,* **37,** 1562 (1965).

34 Sawyer, D. T., and J. C. Roberts, *J. Electroanal. Chem.*, **12,** 90 (1966).

35 Goolsby, A. D., and D. T. Sawyer, *Anal. Chem.*, **40,** 83 (1968).

36 Coetzee, J. F., and I. M. Kolthoff, *J. Amer. Chem. Soc.*, **79,** 6110 (1957).

37 Elving, P. J., and M. S. Spritzer, *Talanta,* **12,** 1243 (1965).

38 Pleskov, V.A., *Usp. Khim.*, **16,** 254 (1947).

39 Coetzee, J. F., and J. M. Simon, *Anal. Chem.*, **44,** 1129 (1972).

40 Kolthoff, I. M., and J. F. Coetzee, *J. Amer. Chem. Soc.*, **79,** 870, 1852 (1957).

41 Coetzee, J. F., and J. J. Campion, *J. Amer. Chem. Soc.*, **89,** 2513, 2517 (1967).

42 Coetzee, J. F., D. K. McGuire, and J. L. Hedrick, *J. Phys. Chem.*, **67,** 1814 (1963).

43 Coetzee, J. F., J. M. Simon, and R. J. Bertozzi,*Anal. Chem.*, **41,** 766 (1969).

44 Brown, G. H., and R. Al-Urfali, *J. Amer. Chem. Soc.*, **80,** 2113 (1958).

45 Kolthoff, I. M., and F. G. Thomas, *J. Phys. Chem.*, **69,** 3049 (1965).

46 Koepp, H. M., H. Wendt, and H. Strehlow, *Z. Elektrochem.*, **64,** 483 (1960).

47 Strehlow, H., "Electrode Potentials in Non-Aqueous Solvents," in J. J. Lagowski, Ed., *The Chemistry of Non-Aqueous Solvents,* Vol. I, Academic Press, New York, 1966, Chap. 3.

48 Strehlow, H., and M. Schneider, *Pure Appl. Chem.*, **25,** 327 (1971).

49 Strehlow, H., and H. Wendt, *Z. Physik. Chem., N. F.*, **30,** 141 (1961).

50 Kolthoff, I. M., and M. K. Chantooni, Jr., *J. Phys. Chem.*, **76,** 2024 (1972).

51 Alexander, R., A. J. Parker, J. H. Sharp, and W. E. Waghorne, *J. Amer. Chem. Soc.*, **94,** 1148 (1972).

52 Cox, B. G., G. R. Hedwig, A. J. Parker, and D. W. Watts, *Aust. J. Chem.*, **27,** 477 (1974).

53 Popovych, O., A. Gibofsky, and D. H. Berne, *Anal. Chem.*, **44,** 811 (1972).

54 Sharp, J. H., and A. J. Parker, *Proc. Roy. Austral. Chem. Inst.*, **39,** 89 (1972).

55 Alfenaar, M., *J. Phys. Chem.*, **79,** 2200 (1975).

56 Gutmann, V., and R. Schmid, *Monatsh.*, **100,** 2113 (1969).

57 Nelson, I. V., and R. T. Iwamoto, *Anal. Chem.*, **35,** 867 (1963).

58 Latimer, W. M., K. S. Pitzer, and C. M. Slansky, *J. Phys. Chem.*, **7,** 108 (1939).

59 Conway, B. E., "Proton Solvation and Proton Transfer Processes in Solution," in J. O'M. Bockris and B. E. Conway, Eds., *Modern Aspects of Electrochemistry,* Butterworth, London, 1964, Vol. III, Chap 2.

60 Laidler, K. J., and C. Regis, *Proc. Roy. Soc.*, **A241,** 80 (1957).

61 Popovych, O., *Crit. Rev. Anal. Chem.*, **1,** 73 (1970).

62 Popovych, O., "Transfer Activity Coefficients (Medium Effects)," in I. M. Kolthoff and P. J. Elving, Eds., *Treatise on Analytical Chemistry,* Part I, Vol. I, 2nd ed., Wiley, New York, 1978, Chap. 12.

63 Bockris, J. O'M., and A. K. N. Reddy, *Modern Electrochemistry*, Vol. II, Plenum Press, New York, 1970, Chap. 7.

64 Delahay, P., *Double Layer and Electrode Kinetics,* Wiley-Interscience, New York, 1965.

65 Devanathan, M. A. V., *Chem. Rev.*, **65,** 635, (1965).

66 Grahame, D. C., *Chem. Rev.*, **41,** 441 (1947).

67 Mohilner, D. M., "The Electrical Double Layer Part I. Elements of Double Layer Theory," in A. J. Bard, Ed., *Electroanalytical Chemistry, A Series of Advances,* Vol. I, Marcel Dekker, New York, 1966.

68 Parsons, R., "Equilibrium Properties of Electrified Interphases," in J. O'M Bockris, Ed., *Modern Aspects of Electrochemistry,* No. 1, Butterworths, London, 1954, Chap. 3.

69 Parsons, R., "The Structure of the Electrical Double Layer and Its Influence on the Rates of Electrode Reactions," in P. Delahay, Ed., *Advances in Electrochemistry and Electrochemical Engineering,* Vol. I, Wiley-Interscience, New York, 1961, Chap. 1.

70 Payne, R., "The Electrical Double Layer in Nonaqueous Solutions," in P. Delahay, Ed., *Advances in Electrochemistry and Electrochemical Engineering,* Vol. VII, Wiley-Interscience, New York, 1970, Chap. 1.

71 Payne, R., "Electrode Processes Part I—The Electrical Double Layer," in A. K. Covington and T. Dickinson, Eds., *Physical Chemistry of Organic Solvent Systems,* Plenum Press, London and New York, 1973.

72 Schiffrin, D. J., *Electrochemistry,* **1,** 223 (1970); **2,** 169 (1972); **3,** 94 (1073).

73 Payne, R., *J. Phys. Chem.,* **73,** 3598 (1969).

74 Macdonald, J. R., and C. A. Barlow, *J. Chem. Phys.,* **36,** 3062 (1962).

75 Watts-Tobin, R. J., *Phil. Mag.,* **6,** 133 (1961).

76 Payne, R., *J. Phys. Chem.,* **71,** 1548 (1967).

77 Guoy, G., *Ann. Chim. Phys.,* **9,** (8), 75 (1906).

78 Frumkin, A. N., *Z. Elektrochem.,* **103,** 43 (1923).

79 Grahame, D. C., *Z. Elektrochem.,* **59,** 740 (1955).

80 Garnish, J. D., and R. Parsons, *Trans. Faraday Soc.,* **63,** 1754 (1967).

81 Damashkin, B. B., and R. V. Ivanova, *Zh. Fiz. Khim.,* **38,** 176 (1964).

82 Kim, S. H., T. N. Andersen, and H. Eyring, *J. Phys. Chem.,* **74,** 4555 (1970).

GENERAL REFERENCES

Bard, A. J., Ed., *Electroanalytical Chemistry, A Series of Advances,* Marcel Dekker, New York, 1967– .

Brezina, M., and P. Zuman, *Polarography in Medicine, Biochemistry and Pharmacy,* Interscience, New York, 1958.

Crow, D. R., and J. V. Westwood, *Polarography,* Methuen, London, 1968.

Delahay, P., and C. W. Tobias, Eds., *Advances in Electrochemistry and Electrochemical Engineering,* Wiley-Interscience, New York, 1970– .

Gough, T. A., and M. E. Peover, *Polarography—1964,* Macmillan, London, 1966.

Heyrovsky, J., and P. Zuman, *Practical Polarography,* Academic Press, New York, 1968.

Kolthoff, I. M., and J. J. Lingane, *Polarography,* 2nd ed., Interscience, New York, 1952.

Longmuir, I. S., Ed., *Advances in Polarography,* Vols. I–III, Pergamon Press, Oxford, London, New York, Paris, 1960.

Mann, C. K., and K. K. Barnes, *Electrochemical Reactions in Nonaqueous Systems,* Marcel Dekker, New York, 1970.

Meites, L., "Voltammetry at the Dropping Mercury Electrode," in I. M. Kolthoff, P. J. Elving, and E. B. Sandell, Eds., *Treatise on Analytical Chemistry,* Part I, Vol. IV, Wiley-Interscience, New York, 1963.

Meites, L., *Polarographic Techniques,* 2nd ed., Wiley, New York, 1965.

Milner, G. W. C., *The Principles and Applications of Polarography,* Longmans, Green, New York, 1957.

Müller, O. H., "Polarography," in A. Weissberger, Ed., *Physical Methods of Organic Chemistry,* 3rd ed., Part IV, Interscience, New York, 1960.

Strobel, H. A., *Chemical Instrumentation,* 2nd ed., Addison-Wesley, Reading, Mass., 1973, Chaps. 23, 25, 26.

Zuman, P., *Organic Polarographic Analysis,* Macmillan, New York, 1964.

Zuman, P., *The Elucidation of Organic Electrode Processes,* Academic Press, New York, 1969.

Zuman, P., *Topics in Organic Polarography,* Plenum Press, London, 1970.

Zuman, P., and I. M. Kolthoff, Eds., *Progress in Polarography,* Vol. I, II, Interscience, New York, 1962.

Zuman, P., L. Meites, and I. M. Kolthoff, Eds., *Progress in Polarography,* Vol. III, Wiley-Interscience, New York, 1972.

Zuman, P., and C. L. Perrin, *Organic Polarography,* Interscience, New York, 1969.

CHAPTER TEN

Kinetics

Experimental evidence indicates that the rates of reactions, both organic and inorganic, can vary quite dramatically with the solvent medium. It is the intent of this chapter to examine the various factors that contribute to the rates and mechanisms of reactions and to illustrate the effects with selected examples. Parker (1) has presented an extensive review of solvent effects on rates of bimolecular reactions and that should be consulted for further examples. A discussion of solvent effects on inorganic reactions has been given by Watts (2). Other examples of the interpretation of reaction rates via transfer activity coefficients are given by Parker et al. (3, 4).

One of the goals, apart from the study of fundamental mechanisms of reactions, is to select a solvent for a particular reaction where the reaction time is minimized, the yield is increased, and contaminating side reactions are avoided. Common solvents used for studying reactions in are DMF, DMAC, DMSO, acetone, and HMPT.

10.1 KINETICS OF ORGANIC REACTIONS

Generally, reactions involve either anions, cations, neutral species, or transition states in a particular solvent. Therefore, to understand the relative reaction rates in different solvents it is important to review the interactions (solvation) of the various species with the solvent molecules. Several properties of the solvent, such as solvent structure, dielectric constant, molar polarizability, dipole moment, viscosity, acidity, basicity, and hydrogen-bonding capability are relevant, as are properties such as charge density, polarizability, and size of the solute species. For example, a typical bimolecular reaction between an anion and an uncharged species can be represented by

$$Y:^{-} + RX \rightleftharpoons [YRX]^{\ddagger} \rightarrow YR + X^{-} \qquad (10.1)$$

where $[YRX]^{\ddagger}$ represents the transition state.

Hydrogen bonding is an important contributor to variations in reaction rates. For example, many bimolecular reactions, such as 10.1, are almost a million times faster in a weakly dipolar aprotic solvent like nitromethane than they are in the protic solvent *N*-methylformamide (NMF). However, other factors, such as dispersion forces, ion–dipole, and dipole–dipole interactions, must also be taken into account, as well as complexing of donors with acceptors, for example, I_2–DMSO. Solvent structure making and breaking also must be considered in the overall picture, and this effect is related to the ability of solvent molecules to donate and accept hydrogen bonds. Also one has to consider how effectively a solvent can align its dipolar molecules around a solute molecule for maximum interaction.

In terms of solvation phenomena, one must consider the difference between the solvation free energies of the reactants and the transition state (or other complex), as the change in rate with medium is a consequence of the differences in solvation free energies. This difference in solvation free energy is conveniently expressed in terms of a *medium effect* or *transfer activity coefficient*, discussed in Chapter 5, which is given by

$$_{s}\bar{G}_{i}^{o} = {}_{o}\bar{G}_{i}^{o} + RT \ln {}^{o}\gamma_{i}^{s} \qquad (10.2)$$

where ${}^{o}\gamma_{i}^{s}$ is the medium effect for species i, and o and s represent a reference solvent and the solvent to which the transfer is made, respectively. Transfer activity coefficients† are particularly useful for examining the reaction rates for reactions of type 10.1, that is, bimolecular nucleophilic attack by anions, reactions whose rates are particularly sensitive to solvent change. Some representative values of transfer activity coefficients are given in Table 10.1, in which the reference solvent is methanol. A negative value of the transfer activity coefficient indicates the species i is more strongly solvated in the solvent than it is in methanol, whereas a positive value indicates weaker solvation.

An activity coefficient rate equation can be derived, using absolute rate theory to express the rate constants of a reaction such as 10.1 in two different

† Parker uses the term "solvent activity coefficient."

solvents as a function of transfer activity coefficients:

$$\log \frac{k^s}{k^o} = \log {}^o\gamma^s_{Y^-} + \log {}^o\gamma^s_{RX} - \log {}^o\gamma^s_{YRX^\ddagger} \tag{10.3}$$

where k^s and k^o are the rate constants in the solvent, s, and the reference solvent, o. This rate equation indicates the contribution to reaction rates from changes in solvation of the reactant anion, Y^-, the reactant nonelectrolyte, RX and the anionic transition state, $YRX^\ddagger$. In addition, a comparison of values for $\log {}^o\gamma^s_i$ and $\log {}^o\gamma^s_{YRX^\ddagger}$ can lead to postulations concerning the structure and charge distribution in the transition state $YRX^\ddagger$.

Examination of Table 10.1 brings out several differences between cation

Table 10.1 Some Representative Values of Transfer Activity Coefficients at 25°C (Reference Solvent: Methanol)[a]

Solute	$\log^{MeOH}\gamma^s_i$					
Solvent(s)	H_2O	DMF	DMA	DMSO	CH_3CN	TMS
Xenon	1.7			1.1		
CH_3I	1.4	−0.5		−0.5	−0.4	
n-BuBr		−0.1		0.1	−0.2	
$(C_5H_5)_2Fe$	3.6				−0.3	
I_2	2.3	−1.8		−4.1	−0.2	
Cl^-	−2.5	6.5	7.8	5.5	6.3	5.8
Br^-	−2.1	4.9	5.9	3.6	4.2	
I^-	−1.5	2.6	3.0	1.3	2.4	2.4
N_3^-	−1.8	4.9	6.2	3.5	4.7	5.4
SCN^-	−1.2	2.7	3.2	1.4	2.6	2.6
Ph_4B^-	4.1	−2.7	−2.7	−2.6	−1.6	−2.0
ClO_4^-	−1.9	−0.4		−0.3		
$AgCl_2^-$	−3.3	−0.5	−0.3	−1.3	−0.2	−1.9
I_3^-	2.2	−2.0	−3.0	−3.6	−0.4	
Ag^+	−0.8	−5.1	−6.6	−8.2	−6.3	
Na^+		−3.9		−3.6	1.4	
K^+	−1.5	−3.7		−4.5	−0.8	
Me_3S^+		−3.1	−3.6		−1.6	
Ph_4As^+	4.1	−2.7	−2.7	−2.6	−1.6	
cis-$[CoCl_2(en)_2]^+$	−2.2	−6.0	−6.2	−6.8		−4.3
trans-$[CoCl_2(en)_2]^+$	−0.9	−4.2	−4.2	−4.7		−2.9

[a] The original reference uses the term "solvent activity coefficient."
Reprinted with permission from D. W. Watts, "Reaction Kinetics and Mechanisms" in *Physical Chemistry of Organic Solvent Systems,* A. K. Covington and T. Dickinson, Eds. Plenum Press, New York, 1973. Chap. 6.

and anion solvation effects. First, in the strongly dipolar aprotic solvents (DMSO, CH_3CN, etc.) the cations, including the *trans*- and *cis*-$[CoCl_2(en)_2]^+$, which possess hydrogen-bond donor sites are better solvated than the anions. On the other hand, anions, (F^-, OR^-, Cl^-, NR_2^-) are solvated fairly extensively in the protic solvents, via hydrogen-bond donor solvation. Large polarizable anions, such as picrate, perchlorate, SCN^-, I^-, I_3^-, and $AgCl_2^-$, are better solvated by the strongly dipolar solvents through dipole-induced dipole and dispersion interactions, as the dipolar aprotic solvents are much more polarizable than the protic solvents. Many tight transition-state anions, such as $YRX^{-\ddagger}$, possess dispersed negative charge and are therefore weak hydrogen-bond acceptors and consequently not very well solvated by protic solvents.

Relative to structural aspects, some of the protic solvents are highly associated through hydrogen bonding and therefore possess a fairly open structure. However, if the solvent molecules are small, then large anions, such as BPh_4^-, I_3^-, and some organic transition states, cannot fit into these structures. This type of interaction accounts in part for the positive transfer activity coefficients of large anions in water and formamide, when methanol or dipolar aprotic solvents are the reference media. Consequently there are several competing features in play when dealing with solvent effects on reaction rates.

For a reaction such as 10.1, the transfer activity coefficient of the transition state, ${}^{o}\gamma^{s}_{YRX^{\ddagger}}$, can be obtained by application of Eqn. 10.3, if the rate constant data are known. A knowledge of the transfer activity coefficient for the transition state is very important as an aid in understanding the mechanism of the reaction. To do this, comparisons are made between the calculated transfer activity coefficients for the transition state and values for either stable model species or comparable transition states. Parker (1) presents an extensive tabulation on the protic–dipolar aprotic solvent effects on the rates (i.e., $\delta_s \log k^s/k^o$) of some bimolecular substitution reactions between anions and molecules, including the reactant- and transition-state solvation.

The data for the bimolecular nucleophilic displacement reaction:

$$n\text{-BuI} + X^- \rightleftarrows (n\text{-BuIX}^-)^{\ddagger} \rightleftarrows n\text{-BuX} + I^- \qquad (10.4)$$

are summarized in Table 10.2, all referred to methanol. It is seen that all the reactions are faster in the dipolar aprotic solvents because the reactant anion X^- is much more solvated (stabilized) by protic than by dipolar aprotic solvents and this solvation obviously outweighs any effects due to the solvation of the transition-state anion or the reactant molecule, both of which are very small in this case. The data given by Parker (1) supports those trends for a whole range of species for reactions of this type in that the solvation of reactant anions is the dominant factor in determining rates of substitution reactions. However, to gain a more complete understanding of

Table 10.2 Bimolecular Substitution for the Reaction n-BuI + X^- $\rightleftarrows$ (n-BuI X^-)‡ $\rightleftarrows$ n-BuX + I^-. Reference Solvent: Methanol

	Solvent	$\log k^S/k^M$	$\log {}^M\gamma^S_{RX}$	$\log {}^M\gamma^S_{Y^-}$	$\log {}^M\gamma^S_{\ddagger^-}$
$X = Cl^-$	DMF	5.2	−0.3	6.5	1.0
	DMSO	4.6	0.0	5.5	0.9
$X = Br^-$	DMSO	3.6	0.0	3.6	0.0
$X = I^-$	CH_3CN	1.8	−0.3	2.4	0.3
	Me_2CO	3.1	—	—	—
$X = SCN^-$	DMF	1.9	−0.3	2.7	0.5
	DMSO	1.7	0.0	1.4	−0.3
	HMPT	3.3	—	3.4	—
$X = NO_2{}^-$	DMSO	3.8	0.0	3.7	−0.1
$X = N_3{}^-$	DMF	4.0	−0.3	4.9	0.6
	DMSO	3.5	0.0	3.5	0.0
	CH_3CN	3.1	−0.3	4.7	1.3

the mechanism of a bimolecular reaction it is useful to examine the detailed differences in the solvation of the transition-state anions. An example of this detailed approach is provided by Parker (1) and the highlights will be discussed here.

Consider a bimolecular substitution at a saturated carbon atom of type 10.1. In general, the nucleophile Y can consist of an anion or an uncharged species while the substrate RX can be either a cation, anion, or an uncharged species. The transition state is felt to exist as a species having a spectrum of structures (B) lying between the structures A and C shown below:

R R R R R R

Y÷C÷X $\overset{\delta-}{Y}$:—C—:$\overset{\delta-}{X}$ $\bar{Y}$: C^+ :$\bar{X}$

R R

A B C

Structure A is described as a *tight* structure, as the bond-forming tendency of Y and C is synchronous with the bond-breaking tendency of X and C. On the other hand, structure C is referred to as a *loose* structure, as the bond-breaking effect is more pronounced than the bond-forming effect. Relative to the actual transition state, there will exist a degree of *tightness* or *looseness*, as these structures cover an intermediate range between those represented by A and C. The best evidence for the existence of the transition

Table 10.3 Solvation of S_N2 Transition State Anions at 25°C. Loosening Factors[a]

Transition state $YCR_3X^{-\ddagger}$ [c]	R	R′	$\delta_R \log {}^E\gamma_{\ddagger}^{DMF}$	$\delta_R \log {}^E\gamma_{R_3CBr}^{DMF}$	$\delta_R \log k^{DMF}/k^E$
			Electronic Effects[a]		
N_3(+5.0)—C(H_2)(p-R-C_6H_4)—Br^-(+5.0)	NO_2		+0.2	−1.2	3.6
	H		+2.2	−0.2	2.6
	OMe		+3.0	0.0	2.0
			Steric and Electronic[b]		
	R	R′	$\delta_R \log {}^M\gamma_{\ddagger}^{DMF}$	$\delta_R \log {}^M\gamma_{RR'CHBr}^{DMF}$	$\delta_R \log k^{DMF}/k^M$
N_3(+4.9)—C(H)(R)(R′)—Br^-(+4.9)	H	H	+0.7	−0.3	3.9
	H	n-Pr	+1.5	0.0	3.4
	H	i-Pr	+1.6	0.0	3.3
	Me	Me	+2.1	−0.1	2.7
$Br\text{-}Ag\text{-}Br^-$ [d]	—		−1.2	—	

H H
N_3—C—X^-
(+4.9) H

Leaving Group[b]

X	$\delta_X \log {}^{M}\gamma_{\ddagger}^{DMF}$	$\delta_X \log {}^{M}\gamma_{CH_3X}^{DMF}$	$\delta_X \log {}^{M}\gamma_{X^-}^{DMF}$	$\delta_X \log k^{DMF}/k^{M}$
I	−0.2	−0.5	+2.6	4.6
Br	+0.7	−0.3	+4.9	3.9
OTS	+2.3	−0.6	+3.5	2.0
$OPO(OMe)_2$	+2.7	−0.4	—	1.8

H H
Y—C—I^-
(+2.6)
H

Entering Group[b]

Y	$\delta_Y \log {}^{M}\gamma_{\ddagger}^{DMF}$	$\delta_Y \log {}^{M}\gamma_{Y^-}^{DMF}$
SCN	0.0	2.7
Cl	+0.1	6.5
$OCOCH_3$	+1.8	9.2

[a] Reference solvent ethanol: data from Ref. 5.
[b] Reference solvent methanol.
[c] The numbers in parentheses below the transition-state structures are $\log {}^{o}\gamma_{Y^-}^{DMF}$ and ${}^{o}\gamma_{X^-}^{DMF}$, respectively.
[d] This is a "real" anion, not a transition state.

Reprinted with permission from *Chem. Rev.*, **69,** 1 (1969). Copyright 1969. American Chemical Society.

state in a range of structures is that provided in a study of the effects on reaction rates of protic and dipolar aprotic solvents (5).

The nature of Y and X will be a predominant factor to be considered when discussing the relative reaction rates and the relative degrees of tightness and looseness in the transition-state structures. An example of transfer activity coefficients for four sets of S_N2 transition-state anions is given in Table 10.3. Here, the species Y and X are only slightly different. It is seen that as one variable is changed with the others held constant, the transition state becomes progressively looser (i.e., $\delta_R \log {}^m\gamma_{\ddagger}{}^{DMF}$ becomes increasingly positive). The increased positive value is indicative of an increased localization of negative charge on X and Y and thereby of the structure tending toward C.

Several interesting features were developed by analyzing Table 10.3. With reference to electronic effects, it is seen that the aryl substituent at C_α, *p*-OMe, is more effective in stabilizing the positive charge at C_α and thereby producing a looser structure for the transition state. Both steric and electron effects are shown by the effect of methyl substituents at C_α, which also give rise to a looser structure. The values of the transfer activity coefficients for the transition states also increase when the leaving or entering groups are bonded to C_α through oxygen rather than a direct halogen linkage, suggesting a looser transition state. The silver dibromide anion is included in Table 10.3 for comparison to present an example of a model for a very tight transition state. The final column in Table 10.3 gives the ratios of the rate constants, and it can be seen that rates of reactions involving the halide ion are much enhanced by a transfer from a protic to a dipolar aprotic solvent than are the corresponding reactions involving tosylates or phosphates. Also, enhancement of rates is found for reactions involving a substrate with a methyl carbon, compared to those involving a primary or secondary carbon atom. A further conclusion that can be drawn from the data presented in Table 10.3 is that reactions involving substituted benzyl compounds respond to the change in solvent in various ways, depending upon the electron-withdrawing or electron-donating properties of the substituent.

In practice it is found that the rates of S_N2 reactions 10.1 of alkyl chlorides are less affected by acceleration due to dipolar aprotic solvents than are the corresponding reactions involving iodides, which implies that a chloride ion is assisted in leaving by a protic solvent more readily than an iodide ion. These trends can be understood in terms of the solvation of S_N2 transition-state anions as the variables Y and X are changed. Parker et al. (6, 7) examined these effects and the results are summarized in Table 10.4. When the groups Y and X are large, they (and the transition-state anions) are more polarizable and therefore cannot accept hydrogen bonds from the reference solvent methanol as easily. This fact is reflected in the trends of the values for $\delta_X(\log {}^m\gamma_{\ddagger}{}^{DMF})_{Y,CR_3}$ and $\delta_Y(\log {}^m\gamma_{\ddagger}{}^{DMF})_{X,CR_3}$, which become more negative as X and Y become larger.

Table 10.4 Solvation of S_N2 Transition State Anions. Effect of Entering and Leaving Group (Reference Solvent: Methanol at 25°C)

$YCR_3X^{-\ddagger}$	$\delta_X \log k^{DMF}/k^M$	$\delta_X(\log {}^M\gamma_{\ddagger}^{DMF})_{Y,CR_3}$
	Transition State Anion	
$N_3CH_3Cl^-$	3.3	1.2
$N_3CH_3Br^-$	3.9	0.7
$N_3CH_3I^-$	4.6	−0.2
$NCSCH_3Cl^-$	1.4	0.9
$NCSCH_3Br^-$	1.7	0.7
$NCSCH_3I^-$	2.2	0.0
		$\delta_Y(\log {}^M\gamma_{\ddagger}^{DMF})_{X,\ CR_3}$
$ClCH_3Br^-$	—	1.4
$N_3CH_3Br^-$	—	0.7
$NCSCH_3Br^-$	—	0.7
ICH_3Br^-	—	0.2
		$\delta_Y(\log {}^M\gamma_{\ddagger}^{DMSO})_{X,\ CR_3}$
$ClBuI^-$	—	1.4
$BrBuI^-$	—	0.7
N_3BuI	—	0.7
$NCSBuI^-$	—	0.2
	Model Anion, $XAgX^-$	
		$\delta_X(\log {}^M\gamma^{DMF})_{XAgX^-}$
$ClAgCl^-$	—	−0.2
$BrAgBr^-$	—	−1.2
$IAgI^-$	—	−2.7

It is also of interest to compare the rates of S_N2 displacement reactions at a carbonyl carbon with aryl and vinyl carbons (two-step addition–elimination reactions). The solvation of transition-state anions for S_N2 reactions involving these groups is summarized in Table 10.5. It can be seen that the rates of displacement reactions involving a carbonyl carbon (e.g., ester hydrolysis) are much less sensitive to acceleration by dipolar aprotic solvents than are the corresponding S_NAr reactions. The transition state involving a carbonyl oxygen is a good hydrogen-bond acceptor because of the negative

Table 10.5 Solvation of Transition State Anions for S_N2 Reactions at Aryl, Vinyl, Carbonyl, and a Methyl Carbon Atom. Transfer from Methanol to DMF at 25°C

Transition state‡	$\delta R_3C \log k^{DMF}/k^{M}$	$\delta_{R_3C}(\log {}^{M}\gamma_{\ddagger}^{DMF})_{Y,X}$	$\log {}^{M}\gamma_{R_3CX}^{DMF}$
	Set A		
NO_2–(ring)–$C(N_3)(OAr)$, NO_2	3.4	−0.5	−2.0
$(i-Pr)_2P(Cl)(O^-)(OAr)$	2.3	+1.5	—
CH_3—$C(N_3)(O^-)(OAr)$	1.4	+2.6	−0.9
	Set B		
NO_2–(ring)–$C(N_3)(Cl)$, NO_2	4.1	<−0.2	<−1
$RSO_2\bar{H}C{=}C(N_3)(Cl)$	3.2	+0.6	−1.1
N_3——CH_3——Cl^-	3.3	+1.2	−0.4
NO_2=(ring)=O, NO_2, NO_2	—	−0.4	—

Reprinted with permission from *Chem. Rev.*, **69,** 1 (1969). Copyright 1969. American Chemical Society.

charge located on the carbonyl oxygen. These substitutions at aryl, vinyl, and carbonyl carbon are characterized by the fact that bond forming lies well ahead of bond breaking and involve transition states with tight structures, with very little negative charge localized on X or Y. The smaller transition-state anions have much more positive values of $\delta_{R_3C}(\log {}^{M}\gamma_{\ddagger}{}^{DMF})_{Y,X}$ than do the larger more polarizable transition-state anions, where the charge is more dispersed.

A study of S_NAr transition states is also possible and is of value in determining whether S_NAr reactions are two-step addition–elimination reactions or synchronous S_N2-like reactions and whether the process of bond breaking is rate determining. The available data support the idea of a tight S_NAr transition-state anion, in which the leaving group carries very little negative charge and bond breaking has made very little progress.

The above discussion has focused on an anionic transition state. In the review by Parker (1), other types of transition states were considered, including uncharged, cationic, dianions and anionic states for elimination. A selection of transfer activity coefficients for a variety of S_N2 transition states

Table 10.6 Transfer Activity Coefficients[a] for S_N2 Transition States (Reference Solvent: Methanol at 25°C)

Transition State	log ${}^{M}\gamma_{\ddagger}^{DMF}$	Transition State	log ${}^{M}\gamma_{\ddagger}^{DMF}$
$\overset{\delta+}{Me_2S}CH_3\overset{\delta+}{NMe_3}$ (+1)	<-4.4	$\overset{\delta-}{N_3}CH_3\overset{\delta-}{I}$ (−1)	−0.2
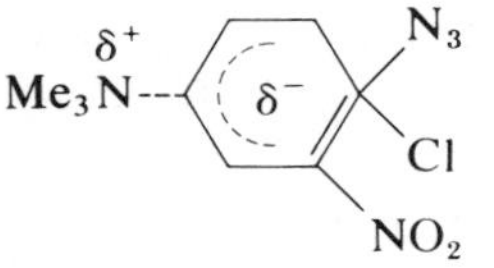 $\overset{\delta^+}{Me_3N}$--(ring, δ^-)(N$_3$)(Cl)–NO_2 (0)	−3.0	$\overset{\delta-}{N_3}CH_3\overset{\delta-}{OTs}$ (−1)	+2.3
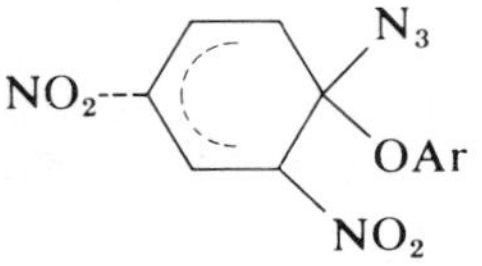 NO_2--(ring)(N$_3$)(OAr)–NO_2 (−1)	−2.0	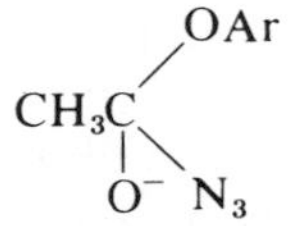 $CH_3C(OAr)(O^-)(N_3)$ (−1)	+2.6
$\overset{\delta-}{N_3}CH_3\overset{\delta+}{SMe_2}$ (0)	−1.3	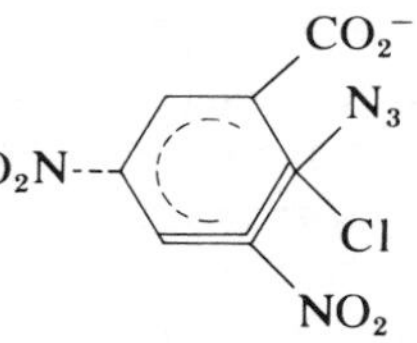 O_2N--(ring)(CO_2^-)(N$_3$)(Cl)–NO_2 (−2)	+2.8

[a] The original reference uses the term "solvent activity coefficients."

for substitution reactions of azide ion is given in Table 10.6. The striking feature is that the solvation of a transition state can bring about a change in rate constant of more than 10^7 due to solvent transfer.

Summary The relative rates of reactions in different solvents are governed by the interactions between the solvent molecules and the reacting species as well as the transition states, all of which can be anionic, cationic or uncharged. Solvent properties such as structure, dielectric constant, molar polarizability, dipole moment, acidity, basicity, and hydrogen-bonding capability are important parameters that determine these interactions. In addition, the charge density, polarizability and size of the solute molecule have to be taken into account. Solvent structure-making and -breaking must also be considered and this effect is related to the ability of solvent molecules to donate and accept hydrogen bonds. Some bimolecular reactions are a million times faster in weakly dipolar aprotic solvents like nitromethane than they are in, for example, *N*-methylformamide.

The changes in reaction rates in different solvents are associated with the differences in solvation free energies of the reactants and the transition state, this difference being expressed as a medium effect, or transfer activity coefficient. A negative value for the transfer activity coefficient indicates that the species under consideration is more strongly solvated in the given solvent than in the reference solvent, whereas a positive value indicates weaker solvation. An activity coefficient rate equation is available, which expresses the rate constants of a bimolecular reaction in two different solvents as a function of transfer activity coefficients. The rate equation indicates the contribution to reaction rates from changes in solvation of the reactant anion, the reactant nonelectrolyte, and the anionic transition state. A comparison of the values is useful as a guide to discussing the structure and charge distribution in the transition state. There are several competing features in play when dealing with solvent effects on reaction rates. For example, in strongly dipolar aprotic solvents, such as DMSO, cations which possess hydrogen-bond donor sites are better solvated than anions. Conversely, anions such as F^- or OR^- are solvated fairly extensively in the protic solvents via hydrogen-bond donor solvation, while large polarizable ions, such as picrate and perchlorate, are better solvated by the strongly dipolar solvents through dipole-induced dipole interactions. Many of the tight transition-state anions, such as $YRX^{-\ddagger}$, possess dispersed negative charge and are therefore weak hydrogen-bond acceptors and consequently not very well solvated by protic solvents.

Knowledge of transfer activity coefficients for the transition state is very important as an aid in understanding the mechanism of a reaction. The relative degree of tightness or looseness of the transition-state structures is a significant factor.

10.2 KINETICS OF INORGANIC REACTIONS

In addition to the study of the kinetics and mechanisms of organic reactions in nonaqueous solvents, the study of inorganic substitution reactions in nonaqueous solvents has led to meaningful discussions in terms of mechanisms. In particular, the solvent-exchange reactions between the bulk solvent and the first solvation sphere have formed the basis of an understanding of substitution reactions at metal centers. This type of reaction for an octahedral complex can be represented by

$$[\mathrm{M(solvent)_6}]^{n+} + \mathrm{X}^{m-} \rightleftharpoons [\mathrm{MX(solvent)_5}]^{(n-m)+} + \text{solvent} \quad (10.5)$$

Also, non-labile octahedral centers, such as Co(III), and Cr(III), square planar complexes of Pt(II) and organometallic complexes have been the subject of several investigations. The techniques used to study these inorganic substitution reactions depend upon the rates of the reactions and include conventional kinetics techniques, T-jump, ultrasonic absorption methods, flow techniques, and NMR relaxation techniques.

Substitution reactions of type 10.5 have been found to possess certain unique characteristics. First, the nature of the ligand X has very little effect on the rate of a reaction of a particular metal in a particular solvent. Second, a limiting value is observed for the rates, which is independent of further increases in the concentration of the ligand. Third, the limiting rate is, within a power of 10, the same as the rate for the solvent exchange process:

$$[\mathrm{M(H_2O)_6}]^{n+} + \mathrm{H_2O^*} \rightleftharpoons [\mathrm{M(H_2O)_5(H_2O^*)}]^{n+} + \mathrm{H_2O} \quad (10.6)$$

Langford and Stengle (8) have discussed the dynamics of ligand substitution reactions in some detail for reactions in both aqueous and nonaqueous solutions. At the outset, some reference should be made concerning terminology in substitution reactions of the type depicted in 10.5. The substitution process involves both the breaking and formation of a bond. There are two possible mechanisms for the formation of the final product, one involving a transition state in which there exists a partial formation of the new bond, referred to as an associative or *a-mode*, and another in which the activation mode is dissociative, or a *d-mode*.

These mechanisms are illustrated by considering the following stoichiometric pathways of substitution:

$\mathrm{MX} \underset{+\mathrm{X}}{\overset{-\mathrm{X}}{\rightleftharpoons}} \mathrm{X} + \mathrm{M} \underset{-\mathrm{Y}}{\overset{+\mathrm{Y}}{\rightleftharpoons}} \mathrm{MY}$	*D*
$\mathrm{MX} + \mathrm{Y} \rightleftharpoons \mathrm{MY} + \mathrm{X}$	*I*
$\mathrm{MX} \underset{-\mathrm{Y}}{\overset{+\mathrm{Y}}{\rightleftharpoons}} \mathrm{MXY} \underset{+\mathrm{X}}{\overset{-\mathrm{X}}{\rightleftharpoons}} \mathrm{MY} + \mathrm{X}$	*A*

where X represents a leaving ligand, Y, an entering ligand, and M, the remainder of the metal complex. A dissociative pathway is represented by D, an associative pathway, by A, and the one-step interchange pathway, by I. The pathway I may be associated with either mode of activation, that is, I_d or I_a. Relative to octahedral substitution, the experimental data indicate that bond breaking (dissociation) is the significant process in the formation of the transition state.

A well-studied example of a ligand-substitution reaction that is consistent with a d-mode activation is that of the Co(III) species in aqueous systems. The possible mechanisms associated with this reaction are given in Fig. 10.1, in which the D path is represented by [I] → [III] → [IV] and the I_d path is represented by [I] → [II] → [IV]. The experimental evidence supports an I_d path mechanism. In Fig. 10.1, K_a represents the equilibrium constant for the formation of the ion pair, k_{-H_2O} is the first-order rate constant for the formation of the five-coordinate intermediate, which is captured by either water k_{+H_2O} or ligand k_{+X}, both of which are second-order processes. The quantity k'_{-H_2O} is the first-order rate constant for the incorporation of X^- into the coordination sphere from the solvation sphere by a dissociative interchange.

Several arguments have been suggested as evidence for an I_d mechanism (8), and these involve a consideration of the relationship between the probability that a ligand occupies an outer-sphere site, the rate of ligand incorporation and relation between the solvent exchange rates, that is, k'_{-H_2O} and k_{+H_2O}. However, Watts (2) has indicated that there is substantial evidence in Co(III) chemistry that those type of considerations are not particularly valid, in so much as the dissociation rate of the ion-pair species will differ from that of the free ion, because the transition state is different and also the differences in nucleophilic characters of ligands will control the tendencies of a ligand to capture a developing dissociative intermediate. In other words, the effect of ion association on the free energy of the reactant and the transition state must be considered, along with the inherent nucleophilicity of the ligand.

$$\underset{[\mathrm{I}]}{\mathrm{Co(NH_3)_5OH_2}^{3+}} \underset{-\mathrm{X}}{\overset{+\mathrm{X}(K_a)}{\rightleftharpoons}} \underset{[\mathrm{II}]}{\mathrm{Co(NH_3)_5OH_2}^{3+}\ldots\ldots \mathrm{X}^-}$$

$$k_{-H_2O} \downarrow\uparrow k_{+H_2O} \qquad\qquad k'_{-H_2O} \downarrow\uparrow k'_{+H_2O}$$

$$\underset{[\mathrm{III}]}{\mathrm{Co(NH_3)_5}^{3+}} \underset{k_{-X}}{\overset{k_{+X}}{\rightleftharpoons}} \underset{[\mathrm{IV}]}{\mathrm{Co(NH_3)_5X}^{2+}}$$

Figure 10.1 Stoichiometric mechanisms consistent with d-mode activation for Co(III). Reproduced with permission from the *Annual Review of Physical Chemistry*, **19**, 193. © 1968 by Annual Reviews Inc.

Table 10.7 Rate Constants at 35°C, Activation Energies, and Entropies of Activation of Some Solvent-Interchange Reactions

Complex[a]	Solvent	k (min^{-1})	E_a kcal mol^{-1}	$\Delta S^{\ddagger}$ cal deg^{-1} mol^{-1}
cis-$[CoCl(DMF)(en)_2]^{2+}$	DMSO	2.74×10^{-3}	24.6	0.6
	H_2O	1.71×10^{-3}	25.2	0.5
	DMA	1.61×10^{-3}	27.4	12.0
cis-$[CoCl(DMSO)(en)_2]^{2+}$	DMSO	4.15×10^{-3}		
	H_2O	3.63×10^{-3}	25.2	1.5
	DMA	2.47×10^{-3}	26.5	9.9
cis-$[CoCl(DMA)(en)_2]^{2+}$	DMSO	2.94×10^{-1}	22.3	1.2
	H_2O	2.13×10^{-1}	23.5	4.7
	DMF	6.27×10^{-1}	30.6	29.5
trans-$[CoCl(H_2O)(en)_2]^{2+}$	DMSO	7.15×10^{-2}	26.9	13.3
	DMF	2.46×10^{-2}	23.8	1.1
trans-$[CoCl(DMF)(en)_2]^{2+}$	DMF	3.24×10^{-2}	24.2	3.0
	DMSO	5.06×10^{-3}		
trans-$[CoCl(DMSO)(en)_2]^{2+}$	DMSO + $2Br^-$	4.1×10^{-3}		
cis-$[CoCl(DMSO)(en)_2]^{2+}$	DMSO	4.65×10^{-3}		
	DMSO	2.65×10^{-3}		
cis-$[CoBr(DMSO)(en)_2]^{2+}$				
cis-$[Co(DMSO)_2(en)_2]^{3+}$				

[a] en = ethylenediamine.

As an example of the chemistry of octahedral cobalt in nonaqueous solvents, Watts (9) determined the rate constants spectrophotometrically for the replacement by the solvent of coordinated molecules for the systems: *cis*-$[CoCl(DMA)(en)_2]^{2+}$ in anhydrous DMF, *trans*-$[CoCl(H_2O)(en)_2]^{2+}$ in acidified DMF; *cis*-$[CoCl(DMA)(en)_2]^{2+}$ and *cis*-$[CoCl(DMF)(en)_2]^{2+}$ in anhydrous DMSO, *trans*-$[CoCl(H_2O)(en)_2]^{2+}$ in acidified DMSO; and *cis*-$[CoCl(DMF)(en)_2]^{2+}$ and *cis*-$[CoCl(DMSO)(en)_2]^{2+}$ in anhydrous DMA. The purpose of this study was to elucidate the mechanisms for these substitution reactions, that is, S_N1 or S_N2. A summary of the rate constants, activation energies, and entropies of activation for some of the solvent-interchange reactions, together with some data for water exchange are given in Table 10.7. It is seen that the rates and activation parameters for the *cis* complexes are solvent dependent and decrease in the order DMA > H_2O > DMSO. In DMF the parameters are anomalously high. The steric course of these reactions is interpreted as proceeding by an S_N1 mechanism, involving trigonal bipyramidal transition states, to yield *cis*- and *trans*-chloro(solvent)bis(ethylenediamine)cobalt(III) ions.

Several revealing factors concerning the mechanisms of ligand substitution are obtained from a study of both equilibria and kinetics in nonaqueous solvents, For example, information concerning transition states, the effect of ion association on mechanisms and the stereochemical course of reactions is often gained by nonaqueous studies. As mentioned earlier, a considerable amount of attention has been devoted to the reactions of the complexes of non-labile cobalt (III) and chromium (III).

An illustration of the approach is provided by the study of Tobe and Watts (10), who investigated the solvolysis and isomerization of dichlorobis(ethylenediamine)cobalt(III) ions in DMSO, containing varying amounts of $(Et)_4NCl$. The objective was to study the role of the solvent in the mechanism of the *cis-trans*-$[CoCl_2(en)_2]^+$ isomerization. In the course of this investigation, the solvent complex $[Co(en)_2DMSOCl_2]^+$ was isolated and tentatively assigned as the *cis* isomer. The other isomer was not isolated.

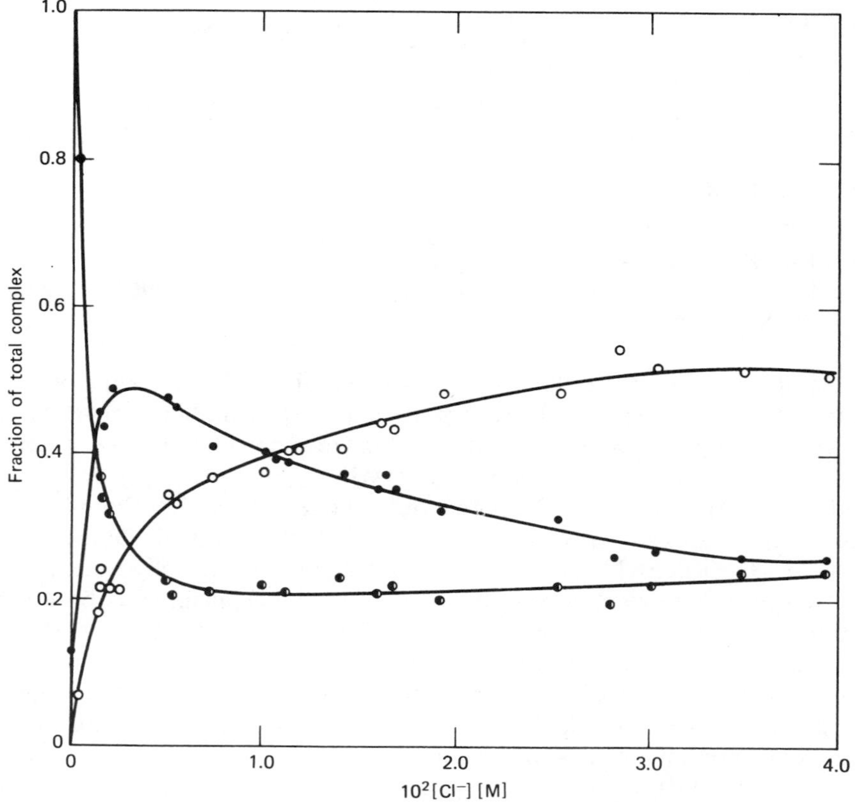

Figure 10.2 Equilibrium composition of solutions in DMSO as a function of added $[Et_4N]Cl$ at 60°C. (○) *cis*-$[Co(en)_2Cl_2]^+$; (●) *trans*-$[Co(en)_2Cl_2]^+$; (◐) *cis*-$[Co(en)_2DMSO\,Cl]^{2+}$. Reprinted with permission from *J. Chem. Soc.*, 2991 (1964).

Table 10.8 Derived Rate Constants for the Solvolysis and Isomerization of *cis*- and *trans*-$[Co(en)_2Cl_2]^+$ in Dimethylsulfoxide at 60°C

$[Cl^-]_{initial}$ (mmol liter^{-1})	$k_C + k_{CT}$ (min^{-1}; $\times 10^2$)	k_C (min^{-1}; $\times 10^2$)	k_{CT} [a] (min^{-1}; $\times 10^3$)
cis-$[Co(en)_2Cl_2]^+$			
1.60	1.01	0.83	2
13.0	1.14	0.8–0.9	2–3
18.2	1.12	0.85	3
29.4	1.25	1.03	2

trans-$[Co(en)_2Cl_2]^+$

$[Cl^-]_{initial}$ (mmol liter^{-1})	$k_T + k_{CT}$ (min^{-1}; $\times 10^{-2}$)	k_T (min^{-1}; $\times 10^2$)	k_{CT} [a] (min^{-1}; $\times 10^3$)
10.2	1.17	1.18	—
14.8	1.24	1.01	2
27.3	1.54	1.09	4

[a] By difference.
Reprinted with permission from *J. Chem. Soc.*, 2991 (1964).

It was found that the composition of the solution at equilibrium was dependent on the chloride-ion concentration and the results of visible absorption spectra demonstrate this trend as shown in Fig. 10.2. The change in the relative amounts of *cis*- and *trans*-dichloro complexes is a reflection of the varying extents of the ion association properties of the two isomers. At high chloride-ion concentrations, the formation of the ion triplet, $[Co(en)_2DMSOCl]^{2+}...2Cl^-$ was advanced to account for the fact that the equilibrium concentration of this solvent complex was independent of chloride-ion concentration at high chloride concentration.

The rates of approach to equilibrium were studied by the same spectrophotometric technique as used in the equilibrium studies. Rates were evaluated starting from each of the three complexes and the rate constants for the various steps were calculated. Several approaches were used to develop the rate constants for the sequence

$$\begin{array}{ccc} & [Co(en)_2DMSOCl]^{2+} & \\ {\scriptstyle k_C}\nearrow\swarrow{\scriptstyle k_{-C}} & & {\scriptstyle k_{-T}}\searrow\nwarrow{\scriptstyle k_T} \\ cis\text{-}[Co(en)_2Cl_2]^+ & \underset{k_{TC}}{\overset{k_{CT}}{\rightleftharpoons}} & trans\text{-}[Co(en)_2Cl_2]^+ \end{array}$$

and the derived rate constants for the solvolysis and isomerization are given in Table 10.8.

In this particular study some qualitative observations were discussed. For example, it was found that the dichlorocomplexes underwent solvolysis and direct isomerization simultaneously, with the solvolysis being more important. These results were interpreted in terms of a unimolecular mechanism in which there is competition between the solvent and the chloride ion for the penta-coordinated intermediate.

An improved understanding of the mechanism of this particular isomerization has been given by Parker et al. (11), using the concept of transfer activity coefficients. The details of this analysis are also discussed thoroughly in Ref. 2. Basically, the transfer activity coefficients were obtained from measuring the solubilities of the chlorides and perchlorates of the *cis* and *trans* isomer cations in the protic solvents water and methanol and in the dipolar aprotic solvents DMF, DMA, TMS, and DMSO. Transfer activity coefficients were estimated for the isomeric cations, the transition states and ion pairs formed between the isomeric cations and the chloride ions (e.g., *cis*-$[CoCl_2(en)_2]^+Cl^-$).

The equilibrium constant ($K = k_c/k_t$) for the isomerization

$$cis\text{-}[CoCl_2(en)_2]^+ \underset{k_t}{\overset{k_c}{\rightleftharpoons}} trans\text{-}[CoCl_2(en)_2]^+$$

has been found to be strongly influenced by ion association and by solvent transfer. The effect has been ascribed to the greater hydrogen-bond donor properties of the *cis* isomer relative to the *trans* isomer. The different hydrogen-bonding abilities of the solvents can give rise to different mechanisms, either involving an intermediate solvent-containing complex, as in DMF and DMSO, or without involving such an intermediate, as in methanol and water. In Table 10.9 are the transfer activity coefficients for the isomeric cation and anion, and the ion pairs (cip and tip), together with a comparison for other cations and anions. It is seen that the symmetrical *trans*-$[CoCl_2(en)_2]^+$ cation is similar in behavior to the potassium cation, that is, without any specific solute–solvent interaction. In contrast, the more polar *cis*-$[CoCl_2(en)_2]^+$ cation has a greater range of transfer activity coefficients, being particularly poorly solvated in methanol. Also the solvation of the ion pairs does not change as significantly as that of the isomeric ions.

There is fairly substantial evidence from a study of anation reactions (where a coordinated solvent molecule is replaced by an anion) to suggest that the mechanism of the kinetics of substitution in Co(III) octahedral complexes is a dissociatively controlled mechanism (S_N1), in which the ionization of a chloride ion from either the *cis*- or *trans*-$[Co(en)Cl_2]^+$ cation is the rate-determining step. With this assumption, the relation between the rate constants in the two solvents, k_c^s/k_c^o, is given by

$$k_c^s/k_c^o = {}^o\gamma_{c^+}^s/{}^o\gamma_{c^\ddagger}^s \tag{10.7}$$

The solvation parameters of the transition states for the isomerization and the solvolysis with DMF as a reference solvent are given in Table 10.10. It

Table 10.9 Solvation of Anions, Cations, and Ion Pairs (Reference Solvent DMA, 25°C)[a]

Solvent	$\log {}^{D}\gamma^{S}c^{+}$	$\log {}^{D}\gamma^{S}t^{+}$	$\log {}^{D}\gamma^{S}Cl^{-}$	$\log {}^{D}\gamma^{S}Ph_4As^{+}$	$\log {}^{D}\gamma^{S}K^{+}$	$\log {}^{D}\gamma^{S}Ag^{+}$	$\log {}^{D}\gamma^{S}cip$	$\log {}^{D}\gamma^{S}tip$
DMA	0.0	0.0	0.0	0.0	—	0.0	0.0	0.0
DMF	+0.2	0.0	−1.3	0.0	0.0a	+1.5	−0.7	−0.3
DMSO	−0.6	−0.5	−2.3	0.0	−1.0a	−1.6	−1.2	−1.8
MeOH	+6.2	+4.2	−7.8	+2.7	+3.7a	+6.6	+0.5	−1.3
H_2O	+4.0	+3.3	−10.3	+6.8	+2.2a	+5.8	—	—
TMS	+1.9	+1.3	−2.0	+0.7	—	+4.6	−0.4	+1.0

[a] Reference solvent is DMF; value for DMA is not available. cip and tip are the ion pairs for the *cis* and *trans* isomers, respectively. Reprinted with permission from *J. Amer. Chem. Soc.*, **90,** 5744 (1968). Copyright 1968. American Chemical Society.

Table 10.10 Solvation of Transition States for Isomerization and Solvolysis in the *cis* and *trans*-$[CoCl_2(en)_2]^+$ System at 25°C (Reference Solvent DMF, k min^{-1})

Solvent (S)	Reaction	$\log k_c$ [a]	$\log k_t$ [a]	$\log^{DMF}\gamma_c^S$	$\log^{DMF}\gamma_t^S$	$\log^{DMF}\gamma_{c^\ddagger}^S$ [b]	$\log^{DMF}\gamma_{t^\ddagger}^S$
DMF	Solvolysis	−2.8	−4.3	0.0	0.0	0.0	0.0
DMSO	Solvolysis	−3.6	−3.8	−0.8	−0.5	0.0	−1.0
H_2O	Solvolysis	−1.8	−2.7	3.8	3.3	2.8	1.6
MeOH	Isomerization	−3.0	−5.9	6.0	4.2	6.0	
$TMSO_2$	Isomerization	−5.4	−7.0	1.9	1.3	4.2	

[a] k_c and k_t are the initial first-order rate constants for removal of *cis*-$[CoCl_2(en)_2]^+$ and *trans*-$[CoCl_2(en)_2]^+$ (that is "*c*" and "*t*," respectively.

[b] $c^\ddagger$ and $t^\ddagger$ are the transition states for the removal of *cis*- and *trans*-$[CoCl_2(en)_2]^+$ respectively.

Reprinted with permission from D. W. Watts, "Reaction Kinetics and Mechanisms" in *Physical Chemistry of Organic Solvent Systems*, A. K. Covington and T. Dickinson, Eds. Plenum Press, New York, 1973. Chap. 6.

is observed that the transition state for solvolysis of the *cis* cation is similarly solvated by DMF and DMSO, while the *trans* cation is more solvated by DMSO than by DMF. Also, the transition states are less solvated by water than by DMF, suggesting that the solvolysis in one or both of these solvents does not possess a rate-determining step based on the model transition state

$$[Co(en)_2Cl]^{+\delta+} \ldots {}^{-\delta}Cl$$

in which a negative charge is developing on the chlorine and a doubly positive charge is developing on the cation.

Langford and Chung (12) have examined the mechanism of complex-formation and solvent-exchange reactions of $Fe(DMSO)_6{}^{3+}$ with SCN^- and sulfosalicylic acid (SSA) as ligands. The mechanism due to Eigen for this process is given by

$$Fe(DMSO)_6{}^{3+} + L^{n-} \underset{}{\overset{K}{\rightleftharpoons}} Fe(DMSO)_6{}^{3+}, L^{n-} \xrightarrow{k_1} Fe(DMSO)_5L^{(3-n)+} + DMSO \qquad (10.8)$$

where K is the equilibrium constant for outer-sphere association and k_1 is the first-order rate constant for outer-sphere to inner-sphere conversion. The values observed for k_1 were as follows:

DMSO, 50 sec^{-1}; SCN^-, 6.3 sec^{-1}; and SSA, 23 sec^{-1}

The fact that the three values are clustered fairly close together indicates that the substitution process is insensitive to the entering ligand and consistent with a dissociative mode of activation.

A further study by Tobe and Watts (13) on the isomerization of dichlorobisethylenediaminecobalt(III) was undertaken in anhydrous DMF and DMA using *cis*- and *trans*-$[Co(en)_2Cl_2]ClO_4$. As in the earlier study, both equilibrium and kinetic data were obtained. It was found that the position of equilibrium between the two isomers was dependent on the concentration of the free chloride ion in solution. It was also apparent that a fairly strong ion pair was formed between the *cis* isomer and chloride and the resulting stabilization led to an increase of *cis* isomer concentration with increasing chloride-ion concentration at equilibrium. Any association occurring with the *trans* isomer was weak.

The rates of approach to equilibrium were found to be first-order with respect to the complex and are shown in Fig. 10.3. It is observed that the rate constant for the conversion of *cis* to *trans*, k_c, shows very little variation with chloride-ion concentration in DMF, whereas the rate constant for the *trans* to *cis* conversion, k_t, is very sensitive to the concentration. These results provide further evidence for the ion-pair-forming capabilities of the *cis* isomer, for even at the lowest concentrations studied (1.5×10^{-3} M), 70% of the *cis* complex is in the form of the ion pair. In addition, a comparison of the rate of exchange of the free *cis* ion with that of the ion pair ($k_c \approx 8 \times 10^{-3} min^{-1}$) indicates that they are roughly equal. The general

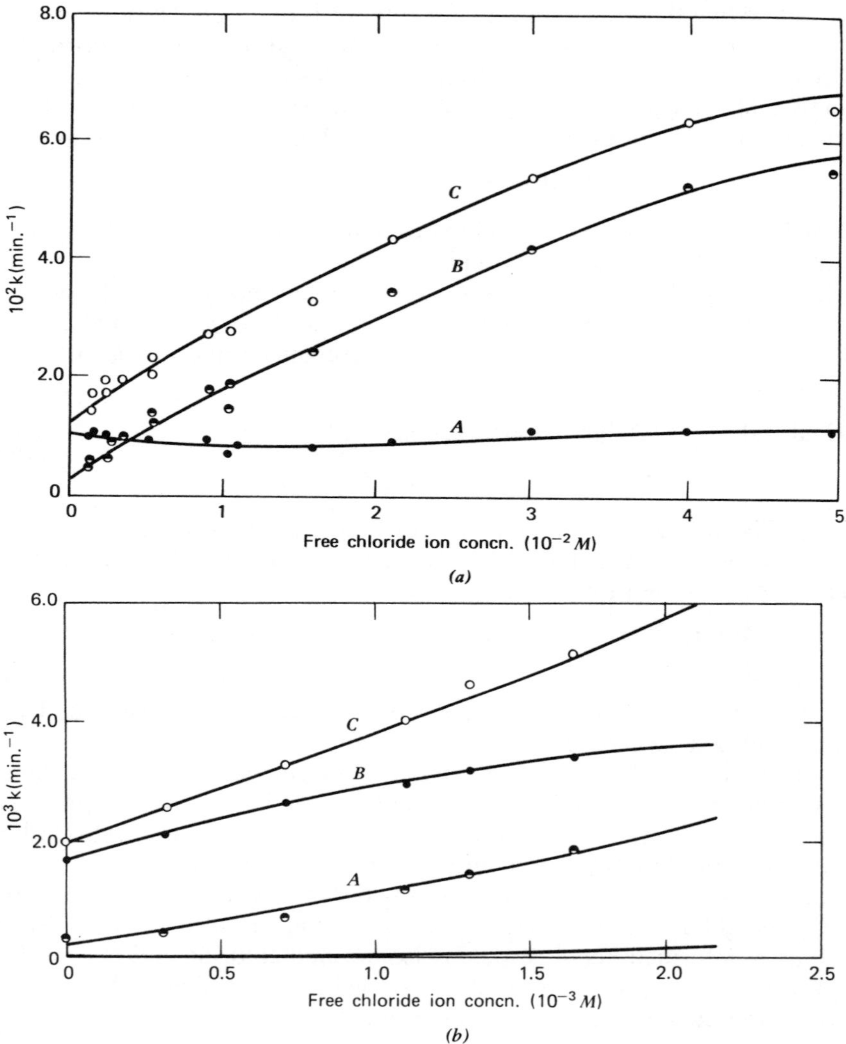

Figure 10.3 First-order rate constants for isomerisation of $[Co(en)_2Cl_2]^+$ in (a) DMF and (b) DMA. *A*, k_c (*cis* → *trans*); *B*, k_t (*trans* → *cis*); *C*, $k_c + k_t$. Reprinted with permission from *J. Chem Soc.*, 4614 (1962).

conclusion from this particular study is that the rate constant for the isomerization of the *cis*–$[Co(en)_2Cl_2]^+$ cation is fairly constant when it changes to the ion pair, but the *trans* isomer isomerizes some 30 times faster than the free ion.

To interpret the actual mechanism of ligand substitution reactions, the detailed nature of the reaction must be considered. Often there are several steps involving association prior to establishment of equilibrium and it is therefore important to consider the free energies both of reactants and transition states. This type of approach is demonstrated in the study by Fitzgerald and Watts (14), who investigated the isomerization of *cis*- and *trans*-dichlorobis(ethylenediamine)cobalt(III) ions, $[CoCl_2(en)_2]^+$, in anhydrous sulfolane, $TMSO_2$. The overall system can be represented by

$$\begin{array}{ccc} cis\text{-}[CoCl_2(en)_2]^+ & \underset{k_t}{\overset{k_c}{\rightleftharpoons}}\ (1) & trans\text{-}[CoCl_2(en)_2]^+ \\ + & & + \\ Cl^- & & Cl^- \\ (2)\ \downarrow\uparrow K_2 & & (3)\ \uparrow\downarrow K_3 \\ cis\text{-}[CoCl_2(en)_2]^+\ldots Cl^- & \rightleftharpoons & trans\text{-}[CoCl_2(en)_2]^+\ldots Cl^- \end{array}$$

where $K_1 = k_c/k_t$, with k_c and k_t being the initial first-order rate constants for the removal of *cis*-$[CoCl_2(en)_2]^+$ and *trans*-$[CoCl_2(en)_2]^+$, respectively. No evidence was found for any solvent-containing complexes, that is, $[CoX(SOL)(en)_2]^{2+}$. As in the other studies for reactions of this type, the position of equilibrium is strongly dependent on the concentration of chloride, as the *cis* isomer is stabilized by ion-pair formation. ($K_{IP,cis} \sim 5000$; $K_{IP,trans} \sim 4$) Also, the rates k_c and k_t are affected by the formation of ion pairs. The solvation of the dissociating anion by sulfolane to form a trigonal bipyramidal intermediate is much weaker than the corresponding solvation by most other solvents, particularly protic solvents. This behavior is reflected in the observation that reactions of this type in sulfolane are much slower than the corresponding reactions in other solvents. For example, the half-life for the isomerization of *cis*-$[CoCl_2(en)_2]^+$ at zero chloride concen-

Table 10.11 Activation Parameters in Sulfolane for the Isomerization of *cis–trans*-Dichlorobis(ethylenediamine)cobalt(III)[a]

	$\Delta H^{\circ\ddagger}$ kcal mol^{-1}			$\Delta S^{\circ\ddagger}$ cal mol^{-1}		
$10^3[Cl^-]$	$\Delta H_c^{\circ\ddagger}$	$\Delta H_{cip}^{\circ\ddagger a}$	$\Delta H_t^{\circ\ddagger}$	$\Delta S_{c\ddagger}^{\circ}$	ΔS_{cip}°	$\Delta S_t^{\circ\ddagger}$
0	25.0			−6.4		
4.3		25.0	20.2		−4.7	−18.4
30.0		25.0			−3.1	

[a] The cip represents the case for the removal of the *cis* isomer under conditions of Cl^- in which formation of the ion pair is complete.

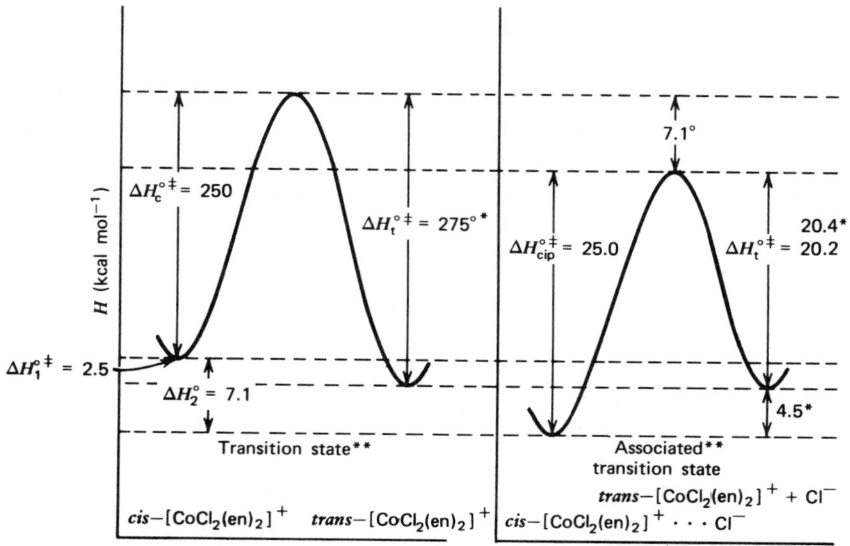

Figure 10.4 Reaction profile for the reactions. (a) *cis*-$[CoCl_2(en)_2]^+ \rightleftarrows$ *trans*-$[CoCl_2(en)_2]^+$. (b) *cis*-$[CoCl_2(en)_2]^+$ $Cl^- \rightleftarrows$ *trans*-$[CoCl_2(en)_2]^+$ $+Cl^-$. (*) Calculated from the other energies, which have been measured independently. (**) In the region of these transition states, there must be at least one dissociative intermediate and one other transition state. Reprinted with permission from *J. Amer Chem Soc.*, **89,** 821 (1967).

tration in methanol is 8.3 min and in DMA it is 350 min at 60°C, compared with 1310 min in sulfolane.

The various activation parameters are given in Table 10.11. The results show that the transition state is stabilized by ion association to the same extent as the reacting *cis*-$[CoCl_2(en)_2]^+$ ion. This effect accounts for the fact that the rate of *cis* removal to form *trans* is much less sensitive to the chloride concentration than is the rate of *trans* removal to form the *cis* isomer. The reaction profile is shown in Fig. 10.4.

Muir and Langford (15) have made a study on the reverse of solvolysis, namely, anation reactions. In particular, they studied the reaction

$$cis\text{-}[Co(en)_2NO_2DMSO]^{2+} + X^- = cis\ [Co(en)_2NO_2X]^{2+} + DMSO \quad (10.9)$$

in DMSO, where $X^- = Cl^-$, NO_2^- or SCN^-. The rates of anation are shown in Fig. 10.5. It is seen that the rates of anation by Cl^- and NO_2^- reach limiting values at low anion concentrations, whereas the reaction rate with SCN^- does not reach a limiting value. The equilibrium constants for forming ion pairs between the complex and Cl^- or NO_2^- are larger than 10^2, while the constant for SCN^- is less than 20. On the basis of the kinetic data, the anation reaction was resolved into two steps:

$$cis\text{-}[Co(en)_2NO_2DMSO]^{2+} + X^- \overset{K}{\rightleftharpoons} cis\text{-}[Co(en)_2NO_2DMSO]^{2+}, X^-$$

$$cis\text{-}[Co(en)_2NO_2DMSO]^{2+}, X^- \xrightarrow{k} cis\text{-}[Co(en)_2NO_2X]^+ + DMSO$$

where K is the equilibrium constant for ion association and k is the first-order rate constant for anation. The observed pseudo-first-order rate constants, k_{obsd}, are related to k and K by

$$k_{obsd} = \frac{kK[X^-]}{1 + K[X^-]} \tag{10.10}$$

Values of k_{obsd} for Cl^-, NO_2^-, and SCN^- are respectively, 5.0×10^{-4}, 1.2×10^{-4} and $> 5.0 \times 10^{-5}$, indicative of a dissociative interchange (I_d) mechanism rather than a conventional dissociative pathway via an intermediate of reduced coordination number (D).

Summary The study of solvent-exchange reactions between the bulk solvent and the first solvation sphere of inorganic ions is important for the understanding of substitution reactions at metal centers. Several revealing factors concerning the mechanisms of ligand substitution are obtained from

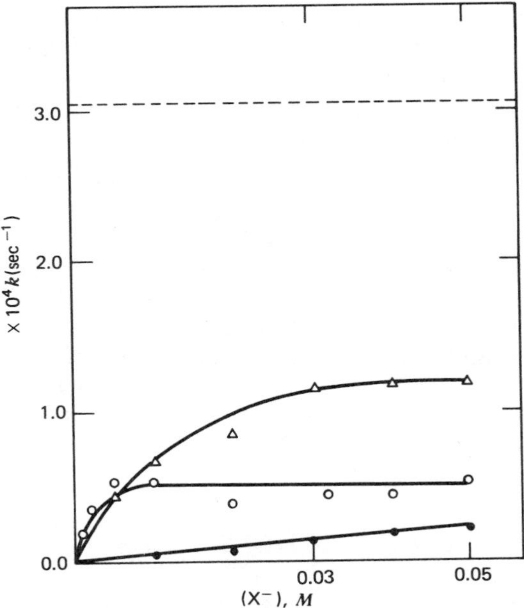

Figure 10.5 Rate of anation of *cis*-$Co(en)_2NO_2$ $DMSO^{2+}$ ion as a function of the concentration of the entering anion X^-. (△) $X^- = NO_2$; (○) $X^- = Cl^-$; (•) $X^- = SCN^-$. The broken line shows the rate of the DMSO-exchange reaction. Reprinted with permission from *Inorg. Chem.*, **7**, 1032 (1968). Copyright 1968. American Chemical Society.

a study of both equilibria and kinetics in nonaqueous solvents. For example, information concerning transition states, the effect of ion association on mechanisms and the stereochemical course of reactions is often gained by nonaqueous studies. Typical non-labile octahedral centers that have been investigated are Co(III) and Cr(III). In addition, square-planar complexes of Pt(II) and organometallic complexes have been studied. The techniques used for the study of exchange rates include temperature-jump, ultrasonic absorption, NMR, and flow methods.

These substitution reactions appear to have several characteristics. For example, the nature of the ligand has very little effect on the rate of a reaction of a particular metal in a given solvent. Also, a limiting value of the rate is attained, which is independent of further increases in concentration of the ligand, and the limiting rate is within a power of 10 the same as the rate for the solvent-exchange process.

The mechanism for the substitution process involves both the breaking and formation of a bond. The two possible mechanisms for the formation of the final product involve either a transition state in which there exists a partial formation of the new bond (an associative or *a-mode*) or a mechanism is which the activation mode is dissociative (a *d-mode*). In addition, a pathway, I, may be associated with either mode of activation. For octahedral substitution, experimental data indicate that bond breaking (dissociation) is the significant process in the formation of the transition state.

In postulating a particular mechanism (dissociative or associative) the effect of ion association on the free energy of the reactant and the transition state must be considered, together with the nucleophilicity of the ligand.

Solvolysis and isomerization of species such as dichlorobis(ethylenediamine)cobalt(III) ions in DMSO containing varying amounts of $(Et)_4NCl$ have been investigated. The change in the relative amounts of the *cis*- and *trans*-dichlorocomplexes is a reflection of the varying extents of the ion association properties of the two isomers. The dichlorocomplexes undergo solvolysis and direct isomerization simultaneously, with the solvolysis being more important. The mechanism of this isomerization can also be discussed using the concept of transfer activity coefficients for the isomeric cations, the transition states and ion pairs formed between isomeric cations and the chloride ions.

10.3 KINETICS OF FAST REACTIONS

We now turn our attention to the study of faster reactions, using relaxation techniques such as NMR, and temperature-jump. In these cases the interest is on the exchange rates of solvent molecules between the coordination sphere of the ion and the bulk solvent and the activation parameters ($\Delta H^{\ddagger}$ and $\Delta S^{\ddagger}$) for these exchanges. Also the mechanisms of these exchanges can be postulated from relaxation data as a function of temperature, that is, what are the factors that control the exchange rate of solvent molecules from the inner coordination sphere of the metal ion and the bulk solvent? A significant point that arises concerns the metal ion–solvent interaction and what type of complex exists. In other words, even if a metal ion–solvent complex can be kinetically distinguished, does this mean that a single complex exists over the entire temperature and composition range of the study or does an equilibrium exist between kinetically distinguishable complexes of different coordination numbers and geometries?

10.3.1 Studies of Fast Reactions Using NMR Techniques In aqueous solutions, the exchange of protons between the free solvent and the coordination sphere is in most cases so rapid that the observed proton relaxation rate is the weighted average of the relaxation rates in the two environments. However, in nonaqueous solvents, such as methanol, proton exchange is slower than in water and by lowering the temperature can be reduced to rates low enough so that separate peaks can be observed for the solvation-shell proton and the free-solvent proton. Rates of exchange of solvent molecules between the bulk solvent and the coordination sphere can be calculated from the widths of the signals. The relative intensities of the signals can also be used to determine solvation numbers.

The basic relations used in the interpretation of NMR data for solvent-exchange reactions were developed by Swift and Connick (16) and are clearly presented by Babiec et al. (17). To gain data on rates of exchange reactions and activation parameters for the exchange processes, the key parameter is the relaxation time. In order to develop the key relationships, it is assumed that there are two environments in which the resonating species (e.g., ^{1}H or ^{17}O) can exist, that is, the coordination sphere of the paramagnetic ion (denoted by M) and the bulk solvent (denoted by A). The ratio of the number of nuclei in the M environment to the number of nuclei in the A environment is given by P_M. Also, for dilute solutions the number of nuclei in the A environment is much larger than the number of nuclei in the M environment. The nuclei can exchange between the two environments at a rate which is characterized by a quantity τ_M, the average residence time of the nuclei in the M environment. The transverse relaxation times in the A and M environments are given by T_{2A} and T_{2M}, respectively, and these relaxation times control the observed linewidths in the NMR spectra. The full width corresponding to the half-height of the absorption line, $\Delta\nu$, yields one observed relaxation time, T_2, from the relation

$$\frac{1}{T_2} = \tau\Delta\nu \tag{10.11}$$

The working equation is given by

$$\frac{1}{T_2} = \frac{1}{T_{2A}} + \frac{P_M[1/T_{2M}(1/T_{2M} + 1/\tau_M) + (\Delta\omega_M)^2]}{\tau_M(1/T_{2M} + 1/\tau_M)^2 + \tau_M(\Delta\omega_M)^2} \tag{10.12}$$

where $\Delta\omega_M$ is the chemical shift of the resonance peak in the M environment relative to the chemical shift in the A environment. The average residence time τ_M is obtained from the quantity $(1/T_2 - 1/T_{2A})$.

For practical purposes, two limiting forms of Eqn. 10.12 can be deduced by making several assumptions. First, if $\Delta\omega_M^2 \gg 1/T_{2M}^2$ and $1/\tau_M^2$, then

$$\left(\frac{1}{T_2} - \frac{1}{T_{2A}}\right) = \frac{P_M}{\tau_M} \tag{10.13}$$

Equation 10.13 shows that the relaxation time is controlled by the rate of chemical exchange which is in accord with the relation

$$\frac{1}{\tau_M} = \frac{kT}{h} \exp\left(\frac{-\Delta H^\ddagger}{RT} + \frac{\Delta S^\ddagger}{R}\right) \tag{10.14}$$

where $\Delta H^\ddagger$ and $\Delta S^\ddagger$ are the enthalpy and entropy of activation for a first-order solvent exchange reaction.

A further limiting condition occurs for the higher temperature region, where the chemical exchange process is faster and $1/\tau_M^2 \gg \Delta\omega_M^2 \gg (1//T_{2M}\tau_M)$, then

$$\frac{1}{T_2} - \frac{1}{T_{2A}} = P_M\tau_M(\Delta\omega_M)^2 \tag{10.15}$$

The activation parameters are again the principal factors in the temperature dependence, but now the reciprocal of the first-order rate constant appears in the relaxation equation.

Some examples of the use of the approach outlined above will now be presented. Breivogel (18) used NMR to determine solvent exchange rates for solvent metal complexes in systems containing the ligand Fe^{3+} in DMF, CH_3CN, and ethanol and for the ligand Ni^{2+} in ethanol. The rate constants and the activation parameters for the exchange processes were evaluated from the measured temperature dependence of the relaxation rates. The results are summarized in Table 10.12. A comparison of the rate constants

Table 10.12 Rate Constants and Activation Parameters for the Exchange of Solvent Molecules Between the First Coordination Sphere of the Metal Ion and the Bulk Solvent

Metal Ion	Solvent	k_1 (sec^{-1} at 25°C)	$\Delta H^\ddagger$ (kcal mol^{-1})	$\Delta S^\ddagger$ (eu)
Fe^{3+}	Ethanol	2.0×10^4	6.2 ± 1.5	-18 ± 5
Fe^{3+}	DMF	33	12.5 ± 1.5	-10 ± 5
Fe^{3+}	AN	<40		
Ni^{2+}	Ethanol	1.1×10^4	10.8 ± 1.5	-4 ± 5

Reprinted with permission from *J. Phys. Chem.*, **73**, 4203 (1969). Copyright 1969. American Chemical Society.

Table 10.13 Comparison of Rate Constants for Solvent Exchange at 25°C for Fe^{3+} and Ni^{2+}

Solvent	$k_{Ni^{2+}}$ (sec^{-1})	$k_{Fe^{3+}}$ (sec^{-1})	$k_{Fe^{3+}}/k_{Ni^{2+}}$
H_2O	2.7×10^4	1.5×10^2	0.0055
DMSO	7.5×10^3	50	0.0067
DMF	7.7×10^3	33	0.0043
AN	3.9×10^3	<40	<0.01
Methanol	1.0×10^3	2.4×10^3	2.4
Ethanol	1.1×10^4	2.0×10^4	1.8

for solvent exchange for Fe^{3+} and Ni^{2+} in several solvents is given in Table 10.13. It is seen that the ions have similar relative exchange rates in all the solvents. However, for methanol and ethanol, the relative exchange rates for Fe^{3+} are between two and three orders of magnitude greater than for the other solvents. This effect was attributed to the steric crowding around the small Fe^{3+} ion. The difference can also be viewed in terms of the bonding existing between the metal ion and the solvent. In the case of DMF, AN, and DMSO, the atom through which the solvent coordinates to the metal ion is bonded to only one atom in the solvent molecule, while in methanol and ethanol the oxygen atom is bonded to two atoms in the solvent.

Thomas and Reynolds (19) obtained proton nuclear magnetic resonance data for Ni^{2+} and Co^{2+} in DMSO solutions over a range of temperatures. Values for the rate constants and activation parameters for the Ni(II) system are given in Table 10.14, together with a comparison for similar data in H_2O, NH_3, CH_3OH, and DMF. To account for the smaller value for $\Delta H^‡$ obtained for the Ni(II)–DMSO system it was suggested that in the substitution reaction involving DMSO the requirement of breaking hydrogen bonds was absent compared to the other solvents. In addition, DMSO is a smaller molecule than DMF. The absence of hydrogen bonding in DMSO was also postulated to account for the higher negative value for $\Delta S^‡$, since the hydrogen bonding has a tendency to oppose the aligning forces of the ion and hence to create disorder around an ion.

An NMR line-broadening study was used to obtain solvent exchange rates for Co(II) and Ni(II) in DMSO and trimethyl phosphate (TMPA) by Angerman and Jordan (20). The solvent exchange rates are listed in Table 10.15. It is possible to measure the ligand exchange rate for the DMSO–Ni(II) system because in all the other systems the exchange rates were too fast to control the proton relaxation times, or in the case of Co(II)–DMSO, the high melting point of DMSO (18.5°C) precludes going to low enough tem-

Table 10.14 Solvent-Exchange Data for Ni(II) at 25°C

Solvent	$10^{-4}k_M$ (sec^{-1})	$\Delta H^‡$ (kcal mol^{-1})	$\Delta S^‡$ (eu)
H_2O	2.7	11.6	0.6
NH_3	4.7	10 ± 1	−3 ± 4
CH_3OH	0.10	15.8	8.0
DMF	0.38	15 ± 0.5	8 ± 2
DMSO	0.75	8 ± 0.7	−16 ± 7

peratures to obtain exchange rates, as generally exchange rates for Co(II) systems are about 10^2 faster than for Ni(II) systems at 25°C. The other exchange rates given in Table 10.15 are minimum values.

In order to explain the relative exchange rates in these systems, Angerman and Jordan (20) assumed that the total activation energy for exchange was made up of a crystal-field contribution, $\Delta H^‡_{CF}$, and a solvation contribution, $\Delta H^‡_{solv.}$, where

$$\Delta H^‡ = \Delta H^‡_{CF} + \Delta H^‡_{solv.} \qquad (10.16)$$

For a given solvent and for metal ions of similar charge and exchanging by the same mechanism, the $\Delta H_{solv.}$ should be constant. If this assumption is correct, then it is apparent that the crystal-field theory can only predict differences in activation energies for different metal ions. To give an example of this treatment, suppose that the activation energy, $\Delta H^‡$, for solvent exchange with Ni^{2+} is taken as an arbitrary standard. It follows that the $\Delta H^‡$ for other metal ions, for example, Co^{2+}, may be calculated using

$$(\Delta H^‡)^{calcd.}_{Co} = (\Delta H^‡)_{Ni,\,obsd.} + (\Delta H^‡_{CF})_{Co} - (\Delta H^‡_{CF})_{Ni} \qquad (10.17)$$

$$= (\Delta H^‡)_{Ni,\,obsd.} + 3.65\,(D_q)_{Co} - 4.80\,(D_q)_{Ni}$$

Calculated and observed values for Co(II) in various solvents using this prediction are given in Table 10.16. The disagreement in the value in ammonia was attributed to some type of octahedral–tetrahedral equilibrium.

Matwiyoff (21) has examined both low-temperature and high-temperature data for Co(II) and Ni(II) perchlorates in DMF. A summary of the results is presented in Table 10.17, where the methyl protons are as shown below:

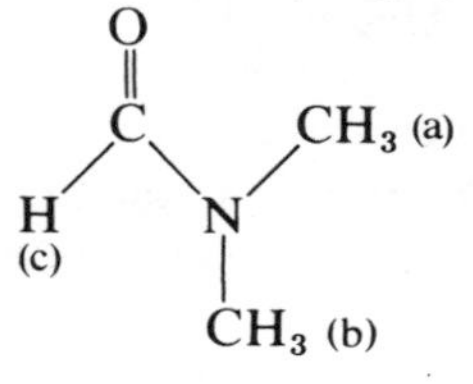

```
          O
          ‖
          C        CH3 (a)
        /   \    /
      H       N
     (c)      |
             CH3 (b)
```

Table 10.15 Solvent-Exchange Rates for Co(II) and Ni(II) in DMSO and TMPA

Compound	Solvent	k_{ex} (25°; sec^{-1})	$\Delta H^{\ddagger}$ (kcal mol^{-1})	$\Delta S^{\ddagger}$ (eu)
$Ni(ClO_4)_2$	DMSO	5.2×10^3	12.1	−1.3
$Ni(ClO_4)_2$	TMPA	$>1.8 \times 10^2$		
$Co(ClO_4)_2$	DMSO	$>1.5 \times 10^4$		
$Co(ClO_4)_2$	TMPA	$>6.3 \times 10^3$		

In the case of $Ni(ClO_4)_2$, it was only possible to detect bulk solvent NMR signals. The temperature studied covered the range −50 to +50°C, in which case at least a twofold change in solvent dielectric constant would be expected. However, the results for the high and low temperature are insensitive to changes in dielectric constant. The results also show that the rate of exchange of the Co(II) complex is more rapid than that of Ni(II). The parameters obtained were found to be consistent with a mechanism where the rate-determining step involves a transition state in which the breaking of the bond to the leaving group (DMF) is much more advanced than the formation of the bond to the entering group (DMF).

A composite of the proton NMR spectra of a solution of $Co(ClO_4)_2$ in DMF is given in Fig. 10.6. Peaks *a*, *b*, and *c* refer to the bulk solvent protons, whereas the primes represent proton peaks in the coordination shell of Co(II). The formyl proton is assigned as c', while the *cis*- and *trans*-methyl protons are assigned as a' and b'. The solvation number, *n*, of cobalt in the complex is obtained by using the ratios of the area of peak *c* to those of a', b', and c'. In this case, a solvation number of 6 was found for the entire

Table 10.16 Observed and Calculated Energies of Activation for Solvent Exchange for Co(II) in Various Solvents

Solvent	$\Delta H^{\ddagger}$ (kcal mol^{-1}) (obs.)	$\Delta H^{\ddagger}$ (kcal mol^{-1}) (calcd.)
H_2O	8.0	8.0
CH_3OH	13.8	13.4
DMF	13.6	13.0
CH_3CN	8.1	7.2
NH_3	11.2	7.3

Table 10.17 Parameters for the Exchange of DMF from the First Coordination Sphere of Co(II) and Ni(II)

Parameter	Formyl proton (c) Low-Temp. Data	Formyl proton (c) High-Temp. Data	Methyl (a)	Methyl (b)
	Co(II)			
k_1 (sec^{-1} at 25°C)	4.0×10^5	3.8×10^5	3.9×10^5	3.9×10^5
$\Delta H^\ddagger$ (kcal)	13.6	13.7	13.4	13.4
$\Delta S^\ddagger$ (eu)	12.6	12.5	12.9	12.9
A/h (Hz)	3.88×10^5	3.88×10^5	4.07×10^4	2.51×10^4
	Ni(II)			
k_1, sec^{-1}		3.8×10^3		
$\Delta H^\ddagger$ (kcal)		15.0		
$\Delta S^\ddagger$ (eu)		8.0		
A/h (Hz)		6.81×10^5		

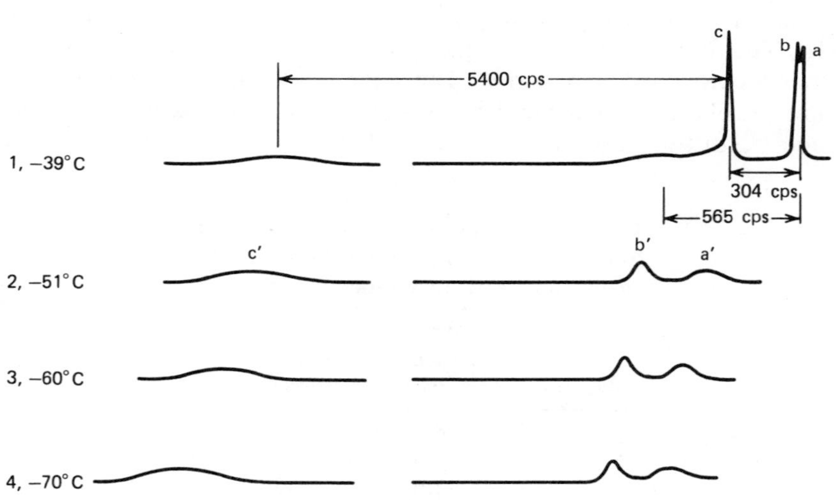

Figure 10.6 Composite proton NMR spectra of a DMF solution of $Co(ClO_4)_2$ as a function of temperature. The magnetic field increases from left to right. Reprinted with permission from *Inorg. Chem.*, **5**, 788 (1966). Copyright 1966. American Chemical Society.

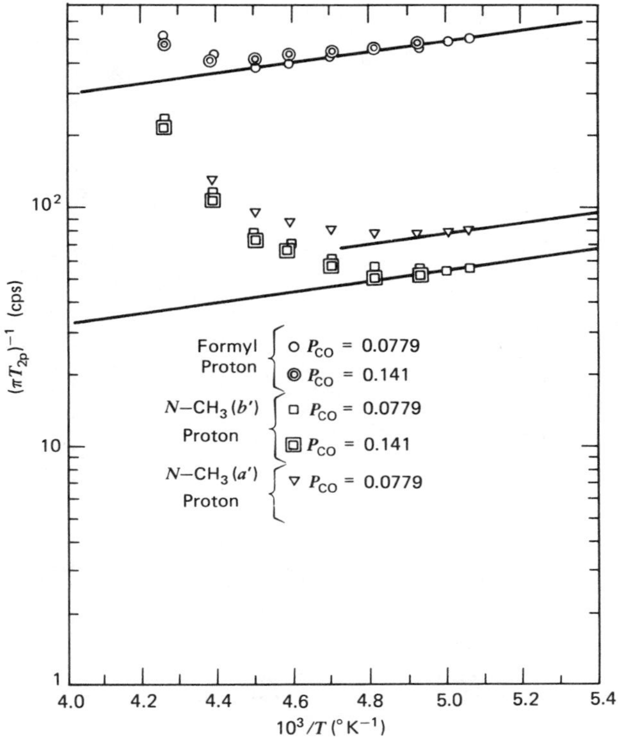

Figure 10.7 Temperature dependence of $(T_{2p})^{-1}$ for the formyl and methyl protons in the complex $Co(DMF)_6^{2+}$. Reprinted with permission from *Inorg. Chem.*, **5**, 788 (1966). Copyright 1966. American Chemical Society.

temperature and composition range. The relaxation of the protons in $[Co(DMF)_6]^{2+}$ was investigated by measuring the width of the lines corresponding to c', b', and a' as a function of temperature (Fig. 10.7) and also the variation in the chemical shift ($\Delta\omega_m$) with temperature (Fig. 10.8). It was found that each set of data in Fig. 10.7 could be fitted to two straight lines. One line represented the data at the higher temperatures and was attributed to line broadening due to the exchange of DMF between the two sites, and a second line represented data at the lower temperatures and is associated with the temperature dependence of the relaxation rate of the proton in the first coordination sphere of Co(II), as at the lower temperatures the chemical exchange is too slow to affect the observed line widths.

10.3.2 Studies of Fast Reactions Using Relaxation Methods Some of the more recent studies of solvent-exchange and ligand-substitution reactions of metal ions have concerned themselves with the so-called fast-substitution reactions, using such techniques as stopped-flow spectrophotometry, tem-

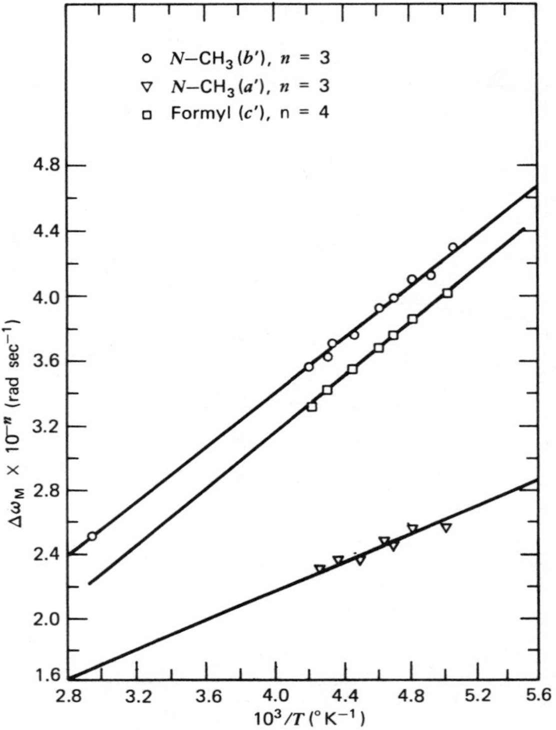

Figure 10.8 Temperature dependence of $\Delta\omega_M$ for the protons of the $Co(DMF)_6^{2+}$ complex in DMF solutions at 60 Mc. Reprinted with permission from *Inorg. Chem.*, **5**, 788 (1966). Copyright 1966. American Chemical Society.

perature-jump and ultrasonic-absorption relaxation methods. An example of this approach is that of several studies by Caldin and Bennetto (22, 23), who have used stopped-flow methods to examine the solvent effects on the kinetics of reactions of Ni(II) and Co(II) ions with 2,2′-bipyridyl and 2,2′,2″-terpyridyl in a variety of solvents. Measurements were made of rate constants and activation parameters. Results obtained for the solvent exchange parameters and the ligand-substitution reactions according to the scheme

$$[M(S)_m]^{2+} + L \overset{K_o}{\rightleftharpoons} [M(S)_m]^{2+}, L \xrightarrow{k_{ex}} [M(S)_{m-1}L], S \qquad (10.18)$$

are given in Table 10.18. Here k_{ex} is the first-order rate constant for exchange of a solvent molecule between the first solvation sphere and the bulk solvent and K_o is the equilibrium constant for the formation of the outer-sphere complex. In the simplest case, the kinetic behavior should be represented by

$$k_f \equiv K_o k_{ex} \tag{10.19}$$

and the value of $n = k_f/K_o k_{ex}$ (see Table 10.18) should be unity in all solvents. However, the results indicate that there is a range of variation of n of more than 200-fold at 25°C and it is not attributable to changes of K_o with the dielectric constant of the solvent. These findings are reasonable evidence that the simple I_d mechanism is inadequate to account for ligand-substitution kinetics in nonaqueous solvents. The treatment by Caldin and Bennetto assumed that the value of K_o (0.1 liter mol^{-1}), the equilibrium constant for outer-sphere complexation, was the same in all solvents.

It was found that the variations of n for the different solvents did not depend on the aprotic or protic nature of the solvents or on the polarity of the solvent molecules. However, a correlation was obtained for a plot of log n vs $\Delta\bar{H}_{vap}$ of the solvent, as shown in Fig. 10.9. A further correlation was obtained for plots of log n vs the fluidity of the solvent (ρ/η), as shown in Fig. 10.10. No correlation was found between either k_f or k_{ex} and $\Delta\bar{H}_{vap}$ or the fluidity. These trends were interpreted in terms of various structural features of the solvent playing a role in the rate-determining step. For example, both the enthalpy of vaporization, $\Delta\bar{H}_{vap}$ and the fluidity of the solvent are related to the "stiffness" of the solvent or its resistance to molecular motion, or the "openness" of the solvent represented by its free volume. The $\Delta\bar{H}_{vap}$ was considered as being related to the energy required to make a hole of molecular dimensions in the liquid (ΔE_h).

The effect of the detailed solvent structure on the mechanism of solvent exchange and ligand substitution was considered from the following view-

Table 10.18 Rate Constants and Activation Parameters for the Reaction Between Ni(II) and Bipyridine in Various Solvents

Solvent	log k_{ex} [a]	$\Delta H^{\ddagger}_{ex}$ [a]	$\Delta S^{\ddagger}_{ex}$ [a]	log k_f [b]	$\Delta H^{\ddagger}_f$ [b]	$\Delta S^{\ddagger}_f$ [b]	log n [c]
Water	4.5	10.8	−2	3.2	12.6	−2	−0.3
D_2O	4.5	10.8	—	3.04	14.0	+2	−0.5
Methanol	3.0	15.8	8	1.92	17.0	+7	−0.1
Ethylene Glycol	3.6	—	—	1.52	16.8	+5	−1.1
DMF	3.9	9.4	−9	2.73	12.7	−3	−0.2
DMSO	3.9	8	−14	1.84	12.6	−8	−1.1
Acetonitrile	3.4	11.7	−4	3.67	6.6	−20	+1.3

[a] Subscript ex represents solvent exchange.
[b] Subscript f represents ligand substitution.
[c] $n = k_f/K_o k_{ex}$, where k_f is the second-order rate constant for the overall forward reaction. The quantity n is a measure of the chance that a ligand molecule will enter the primary solvation shell of the cation at a particular site when a molecule leaves it.

Reprinted from Ref. 24, p. 358, by courtesy of Marcel Dekker, Inc.

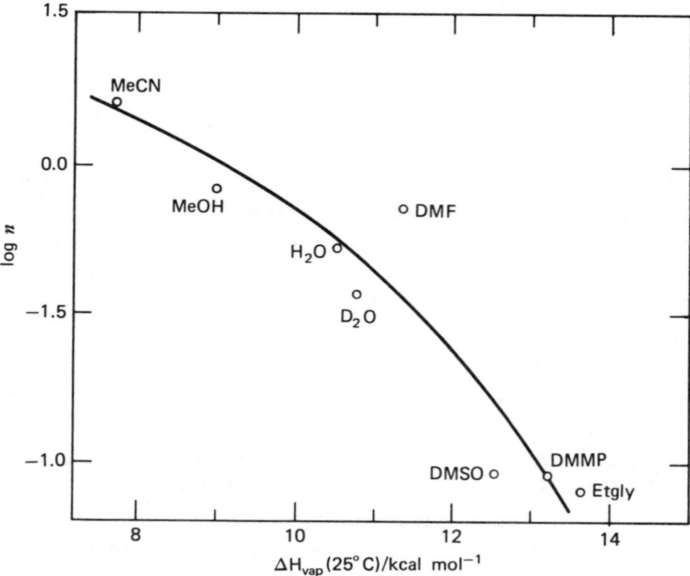

Figure 10.9 Plot of log n at 25°C for the reaction of Ni(II) with bipyridyl in various solvents, against enthalpy of evaporation of solvent (ΔH_{vap}). Reprinted with permission from *J. Chem. Soc.*, **A,** 2191 (1971).

point. The process of the exchange of a solvent molecule between an ion and the bulk solvent, requiring an enthalpy comparable to $\Delta \bar{H}_{vap}$, consisted of the solvent leaving its position in the first solvation shell, passing through a relatively disordered region near the ion and then entering the relatively ordered bulk solvent region. For the formation of the activated complex a solvent molecule S is partly detached from the cation, and a ligand molecule L is in the process of entering the first coordination shell of the cation. In the case of solvent exchange, the ligand molecule is just another solvent molecule. In the absence of any solvent structure, it would be possible to represent the transition states by three-center models as shown:

Ligand Substitution

$L \ldots > \ldots MS_5 \ldots S \rightarrow$

Solvent Exchange

$S \ldots > \ldots MS_5 \ldots S \rightarrow$

However, if the solvent structure is considered, then in the process of solvent exchange the leaving molecule S moves into the structured bulk solvent, while the incoming molecule leaves a similarly structured region. In the case of ligand substitution one has to consider the differences in the interactions between ligand molecules and solvent molecules. Solvent exchange thus requires at least a five-center model and ligand substitution, at least a four-center model, as shown below

Ligand Substitution

```
                              sss
L . . . > . . . MS5 . . . S > . . . sss
                              sss
```

Solvent Exchange

```
sss                                          sss
sss . . . S . . . > . . . MS5 . . . S . . . > . . . sss
sss                                          sss
```

The model based on solvent structure thus predicts that the approach to

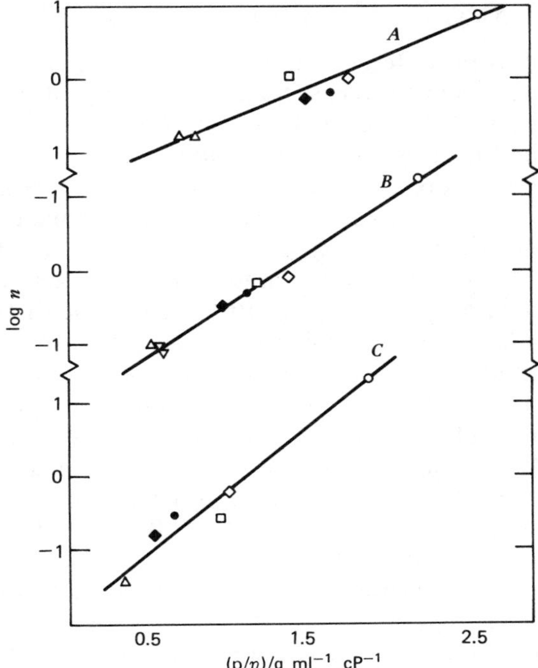

Figure 10.10 Plot of log n for the reaction of Ni(II) with bipyridyl in various solvents, against fluidity, at 25°C. (○) CH_3CN; (◇) MeOH; (□) DMF; (●) H_2O; (◆) D_2O; (△) DMMP; (▲) DMSO; (▽) Etgly; (*A*), 45°C; (*B*), 25°C; (*C*), 5°C. Reprinted with permission from *J. Chem. Soc.*, **A**, 2191 (1971).

Table 10.19 Enthalpy, Entropy, and Free Energy of Activation Differences for Bipyridyl Substitution and Solvent Exchange at Ni^{2+} at 25°C

Solvent	$\Delta\Delta G^{\ddagger} =$ $(\Delta G^{\ddagger}_{f} - \Delta G^{\ddagger}_{ex})$ (kcal mol^{-1})	$\Delta\Delta H^{\ddagger} =$ $(\Delta H^{\ddagger}_{f} - \Delta H^{\ddagger}_{ex})$ (kcal mol^{-1})	$\Delta\Delta S^{\ddagger} =$ $(\Delta S^{\ddagger}_{f} - S^{\ddagger}_{ex})$ (cal mol^{-1} K^{-1})
MeCN	−0.3	−5.1	−15.9
MeOH	+1.4	+1.2	−0.6
DMF	+1.6	+3.3	+5.8
H_2O	+1.8	+1.8	+0.0
DMSO	+2.7	+4.6	+6.3

Reprinted with permission from *J. Chem. Soc.*, **A**, 2191 (1971).

the cation of ligand molecules and of solvent molecules will be affected differently by the solvent structure.

Correlations were also observed for the temperature coefficient of log n and the fluidity and $\Delta\bar{H}_{vap}$ of the solvent using

$$RT\frac{d \ln n}{d(1/T)} = RT\, d \ln \frac{k_f/k_{ex}}{d(1/T)} = \Delta H^{\ddagger}_{f} - \Delta H^{\ddagger}_{ex} \qquad (10.20)$$

where $\Delta H^{\ddagger}_{f} - \Delta H^{\ddagger}_{ex}$ is the difference in the energies of activation for solvent exchange and bipyridyl substitution at Ni(II). A summary of these activation parameters is given in Table 10.19. It is seen that a fairly large effect of solvent structure on the transition state is found so that one must take into account the process of reorganization of the solvent in the region of the ion.

Another striking feature found in the study by Bennetto and Caldin (22, 23) was the apparent direct relation between $\Delta H^{\ddagger}$ and $\Delta S^{\ddagger}$ for both solvent exchange and ligand substitution for various divalent metal ions in a variety of solvents, as shown in Fig. 10.11. This type of behavior was attributed to an enthalpy-entropy balance for changes in solvent structure. It is characteristic of the processes which are dominated by solvation changes that the changes in ΔH are compensated by those in ΔS.

To discuss their results, Bennetto and Caldin (23) proposed an extension of the Frank and Wen model for aqueous solvation to nonaqueous solvents as shown in Fig. 10.12. This particular model invokes three specific regions around an ion; (A), an ordered solvated layer; (B), an intermediate disordered region, and (C) the bulk solvent. Using this model the process of solvent exchange was considered to take place along the following paths:

1 The transfer of solvent molecules from the bulk solvent (C) to a position in the disordered region (B).
2 A solvent molecule can move from region (B) into the first solvation shell of the ion (A).
3 The converse of process (2), i.e., the movement of a solvent molecule

from the solvation sheath (A) to the disordered region (B) can occur, and the metal ion–solvent bond will be broken.

4 A solvent molecule can return to the bulk solvent from the disordered region.

5 Reorganization of the bulk solvent may occur.

In using these concepts to explain the exchange process it is obvious that all five steps can contribute to the values of $\Delta H^{\ddagger}$, $\Delta S^{\ddagger}$, $\Delta G^{\ddagger}$, and in particular steps 4 and 5 may be especially relevant in nonaqueous solvents as compared to aqueous solvents, where step 3 dominates the kinetic process. The effect of the particular ligand on the structure of the solvent is regarded as the crucial point. For example, if the ligand enhances the solvent structure, the heat evolved in step 4 increases and therefore reduces $\Delta H^{\ddagger}$, and $\Delta\Delta H^{\ddagger}$ is positive, whereas if the ligand breaks down the solvent structure, the $\Delta\Delta H^{\ddagger}$ becomes negative. The effect of the ligand on the solvent structure depends in turn on the strength of the solvent–solvent interactions.

As a working hypothesis, it is assumed that when an exchange of a solvent molecule between an ion (region A) and the disordered region (B)

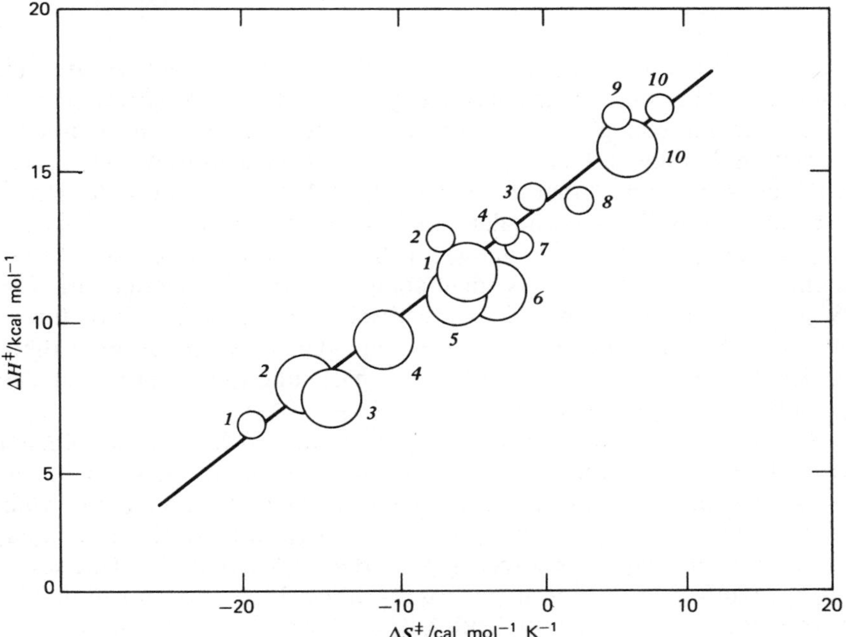

Figure 10.11 Enthalpy–entropy relation for substitutions at Ni^{2+}, *1*, MeCN; *2*, DMSO; *3*, DMMP; *4*, DMF; *5*, H_2O, EtOH; *6*, NH_3; *7*, H_2O; *8*, D_2O; *9*, Etgly; *10*, MeOH. Reprinted with permission from *J. Chem. Soc.*, **A,** 2198 (1971).

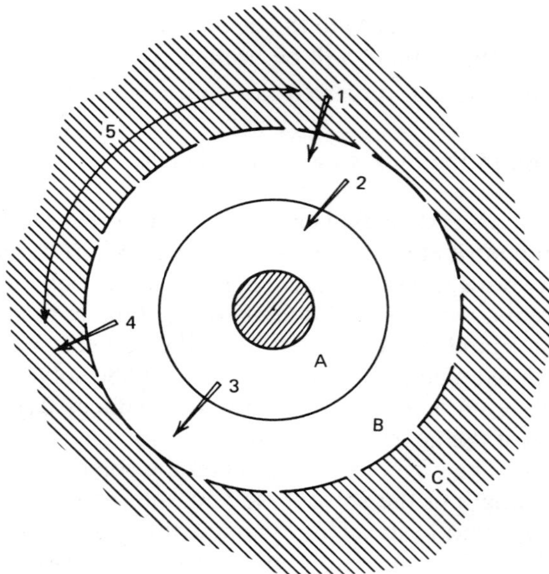

Figure 10.12 Structural model for a solvated ion. Reprinted with permission from *J. Chem. Soc.*, **A**, 2198 (1971).

occurs there is a contribution to $\Delta G^{\ddagger}$ and thus to the reaction rate that is governed by the enthalpy and is given by ΔH_{MS}, that is, the solvent molecule has moved far enough away from the ion to overcome the ion–solvent interaction, but not sufficiently to attain its increased entropy state in region B. It was also assumed that the magnitude of ΔH_{MS} was about the same for all bivalent ions. This particular model implied that the dissociative step required an enthalpy which was about the same for different solvents and an entropy change which was small compared with the overall variation of $\Delta S^{\ddagger}$ and also that the effect of an added solute on the solvent structure was related to the difference between the enthalpies of evaporation of the two species. It should be pointed out that this particular treatment is speculative and subject to further testing.

The more recent work of Coetzee (23a) has shown that the effects observed by Caldin are the result of steric problems of bipyridine and provide a simpler explanation than that discussed above. Coetzee has reexamined the "abnormal" kinetics of the labile complexes of Mn(II), Co(II), and Ni(II) in a variety of nonaqueous solvents. The order of reactivity in all the solvents was phenanthroline > bipyridine > terpyridine. It was also noted that in all the solvents, except dimethylsulfoxide and water, phenanthroline reacts faster than the norm and in several solvents bipyridine and particularly terpyridine react slower than the norm. These "abnormal" kinetics were accounted for in terms of the decreased reactivity of bipyridine and terpyridine in certain solvents being caused by a shift in the rate-determining step to the ring-closure step in the reaction scheme:

$$Ni(H_2O)_6^{2+} + L{-}L \underset{k_{21}}{\overset{k_{12}}{\rightleftharpoons}} \underset{\text{(I) Outer-sphere complex}}{Ni(H_2O)_6^{2+}\cdot LL} \underset{k_{32}}{\overset{k_{23}}{\rightleftharpoons}} \underset{\text{(II) Singly coordinated inner-sphere complex}}{NiL{-}L(H_2O)_5^{2+}} + H_2O$$

$$\underset{k_{43}}{\overset{k_{34}}{\rightleftharpoons}} \underset{\text{(III) Doubly coordinated (chelated) inner-sphere complex}}{Ni\langle L{\frown}L\rangle (H_2O)_4^{2+}} + 2H_2O$$

The repulsion occurring between the 3 and 3′ hydrogen atoms opposes the *cis* configuration and consequently the *trans* configuration is favored. The reason suggested for the inhibited reactivity of bipyridine is that the first bond formation with the metal ion is subject to severe steric hindrance, unless bipyridine in the outer sphere interacts sufficiently strongly with the field of the metal ion to be essentially in the *cis* configuration. For other configurations of bipyridine, the nonreactivity arises from steric hindrance of first-bond formation by the 3′ hydrogen atom of the flanking pyridine ring. The general conclusion arrived at was that first-bond formation of bipyridine occurs predominantly via the *cis* configuration. The nature of the nonaqueous solvent determines the extent to which the *cis* configuration is acquired in the outer sphere. The formation of the *cis* configuration will be promoted by hydrogen bonds donated by the solvent molecules of the inner sphere and by ion–dipole interactions, which in turn will be favored by low electron-donor strength of the solvent and a relatively open inner sphere. In general, the kinetics of a pyridine-type ligand are influenced by its polarity and steric requirements as well as by the electron-donor ability, hydrogen bond-donor ability and steric requirements of the solvent. Some qualitative estimates of the kinetics influence of the various solvent properties are discussed by Coetzee.

Coetzee has made other major contributions to the study of the kinetics of fast ligand-substitution reactions. A fairly complete account is given of his contributions, together with others, in Volume II of *Solute–Solvent Interactions* (24). The objective here is to focus on some specific studies of ligand substitution reactions in acetonitrile for Ni(II) and Co(II) ions with pyridine, 4-phenylpyridine, and isoquinoline and several 1,10-phenanthrolines substituted in the five position (25). Some comparisons are also made with a similar study in DMSO (26). The main goals of these studies were to demonstrate whether the I_d mechanism could adequately account for the reactions of Ni(II) ions, for example, in acetonitrile. The particular ligands were selected to investigate the steric effects on the relative rates and activation parameters involved in ligand substitution. In addition, the various substituted 1,10-phenanthrolines were chosen to examine the effects of electron-donating and electron-withdrawing groups on the kinetic processes.

The I_d or S_N1 mechanism can be represented as

$$\underset{(I)}{Ni(S)_6^{2+}} + L{-}L \underset{k_{21}}{\overset{k_{12}}{\rightleftharpoons}} \underset{(II)}{Ni(S)_6^{2+},L{-}L} \rightleftharpoons$$

$$\underset{(III)}{NiL{-}L(S)_5^{2+},S} \rightleftharpoons \underset{(IV)}{Ni\left<\begin{matrix} L \\ | \\ L \end{matrix}\right.(S)_4^{2+}} + 2S \qquad (10.21)$$

In this mechanism the outer-sphere complex is represented by (II), whereas (III) represents a singly coordinated inner-sphere complex and (IV), a doubly coordinated inner-sphere complex. The rate-determining step in this mechanism is the solvent exchange, process (II). In the majority of cases, both processes (I) and (III) are rapid. The pertinent data are presented in Table 10.20, which includes rate constants and activation parameters for substitution by unidentate as well as multidentate ligands, with comparisons with the corresponding quantities for solvent exchange. If the mechanism is considered to be a "normal" I_d mechanism, then the value of R in Table 10.20 should be unity, where $R = (4/3)\ k_1/k_{12}k_s$. It is seen that for phenanthroline itself and all other derivatives, the values obtained for R are significantly greater than unity. In addition, the differences between the activation energies for ligand substitution and solvent exchange, $\Delta H^\ddagger_L - \Delta H^\ddagger_S$ are ligand dependent and are large and negative, and this behavior appeared to be peculiar to acetonitrile. On the other hand, in aqueous solutions the values of $\Delta H^\ddagger_L - \Delta H^\ddagger_S$ are similar for a variety of ligands, and the value of R is close to unity.

To account for the behavior of multidentate ligands, two factors were considered, the stabilization of the outer-sphere complexes and the steric inhibition of coordination, factors which oppose each other. The origin of the outer-sphere complex stabilization is vague and in the case of acetonitrile was considered to arise from a π-orbital interaction with the nitrile group or an electrostatic interaction of the ligand molecules present in the outer sphere with the polarized acetonitrile molecules present in the inner sphere. The outer-sphere complex stabilization was enhanced by introducing into the ligands, centers of high electron density and the resulting increased stabilization accounted for the decrease in $\Delta H^\ddagger_{L,f}$. The stabilization effect can be either aided or hindered by steric factors, as illustrated by the cumulative stabilizing effect in the phenanthrolines due to the nitrogen atoms located in favorable, fixed positions. In the case of bipyridine and terpyridine the effect of steric factors on the formation of the outer-sphere complex is apparent. For several solvents, the order of rate constants is phenanthroline > bipyridine > terpyridine covering a range of $10^{0.4}$ in water and $10^{1.3}$ in the three nonaqueous solvents, acetonitrile, methanol, and DMSO.

Table 10.20 Comparison of Rate Constants and Activation Parameters for Ligand Substitution in Acetonitrile with Those for Solvent Exchange

Ligand	k_s or k_1 [a]	R	$\Delta H^\ddagger$	$\Delta S^\ddagger$
	Metal Ion—Nickel(II)			
Acetonitrile	2.7×10^3	—	ca.15	ca. +11
Nitrate ion	1.9×10^5	1	17.5	—
Trifluoroacetate ion	1.5×10^5	1	—	—
p-Toluenesulfonate ion	1.3×10^5	1	17.0	—
Ammonia	ca.3×10^3	1	—	—
Pyridine	8.3×10^2	0.4	14.7	+4
4-Phenylpyridine	9.7×10^2	0.5	11.2	−7
Isoquinoline	1.24×10^3	0.6	11.9	−5
2,2′-Bipyridine	4.1×10^3	2.0	6.5	−20
2,2′,2″-Terpyridine	2.20×10^3	1.1	8.4	−15
1,10-Phenanthroline	5.0×10^4	25	4.7	−21
5-Nitrophenanthroline	1.23×10^4	6	5.1	−23
5-Chlorophenanthroline	2.32×10^4	11	6.5	−17
5-Methylphenanthroline	6.5×10^4	32	5.5	−18
5,6-Dimethylphenanthroline	7.3×10^4	36	3.5	−25
2,9-Dimethylphenanthroline	4.3×10^2	0.21	10.1	−13
	Metal Ion—Cobalt(II)			
Acetonitrile	3.2×10^5	—	11.4	+5
Pyridine	1.16×10^5	0.5	10.5	0
4-Phenylpyridine	9.5×10^4	0.4	8.7	−7
Isoquinoline	1.11×10^5	0.5	7.0	−12
	Metal Ion—Iron(II)			
Acetonitrile	5.5×10^5	—	9.7	0
4-Phenylpyridine	2×10^6	5	7	−7

[a] Units: sec^{-1} for k_s, $l \cdot mol^{-1}\ sec^{-1}$ for k_1.
Temperature = 25°C, except where noted otherwise.

Summary In some solutions, the exchange of solvent molecules between the bulk solvent and the coordination sphere of the charged species is so rapid that special methods have to be used to study the kinetics. Relaxation techniques, such as NMR and temperature-jump, are normally used. The parameters of interest are normally the rate constants and the activation parameters, $\Delta H^\ddagger$ and $\Delta S^\ddagger$ for the exchange process, and are obtained from relaxation times. Relaxation data obtained as a function of temperature provide valuable information concerning the exchange mechanism. NMR spectra show that proton exchange in nonaqueous solvents such as methanol is slower than in water and by lowering the temperature can be reduced to rates low enough so that separate peaks can be observed for the solvation-shell proton and the free-solvent proton. Rates of exchange of solvent molecules between the bulk solvent and the coordination sphere can be calculated from the widths of the signals, while the relative intensities of the signals can be used to determine solvation numbers. The key relationships are developed assuming that there are two environments in which the resonating species can exist, namely, the coordination sphere of the paramagnetic ion and the bulk solvent. The nuclei can exchange between the two environments at a rate that is characterized by a quantity, τ_m, the average residence time of the nuclei in the M environment. Differences in solvent exchange rates for ions, such as Fe^{3+} and Ni^{2+}, in a variety of solvents, can be attributed to steric crowding around the small ion or can be viewed in terms of the bonding that exists between the metal ion and the solvent, that is, depending on whether the ion is bonded to one or two atoms in the solvent molecule. The presence or absence of hydrogen bonding in a solvent can have a significant effect on the activation parameters.

NMR line-broadening studies are often used to obtain solvent exchange rates. To explain relative exchange rates in systems involving Ni(II) or Co(II) in DMSO it is assumed that the total activation energy for exchange is made up of a crystal field contribution and a solvation contribution.

As an example of the use of the stopped-flow technique for studying fast reactions, solvent effects on the kinetics of reactions of Ni(II) and Co(II) ions with 2,2′-bipyridyl and 2,2′,2″-terpyridyl were measured and rate constants and activation parameters established. The effects observed have been attributed to a steric problem with the bipyridine. The order of reactivity in all the solvents is phenanthroline > bipyridine > terpyridine. In all solvents except DMSO and water, phenanthroline reacts faster than the norm, and in several solvents bipyridine and particularly terpyridine react slower than the norm. These "abnormal" kinetics are accounted for in terms of the decreased reactivity of bipyridine and terpyridine in certain solvents being caused by a shift in the rate-determining step to the ring-closure step in the proposed reaction scheme. In general the kinetics of a pyridine type ligand are influenced by its polarity and steric requirements as well as by the electron-donor ability, hydrogen bond-donor ability, and steric requirements of the solvent.

LITERATURE CITED

1 Parker, A. J., *Chem. Rev.*, **69**, 1 (1969).

2 Watts, D. W., "Reaction Kinetics and Mechanisms," in *Physical Chemistry of Organic*

Solvent Systems, A. K. Covington and T. Dickinson, Eds., Plenum Press, New York, 1973, Chap. 6.

3 Parker, A. J., *Pure Appl. Chem.,* **25,** 345 (1971).

4 Parker, A. J., and R. Alexander, *J. Amer. Chem. Soc.,* **90,** 3313 (1968).

5 Parker, A. J., and E. C. F. Ko, *J. Amer. Chem. Soc.,* **90,** 6447 (1968).

6 Parker, A. J., *Quart. Rev. (London),* **16,** 163 (1962).

7 Alexander, R., E. C. F. Ko, A. J. Parker, and T. J. Broxton, *J. Amer. Chem. Soc.,* **90,** 5049 (1968).

8 Langford, C. H., and T. R. Stengle, *Ann. Rev. Phys. Chem.,* **19,** 193 (1968).

9 Lantzke, I. R., and D. W. Watts, *J. Amer. Chem. Soc.,* **89,** 815 (1967).

10 Tobe, M. L., and D. W. Watts, *J. Chem. Soc.,* 2991 (1964).

11 Fitzgerald, W. R., A. J. Parker, and D. W. Watts, *J. Amer. Chem. Soc.,* **90,** 5744 (1968).

12 Langford, C. H., and F. M. Chung, *J. Amer. Chem. Soc.,* **90,** 4485 (1968).

13 Tobe, M. L., and D. W. Watts, *J. Chem. Soc.,* 4614 (1962).

14 Fitzgerald, W. R., and D. W. Watts, *J. Amer. Chem. Soc.,* **89,** 821 (1967).

15 Muir, W. R., and C. H. Langford, *Inorg. Chem.,* **7,** 1032 (1968).

16 Swift, T. J., and R. E. Connick, *J. Chem. Phys.,* **37,** 307 (1962).

17 Babiec, J. S., C. H. Langford, and T. R. Stengle, *Inorg. Chem.,* **5,** 1362 (1966).

18 Breivogel, F. W., Jr., *J. Phys. Chem.,* **73,** 4203 (1969).

19 Thomas, S., and W. L. Reynolds, *J. Chem. Phys.,* **46,** 4164 (1967).

20 Angerman, N. S., and R. B. Jordan, *Inorg. Chem.,* **8,** 2579 (1969).

21 Matwiyoff, N. A., *Inorg. Chem.,* **5,** 788 (1966).

22 Bennetto, H. P., and E. F. Caldin, *J. Chem. Soc., A,* 2191 (1971).

23 Bennetto, H. P., and E. F. Caldin, *J. Chem. Soc., A,* 2198 (1971).

23a Coetzee, J. F., "The Role of the Solvent in the Ligand Substitution Dynamics of Labile Complexes," in *Protons and Ions Involved in Fast Dynamic Phenomena,* Elsevier, 1978.

24 Coetzee, J. F., "Ligand Substitution Kinetics of Labile Metal Complexes in Nonaqueous Solvents," in *Solute–Solvent Interactions,* Vol. II, J. F. Coetzee and C. D. Ritchie, Eds., Marcel Dekker, New York, 1976.

25 Chattopadhyay, P. K., and J. F. Coetzee, *Anal. Chem.,* **46,** 2014 (1974).

26 Chattopadhyay, P. K., and J. F. Coetzee, *Inorg. Chem.,* **12,** 113 (1973).

GENERAL REFERENCES

Coetzee, J. F., "Ligand Substitution Kinetics of Labile Metal Complexes in Nonaqueous Solvents," in *Solute–Solvent Interactions,* Vol. II, J. F. Coetzee and C. D. Ritchie, Eds., Marcel Dekker, New York, 1976, Chap. 14.

Watts, D. W., "Reaction Kinetics and Mechanisms," in *Physical Chemistry of Organic Solvent Systems,* A. K. Covington and T. Dickinson, Eds., Plenum Press, New York, 1973, Chap. 6.

CHAPTER ELEVEN

Some Specialized Applications of Nonaqueous Solvents

11.1 APPLICATION OF NONAQUEOUS SOLVATION OF IONS TO HYDROMETALLURGY INCLUDING SOLVENT EXTRACTION

Some interesting ideas and predictions on the applications of ion-solvent interactions to the chemistry of metals such as copper and silver are discussed by Parker (1, 2). These papers demonstrate clearly how the knowledge gained from free energies of transfer of ions from one solvent system to another can be used for such processes as electrorefining and electrowinning. This particular study focuses attention on the use of nitriles (acetonitrile, acrylonitrile, and 2-hydroxycyanoethane) in hydrometallurgical processes. First, let us review the behavior of a series of ions in both acetonitrile and acetonitrile–water mixtures by examining the free energies of transfer from water to these systems. The data are summarized in Table 11.1. In earlier chapters, we have observed that generally cations are more strongly solvated in water than they are in the more weakly basic solvent, acetonitrile. However, cations such as Cu^+, Ag^+, and Au^+ have the ability of back-donation of their electrons into a π^* antibonding orbital of the nitrile group with a result that the solvation of cuprous copper (Cu^+) and Ag^+ is stronger in acetonitrile than in water. The negative values for the free energies of transfer in the mixed solvent systems suggest that the Cu^+ and Ag^+ are preferentially solvated by acetonitrile in these systems. Additional evidence for this is provided by NMR spectra.

An important equilibrium in the chemistry of copper is

$$Cu^{2+} + Cu \rightleftharpoons 2\,Cu^+ \tag{11.1}$$

It is found that the presence of acetonitrile strongly influences the position of this equilibrium, shifting it from 10^{-6} *M* in water to 10^8 *M* in the presence of small amounts of acetonitrile. The variation in the equilibrium constant for a large range of acetonitrile–water mixtures ranges from 10^8 to 10^{11} *M*, whereas in anhydrous acetonitrile the value rises sharply to 10^{20} *M*,

Table 11.1 Free Energies of Transfer (ΔG_t kcal g ion^{-1}) of Cations from Water to Acetonitrile and to Acetonitrile–Water Mixtures at 25°C [Assumption ΔG_t Ph_4As^+ = ΔG_t Ph_4B^-]

Cation M^+	ΔG_t M^+ 0.1 AN/H_2O	ΔG_t M^+ AN
H^+	0.1	+11
Na^+	0.0	+5
Cu^{2+}	−0.5	+12
Fe^{2+}	0.8	+18
Fe^{3+}	0.0	+48
Cu^+	−10	−12
Ag^+	−1.7	−4

Reprinted with permission from *Search*, **4**(10), 426 (1973), the Journal of the Australian and New Zealand Association for the Advancement of Science Incorporated. This table was originally reported in the Ph.D. thesis of W. E. Waghorne at ANU.

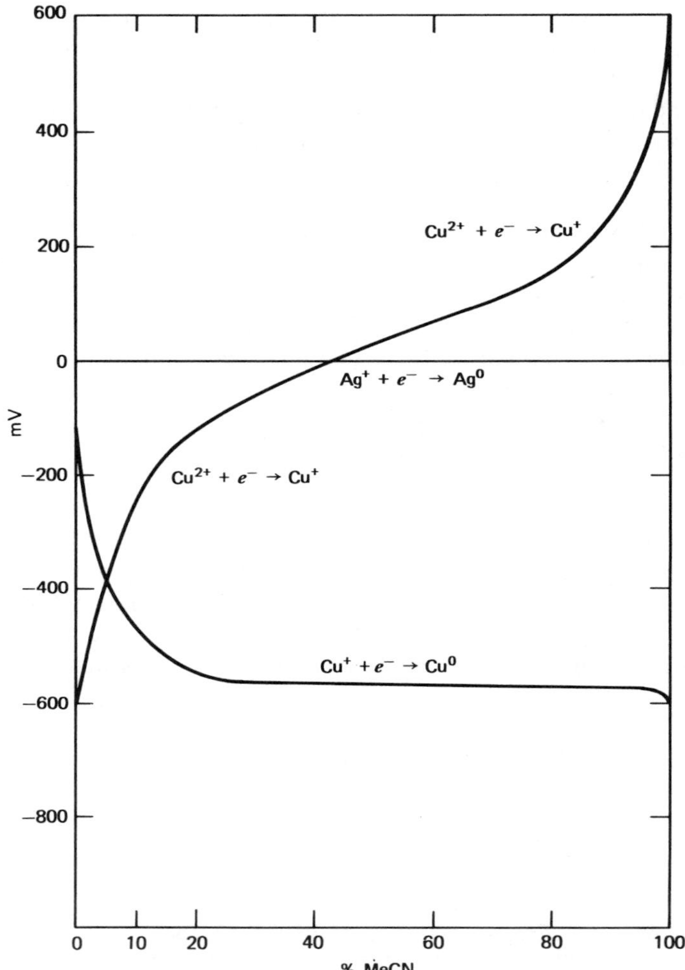

Figure 11.1 Reduction potentials of copper sulfates in acetonitrile–water mixtures containing 0.01 *M* H_2SO_4 vs a Ag/Ag_2SO_4 reference electrode. Reprinted from Ref. 1 with permission of the Royal Australian Chemical Institute.

reflecting the poor solvation of Cu^{2+} in acetonitrile. Another point that is important in the extractive metallurgical processes is that at pH < 5 the variation of the equilibrium constant is more or less independent of acidity.

When discussing the stabilities of ions such as Ag^+ and Cu^+ in various media, it is also informative to examine the reduction potentials of the ions. The reduction potentials of copper sulfates in acetonitrile–water mixtures containing 0.01 *M* H_2SO_4 measured against a $Ag \mid Ag_2SO_4$ reference electrode are shown in Fig. 11.1. The differences in the solvations of Ag^+, Cu^+, and Cu^{2+} in acetonitrile and water account for the ease or difficulty of reducing these ions to the metals. It is seen that it is more difficult to reduce Ag^+ and Cu^+ to metallic silver and copper, respectively, but it is easier to reduce cupric to cuprous sulfate as the proportion of acetonitrile in the solvent increases. The disproportionation reaction between the two oxidation states of copper:

$$Cu^+ + Cu^+ \rightleftharpoons Cu^{2+} + Cu \qquad (11.2)$$

is influenced by the amount of acetonitrile present in an acetonitrile–water mixture. If the amount of acetonitrile present exceeds 6 moles per mole of cuprous ion, then it is found that cuprous sulfate is stable, but if the amount of acetonitrile is less than this, the disproportionation of cuprous sulfate to yield cupric sulfate and metallic copper occurs (Eqn. 11.2). A further consequence of the effect of redox potentials is that it is possible to oxidize silver with acidified cupric sulfate in the presence of high proportions of acetonitrile. We will now consider some applications of this information to hydrometallurgy.

11.1.1 Separation of Copper, Silver, and Lead Often it is required to separate the metals copper, silver, and lead from a blast-furnace residue. This separation can be achieved by either adding acetonitrile or boiling it off and thus switching the various redox equilibria backward and forward. Consideration of Fig. 11.1 indicates that it is possible to oxidize both copper and silver using acidified cupric sulfate in a 50% acetonitrile–water mixture. Lead is not attacked by this system, due to the protective coating of lead sulfate formed on its surface. Reaction (11.1) occurs readily at room temperature in the 50% solvent system, while the reaction

$$Cu^{2+} + Ag \rightleftharpoons Ag^+ + Cu^+ \qquad (11.3)$$

also occurs, although less readily. Note that in pure water, reaction (11.1) is reversed. Additionally it is seen in Fig. 11.1 that the reaction

$$Ag^+ + Cu \longrightarrow Ag + Cu^+ \qquad (11.4)$$

occurs in 50% acetonitrile–water. A possible scheme for the separation of the metals copper, silver, and lead using the variation of equilibria with solvent composition is shown in Fig. 11.2.

11.1.2 Electrowinning An important industrial process in the copper industry is the production of pure, high-quality copper cathode material from impure copper powder or cement–copper, produced by adding iron to cupric sulfate solutions. A method involving the use of an acetonitrile–water–sulfuric acid mixture with cuprous sulfate as the electrolyte has been suggested by Parker (1). The scheme is outlined in Fig. 11.3. The electrolysis reactions

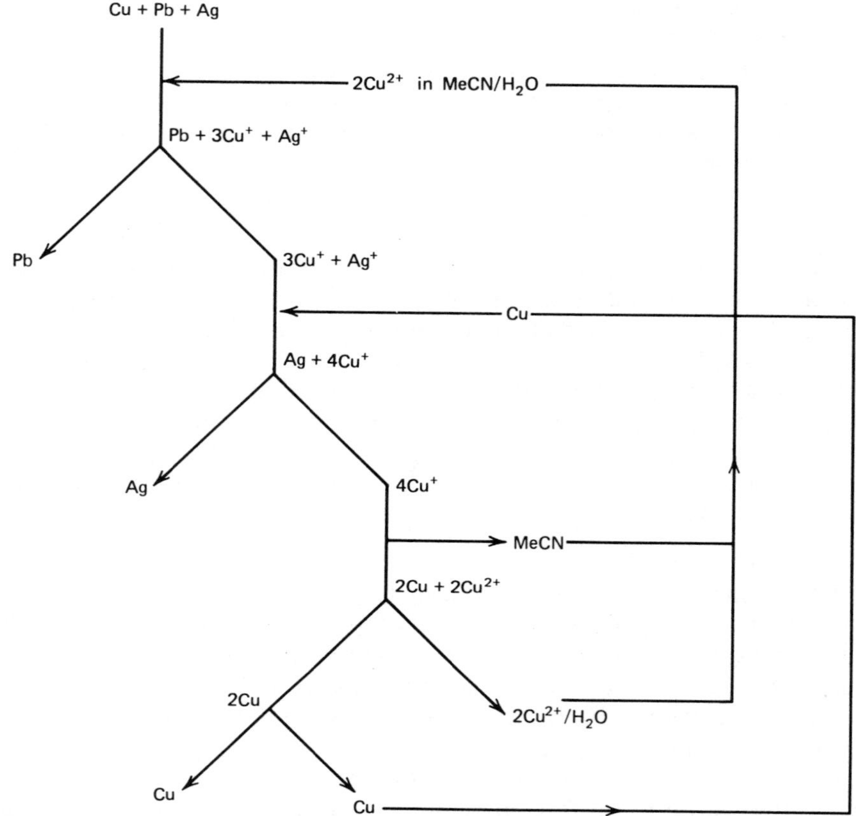

Figure 11.2 Flow diagram to show the separation of copper, silver, and lead in nonaqueous media. Reprinted from Ref. 1. with permission of the Royal Australian Chemical Institute.

are

$$\text{Anode:} \qquad Cu^{+} \longrightarrow Cu^{2+} + e \tag{11.5}$$

$$\text{Cathode:} \qquad Cu^{+} + e \longrightarrow Cu \tag{11.6}$$

It is found that if the composition of the electrolyte is carefully chosen, the electrolysis proceeds efficiently for about 25% disproportionation without a diaphragm and to about 50% disproportionation with a diaphragm separating the anode and the cathode. The advantage of this method is that it involves a one-electron process at 0.5–0.6 V, and consequently the power consumption is less than in methods involving a two-electron process, such as adding iron to cupric sulfate solutions. The cupric sulfate formed in the nonaqueous method is recycled by passing it through a tower containing particulate copper, which dissolves to form additional cuprous solution.

11.1.3 Electrorefining of Copper A further application of the enhanced stabilization of cuprous ion in nitriles is in the electrorefining of copper. In the method proposed by Parker (1), a crude copper anode is converted to a pure copper cathode by using an electrolyte system comprising cuprous sulfate in acetonitrile or 2-hydroxycyanoethane, together with water and sulfuric acid plus some additives. In the traditional method, an acidified cupric sulfate solution is electrolyzed. The electrochemical reactions for these two methods are

Anode Cathode

$$Cu \xrightarrow{-2e} Cu^{2+} \xrightarrow{+2e} Cu \quad \text{(cupric electrolyte)} \tag{11.7}$$

$$Cu \xrightarrow{-e} Cu^{+} \xrightarrow{+e} Cu \quad \text{(cuprous electrolyte)} \tag{11.8}$$

As discussed in the electrowinning process above, the use of the cuprous electrolyte involving a one-electron process leads to a lower power cost and a faster refining process. However, these advantages have to be weighed against some disadvantages, such as the higher capital cost of the solvent and the need to use a kerosene layer to protect the solutions from oxidation by air. The electrorefining of silver could be achieved in a similar manner.

11.1.4 Leaching of Cuprous-Sulfide Ores Acetonitrile–water mixtures can be applied to gain copper from its ores, cuprous sulfide (digenite, chalcocite, or bornite), using cupric sulfate as an oxidizer. The combined processes of

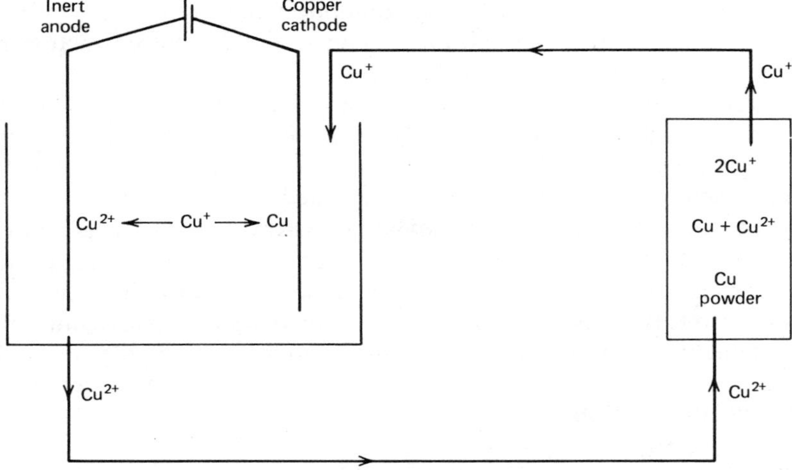

Figure 11.3 Powder refining of copper using an electrochemical cell with acetonitrile–water; 0.6 V at 10 A ft^{-2}. Reprinted from Ref. 1 with permission of the Royal Australian Chemical Institute.

leaching and distillation are summarized below:

$$\text{Leach} \quad Cu^{2+} + Cu_2S + 4AN \rightleftharpoons CuS^* + 2CuAN_2^+ \tag{11.9}$$

$$\text{Distil} \quad 2CuAN_2^+ \longrightarrow Cu + Cu^{2+} + 4AN \tag{11.10}$$

$$\text{Net} \quad Cu_2S \longrightarrow Cu + CuS^* \tag{11.11}$$

where CuS^* is an activated copper sulfide. Equation 11.9 is another example of a case in which a dramatic change in equilibrium constant occurs when one changes the medium from water to an acetonitrile–water mixture. In this case, the equilibrium constant for reaction 11.9 in water is $10^{-13.3}$ M, while in a solution containing 0.3 mole fraction of acetonitrile, the equilibrium constant is increased to 10^3 M, reflecting again the enhanced solvation of the cuprous ion by acetonitrile. Copper can also be leached from the activated copper sulfide using sulfuric acid and oxygen, with cupric sulfate and acetonitrile as catalyst, as shown in Eqn. 11.12:

$$CuS^* + H_2SO_4 + \tfrac{1}{2}O_2 \xrightarrow[AN]{Cu^{2+}} CuSO_4 + S + H_2O \tag{11.12}$$

The copper can be obtained by reduction with zinc powder (2).

11.1.5 General Applications of Solvent Extraction Some very interesting practical applications of solvent extraction for isolating metals from effluents and waste residues, concentrated leach liquors and scrap materials are discussed by Bailes et al. (3). Solvent extraction had its early application in the separation of uranium from its ore. However, the consideration of solvent extraction for isolating such metals as copper and nickel has to be viewed carefully from the point of view of the economics of the process. In this case, loss of the extracting solvent by evaporation, chemical breakdown, or solubility of the solvent in the various streams of a solvent extraction plant become critical features for consideration.

A large variety of solvent extractants are presently being considered for extractive metallurgical processes. A summary of some of the candidate solvents is given in Table 11.1. Because many of the solvents are fairly viscous when pure, a diluent is normally added to the extracting solvent to lower the viscosity and thereby alleviate mass-transfer problems and interfacial problems. In some cases, a third component (a modifier) may be added to the organic phase to prevent salt precipitation or third-phase formation.

The various types of extractants may be conveniently divided into several types, namely, solvating, acidic, chelating, and ionic. Examples of all these types are given in Table 11.2. A brief discussion of the different types of extracting solvents follows.

Solvating Extractants A solvent that readily solvates the hydrogen ion, that is, a basic extractant, can be used to extract a metal. For example, tributylphosphate (TBP) has been used extensively for the treatment of nuclear

Table 11.2 Commercial Extractants, Their Chemical Structures and Applications (3)

Extractant Classification and Name	Structure	Known Applications
1. Solvating tributyl phosphate (TBP)	$(C_4H_9O)_3P{=}O$	Uranium from nitric acid leach liquors, separation of uranium, plutonium and fission products
		Separation of Zr/Hf from nitric acid liquors
		Used in conjunction with amine circuit in separation of iron (as $HFeCl_4$) from nickel and cobalt
Methyl isobutyl ketone (MIBK)	$CH_3(C_4H_9)C{=}O$	Extraction of niobium from H_2SO_4/HF mixtures
		Separation of Zr/Hf from H_2SO_4/SCN mixtures
Dibutyl carbitol (Butex, Hexone)	C_4H_9 O CH_2 CH_2O CH_2O C_4H_9	Reprocessing of nuclear fuels
		Extraction of Au from HCl/HNO_3 mixtures

Table 11.2 (*Continued*)

Extractant Classification and Name	Structure	Known Applications
2. Acidic		
Naphthenic acids	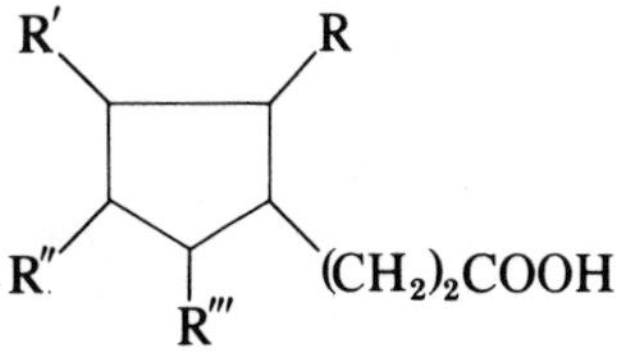 Cyclopentane ring with substituents R′, R, R″, R‴ and $(CH_2)_2COOH$	No known commercial application
Versatic acids	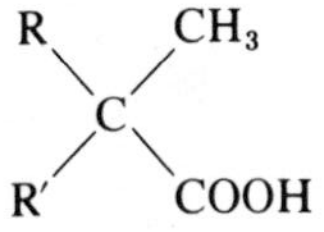 R, R′, CH_3 and COOH attached to a central C	V_{911} is used to separate Co/Ni Extraction of rare earths
D2EHPA (Di-2-ethylhexyl phosphoric acid)	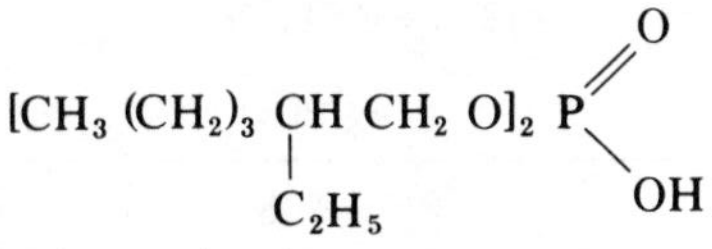 $[CH_3(CH_2)_3CH(C_2H_5)CH_2O]_2P(=O)OH$	Rare earths extracted from H_2SO_4 Separation of Co/Ni Treatment of uranium in acid sulfate leach liquors Extraction of vanadium as vanadyl from reducing acid–sulfate liquors

3. Chelating

LIX 63	$CH_3(CH_2)_3CH(C_2H_5)CH(OH)C(=NOH)CH(C_2H_5)(CH_2)_3CH_3$	
LIX 65N (*anti* form)	C_9H_{19}–substituted 2-hydroxyphenyl–C(=N–OH)–phenyl	
LIX 64N	Mixture of 65N and 63	Extraction of copper from sulfuric acid or ammonia leach liquors
P17	C_9H_{19}–substituted 2-hydroxyphenyl–C(=N–OH)–CH_2–phenyl	Proposed alternative to 64N

Table 11.2 (*Continued*)

Extractant Classification and Name	Structure	Known Applications
P50	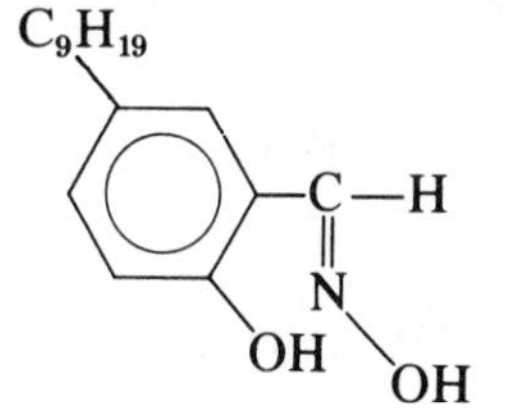 C_9H_{19}; C—H; N; OH; OH	Proposed alternative to 64N
SME 529	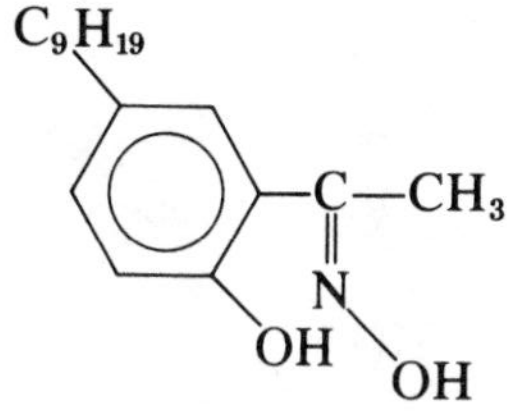 C_9H_{19}; C—CH_3; N; OH; OH	Proposed alternative to 64N
Kelex 100	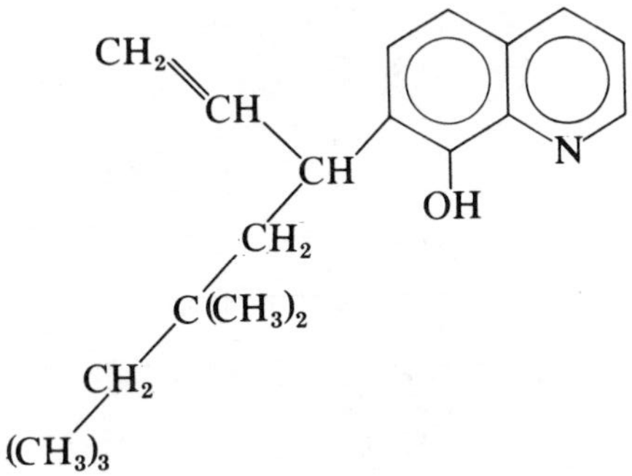 CH_2; CH; CH; CH_2; $C(CH_3)_2$; CH_2; $(CH_3)_3$; OH; N	Proposed extractant for copper from leach liquors

Polyol	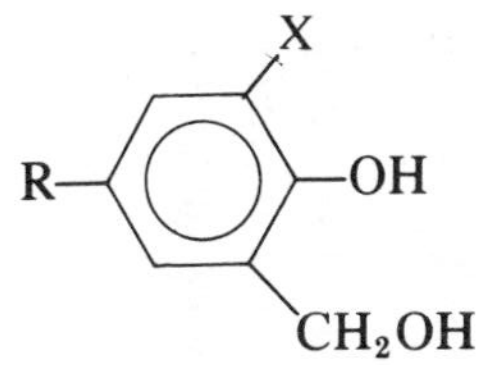 R—C_6H_2(X)(OH)(CH_2OH)	Removal of boron from brines
4. Ionic		
(*a*) Primary amine, Primene JMJ	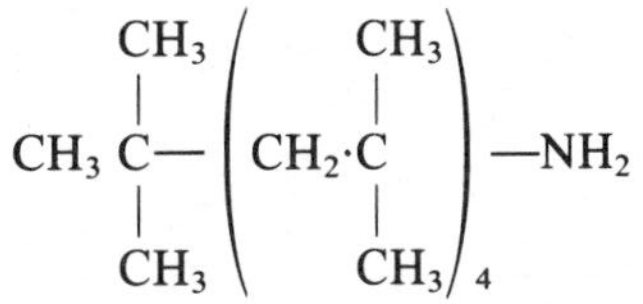 $CH_3C(CH_3)_2—(CH_2C(CH_3)_2)_4—NH_2$	
(*b*) Secondary amine, LA-1, LA-2	$R_2^{IV}NH$	
(*c*) Tertiary amines		Extraction of tungsten
Alamine 336	$R_3^{IV}N$	Extraction of uranium, reprocessing of nuclear fuel
Adogen 336	$R_3^{IV}N$	Used in one circuit on the Falconbridge process for nickel
(*d*) Quaternary amines		
Aliquat 336	$(R_3^{IV}NCH_3)^+ Cl^-$	Extraction of vanadium
Adogen 464	$(R^{IV} = C_8–C_{10})$	

fission products, as shown below

$$2TBP_{org} + [UO_2(NO_3)_2]_{aq} \rightleftharpoons [UO_2(NO_3)_2 \cdot 2TBP]_{org} \qquad (11.13)$$

The effective solvating ability of the typical basic extracting solvents follows the order

$$R_3N > R_3PO > (RO)_3PO > R_2CO > R_2O$$

Acidic Extractants Acidic extractants function by exchange of an acidic hydrogen for a metal, thus forming a metal salt, that is,

$$M^{n+}_{aq} + nHR_{org} \rightleftharpoons [MR_n]_{org} + nH^+_{aq} \qquad (11.14)$$

where HR_{org} is the organic acid and M^{n+} is the metal ion. The dialkylphosphoric acids are more effective than the carboxylic acids. In addition, the extraction process is dependent both on the pH of the solution and on the stability constant of the complex MR_n. In fact, by changing the pH in a solvent feed, one metal may be separated from another. Both naphthenic and versatic acids have been proposed for extraction of transition metals.

Chelating Extractants Chelating extractants are essentially molecules consisting of ring structures which form chelates with a metal ion. These molecules may be either charged or neutral. Ring stability is normally maximized if the ring contains six or seven members. Classic examples are the substituted 8-hydroxyquinolines and the so-called LIX extractants, which are derivatives of the salicylaldoximes, which have been used in the extraction of copper. The simplified process may be represented as

$$Cu^{2+}_{aq} + 2HR_{org} \rightleftharpoons [CuR_2]_{org} + 2H^+_{aq} \qquad (11.15)$$

Most of the chelating extractants are specific for the transition metals.

Ionic Extractants Ionic extractants are those that carry within an ion pair a labile cation or anion that will exchange with the appropriate metal species in the aqueous phase, somewhat similar to a liquid ion exchanger. Specific examples are the quaternary amines [Aliquat 336 and Adogen 364], which are capable of reacting with strong alkali leach liquors, so that chromates, vanadates, molybdates, and tungstates can be extracted, as these metals are capable of forming anionic species under such conditions.

11.2 NONAQUEOUS-ELECTROLYTE BATTERIES

11.2.1 Organic Systems The application of organic electrolyte systems to both primary (nonrechargeable) and secondary (rechargeable) batteries stems from the fact that the reactive metals such as lithium, sodium, and magnesium can be used as the negative electrodes, and thereby fairly high energy densities (W hr lb^{-1}) can be achieved. Also, aqueous battery systems are ineffective at low temperatures due to increased polarization and very low conductivities. Both the increased cell voltage attainable with, for example, the lithium electrode and its low equivalent weight are the important contributing factors to the high energy density.

Strictly speaking, the broad area of nonaqueous battery systems can be conveniently divided into several classes, namely,

- Molten-salt batteries–high-temperature systems (e.g., Li/Cl_2 in molten LiCl).
- Organic or inorganic solvents, which are rendered conducting by dissolved salts.
- Solid electrolytes (e.g., β-alumina).

In this section, only the ambient-temperature organic or inorganic solvent systems will be discussed. Some important applications of nonaqueous battery systems are as follows:

- Vehicle propulsion.
- Load leveling—off-peak power storage.
- Cardiac pacemakers.
- Electronic equipment.

Any organic solvent selected for study as a candidate for a nonaqueous battery must meet certain fairly stringent requirements. First, the solvent must possess a good liquid range, that is, a low freezing point and high boiling point. Second, the fact that a sufficient quantity of electrolyte must dissolve in the solvent to produce a high conductivity normally requires that the solvent have a fairly high dielectric constant. Also, the solvent should have a low vapor pressure to minimize possible explosions and should be both thermally stable and unreactive toward the electrode materials. A further requirement is that the electrode processes occur at a useful rate in the solvent. In addition, a suitable solvent must be both inexpensive and nontoxic. A large variety of solvents have been considered for battery applications and some of the more popular organic solvent candidates are propylene carbonate, DMSO, DMF, acetonitrile, butyrolactone, methyl formate, and nitromethane. Some inorganic solvents include liquid SO_2, sulfuryl chloride and thionyl chloride.

The most important requirements of the electrolyte (solute) are high solubility, high electrical conductivity, inertness to chemical attack, ease of handling and low cost. For secondary batteries, the electrolyte must be stable over the voltage range required for recharging. The most frequently employed solutes that meet these requirements are

- Simple salts $LiClO_4$, $Mg(ClO_4)_2$
- Lewis acids $AlCl_3$, BF_3, $ZrCl_4$ or $AlCl_3$ + LiCl
- Complex fluorides $NaPF_6$, KBF_4, $NaAsF_6$
- Complex salts $(C_2H_5)_4NClO_4$

The most frequently used anodes and cathodes are shown in Table 11.3. The negatives comprise mostly those elements in the upper left-hand corner of the periodic table, while the positives are generally halides, oxides, or sulfides of the transition metals. An extensive amount of information is needed to apply value judgments on candidate systems for batteries and the range of this information is depicted in Table 11.4.

One of the more complex problems encountered with nonaqueous battery systems is the effect of impurities and, in particular, water. For example, for the battery system involving a lithium anode in a KSCN–PC system, the discharge characteristics or change in anode potential with change of water content is only very slight, but much more serious problems are encountered when attempting to charge lithium in propylene carbonate containing about 200 ppm of moisture. For successful application when using lithium electrodes, it is essential to maintain the system sealed from the surrounding atmosphere, and all studies concerning such systems have been performed in glove boxes with a carefully controlled atmosphere. Another problem encountered with lithium electrodes is that of the stability of lithium in organic solvents. In some cases, slow gassing has been observed. Corrosion problems are also significant in some nonaqueous battery systems,

Table 11.3 Common Electrode Couples Used in Nonaqueous Batteries

Anodes	Cathodes
Li	$CuCl_2$
Mg	CuF_2
Al	CuS
Ca	AgF_2
	AgCl
	AgS
	NiF_2
	$NiCl_2$
	CoF_3
	CF

Table 11.4 Information and Studies Needed on Candidate Systems for Nonaqueous Batteries

Type of Information	Examples
Thermodynamic	Equilibrium emf, heats of solution, vapor pressures, activities
Transport	Electrical conductance, viscosity, diffusion
Electrochemical	Polarography, cyclic voltammetry; charge and discharge curves; corrosion, electrode kinetics
Spectroscopic	ion pairing, solution structure (species); NMR, laser Raman, infrared

particularly when complex fluorides are used as the electrolytes, as, for example, in the system KPF_6–propylene carbonate. The organic solvents DMSO and *N*-methyl-2-pyrrolidone appear to be compatible with the lithium electrode, whereas nitromethane is apparently susceptible to reduction on battery charging. Both zinc and magnesium anodes have been used in dry-cell batteries involving acetonitrile and propionitrile, but the low-temperature performance of magnesium is affected by film formation and long activation times. As an example of the characteristics of an organic battery system, the Li/CF (carbon monofluoride) battery has a capacity of 5 ampere-hr, an operating voltage of 2.4 V and an energy density of 10 W hr lb^{-1}.

11.2.2 Inorganic Systems Currently there is considerable interest in both primary and secondary batteries which involve inorganic solvents such as liquid SO_2, thionyl chloride, sulfuryl chloride, and phosphorus oxychloride. These systems contain dissolved salts such as lithium tetrachloroaluminate and possess the interesting feature that the electrolyte solution is also the active cathode and is reduced on discharge. The external electric circuit is completed with an inert current collector inserted into the electrolyte as the positive pole of the cell.

The Li/SO_2 battery is being considered for military applications, as for example in radio sets and missile launching. The battery consists of a lithium anode, SO_2 reactant in acetonitrile and a carbon-containing cathode. For these applications, both thermal stability and low-temperature capabilities are important (e.g., in Norway on the average there are 245 days per year where the temperature dips below freezing). A further problem is the loss of capacity with time. One of the more controversial aspects of the Li/SO_2 battery at present is that of safety, as some explosions have been reported during initial tests, due to a voltage reversal problem. Apparently, the ratio of $Li:SO_2$ is important. The electrochemical reactions for the Li/SO_2 battery can be represented as

$$\text{Anodic} \quad 2Li \rightarrow 2Li^+ + 2e^- \tag{11.16}$$

$$\text{Cathodic} \quad 2SO_2 + 2e^- \rightarrow S_2O_4^{-2} \tag{11.17}$$

$$\text{Overall} \quad 2Li + 2SO_2 \rightleftarrows Li_2S_2O_4 \tag{11.18}$$

In addition, there are side reactions occurring between lithium and acetonitrile

$$2Li + 2CH_3CN \rightleftarrows (CH_2\text{—}CN)^- + Li^+ + CH_4 + LiCN \tag{11.19}$$

$$CH_3CN + (CH_2\text{—}CN)^- \rightleftarrows (CH_3\text{—}CN\text{—}CH_2CN)^- \tag{11.20}$$

Apparently, considerable improvement is gained relative to the explosion hazard if precautions are taken not to discharge the cells into the voltage reversal zone at rates exceeding 1 ampere. Obviously, the need exists for more extensive research with this system.

The $Li/SOCl_2$ cell is being considered for application in cardiac pace makers. The overall electrode reaction for this cell has been represented by two possibilities, namely,

$$4Li + 2SOCl_2 \longrightarrow SO_2 + 4LiCl + S \tag{11.21}$$

or

$$8Li + 3SOCl_2 \longrightarrow 2S + Li_2SO_3 + 6LiCl \tag{11.22}$$

Some of the present problems concerning this particular battery include self discharge, corrosion resistance, and hermeticity. Both nickel alloys and stainless steels have been found to be stable in $SOCl_2$. Self-discharge occurs at the rate of about 2% per year, and hermiticity has been maintained up to pressures of 5700 psi. Apparently, more than 20,000 cells using this system have been implanted over a four-year period. However, some problems have been reported concerning the reduction of $SOCl_2$ on nickel and stainless steel.

11.3 ELECTRODEPOSITION OF METALS FROM NONAQUEOUS SOLUTIONS

The need to deposit an adherent layer of a metal such as aluminum on a substrate such as steel represents a significant part of many industrial applications designed to minimize problems due to corrosion. For structural applications, the electrodeposition of aluminum on steel is very important, as aluminum is resistant to ordinary atmospheric corrosion and is little attacked by concentrated acids such as nitric and acetic. Sulfur compounds in ordinary petroleum will not corrode aluminum tubes to any appreciable extent, even at high temperatures. However, pure aluminum has a low tensile strength and a low elastic limit and therefore is very unsatisfactory as a structural material. A further example is found with molybdenum, which is used for several specialized applications at high temperature, but a composite coating of electrodeposited nickel and chromium can be applied to protect molybdenum from oxidation at elevated temperatures. In secondary batteries containing nonaqueous solvent systems, lithium anodes are often made by electrodeposition of lithium, as, for example, in the deposition of lithium from a solution of LiCl + $AlCl_3$ in propylene carbonate. About 30 metals can be deposited satisfactorily from aqueous solutions, but for a fairly substantial number of metals, such as aluminum, magnesium, and the alkali metals, it is not possible to use aqueous solutions and this is where nonaqueous electrolyte solutions can be used to advantage.

In this section, some of the basic principles of electrodeposition are reviewed and illustrated by some representative examples. One of the most informative discussions of this subject can be found in the general survey presented by Brenner (4). An important concept in studying the electrodeposition process is the deposition potential. For successful deposition, a metal usually requires a potential that is less negative than its standard potential. However, in aqueous solutions, the potentials which can be established are limited by the discharge of hydrogen, which usually, because of the phenomenon of overvoltage, occurs at a potential less negative than its equilibrium value, sometimes by as much as 1 V. It is the large overvoltage for the discharge of hydrogen that makes it possible to electrodeposit a metal such as zinc from an aqueous solution, although zinc is less noble than hydrogen. In other words, hydrogen is discharged less readily on a zinc cathode that is zinc itself. In the case of the electronegative metals such as Mg, Be, and Al, their electrodeposition cannot be effected from aqueous electrolytes since in such solutions hydrogen is reduced at a much more positive potential than the metal ions, and hence the entire current will be devoted to hydrogen evolution.

However, the preferential discharge of hydrogen fails to explain why some metals cannot be deposited from aqueous media. For example, tungsten, molybdenum, and germanium cannot be deposited in the pure state from aqueous solutions even though their theoretical electrode potentials for reduction from acid solution are close to zero:

$$WO_3 + 6H^+ + 6e^- \longrightarrow W + 3H_2O \qquad E^\circ = -0.09\ \text{V}$$

$$Ge^{2+} + 2e^- \longrightarrow Ge \qquad E^\circ = 0$$

It therefore appears that the electrode potentials are not the only factors to be considered for the process of electrodeposition, but other factors such as the reactivity of the ions at the cathode should also be taken into account. It is convenient to divide the metals into two classes, namely, class I, which contains those metals that cannot be deposited from aqueous solutions because their deposition potentials are too negative, and class II, which contains metals which have a deposition potential theoretically attainable in aqueous solution, but where the ions do not react at the cathode. It should be noted that although the metals, W, Mo, and Ge cannot be deposited from aqueous solution, alloys containing appreciable contents of these metals can be deposited from aqueous solutions.

For metals in class I, such as the alkali metals, the problem of preferential discharge of hydrogen can be readily overcome by using an organic solvent in which the hydrogen atom is more tightly bound than it is in water. However, for metals in class II, the necessary conditions for electrodeposition are not easy to predict because of the limited knowledge available on the types of ions most likely to react at the cathode.

Organic solvents possess several key features which make them important candidates for the study of electrodeposition. First, a large variety of complex ions can exist in organic solvent systems relative to aqueous solution, where hydrolysis of many complex ions takes place. Second, the fact that most organic solvents require a higher decomposition voltage than that required for water means that the attainment of higher voltages makes it feasible to carry out reactions of those ions in class II. The guidelines for predicting the most suitable conditions for electrodeposition are rather vague, but it is fairly clear that the chemical nature of the solvents and solutes are key factors in determining whether a particular solution will conduct or whether a metal can be deposited from it. Only certain classes of solutes and solvents form plating baths and an important criterion is that a loose ionic complex must form between the solute and the solvent. In the absence of the complex, no conductivity occurs, whereas if the complex is too stable, conductivity may occur, but not metal deposition. We will consider some important features required of both solutes and solvents for the process of electrodeposition of Al, Be, Mg, Ti, and Zr, which are the main metals presently under consideration.

The solutes that have been used are for the most part simple compounds of low molecular weight and include halides, hydrides, borohydrides and organometallic compounds. For example, aluminum plating has used a mixture of $AlCl_3$ and AlH_3. It should be noted that these solutes do not contain oxygen or nitrogen. To illustrate this point, it is observed that solutions of magnesium perchlorate, $Mg(ClO_4)_2$ do not yield metallic deposits on electrolysis. It has been found that dimethylberyllium does yield a metallic deposit in ethyl ether, whereas the phenyl derivative does not form a conducting solution in ether. Also, in ether, the borohydrides of Be, Al, and

Mg yield metallic deposits, whereas those of Ti and Zr are not even conductive.

The selection of a suitable solvent is governed by the stability of the coordination compound it forms with the solute. As indicated earlier, the compound formed must not be too stable. This condition eliminates the possibility of nonreactive liquids, such as hydrocarbons and some of their halogen derivatives. Although alcohols, ethers, ketones, acids, acid-anhydrides, amines, amides, nitriles, benzene, and toluene all form coordination compounds with the solutes listed earlier, it has been found that electrodeposits were only found for baths containing either ethers or aromatic hydrocarbons. It would therefore appear that all the other solvents formed a complex with the solute that was too stable. In the case of the ether solvents, the oxygen atom coordinates sufficiently but not too strongly with the various types of metallic solutes considered. In particular, ethyl ether yielded the best plating baths. These observations apply to the metals Mg, Al, Be, Ti, and Zr, but, with the exception of the alkali metals and a few others, for example, Ge, the conclusions are generally valid. Some of the alkali metals can be deposited from solutions of their salts in alcohols and ketones.

It should be pointed out that there were several earlier studies of electrodeposition of metals that were recorded, but in the majority of cases the deposits were impure and no attempts were made to analyze for purity. The most successful electrodeposition is that of aluminum, which has also attracted the greatest amount of attention and research. Aluminum can be obtained at a high efficiency of cathodic current, in high purity and with good physical properties. The system that is most popular consists of $AlCl_3$ dissolved in ether, with the addition of some $LiAlH_4$. The ionizing complex in this system is probably $Al_2Cl_6{}^+ AlH_4{}^-$. Deposits several hundredths of an inch thick were obtained from this bath. To prevent the formation of nodules, methyl borate was used as an additive. Also, it was found that the bath could tolerate the presence of oxygen, but not moisture. The hydride bath appears to be specific for the plating of aluminum. On the other hand, attempts to electrodeposit beryllium have focused on metal-alkyl baths using, for example, solutions of beryllium dimethyl and $BeCl_2$. It was found that if conditions were optimized, Be could be deposited with 95% purity. Borohydride baths containing borohydrides of Al, Mg, or Be in ether solution have been used to deposit alloys of the metals with boron. In general, the metals which separate from nonaqueous media are less pure than those obtained from aqueous solutions. This phenomenon may be attributed to either incomplete reduction or adsorption of the organic matter on the surface of the metal or a combination of both.

In conclusion, it may be seen that advances in nonaqueous electrodeposition have been slow, although many studies have been devoted to the evaluation of a variety of solvents and solutes. Most of the success has been with aluminum and beryllium, with some promising results for the deposition of alloys such as Be–B, Mg–B, Mg–Al, Ti–Al, and Zr–Al.

11.4 MISCELLANEOUS APPLICATIONS OF NONAQUEOUS SOLVENTS

11.4.1 Chemistry in Ionizing Solvents Ionizing solvents, such as liquid ammonia, liquid hydrogen fluoride, sulfuric acid, glacial acetic acid, liquid sulfur dioxide, and liquid dinitrogen tetroxide, have been used to study a variety of interesting phenomena. Details of the chemistry and physical interactions in these solvents have been described in Chapter 3. The types of studies encountered include solvolysis, oxidation–reduction reactions, complex-formation reactions, organic reactions (in e.g., liquid SO_2), disproportionation reactions, and reactions of metals dissolved in media such as liquid ammonia.

For example, the formation of precipitates is a function of the solubility of the substances involved in the reaction. Thus, in the reaction

$$2AgNO_3 + BaCl_2 \longrightarrow 2AgCl + Ba(NO_3)_2 \qquad (11.23)$$

silver chloride is precipitated in aqueous media, whereas barium chloride is precipitated in liquid ammonia.

The formation of salts by the process of neutralization depends on the choice of a solvent in which both the acid and base can exist. For instance the sodium salt of urea $H_2NCONHNa$ cannot exist in water, because even the strongest base cannot remove a proton from urea in water, but the salt can be prepared in liquid ammonia.

As an example of different solvolysis reactions consider the case of sulfuryl chloride in water, liquid ammonia, and glacial acetic acid. The differing behaviors are illustrated below:

$$SO_2Cl_2 + 2H_2O \longrightarrow H_2SO_4 + 2HCl \qquad (11.24)$$

$$SO_2Cl_2 + 4NH_3 \longrightarrow SO_2(NH_2)_2 + 2NH_4^+ + 2Cl^- \qquad (11.25)$$

$$SO_2Cl_2 + 4CH_3COOH \longrightarrow SO_2(OOCCH_3)_2 + 2CH_3C(OH)_2^+ + 2Cl^- \qquad (11.26)$$

Liquid ammonia has been found to be a very powerful solvent for the alkali metals, with the resulting solutions being excellent conductors of electricity. These metal-ammonia solutions are strong reducing agents and can be used to reduce complex transition metal ions as, for example,

$$MnO_4^- + e \longrightarrow MnO_4^{2-} \qquad (11.27)$$

Some ionizing solvents are useful for effecting anodic oxidation reactions. In particular the very high potential required for the anodic reaction in liquid hydrogen fluoride

$$F^- \rightarrow \tfrac{1}{2}F_2 + e \tag{11.28}$$

makes the solvent an excellent medium for anodic oxidation reactions, particularly for the case of synthesis of fluorine-containing organic molecules.

11.4.2 Influence of Nonaqueous Solvents on Reaction Rates and Reaction Mechanisms A very important application of nonaqueous solvents is the large range of physical properties available for protic and dipolar aprotic solvents which makes it possible to vary the rates of chemical reactions by several orders of magnitude. The acceleration of reaction rates is due mainly to differences in the stabilization of the reactants and the activated complex by the solvent. This means that a whole range of rates for a given type of reaction can be obtained without having to vary the reactants. This in turn makes it possible to reduce production costs for commercial reactions by reducing reaction times.

It is well known that most anions are much less solvated in dipolar aprotic solvents than in protic solvents, while the reverse is true of the polarizable charged transition states. The results of these trends are reflected in the fact that the bimolecular reactions of anions which pass through a large polarizable transition state containing that anion are much faster in dipolar aprotic solvents than in protic solvents. On the other hand, bimolecular substitutions that do not involve anions are much less sensitive to solvent type, provided that the dielectric constant does not change appreciably.

Important organic syntheses can be conducted with advantage in nonaqueous solvents. These syntheses include anionic polymerization, nitrile synthesis, and the synthesis of fluoro compounds.

A great deal of attention has been focused on the understanding of the effect of solvents on reaction rates and reaction mechanisms and detailed accounts can be found in the general references at the end of this chapter and in Chapter 10. The major factors considered in a large variety of theories and models concern themselves with electrostatic effects, the effects of external pressure, viscosity of the solvent, selective solvation (mixed solvents), hydrogen bonding, cage effects, ionization effects of the solvent, effects of solvolysis by the solvent and effects of the acidity of the medium. Much of the discussion has focussed on the different extent of solvation by different solvents for the reactants, the activated complex and the products.

Substitution nucleophilic second-order (S_N2) reactions are very sensitive to changes from protic to dipolar aprotic solvents. Any consideration of solvent effects on rates or equilibria must consider transfer† activity coefficients of reactants, transition states, and products. Several linear free energy relationships have been formulated to correlate solvolysis rates.

† The original literature citation uses the term "solvent activity coefficient."

In conclusion, it is obvious that a careful study of the detailed mechanism and rates of reactions in nonaqueous solvents is essential to improve both yields and production costs.

11.4.3 Nonaqueous Titrations Nonaqueous acid–base titrations have been discussed in some detail in Section 6.3.

SUMMARY

Nonaqueous solution chemistry has several important applications of technological interest. For instance, the fundamentals of the solvent-extraction process are based on the free energies of transfer of metallic ions from one solvent to another. The chemistry of copper involves a disproportionation reaction between Cu^{2+}, metallic copper, and Cu^{+}, and the presence of acetonitrile influences the position of this equilibrium. In fact the variation of the equilibrium constant for a large range of acetonitrile–water mixtures ranges from 10^{8}–10^{11} M, whereas in anhydrous acetonitrile the value rises sharply to 10^{20} M, reflecting the poor solvation of Cu^{2+} in acetonitrile. The differences in the solvation energies of Ag^{+}, Cu^{+}, and Cu^{2+} in acetonitrile and water account for the ease or difficulty of reducing these ions to the metals. The use of mixed solvents for the separation of metals from a furnace residue can be achieved by making use of the shifting equilibria. In addition the enhanced stabilization of the cuprous ion in nitriles is the basis for the production of high-purity copper in the electrowinning process and in the electrorefining of copper. Further practical applications of solvent extraction involve the isolation of metals from effluents and waste residues, concentrated leach liquors, and scrap material. A large variety of solvent extractants (solvating, acidic, chelating, and ionic) are presently being considered for extractive metallurgical processes.

Nonaqueous-electrolyte batteries have received considerable interest over the past decade for possible use in vehicle propulsion and for use in energy storage systems. Reactive metals, such as lithium, sodium, and magnesium, can be used as anode materials and, therefore, fairly high energy densities can be obtained. The nonaqueous solvents that are good candidates for these battery systems must satisfy certain requirements, such as high boiling point, low freezing point and vapor pressure, and a fairly high dielectric constant. With these restrictions, the most popular choices of organic solvents have been propylene carbonate, DMSO, DMF, acetonitrile, and nitromethane. The choices among inorganic solvents include liquid SO_2, sulfuryl chloride, and thionyl chloride. The most important requirements of electrolytes in nonaqueous-battery systems are high solubility and electrical conductivity, and the solutes that are most frequently used are alkali-metal perchlorates, Lewis acids, such as $AlCl_3$ or BF_3, complex fluorides, such as $NaPF_6$, or complex salts, such as $(C_2H_5)_4NClO_4$. Impurities, particularly water, have a significant effect on the behavior of nonaqueous batteries. Batteries using inorganic solvents, such as $SOCl_2$, contain dissolved salts, such as lithium tetrachloroaluminate, and possess the interesting feature that the electrolyte solution is also the active cathode and is reduced on discharge. Several problems presently exist with these batteries, including self-discharge, corrosion resistance, and hermeticity, as well as safety and hazards.

A further application of nonaqueous electrolyte solutions is in the electrodeposition of metals. Of chief concern in this case are aluminum, magnesium, and the alkali metals, where the problem of preferential discharge of hydrogen can be overcome with the use of nonaqueous solvents. For successful deposition a metal usually requires a potential that is less negative than its standard potential. In addition the reactivity of the ions at the cathode should also be taken into account.

Only certain classes of solutes and solvents form plating baths and an important criterion is that a loose ionic complex must form between the solute and the solvent. The solutes most commonly used include halides, hydrides, borohydrides, and organometallic compounds. The selection of a suitable solvent is governed by the stability of the coordination compound it forms with the solute. Ethers and aromatic hydrocarbons are the solvents that satisfy the requirement of forming a relatively weak complex with the solute. The most successful electrodeposition is that of aluminum, with the most popular system consisting of $AlCl_3$ dissolved in ether, with the addition of some $LiAlH_4$. Some success has been achieved with beryllium.

Ionizing solvents, such as liquid ammonia, liquid hydrogen fluoride, and liquid sulfur dioxide, are of particular interest for studying solvolysis, oxidation-reduction, and complex-formation reactions, as well as organic reactions and reactions of metals.

As discussed in Chapter 10, a very important application of nonaqueous solvents is to take advantage of their ability to vary the rates of chemical reactions to significant extents and a great deal of attention has been focused on the understanding of the effect of solvents on reaction rates and mechanisms.

LITERATURE CITED

1 Parker, A. J., *Proc. Roy. Aust. Chem. Inst.*, **39,** 163 (1972).

2 Parker, A. J., *Search,* **4 (10),** 426 (1973).

3 Bailes, P. J., C. Hanson, and M. A. Hughes, *Chem. Eng.*, **83 (18),** 86 (1976).

4 Brenner, A., *J. Electrochem. Soc.*, **103,** 652 (1956).

GENERAL REFERENCES

Solvent Extraction

Alders, L., *Liquid-Liquid Extraction, Theory and Practice,* 2nd ed., Elsevier, Amsterdam, 1959.

Hanson, C., Ed., *Recent Advances in Liquid-Liquid Extraction,* Pergamon Press, Oxford, 1971.

Kertes, A. S., and Y. Marcus, Eds., *Solvent Extraction Research,* Wiley, New York, 1969.

Marcus, Y., and A. S. Kertes, *Ion Exchange and Solvent Extraction of Metal Complexes,* Wiley-Interscience, London, 1969.

Ritcey, G. M., Ed., Proceedings International Solvent Extraction Conferences, Soc. of Chemical Industry, London; *Advances in Extractive Metallurgy,* publication of Min. and Met., London, 1974.

Queneau, P., Ed., *Extractive Metallurgy of Copper, Nickel and Cobalt,* Interscience, New York, 1961.

Starcy, J., *The Solvent Extraction of Metal Chelates,* Pergamon Press, Elmsford, N.Y., 1964.

Treybal, R. E., *Liquid Extraction,* 2nd ed., McGraw-Hill, New York, 1963.

Nonaqueous-Electrolyte Batteries

Freund, J. M., and W. C. Spindler, "Low-Temperature Nonaqueous Cells," in *The Primary Battery,* Vol. 1, G. W. Heise and N. Corey Cahoon, Eds., Wiley, New York, 1971, Chap. 10.

Jasinski, R., *High-Energy Batteries,* Plenum Press, New York, 1967.

Electrodeposition

Brenner, A., "Electrolysis of Nonaqueous Systems," in *Advances in Electrochemistry and Electrochemical Engineering,* P. Delahay and C. W. Tobias, Eds., Vol. V, *Electrochemical Engineering,* C. W. Tobias, Ed., Wiley-Interscience, New York, 1967.

Reaction Rates and Mechanisms

Amis, E. S., and J. F. Hinton, *Solvent Effects on Chemical Phenomena,* Academic Press, New York, 1973, Chap. 5.

Dack, M. R. J., *Chem. Brit.,* **6,** 347 (1970).

Parker, A. J., *Quart. Rev.,* **16,** 163 (1962).

Ritchie, C. D., "Solvent Effects on Some Simple Organic Reactions," in *Solute-Solvent Interactions,* Vol. II, J. F. Coetzee and C. D. Ritchie, Eds., Marcel Dekker, New York, 1976, Chap. 12.

Nonaqueous Titrations

Fritz, J. S., *Acid–Base Titrations in Nonaqueous Solvents,* Allyn and Bacon, Boston, Mass., 1973.

Ionizing Solvents

Jander, J., and C. Lafrenz, *Ionizing Solvents: Chemical Topics for Students,* W. Foerst and H. Grunewald, Eds., John Wiley & Sons, Ltd., London, 1970. (English translation.)

Waddington, T. C., *Nonaqueous Solvents,* Appleton-Century-Crofts, Educational Division, New York, 1969.

Index

ALVERNO COLLEGE LIBRARY
Nonaqueous solution chemistry